Resolution of Equations in Algebraic Structures

Volume 1

Algebraic Techniques

Resolution of Equations in Algebraic Structures

Volume 1

Algebraic Techniques

Edited by

Hassan Aït-Kaci

ACA Systems Technology Laboratory
Programming Languages Group
Microelectronics and Computer Technology Corporation
Austin, Texas

Maurice Nivat

LITP
Université Paris VII
Paris, France

ACADEMIC PRESS, INC.
Harcourt Brace Jovanovich, Publishers
Boston San Diego New York
Berkeley London Sydney
Tokyo Toronto

ACADEMIC PRESS, INC.
1250 Sixth Avenue, San Diego, CA 92101

United Kingdom Edition published by
ACADEMIC PRESS INC. (LONDON) LTD.
24–28 Oval Road, London NW1 7DX

Library of Congress Cataloging-in-Publication Data

Resolution of equations in algebraic structures / edited by Hassan Aït-Kaci, Maurice Nivat.
p. cm.
Includes bibliographies and index.
Contents: v. 1. Algebraic techniques – v. 2. Rewriting techniques.
ISBN 0-12-046370-9 (v. 1). – ISBN 0-12-046371-7 (v. 2)
1. Algebra, Abstract. 2. Equations–Numerical solutions. I. Aït-Kaci, Hassan, Date- II. Nivat, M.
QA6162.R47 1989
512'.02–dc 19 88-21727
CIP

89 90 91 92 9 8 7 6 5 4 3 2 1
Printed in the United States of America

Contents

Contents of Volume 2: Rewriting Techniques

Contributors

Numbers in parentheses indicate the pages on which the authors' contributions begin.

Egidio Astesiano (1), *Dipartimento di Matematica, Università de Genova, via L. B. Alberti 4, I-16132 Genova, Italy*

Shang-Ching Chou (33), *Department of Computer Science, University of Texas, Austin, Texas 78712*

Bruno Courcelle (93), *Laboratoire d'Informatique Université Bordeaux–I, 351 cours de la Libération, F-33405 Talence, France*

Desmond Fearnley-Sander (127), *Department of Mathematics, University of Tasmania, GPO Box 252 C, Hobart, Australia 7001*

Jean Gallier (151), *Department of Computer and Information Science, University of Pennsylvania, Philadelphia, Pennsylvania 19104*

Joseph A. Goguen (217) *Oxford University Computing Laboratory, Programming Research Group, 8–11 Keble Road, Oxford, OX1 3QD, England*

Irène Guessarian (263), *LITP, UER de Mathématique, Université de Paris VII, 2 place Jussieu, F-75251 Paris Cédex 05, France*

Paul Mendler (369), *Computer Science Department, Manchester University, Manchester M13 9PL, England*

Ugo Montanari (293), *Istituto di Scienze della Informazione, Università di Pisa, corso Italia 40, I-56100 Pisa, Italy*

Jeffrey F. Naughton (321), *Department of Computer Science, Princeton University, Princeton, New Jersey 08540*

Maurice Nivat (351), *LITP, Université de Paris VII, 2 place Jussieu, F-75251 Cédex 05, France*

Prakash Panangaden (369), *Department of Computer Science, Cornell University, Upson Hall, Ithaca, New York 14853*

Andreas Podelski (351), *LITP, Université de Paris VII, 2 place Jussieu, F-75251 Cédex 05, France*

Stan Raatz (151), *Department of Computer Science, Rutgers University, New Brunswick, New Jersey 08901*

Domenico Sacca (411), *viale della Resistenza 198, I-87306 Renede, Italy*

Yehoshua Sagiv (321), *Department of Computer Science, Hebrew University of Jerusalem, Givat Ram Campus, Jerusalem, Israel*

William F. Schelter (33), *Department of Mathematics, University of Texas, Austin, Texas 78712*

Michael I. Schwartzbach (369), *Computer Science Department, Aarhus University, Ny Munkegade, DK-8000 Aarhus C, Denmark*

Marcello Sgamma (293), *Olivetti DOR, I-56100 Pisa, Italy*

Wayne Snyder (151), *Department of Computer and Information Science, University of Pennsylvania, Philadelphia, Pennsylvania 19104*

Martin Wirsing (1), *Fakultät für Mathematik und Informatik, Universität Passau, Postfach 2540, D-8390 Passau, Federal Republic of Germany*

Jin-Gen Yang (33), *Department of Computer Science, University of Texas, Austin, Texas 78712*

Carlo Zanziolo (411), *Microelectronics and Computer Technology Corporation, ACA Systems Technology Laboratory, 3500 West Balconnes Center Drive, Austin, Texas 78759*

Foreword

Equations are pervasive in computer science. They appear in a wide variety of algebraic structures, and in a rich diversity of applications. Therefore, formal methods for establishing the existence of solutions to equations or for effectively finding such solutions constitute a body of scientific knowledge of fundamental importance. For as young a discipline as computer science, growing rapidly from several shoots into a multitude of branches, the danger of losing sight of the generality of equational approaches is real. Thus, it is important to recognize that results obtained in one area of computer science could be applied in another. The example of fixed-point equations is blatant, as these manifest themselves virtually everywhere from programming language semantics and domain theory to deductive databases. Surveying the state of the art in equation resolution reflects more than just a desire to avoid duplication of effort: it is bound to shed light on those general structures and methods that may be abstracted from each idiosyncratic application. This is not only bound to foster cross-fertilization and systematic understanding of the many equational structures and concepts found to be valuable in separate branches of computer science, but it is also a vital and necessary step in the concretization of the discipline that will enable it to endure as a science.

Having come to this realization, we decided, on the occasion of a visit of Maurice Nivat to MCC in 1986, to do something about it. We agreed to organize a colloquium to which we would invite researchers, working with equations, from diverse areas in computer science. The idea was to obtain from the best researchers the solid and stable results of the equational approaches they had devised or mastered in particular applications. Although the aim was to bring together experts from several areas of computer science, we realized that for the size of the meeting to stay within reason, as well as for exchanges to be feasible and bear fruit, it would be wise to limit ourselves to a restricted notion of equations and algebraic structures—namely, those relevant to symbolic computation and the foundation of programming. In May of 1987, under the generous sponsorship of MCC in Austin, Texas, and some partial contribution from INRIA in Paris, France, the Colloquium on Resolution of Equations in Algebraic Structures (CREAS) took place in Lakeway, Texas, with thirty outstanding participants from Australia, France, Germany, Italy, and the United States. For three overloaded days, participants exposed and exchanged a wealth of results, ideas, and prospects.

The meeting was unanimously judged to be a success. Therefore, it became our duty as organizers to edit a book containing the highlights of the contributions to CREAS, asking each willing participant to write a special original piece. Gathering twenty-two contributions from some of the best researchers working in equation solving, the book would be meant as a reference, a compendium of results, methods, algorithms, and state of the art in resolution of equations in algebraic structures. It would be destined to go on the shelves of computer science libraries, university departments, and research laboratories, as well as to serve as a fine textbook for graduate students coming to this field of research.

The resulting collection of papers came to a size too large to fit in one volume. Therefore, it was decided to divide it into two volumes—one of which you are holding in your hands. The split was made easy, as a conceptual line appeared naturally that created a balanced partition between algebraic techniques and, more specifically, rewriting techniques. Thus, although both volumes share the common title *Resolution of Equations in Algebraic Structures*, Volume 1 is subtitled *Algebraic Techniques* and Volume 2 is subtitled *Rewriting Techniques*. More than just a convenient coincidence, the separation of the specific area of rewriting techniques for equational problems translates the historical fact that a great deal has happened lately in this field, which has made it one of the most active fields of research in symbolic computation. Thus it is only natural to devote a well-delineated volume to these among all equation-resolution techniques.

Each volume's contents are organized in alphabetical order by first authors. In each, an introduction reviews each contribution in order of appearance, giving an informal summary of the work and results reported. The purpose of these reviews is to serve as a quick first-pass reference to the reader, putting the contribution in context, stressing the significance of the work.

Finally, we would like to express some thoughts of acknowledgement. We are, of course, indebted to the Advanced Computer Architecture program of MCC for their generous financial help for organizing CREAS, as well as for providing their efficient logistic support before, during, and after the colloquium. In particular, CREAS would not have been without the encouragement and support of Woody Bledsoe and Bob Boyer. Many thanks also to INRIA for agreeing to pay for the travel of their participating researchers. We owe Denise White (the charming and ever-smiling MCC coordinator of CREAS) a great deal for taking care of every detail and person without ever pulling out a shotgun. We were lucky to find at Lakeway Inn in Lakeway, Texas, two extremely qualified and helpful professionals, Jo Ann Freeman and Lisa Parker, who provided a most propitious environment for CREAS. Naturally, we are especially grateful to the contributors of this volume for gracefully taking the (short) time to write and polish their articles, and for bending to our editorial whims. Finally, we acknowledge the kind help and assistance of Sari Kalin and Alice Peters, of Academic Press Boston. In addition, the first editor would like to thank Carlo Zaniolo, director of the Languages Group of MCC's ACA Systems Technology, and the members of the LIFE project for their understanding and patience during the interminable time he spent editing these volumes. Most of all, he needs to express his guilty gratitude to his loving and patient family for having foolishly stuck his foot into an unexpected vortex of never-ending obligations.

We wish you, the reader, a pleasant time sharing the excitement felt by all who attended CREAS, and hope that we have contributed in giving you the means and tools to pursue your own research, solving equations in algebraic structures.

Hassan Aït-Kaci
Maurice Nivat

A Preview of Volume 1: Algebraic Techniques

In *Bisimulation in Algebraic Specifications*, Astesiano and Wirsing expose an extension of Milner's notion of observational equivalence of processes in concurrent systems to more abstract algebraic specifications of transition systems. They build their results upon the notion of *partial abstract types* as an appropriate tool for expressing hierarchies of type enrichments from primitive specifications. Working from hierarchical specifications consisting of conditional equation axioms, they construct a complete lattice of simulation congruences and corresponding fully abstract quotient models. They study in detail the special case of the complete lattice of simulation congruences associated with the ground atomic theory of hierarchical specification, retrieving as the lattice's maximum element Milner's strong bisimulation for CCS.

In *Characteristic Sets and Gröbner Bases in Geometry Theorem Proving*, Chou, Schelter, and Yang report work done in algebraic methods for geometry theorem proving. Geometric objects are described by polynomial equations in cartesian coordinates, translating the necessary and sufficient constraints that characterize them. Proving a theorem, also expressed as a polynomial equation, consists of verifying that it is a consequence of the hypothesis equations. The authors survey and compare in depth two essential methods, both based on symbolic polynomial manipulation and both effectively mechanizable. The first one, mainly developed by the Chinese mathematician Wu Wen-Tsün, uses triangulation and polynomial pseudo-division. The other, explored by Chou and Schelter among others, employs a rewrite rule method in conjunction with Buchberger's algorithm for finding a Gröbner basis in polynomial ideal theory. They discuss and illustrate the effectiveness of these methods on numerous nontrivial theorems of planar Euclidian geometry.

In *On Recognizable Sets and Tree Automata*, Courcelle exposes the foundations of an algebraic theory of recognizability as opposed to a more familiar by less precise and less general notion of algorithmic recognizability. The former elegantly characterizes a recognizable subset as a partition of a congruence of finite index, whereas the latter relies on a cumbersome notion of automaton. Courcelle studies in detail algebraic recognizability of subsets of free T-algebras, or equational theories, thus covering the cases of useful discrete structures like those of words, terms, finite trees, and finite graphs. (Although not illustrated in this contribution, the fundamental results of this notion of recognizability are not limited to discrete objects.) Looking at equational presentations as term rewriting systems, syntactic properties for recognizability in free T-algebras are formulated. In addition, Courcelle extends the familiar determinization nondeterministic (word or tree) automata to special equational theories called *bilinear*. These cover all interesting cases of the discrete structures considered, and allow a connection with an algorithmic notion of recognizability.

In *Idea of a Diagram*, Fearnley-Sander proposes a language for geometric descriptions. He elaborates on the concept of a *Grassmann Geometry* to construct a high-level symbolic language in which diagrams are represented as sets of polynomial equations. Grassmann geometry is essentially a first-order language for a *Grassmann Algebra*, which is a ring structure over the union of a field and an affine space. Such a structure allows at once the manipulation of scalar quantities (elements of the field), points (elements of the affine space), and vectors (as affine point differences). Thus, both topological and scalar constraints can be expressed as polynomial equations. Explicit coordinate manipulation is conveniently avoided, yielding elegant abstract descriptions. Coordinate-oriented specifications are possible however, reducing Grassmann geometry to polynomial algebra. Properties of diagrams in Grassmann geometry corresponding to elementary geometry theorems are provable using simple equational reasoning. Computationally, the approach is for the most part directly mappable to Horn logic programming à la Prolog. A few additional inference rules like superposition also may be needed. This approach to abstract computational geometry has undeniable simplicity and elegance which open intriguing prospects.

In *Rigid E-Unification and Its Application to Equational Matings*, Gallier, Raatz, and Snyder elaborate on the concept of *matings* in theorem proving due to Peter Andrews, and independently investigated by Wolfgang Bibel. The method of matings is interesting as it does not require, contrary to resolution-based procedures, that formulae be expanded out in disjunctive normal forms. Rather it keeps formulae in negative normal form (nnf), thus both avoiding exponential growth of the size as well as taking advantage of their global structures. Through such a formula which has no quantifiers, *vertical paths* can be threaded which are sets (conjections) of literals obtained by going down the formula's syntactic structure, merging subpaths at conjunctions and choosing subpaths at disjunctions. Andrews observed that a quantifier-free formula in nnf is unsatisfiable if and only if all its vertical paths are unsatisfiable. Thus, unsatisfiability of a universally quantified formula is checked by attempting to close all vertical paths beginning by erasing quantifiers, and then, if that fails, splitting quantified subformulae and iterate. A mating consists of a set of unifiable pairwise complementary literals spanning all vertical paths. Finding one comes in handy in the path-closing step, as it signifies that its unifying substitution closes all paths. The authors' contribution is a nontrivial extension of the method to theories with equality. They introduce the notion of *equational matings*, extending the notion of unsatisfiability to deal directly with E-unifiers—i.e., unifiers modulo equations. They derive rigorous soundness and completeness results for their method as well as establishing its NP-completeness. This contribution is a self-contained and thorough study.

In *What Is Unification?*, Goguen develops a very general theory of unification using a category theoretic framework. This approach transcends conventional unification of first-order terms and gives abstract foundations to the quintessential process of equation solving. The notion of a *substitution system*, where objects are sets of variables and morphisms are substitutions, is introduced as a category. An *equation* is simply a pair of arrows. A (most general) *solution* is an equalizer. These ideas generalize to systems of equations, and equations with solution sets. The idea is amazingly simple and powerful. Thus, Goguen proceeds to illustrate the generality of his theory to various substitution systems including (of course) classical term unification, E-unification, and order-sorted unification. In addition, he discusses several and diverse instances that may be viewed as substitution systems, such as infinite term unification, fixed-point equations, domain equations, polymorphic-type inference, and unification grammars. This contribution is a ground-breaking step toward a truly general unification theory. It is self-contained as far as category theory is concerned, and thus is accessible to a broader audience than might be presumed.

In *Some Fixpoint Techniques in Algebraic Structures and Applications to Computer Science*, Guessarian gives a brief tutorial on various fixpoint theorems in ordered algebraic structures and then illustrates their applications to a few areas of computer science. She touches upon denotational semantics, domain equations, algebraic semantics of recursive programs, inductive proof techniques for program verification, semantics of logic programs, and recursive query evaluation in deductive databases. This is a lucid and to-the-point tutorial which shows the universality of algebraic equations in computer science.

In *Canonical Representatives for Observational Equivalence Classes*, Montanari and Sgamma consider the problem of defining observation criteria for nondeterministic systems. The model chosen is (labeled) transition systems. In particular, the notion of *bisimulation* due to Park and Milner is examined: bisimulations are expressed as fixpoints of a noncontiuous functional, and the *observational equivalence* is defined as the maximal bisimulation. It is emphasized that an extension is needed to encompass more general notions of observation than event sequences. The case of partial ordering observations is considered, where bisimulation can be used to define a "truly concurrent" semantic for CCS and a meaningful observation equivalence for Petri nets. A more general notion of bisimulation is presented, where observations may be taken from any partial ordering. An independent, algebraic notion of simplification is given, such that a unique, reduced canonical representative exists. Finally, the two notions are proved to coincide, i.e., two systems are observationally equivalent if and only if they have the same canonical representative. The results apply to both standard and partial ordering semantics of (unguarded) CCS.

In *Minimizing Expansions of Recursions*, Naughton and Sagiv address the issue of controlling the proliferating of redundant computations in function-free Horn clause recursion (Datalog). Intentional rules in a rational database extending the more limited classical relational algebra give rise, when recursive, to nontrivial optimization questions. Most research in this area has concentrated on efficient evaluation of recursive relations. Here, the authors study the less often trodden path of detecting redundancy. They limit themselves to the particular but not so trivial case of programs containing a single *linear recursive* rule; that is, a rule in which the head predicate has no repeated arguments and occurs exactly once in the rule body along with extensional predicates only. Such a recursive relation's *expansion* is simply the set of strings of predicates obtained by unfolding the relation's definition. In expanding a recursive relation, a same predicate may be unnecessarily computed over and over again. Also, a recursive relation may not need to be recursive (bounded recursion). Detecting bounded recursion for arbitrary Datalog programs has only recently been proven undecidable by Gaifman, Maierson, Sagiv, and Vardi, although it has been shown feasible under some sufficient conditions and/or restrictive criteria by several authors. Naughton and Sagiv observe that detecting bounded recursion is a particular case of detecting redundant predicates, thus establishing that the latter problem is also undecidable. They also give the criteria whereby it is possible to transform the recursive programs to an equivalent minimally redundant one. They define the notion of an *argument/value* (A/V) *graph* through which they relate the patterns of the variables appearing in the strings of an expansion to the structure of the rules. They survey known results and establish several new ones, relating detection of redundancy and boundedness to various types of paths in the (A/V) graph of the rules. This contribution is a clear account of problems and solutions for the removal of redundancy in deductive databases.

In *Tree Monoids and Recognizability of Sets of Finite Trees*, Nivat and Podelski lay the foundation of a new theory of tree monoids for tree automata in analogy with the theory of word monoids that underlies word automata. Until now, no such theory has been devised as transitions in tree automata could not be seen, in contrast with word automata, as finitely generated as a transition monoid. The difficulty is that the algebraic structure of transitions is, by construction, a homomorphic image of the free monoid of words, and no such structure was found to exist obviously on trees. However, Nivat and Podelski uncover the right monoid structure on finite labeled *pointed trees*—trees with a distinguished *border node* lying immediately outside the tree—endowed with a composition law that, given $\langle t_1, f_1 \rangle$ and $\langle t_2, f_2 \rangle$ two pairs of pointed trees, yields the pair $\langle t_1[f \leftarrow t_2], f_1 f_2 \rangle$. Thus, they establish the existence of a syntactic congruence for a given pointed tree language L and a syntactic monoid is obtained as the quotient of the monoid of pointed trees by the syntactic congruence. Such a tree language L is recognizable by an ascending finite tree automaton if and only if its syntactic congruence has a finite index. This analogy with finite word automata goes further, as existence of a minimal tree automata is determined among the recognizers of L whose states are precisely identified with the Nerode equivalence classes associated with L. This short contribution is an example of beautiful mathematics, and shows that the simplicity and elegance gracing the theory of finite word automata—thought too simple for the more complex theory of tree automata—can be retained nonetheless for the latter.

In *Recursively Defined Types in Constructive Type Theory*, Panangaden, Mendler, and Schwartzbach give a category theoretic view of recursive type equations in constructive type theory. It is well known that a (constructive) deductive system may be seen as a category where objects are propositions and arrows are proofs. Logical connectives have direct category theoretic counterparts; namely, conjunction corresponds to product and disjunction to coproduct. In addition, functional abstraction corresponds to an adjointness condition known as *exponentiation*, translating the fact that for any proof may be viewed as a proposition. This connection is known as (higher-order) *Categorical Logic*. The authors build on this connection and the well-known proposition-as-type principle to explicate type constructs of intuitionistic logic in category theoretic terms. This generalization allows them then to justify recursive (least solution) and infinite (greatest solution) type constructs as limits and colimits of countable chains and cochains. This contribution is a pleasant and self-contained reading. It is well motivated and explains the powerful connections between constructive proof theory and category theory.

In *Rule Transformation Methods in the Implementation of Logic Languages*, Sacca and Zaniolo present a unifying approach to several techniques developed for the efficient evaluation of recursive rules defining intentional relations in a logic-based language. In function-free Horn logic, computation amounts to solving fixed-point equations by unfolding the rules. Sacca and Zaniolo use an approach based on rule rewriting scripts, called methods, that transform programs that, as such, are not implementable safely and efficiently by fixed-point computations into equivalent programs that are. The choice of these methods is driven by the patterns of bound and free arguments in the given recursive goal, which is assumed to be known at compilation time. One method, known as the *differential fixed-point* method, applies to the situations where no argument is bound in the recursive goal, and yields a more efficient implementation of the least fixed-point computation. The remaining methods are designed to use the initial bindings of the arguments in the recursive goal to improve the efficiency and, often, to guarantee the termination of the computation. The solution consists of implementing the recursive predicate as a cascade of two fixed-point computations. Various methods proposed along these lines are studied and compared, including the *magic set* method, the *minimagic* method, the *counting* method, and the *magic counting* method. This contribution is an eye-opening exposition of tools for efficient recursive query evaluation in logic databases.

1 Bisimulation in Algebraic Specifications[1]

EGIDIO ASTESIANO

Dipartimento di Matematica
Università di Genova
Genova, Italy

MARTIN WIRSING

Fakultät für Mathematik und Informatik
Universität Passau
Passau, West Germany

1. Introduction

Over the last ten years algebraic specifications techniques have been successfully applied to the description of data structures and abstract requirement specifications. Abstract data types are included (in a more or less pure form) in programming languages such as Ada, CLU, MODULA, CIP-L or ML in order to provide a syntactic frame for the modular and abstract definition of programming systems. Hence, it appears to be consequent to describe also programming languages by these techniques. Main steps in this direction are Nivat's infinite magmas (1975), the ADJ-approach using continuous algebraic theories (Wagner, 1978), the abstract semantic algebras of Peter Mosses (1982), the ACP-framework of Jan Bergstra and his team (cf. e.g., Bergstra, 1984) and the CIP-project at Munich (Bauer et al., 1985). In this paper we consider the hierarchical approach to abstract data

[1]Work partially supported by the ESPRIT-project METEOR and by CNR (Italy).

ISBN 0-12-046370-9

types based on partial algebra semantics (Broy, 1982) as it has been pursued in the CIP-project at Munich (Bauer et al., 1985). In this framework deterministic and non-deterministic as well as applicative and procedural programming language constructs have been formalized (cf. Broy (1980), Broy (1983)).

These algebraic techniques have also been applied to formalize a parameterized schema for the specification of concurrent systems in the framework of the SMoLCS methodology (Astesiano (1987a), (1987b)) which has been used to give the dynamic semantics of full Ada (Astesiano (1986), (1987a)). In particular, the semantics of the intermediate language, which formalizes the concurrent structure of Ada, has been given by an algebraic specification where processes are viewed as labelled transition systems, thus generalizing an idea underlying CCS (Milner (1980), (1983)). Within this approach, in order to give processes a sufficiently abstract semantics, we have been led to extend to algebraic specifications of transition systems the notion of bisimulation, introduced by Park (1981) and Milner (1983) in connection with nondeterministic automata and labelled transition systems. Bisimulation essentially means that two processes can perform the same observable actions. In this paper we provide a generalization of that notion for hierarchical algebraic specifications, illustrating our concepts by an example CCSO, which is a subset of CCS. Let us now give a more detailed overview of our work.

Given an algebraic specification T of a programming language (such as the example CCSO in Section 1), in the partial algebra approach (Broy, 1982) one considers as semantics the class of all so-called minimally defined models of T (that are those models in which a term t is defined only if the definedness of t is provable by the axioms of T). Under reasonable assumptions (cf. e.g., (Broy (1984), (1982)), this class is not empty and contains a terminal model, the so-called weakly terminal model which is fully abstract in the sense of Milner (1972). Moreover, the minimal definedness of a term corresponds to the least fixpoint approach to semantics. As a consequence, the algebraic semantics yields for deterministic languages a class of least fixpoint models where some of them describe the operational behaviour of programs whereas others describe the functional behaviour. In particular, the weakly terminal models are fully abstract models for the functional behaviours (Broy, 1983).

In the case of concurrent languages, one can apply the same technique (Broy, 1982). For example, in the case of CSP one gets also a non-empty class of minimally defined models with a fully abstract weakly terminal model. But this model is isomorphic to the initial model and describes an operational behaviour which distinguishes even more programs than the concept of bisimulation of Milner (1980, 1983). For the semantics a more abstract model is needed. In (Astesiano, 1986) this is obtained by a second step where the algebraic specification is factorized using a socalled behavioural abstraction operation.

In this paper, a more direct approach is pursued: the notion of strong bisimulation (Park (1981), Milner (1980), Milner (1983)) is generalized to arbitrary signatures. For example, a transition relation is a particular case of an "output" function with result in the "primitive" part of the hierarchical specification.

In this way we obtain the concept of "simulation congruence" relative to a given hierarchical signature Σ and a given fixed congruence $\approx$ (which often is chosen to be the ground atomic theory of an algebraic specification T with signature Σ). It is shown that every simulation congruence is fully abstract and that the set of simulation congruences (wrt $\approx$) forms a complete lattice, the coarsest element of which corresponds to strong bisimulation (in the special case of CCS).

Then these results are applied to hierarchical specifications T with conditional equations as axioms. It is shown that under reasonable syntactic restrictions of the form of the axioms of T the simulation algebras (which correspond to the simulation congruences) are models of T. Thus every hierarchical algebraic specification (under some assumptions) possesses a complete lattice of (isomorphism classes of) fully abstract models which simulate the ground atomic theory of T. The weakly terminal models form the smallest element of this lattice whereas the bisimulation algebras form the coarsest element.

The paper is organized as follows. After presenting in the first section an example (CCSO) of algebraic specification, we introduce in the second the simulation congruences associated to a given congruence and show in the third that they form a complete lattice. Finally in the last section we consider the lattice of simulation congruences associated to the ground atomic theory of a hierarchical specification.

2. Algebraic Specification of Concurrent Systems: An Example

We refer to the appendix for basic concepts and notations on algebraic specifications. However, for reader's convenience, we recall here the two main concepts.

A specification, also called a **partial abstract type**, $T = (\Sigma, E)$ consists of a signature $\Sigma = (S, F)$ of sorts $s \in S$ and operation symbols $f : s_1 \times \cdots \times s_n \rightarrow s_{n+1} \in F$ (where $s_1, \ldots, s_{n+1} \in S$) and of a set E of positive conditional formulas, i.e. formulas which are first-order equivalent to formulas of the form

$$(D(u_1) \wedge u_1 = v_1) \ldots \wedge (D(u_n) \wedge u_n = v_n) \Rightarrow e, \tag{1}$$

where e is either $D(t)$ (expressing the definedness of the term t) or $t = t'$ and u_i, v_i, t, t' are Σ-terms (possibly containing free variables) and $n \geq 0$.

A **hierarchical specification** $T = (\Sigma, E, P)$ is a specification (Σ, E) in which the **primitive** specification P is a hierarchical specification (Σ_P, E_P, P') or a specification (Σ_P, E_P) and where $(\Sigma_P, E_P) \subseteq (\Sigma, E)$. Whenever (Σ_P, E_P) is the union of some specifications (Σ_{P_i}, E_{P_i}) $(i = 1, \ldots, n)$ we say that (Σ, E) is based on (Σ_{P_i}, E_{P_i}) $(i = 1, \ldots, n)$.

We denote by

$S^+ =_{\text{def}} S - S_p$	the set of 'non-primitive sorts'
$F^+ =_{\text{def}} F - F_p$	the set of 'non-primitive operation symbols'
$E^+ =_{\text{def}} E - E_p$	the set of 'non-primitive axioms'
$W(\Sigma, x_1, \ldots, x_n)$	the set of 'contexts' of $x_1, \ldots x_n$.

Informally, the primitive specification P represents the 'observable part' of the hierarchical specification whereas the non-primitive part $((S^+, F^+), E^+)$ is considered as 'black box', the characteristic properties of which are given by the non-primitive axioms E^+.

In the following we write often W(SPEC) instead of W(Sig(SPEC)) for a specification SPEC with signature Sig(SPEC).

As a first but typical example we consider the following specification of a simple nondeterministic sublanguage of CCS. The specification CCSO is hierarchically based on given specifications BOOL for the truth-values '*tr*' and '*fa*' and ACTION for a data type of 'actions' which is supposed to contain a constant τ for 'invisible' actions and a unary function symbol c ('complement') such that $c(c(a)) = a$ for all actions a and $c(\tau) = \tau$.

CCSO defines a sort *ccs* of 'programs' containing a constant *nil*, a binary composition '$\blacksquare$' of actions with programs and function symbols '+' (for nondeterministic choice) and '|' (for parallel composition). The semantics of programs is given by a ternary boolean function symbol '$\rightarrow$' where $(p \xrightarrow{a} p') = tr$ indicates that there is a transition from program p to program p' when executing action a. In the following we use an infix notation, where dots stand for arguments.

type `CCSO` $\equiv$
  based on `ACTION, BOOL`
  sort *ccs*
  $nil \; : \quad \rightarrow ccs$
  $.\blacksquare. \; : \quad action \times ccs \rightarrow ccs$
  $.+. \; : \quad ccs \times ccs \rightarrow ccs$
  $.|. \; : \quad ccs \times ccs \rightarrow ccs$
  $.\xrightarrow{.}. \; : \quad ccs \times action \times ccs \rightarrow bool$
  axioms `for all` a `of sort` *action* `and all` p, q, p', q' `of sort` *ccs*
  `(`$\blacksquare$`1)` $(a \blacksquare p \xrightarrow{a} p) = tr$
  `(+1)` $(p \xrightarrow{a} p') = tr \Rightarrow (p + q \xrightarrow{a} p') = tr$
  `(+2)` $(q \xrightarrow{a} q') = tr \Rightarrow (p + q \xrightarrow{a} q') = tr$
  `(C1)` $(p \xrightarrow{a} p') = tr \Rightarrow (p|q \xrightarrow{a} p'|q) = tr$
  `(C2)` $(q \xrightarrow{a} q') = tr \Rightarrow (p|q \xrightarrow{a} p|q') = tr$
  `(C3)` $(p \xrightarrow{a} p') = tr \wedge (q \xrightarrow{c(a)} q') = tr \Rightarrow (p|q \xrightarrow{\tau} p'|q') = tr$
  `(D)` $D(nil) \wedge D(p + q)$

The specifications ACTION and BOOL are the primitive specifications of CCSO. The only non-primitive sort is '*ccs*', the non-primitive operation symbols are '*nil*', '.$\blacksquare$.', '.+.', '.|.', '$\rightarrow$' and the non-primitive axioms are ($\blacksquare$1)—(D). The terms $(a \blacksquare x \xrightarrow{a} y)$, $(x|nil \xrightarrow{\tau} y|x) \in W(\text{CCSO}, x, y)|_{bool}$ are examples of 'contexts of primitive sort'.

Let ACTION and BOOL be monomorphic specifications. From the form of the axioms one may easily see that CCSO admits an initial model, I say, and a weakly terminal model, Z say. Both models are minimally defined. The equality in I is characterized by the provable equality $=_{\mathrm{CCSO}}$ and the equality in Z by the equality in all contexts. However (surprisingly) I and Z are isomorphic; the primitive contexts distinguish all terms which are not provably equal. For example, the terms $p+q$ and $q+p$ (for $p, q \in W(\mathrm{CCSO})_{ccs}$ such that $\mathrm{CCSO} \vdash p = q$ does **not** hold) can be distinguished by the context $\mathrm{C}[x] \equiv (a \blacksquare nil|x \xrightarrow{a} nil|p+q)$ since $\mathrm{C}[p+q] = tr$ is provable according to axiom ($\blacksquare$1) whereas $\mathrm{C}[q+p] = tr$ is not provable (cf. corollary 0). We summarize in a proposition the relevant properties.

Proposition 1. *Let I and Z denote respectively the initial and weakly terminal model of CCSO which exist. Then*

(0) *for all $t \in W(CCSO)$,*
$I \models D(t)$ iff $CCSO \vdash D(t)$ (i.e. I is minimally defined), and
$Z \models D(t)$ iff $CCSO \vdash D(t)$ (i.e. Z is minimally defined);

(1) *for all $t_1, t_2 \in W(CCSO)$ s.t. $CCSO \vdash D(t_1) \wedge D(t_2)$,*
$I \models t_1 = t_2$ iff $t_1 =_{\mathrm{CCSO}} t_2$ (i.e. $CCSO \vdash t_1 = t_2$);

(2) *for all $p, q \in W(CCSO)|_{ccs}$,*
$Z \models p = q$ iff for all contexts $c \in W(CCSO, x : ccs)|_{bool}$ and all $b \in W(CCSO)|_{bool}$,
$c[p] =_{\mathrm{CCSO}} b \Longleftrightarrow c[q] =_{\mathrm{CCSO}} b$.

Moreover, the class MDEF(CCSO) of minimally defined models of CCSO satisfies the following properties: for every model $M \in$ MDEF(CCSO),

(3) *for all $t \in W(CCSO)$, $M \models D(t)$ iff $CCSO \vdash D(t)$.*

(4) *$M \models D(nil)$, i.e. nil is defined.*

(5) *for all $p, q \in W(CCSO)|_{ccs}$ and $a \in W(CCSO)|_{action}$, s.t. $CCSO \vdash D(p) \wedge D(q)$ and $ACTION \vdash D(a)$,*

   (a) *$M \models D(p+q) \wedge D(a \blacksquare p) \wedge D(p|q)$ (i.e. $p+q$, $a \blacksquare q$, $p|q$ are defined);*

   (b) *if p and q do not contain any subterm of sort bool, then $CCSO \vdash p=q$ iff $CCSO|_{\mathrm{ACTION}} \vdash p=q$ iff $M \models p=q$, where $CCSO|_{\mathrm{ACTION}}$ is the specification obtained from CCSO by forgetting the axioms ($\blacksquare$ 1)–(D) (i.e. iff the equality of p and q can be proven only using the axioms of ACTION).*

Proof sketch. For the existence of the initial and terminal model and properties 0, 1 and 2, see (Broy, 1982);

(3) Obvious, since $M \models D(t)$ iff $I \models D(t)$;

(4)(5a) $D(nil)$ and $D(p+q)$ hold according to axiom (D); the definedness of $a \bullet p$ follows from $(a \bullet p \xrightarrow{a} p) = tr$ because of the strictness of the operation '$\rightarrow$'. Similarly the definedness of $p|q$ follows from $(a \bullet p)|q \xrightarrow{a} (p|q) = tr$ by the strictness of the operation '$\rightarrow$';

(5b) by the form of the axioms it is easy to see that for defined ground terms $p, q \in W(\text{CCSO})|_{ccs}$ without boolean subterms, a nontrivial equality $p = q$ can be proven only using identities proven in ACTION. It remains to show that $M \models p = q$ implies CCSO $\vdash p = q$, i.e. that $p = q$ holds in I. Assume $p \neq q$ in I and $p = q$ in M. Let $a \neq \tau$. Then, since $(a \bullet nil|p \xrightarrow{a} nil|p) = tr$ holds in I and hence in M, we have also that $(a \bullet nil|q \xrightarrow{a} nil|p) = tr$ holds in M and hence $(a \bullet nil|q \xrightarrow{a} nil|p)$ is defined in M. On the other hand, because of $p \neq q$ in I the equation $(a \bullet nil|q \xrightarrow{a} nil|p) = tr$ is not provable in CCSO; thus the definedness of $(a \bullet nil|q \xrightarrow{a} nil|p)$ is not provable in CCSO and therefore, because of (3), $(a \bullet nil|q \xrightarrow{a} nil|p)$ is undefined in M: contradiction. □

Corollary 1. *The initial model I of CCSO and the weakly terminal model Z of CCSO are isomorphic (recall that BOOL and ACTION are monomorphic).*

Proof. By 0 and 5b of Proposition 1. □

Hence the congruence between CCSO-programs induced by the weakly terminal model is the same as the one induced by the initial model. The weakly terminal model distinguishes e.g. between the terms $p+q$ and $q+p$ even if intuitively they can perform the same actions.

3. Simulation

In order to define a congruence between CCSO-programs where all those programs are identified which can perform the same actions, Milner's idea was to allow the addition of congruences on sort ccs.

The most basic of such congruences, called strong bisimulation, can be formulated, adopting Park's elegant approach (Park, 1981), in the following way.

First let us assume, as in the original CCS, that CCSO is just a labelled transition system, i.e. we assume that ACTION is just a set and then we consider an equality $p \xrightarrow{a} p' = tr$ as defining a labelled transition relation $\xrightarrow{a}$.

Let R be a binary relation on $W(\mathrm{CCSO})|_{ccs}$ and R^s be its converse. Then we define two transformations H_{MP} and F_{MP} by

$$\begin{aligned} H_{MP}(R) &=_{\mathrm{def}} \{\langle p, q\rangle : \forall p', a : \text{ if } p \xrightarrow{a} p' \text{ then } \exists q' \text{ s.t. } q \xrightarrow{a} q' \\ &\qquad\qquad \text{and } p' R q'\} \\ F_{MP}(R) &=_{\mathrm{def}} H_{MP}(R) \cap H_{MP}(R^s). \end{aligned} \tag{2}$$

We define the **strong bisimulation**, denoted by $\approx_{MP}$, as the maximum of the fixpoints of F_{MP}, i.e. $\approx_{MP} =_{\mathrm{def}}$ max fixpoint F_{MP}. It can be shown that, F_{MP} being monotonic, such maximum fixpoint exists and we have two equivalent characterizations:

$$\approx_{MP} = \cup\{R | R \subseteq F_{MP}(R)\} \tag{3}$$

and

$$\approx_{MP} = \approx^{\alpha}_{MP} \quad \text{for some ordinal } \alpha < \omega_1, \tag{4}$$

where

$$\approx^{0}_{MP} = W(\mathrm{CCSO})|_{ccs}{}^{2} \text{ and } \approx^{\lambda}_{MP} = F_{MP}\,(\bigcap_{\mu<\lambda} \approx^{\mu}_{MP}) \quad \text{for any ordinal } \lambda. \tag{5}$$

[Here ω_1 can be chosen since $W(\mathrm{CCSO})|_{ccs}{}^{2}$ is countable.] Every relation R, s.t. $R \subseteq F_{MP}(R)$ is called a **bisimulation relation**. In (Milner, 1983) it is shown that $\approx_{MP}$ is a congruence on terms of sort *ccs* w.r.t. the operations $+, \blacksquare$, and $|$.

In the following we will provide a theory for generalizing that concept to the case of algebraic specifications of transition systems and, even more generally, to the case of algebraic specifications with observations.

3.1. Preliminaries on Hierarchical Congruences with Partial Interpretations

Hierarchical types and hierarchical congruences with partial interpretations are considered as in (Broy (1984), (1982)).

Assume that Σ is a signature; then a **partial Σ-congruence**, $\sim$ say, consists of a congruence relation on $W(\Sigma)^2$, equally denoted by $\sim$, and a definedness relation $D^{\sim} \subseteq W(\Sigma)$ which is

(1) compatible with $\sim$, i.e. for all $t, t' \in W(\Sigma) : t \sim t' \Rightarrow (D^{\sim}(t) \Rightarrow D^{\sim}(t'))$;
(2) strict, i.e. for all $c \in W(\Sigma, x{:}s), t \in W(\Sigma)|_s : D^{\sim}(c[t]) \Rightarrow D^{\sim}(t)$;
(3) admitting only unique undefined elements, i.e. for all $t, t' \in W(\Sigma)$ of the same sort:

$$\neg D^{\sim}(t) \wedge \neg D^{\sim}(t') \Rightarrow t \sim t'. \tag{6}$$

Partial Σ-congruences can be compared using the following (partial) ordering relation $\leq$:

$$\begin{aligned} \sim_1 \leq \sim_2 \quad & \text{iff for all } t, t' \in W(\Sigma) : \\ & (\leq 1) D^{\sim_1}(t) \Rightarrow D^{\sim_2}(t) \text{ and} \\ & (\leq 2) D^{\sim_1}(t) \wedge t \sim_1 t' \Rightarrow t \sim_2 t'. \end{aligned} \tag{7}$$

This relation corresponds to the existence of a total Σ-homomorphism (cf. the appendix) from a term generated Σ-algebra A associated with $\sim_1$ to a term generated Σ-algebra associated with $\sim_2$.

For a hierarchical signature Σ over a primitive signature $\Sigma_P = (S_P, F_P)$, a partial Σ-congruence $\sim$ is called **hierarchical w.r.t.** Σ_P iff for all terms $t \in W(\Sigma)$ of a primitive sort $s \in \Sigma_P$ with $D^{\sim}(t)$ there exists a primitive term $t_P \in W(\Sigma_P)$ such that $t \sim t_P$.

The congruence $\sim$ is called **hierarchical w.r.t. a primitive type** $P = (\Sigma_P, E_P)$ if $\sim$ is hierarchical w.r.t. Σ_P and the axioms E_P are satisfied by $\sim$.

Let $\Sigma = (S, F, \Sigma_P)$ be hierarchical w.r.t. $\Sigma_P = (S_P, F_P)$. Then we consider the following sets of sorts and operation symbols:

$$\begin{aligned} F_{\text{out}} &=_{\text{def}} \{f \in F^+ : \text{ range } (f) \in S_P\} && \text{'output' functions,} \\ F_g &=_{\text{def}} F - F_{\text{out}} && \text{generating functions.} \end{aligned} \tag{8}$$

Informally, F_{out} denotes the set of all non-primitive operation symbols the range of which is 'observable', i.e. the range of which is a primitive sort. The 'observable' data are defined by the primitive specification P. Hence, the 'output' operations do not contribute to the construction of the data; they are not 'generating' functions. Note that the above definition of generating functions is a very crude one; it is sufficient for the purpose of the paper but for practical applications a more subtle definition is appropriate (cf. e.g. (Hußmann, 1985)).

Let $x_1, \ldots, x_n$ be free variables; then we define

$G(\Sigma, x_1, \ldots, x_n) =_{\text{def}} W(\langle S, F_g \rangle, x_1, \ldots, x_n)$
as generating terms of the hierarchical structures,

$DC_\Sigma(x_1, \ldots, x_n) =_{\text{def}} \{D(t) : t \in G(\Sigma, x_1, \ldots, x_n)|_{S^+}\}$
as "definedness contexts" of generating terms of nonprimitive sort,

$$\begin{aligned} OC_\Sigma(x_1, \ldots, x_n) = & DC_\Sigma(x_1, \ldots, x_n) \cup \\ & \{f(t_1, \ldots, t_k) : f : s_1 \times \ldots \times s_k \to s \in F_{\text{out}} \\ & \text{and for } i = 1, \ldots, k \\ & t_i \in G(\Sigma, x_1, \ldots, x_n)|_{s_i}\} \end{aligned}$$
as "observable contexts," if $x_1, \ldots, x_n$ are of nonprimitive sort.

We write $G(\Sigma)$ for the set of all ground (i.e. variable-free) generating terms and $t[u_1/v_1, \ldots, u_n/v_n]$ for the simultaneous replacement of (the terms or variables) $v_1, \ldots, v_n$ by $u_1, \ldots, u_n$ in t. As for $W(\Sigma)$ we write often $G(\text{SPEC})$ instead of $G(\text{Sig}(\text{SPEC}))$ and similarly for OC_Σ.

Example CCSO. Here $F_{\text{out}} = \{\to\}$ and hence $G(\text{CCSO}) = \{t \in W(\text{CCSO})|_{ccs} : t$ does not contain '$\to$' $\}$. For example the terms nil, $a \centerdot nil$, $a \centerdot nil + (b \centerdot nil | c \centerdot nil)$ are in $G(\text{CCSO})$ and the terms $(x_1 | x \xrightarrow{a} x_2 | x_3), D(a \centerdot x)$ are examples for observable contexts in $OC_{\text{CCSO}}(x, x_1, x_2, x_3)$. On the other hand, assume for the moment that there exists an injection function $in : bool \to action$ from BOOL into ACTION (i.e. there exist 'boolean actions'). Then $t \equiv in(a \centerdot x \xrightarrow{a} x)$ is a context of primitive sort which is not in $OC_{\text{CCSO}}(x)$; similarly, $t \centerdot nil$ is a non-primitive term which is not in $G(\text{CCSO})$.

3.2. Simulation Congruences

Assume that (Σ, E, P) is hierarchical w.r.t. $P = (\Sigma_P, E_P)$ and that $\approx$ is a hierarchical congruence w.r.t. P. We define Obs $=_{\text{def}} \{t : t \in W(\Sigma_P) \text{ and } D^{\approx}(t)\} \cup \{true\}$, where $true \notin W(\Sigma)$, to be the **set of observable terms**. The special constant symbol '*true*' is used in order to have $D(t) \approx true$ standing for $D^{\approx}(t)$, so that many clauses can have a more compact notation, as we are going to see in the following section.

Definition 1. Let $\sim$ be a binary relation on $G(\Sigma)$, i.e. $\sim \subseteq G(\Sigma)^2$. Then we define the induced equivalence $\sim^*$ on $W(\Sigma, x)^2$ and hence on $OC_\Sigma(x)^2$ as follows:

(1) $\sim^*$ is reflexive and transitive;
(2) if $c \in W(\Sigma, x), u, v \in G(\Sigma)$, u is a subterm of c and $u \sim v$, then $c[v/u] \sim^* c$. □

Note that if $\sim$ is symmetric then $\sim \subseteq \sim^*$ and $\sim^*$ is a Σ-congruence (in the usual sense of (total) Σ-congruences).

As example take for $\sim$ the smallest equivalence relation on $G(\text{CCSO})$ which contains for all $p, q \in G(\text{CCSO})$ $p + p \sim p$ and $p + q \sim q + p$. Then e.g. the following holds (cf. the example above Proposition 1):

$$(a \,\blacksquare\, nil|x \xrightarrow{a} nil|p + q) \sim^* (a \,\blacksquare\, nil|x \xrightarrow{a} nil|q + p) \text{ and}$$
$$(x|q + q \xrightarrow{a} p'|q) \sim^* (x|q \xrightarrow{a} p'|q + q)$$
$$\text{but } (a \,\blacksquare\, nil|x \xrightarrow{a} nil|x) \sim^* tr \text{ does not hold since}$$
$$\sim^* \text{ is defined independently of } =_{\text{CCSO}}.$$

Definition 2. Let $\approx$ be a hierarchical Σ-congruence. A ternary relation $S_\approx$ is called **simulation predicate** if for any relation $\sim \subseteq G(\Sigma)|_{S^+}{}^2$ and all $u, v \in G(\Sigma)|_{S^+}$ the following holds.

$S_\approx(\sim, u, v)$ iff for all contexts $c \in OC_\Sigma(x{:}s)$, where $s \in S^+$, and all $r \in$ Obs, there exists a context c' with $c' \sim^* c$ such that $c[u] \approx r \Rightarrow c'[v] \approx r$. □

Note that, because of the addition of the symbol *true* to Obs, $c[u] \approx r$ may have the form $D(t) \approx true$, whenever $D(t) \in DC_\Sigma(x)$.

Definition 3. Let $\approx$ be a hierarchical Σ-congruence. Then the set $S(\approx) \subseteq \mathcal{P}(G(\Sigma)|_{S+}{}^2)$ of **simulation relations** wrt. $\approx$ is defined as follows:

$$S(\approx) =_{\text{def}} \{\sim \subseteq G(\Sigma)|_{S+}{}^2 : \sim \text{ is symmetric and}$$
$$\forall u, v \in G(\Sigma)|_{S+} : u \sim v \text{ iff } S_{\approx}(\sim, u, v)\}. \square$$

Fact 1. For every $\sim \in S(\approx), \approx|_{S+} \subseteq \sim$.

Proof. If $u \approx v$ holds, then for all contexts c, $c[u] \approx c[v]$ holds as well. Hence for all observations r, we have $c[u] \approx r \Rightarrow c[v] \approx r$ and vice versa with u and v exchanged. Therefore, $S_{\approx}(\sim, u, v)$ holds for any $\sim \in S(\approx)$. $\square$

Example CCSO. In order to relate our theory to the usual notion of bisimulation in CCS that we have recalled at the beginning of section 2, it is useful to consider an equivalent definition of simulation relation as follows.

Definition $3'$. Let $\approx$ be a hierarchical Σ-congruence. Then we denote by $H_{\approx}$ the transformation on the set of binary relations on $G(\Sigma)|_{S+}$, defined by:

$$uH_{\approx}(\sim)v \text{ iff } S_{\approx}(\sim, u, v).$$

Denote by $\sim^s$ the converse of $\sim$ and define $F_{\approx}(\sim) =_{\text{def}} H_{\approx}(\sim) \cap H_{\approx}(\sim^s)$. Then a simulation relation is a fixpoint of $F_{\approx}$ and the set $S(\approx)$ of simulation relations w.r.t. $\approx$, is defined by

$$S(\approx) =_{\text{def}} \{\sim : \sim = F_{\approx}(\sim)\}. \square$$

Consider now the given specification CCSO and denote by $=_{\text{CCSO}}$ the congruence generated by provability, i.e. $t =_{\text{CCSO}} t'$ iff $\text{CCSO} \vdash t = t'$. Recall that $=_{\text{CCSO}}$ corresponds to the equality in the initial (and weakly terminal) model of CCSO. We want to make explicit $F_{=_{\text{CCSO}}}$ and thus the simulation predicate.

Recall that $S^+ = \{ccs\}, F_{\text{out}} = \{\rightarrow\}$ and hence

- $G(\text{CCSO}, x_1, \ldots, x_n)|_{S+} = G(\text{CCSO}, x_1, \ldots, x_n)|_{ccs} =$ $\{t \in W(\text{CCSO}, x_1, \ldots, x_n)|_{ccs} : t \text{ does not contain `}\rightarrow\text{'}\}$,

- for x of sort ccs,
  $DC_{\mathrm{CCSO}}(x) = \{D(a \,\blacksquare\, p), D(p+q), D(p|q) :$
  $a \in W(\mathrm{CCSO})|_{action}, p, q \in G(\mathrm{CCSO}, x)|_{ccs}\}$,
  $OC_{\mathrm{CCSO}}(x) = \{p \xrightarrow{a} q : p, q \in G(\mathrm{CCSO}, x)|_{ccs},$
  $a \in W(\mathrm{CCSO})|_{action}\} \cup DC_{\mathrm{CCSO}}(x)$.

Thus, when dealing with simulation predicates, we have only to consider binary relations on ground terms of sort ccs and precisely those terms, which do not contain '$\rightarrow$'. Notice also that, because of the form of the axioms of CCSO, when instantiating the clause $c[u] \approx r \Rightarrow c'[v] \approx r$ of the definition of $S(\sim, u, v)$, we have only to consider either $r = tr$ or $r = true$; moreover, $D(t) =_{\mathrm{CCSO}} true$ is by definition equivalent to $\mathrm{CCSO} \vdash D(t)$.

Thus we have, for any $\sim \subseteq G(\mathrm{CCSO})|_{ccs}{}^2$, $H_{=_{\mathrm{CCSO}}}(\sim) = \{ \langle p, q \rangle \in G(\mathrm{CCSO})|_{ccs}{}^2$: for any $a \in W(\mathrm{CCSO})|_{action}$ and for any $c1[x], c2[x], c[x] \in G(\mathrm{CCSO}, x : ccs)|_{ccs}$, there exist $c1'[x], c2'[x]$, $c'[x] \in G(\mathrm{CCSO}, x : ccs)|_{ccs}$ with $c1'[x] \sim^* c1[x]$, $c2'[x] \sim^* c2[x]$, $c'[x] \sim^* c[x]$ s.t.

$$\begin{aligned}
&(\mathrm{H1})\ c1[p] \xrightarrow{a} c2[p] =_{\mathrm{CCSO}} tr \Rightarrow c1'[q] \xrightarrow{a} c2'[q] =_{\mathrm{CCSO}} tr,\\
&(\mathrm{H'2})\ \mathrm{CCSO} \vdash D(a \,\blacksquare\, c[p]) \Rightarrow \mathrm{CCSO} \vdash D(a \,\blacksquare\, c'[q]),\\
&(\mathrm{H'3})\ \mathrm{CCSO} \vdash D(c1[p] + c2[p]) \Rightarrow \mathrm{CCSO} \vdash D(c1'[q] + c2'[q]),\\
&(\mathrm{H'4})\mathrm{CSO} \vdash D(c1[p]|c2[p]) \Rightarrow \mathrm{CCSO} \vdash D(c1'[q]|c2'[q]) \ \}
\end{aligned} \tag{9}$$

This characterization of $H_{=_{\mathrm{CCSO}}}(\sim)$ can be simplified.

First note that H1 implies H2, where

$$(\mathrm{H2}) \quad \mathrm{CCSO} \vdash D(p) \Rightarrow \mathrm{CCSO} \vdash D(q). \tag{10}$$

Indeed if we assume $D(p)$ (here we drop CCSO $\vdash$), then for $a \in G(\mathrm{CCSO})|_{action}$ s.t. $D(a)$, we have $a \,\blacksquare\, p \xrightarrow{a} p = tr$ and hence, because of H1, $a \,\blacksquare\, q \xrightarrow{a} q = tr$ and thus, because of strictness, $D(q)$ holds.

On the other hand, H2 implies H$'$2, H$'$3 and H$'$4.

Indeed by structural induction it can be seen that, for $c[x] \in G(\mathrm{CCSO}, x)|_{ccs}$ with x of sort ccs, if $D(c[p])$ and $D(q)$ hold then $D(c[q])$ also holds; this is based on the fact that in CCSO from $D(a)$, $D(p)$, $D(q)$, we can infer, reasoning as above, that $D(a \,\blacksquare\, p)$, $D(p+q)$ and $D(p|q)$ hold. Hence we have

$$\begin{aligned}
H_{=_{\mathrm{CCSO}}}(\sim) = \{ \langle p, q \rangle \in G(\mathrm{CCSO})|_{ccs}{}^2 :\\
\mathrm{H1} \text{ and } \mathrm{CCSO} \vdash D(p) \Rightarrow \mathrm{CCSO} \vdash D(q) \ \}.
\end{aligned}$$

Now the relationship of this transformation and thus of $F_{=\text{CCSO}}$ to the transformations H_{MP} and F_{MP} introduced at the beginning of section 3 for defining Milner's and Park's strong bisimulation, can be better understood if we define the analogous of F_{MP}, still denoted by F_{MP}, for the algebraic specification CCSO.

Let us then define H_{MP} by $H_{MP}(\sim) = \{ \langle p, q \rangle \in W(\text{CCSO})|_{ccs}{}^2$: for any $a \in W(\text{CCSO})|_{action}$ and for any $p' \in G(\text{CCSO})|_{ccs}$, $p \xrightarrow{a} p'_{=\text{CCSO}}\, tr \Rightarrow [\exists q', q' \sim p' \text{ s.t. } q \xrightarrow{a} q'_{=\text{CCSO}}\, tr] \}$.

It is clear from the above characterization of $H_{=\text{CCSO}}(\sim)$ that, for any $\sim$, $F_{=\text{CCSO}}(\sim) \subseteq F_{MP}(\sim)$. We know from Milner's work (1980, 1983) that the greatest fixpoint $\approx_{MP}$ of F_{MP} is a congruence over the terms of sort *ccs* (w.r.t. the operations $\blacksquare$, +, | and *nil*). However from those results, we cannot conclude that it satisfies the axioms of the specification: if $p \xrightarrow{a} p'_{=\text{CCSO}}\, tr$ and $q' \approx_{MP} p'$, then we are not authorized to assert that $p \xrightarrow{a} q'_{=\text{CCSO}}\, tr$, while this assertion corresponds exactly to our intuition on strong bisimulation. Hence we need to extend $\approx_{MP}$ by closure w.r.t. the other operations of the signature, in order to have the axioms satisfied by $\approx_{MP}$.

In the rest of the paper we will provide a characterization of such a bisimulation congruence as the greatest element of a lattice of congruences generated by the fixpoints of $F_{=\text{CCSO}}$, in such a way that the axioms are satisfied by that congruence and thus obtaining a fully algebraic counterpart of the treatment of Milner and Park. But our technique will deal with hierarchical specifications and not only with algebraic labelled transition systems.

However note that the definition, later in Proposition 2, of the greatest bisimulation congruence corresponds roughly to the definition of $\approx_{MP}$ as $\approx^{\alpha}_{MP}$ for some $\alpha < \omega_1$ plus the provision of closure by congruence (see Definition 5).

For proving further properties of simulation relations another more syntactic characterization is helpful.

Definition 4. Let $C[x, x_1, \ldots, x_n]$, also denoted by C when there is no ambiguity, be the following set of contexts of primitive sort:

$$C =_{\text{def}} \{c \in OC_{\Sigma}(x, x_1, \ldots, x_n)|_P : x, x_1, \ldots, x_n \text{ are of non-primitive sor}$$
$$\text{and for all non-primitive subterms } t \text{ of } c \text{ either } x \text{ is subterm of } t \text{ c}$$
$$t \equiv x_i \text{ for some } i \in \{1, \ldots, n\}\}. \square$$

Hence the elements of C are contexts of primitive sort which do not contain composed "x-free" non-primitive terms; in particular they do not contain non-primitive ground terms.

Examples from CCSO.

$$x_1|x \xrightarrow{a} x_2 \in C,$$
$$x_1|x \xrightarrow{a} x_3|x_4 \notin C,$$
$$(a \blacksquare x_1)|x \xrightarrow{a} x_1|x \notin C.$$

Lemma 1. *Let $\approx$ be a hierarchical Σ-congruence and $\sim \subseteq G(\Sigma)|_{S^+}{}^2$. Then for all $u, v \in G(\Sigma)|_{S^+}$ the following statements are equivalent:*

(1) $S_{\approx}(\sim, u, v)$
(2) $\forall c \in C[x, x_1, \ldots, x_n], p_1, \ldots, p_n \in G(\Sigma)|_{S^+}, r \in Obs$: $c[u, p_1, \ldots, p_n] \approx r \Rightarrow \exists q_1, \ldots, q_n \in G(\Sigma)|_{S^+} : p_i \sim q_i$ for $i = 1, \ldots, n$ and $c[v, q_1, \ldots, q_n] \approx r$.

Proof.

(1) $\Rightarrow$ (2)

Let $c \in C[x, x_1, \ldots, x_n]$, $p_1, \ldots, p_n \in G(\Sigma)|_{S^+}$ and $r \in$ Obs be given such that $c[u, p_1, \ldots, p_n] \approx r$ holds. Then $c[x, p_1, \ldots, p_n]$ is an observable context. Because of (1) there exists c' such that $c' \sim^* c[x, p_1, \ldots, p_n]$ and $c'[v] \approx r$ and, due to the definition of $\sim^*$, the equivalence can only be obtained by replacement of ground generating terms of non-primitive sort in $c[x, p_1, \ldots, p_n]$. But the only such ground terms are $p_1, \ldots, p_n$ (which by definition of C are maximal). Therefore c' has the form $c[x, q_1, \ldots, q_n]$ for some terms $q_1, \ldots, q_n$ with $p_i \sim q_i$ for $i = 1, \ldots, n$. This implies $c[v, q_1, \ldots, q_n] \approx r$.

(2) $\Rightarrow$ (1)

Let $c[u] \approx r$ for $c \in OC_{\Sigma}(x)$ and $r \in$ Obs. Then let $p_1, \ldots, p_n$ be the maximal ground terms of non-primitive sort in c, i.e. $c \equiv c''[x, p_1, \ldots, p_n]$ where $c'' \in C[x, x_1, \ldots, x_n]$.

Because of (2) there exist $q_1, \ldots, q_n \in W(\Sigma)|_{S^+}$ such that $q_i \sim p_i$ for $i = 1, \ldots, n$ and $c''[v, q_1, \ldots, q_n] \approx r$. Hence for $c' =_{\text{def}} c''[x, q_1, \ldots, q_n]$ we have $c'[v] \approx r$ and $c'' \sim^* c''[x, p_1, \ldots, p_n] \equiv c$. □

Fact 2. Every $\sim \in S(\approx)$ is an equivalence relation.

Proof. Reflexivity and symmetry are obvious and we prove now transitivity. Let for $u, v, w \in G(\Sigma)|_{S^+}{}^2$ of the same sort s, $u \sim v$ and $v \sim w$ hold; in order to show $u \sim w$, consider any context $c \in OC_\Sigma(x{:}s)$ and any $r \in$ Obs. Because of $S_\approx(\sim, u, v)$ and $S_\approx(\sim, v, w)$ there exist contexts c' and c'' with $c \sim^* c'$ and $c' \sim^* c''$ such that $c[u] \approx r$ implies $c'[v] \approx r$ which in turn implies $c''[w] \approx r$. Thus $S_\approx(\sim, u, w)$ holds. □

Fact 3. For every $\sim \in S(\approx)$ and all $t, t' \in W(\Sigma_P), e \in W(\Sigma, x)|_{S^+}$, with t, t', x of the same sort:

$$t \approx t' \Rightarrow e[t/x] \sim e[t'/x].$$

Proof. $t \approx t'$ implies $e[t/x] \approx e[t'/x]$. Because of Fact 1 $\approx|_{S^+} \subseteq \sim$ and hence $e[t/x] \sim e[t'/x]$ holds as well. □

For any $\sim \in S(\approx)$ denote by $\sim_P$ the union of $\sim$ and $\approx|_{\Sigma_P}$:

$$\sim_P =_{\text{def}} \sim \cup \approx|_{\Sigma_P}. \tag{11}$$

Because of the disjointness of both relations and the previous Fact 2, $\sim_P$ is an equivalence relation. It has the following properties.

Fact 4. For any $\sim \in S(\approx)$ and $t, t', t_1, \ldots, t_n \in G(\Sigma)$ with t, t' of sort s and t_i of sort s_i, the following holds :

(s1) for all $f : s_1 \times \ldots \times s_n \to s \in F_{\text{out}}$, $t_1, \ldots, t_n, u \in G(\Sigma)$, $r \in$ Obs $|_s$, $i \in \{1, \ldots, n\}$, $t_i \sim_P u \wedge f(t_1, \ldots, t_n) \approx r \Rightarrow$ $\exists u_1, \ldots, u_{i-1}, u_{i+1}, \ldots, u_n \in G(\Sigma)$ s.t. $t_j \sim_P u_j$ for $j \neq i \wedge$ $f(u_1, \ldots, u_{i-1}, u, u_{i+1}, \ldots, u_n) \approx r$;

(s2) for all $t, t' \in G(\Sigma)$ of same sort: $D(t) \approx true \wedge t \sim_P t' \Rightarrow$ $D(t') \approx true$. □

As a consequence of this fact every $\sim \in S(\approx)$ can be extended in a natural way to a hierarchical partial Σ-congruence (w.r.t. $\approx$) which will be denoted by $\langle\sim\rangle_\approx$.

Definition 5. For $\sim \in S(\approx)$, let $\langle\sim\rangle_\approx$be the smallest (w.r.t $\leq$) partial Σ-congruence generated as follows:

(sc0) for all $u, v \in G(\Sigma) : u\langle\sim\rangle_\approx v$ iff $u \sim_P v$;

(sc1) for all $f : s_1 \times \ldots \times s_n \rightarrow s \in F_{\text{out}}, t_i \in G(\Sigma)|_{s_i}$ for $i = 1, \ldots, n$ and $r \in \text{Obs}|_s : f(t_1, \ldots, t_n)\langle\sim\rangle_{\approx} r$ iff $\exists t'_i \in G(\Sigma)|_{s_i}$ for $i = 1, \ldots, n : t_1\langle\sim\rangle_{\approx} t'_1 \wedge \ldots \wedge t_n\langle\sim\rangle_{\approx} t'_n \wedge f(t'_1, \ldots, t'_n) \approx r$;

(sc2) for all $t \in W(\Sigma) : D(t)\langle\sim\rangle_{\approx} \mathit{true}$ iff $\exists t' \in G(\Sigma) : t'\langle\sim\rangle_{\approx} t \wedge D(t') \approx \mathit{true}$.

Congruences satisfying (sc0)–(sc2) are called **simulation congruences**:

$SC(\approx) =_{def}$

$\{ \langle\sim\rangle_{\approx} : \langle\sim\rangle_{\approx}$ is a partial Σ-congruence satisfying (sc0), (sc1) and (sc2) for some $\sim \in S(\approx)$ $\}$. □

Fact 5. For each $\langle\sim\rangle_{\approx} \in SC(\approx)$, the following holds:

(0) $\langle\sim\rangle_{\approx}$is hierarchical and, restricted to $G(\Sigma)|_{S^+}{}^2$ it is a simulation relation w.r.t. $\approx$;

(1) the definedness of generating terms is fixed by $\approx$, i.e. for all $t \in G(\Sigma) : D(t)\langle\sim\rangle_{\approx} \mathit{true}$ iff $D(t) \approx \mathit{true}$;

(2) $\langle\sim\rangle_{\approx}$ *bisimulates* $\approx$, i.e. for all $t, t' \in G(\Sigma) : t\langle\sim\rangle_{\approx} t'$ implies that for all $t_1, \ldots, t_n \in G(\Sigma)$, $r \in$ Obs and $c \in C[x, x_1, \ldots, x_n]$: $c[t/x, t_1/x_1, \ldots, t_n/x_n] \approx r \Rightarrow \exists t'_1, \ldots, t'_n \in G(\Sigma){:}t_1\langle\sim\rangle_{\approx} t'_1 \wedge \ldots \wedge t_n\langle\sim\rangle_{\approx} t'_n \wedge c[t'/x, t'_1/x_1, \ldots, t'_n/x_n] \approx r$;

(3) $\langle\sim\rangle_{\approx}$is *fully abstract*, i.e. for all $t, t' \in W(\Sigma)$ of the same sort s: $t\langle\sim\rangle_{\approx} t'$ iff for all $c \in W(\Sigma, x{:}s)|_{S_P}$, $c[t/x]\langle\sim\rangle_{\approx} c[t'/x]$.

Proof.

(0) is obvious from the definition of $\langle\sim\rangle_{\approx}$.

(1) "⇐" follows from (sc2) by reflexivity of $\langle\sim\rangle_{\approx}$. "⇒" Let $t \in G(\Sigma)$ and $D(t)\langle\sim\rangle_{\approx} \mathit{true}$. Then there exists $t' \in G(\Sigma)$ s.t. $t\langle\sim\rangle_{\approx} t'$ and $D(t') \approx \mathit{true}$. Because of (sc0) $t \sim_P t'$ holds and hence by (s2) $D(t) \approx \mathit{true}$.

(2) W.l.o.g. let t and t' in $G(\Sigma)$ be of non-primitive sort. Then $t\langle\sim\rangle_{\approx} t'$ implies $t \sim t'$. Because of $\sim \in S(\approx)$, it is sufficient to apply lemma 1.

(3) "⇒" is obvious since $\langle\sim\rangle_{\approx}$is a congruence. "⇐" is obvious for t, t' of primitive sort. Since $\langle\sim\rangle_{\approx}$ is hierarchical (cf. 0) we can obviously consider $t, t' \in G(\Sigma)$ be of non-primitive sort. In order to apply Lemma 1, let $c \in C[x, x_1, \ldots, x_n]$ and $u_1, \ldots, u_n \in G(\Sigma)$ of the same sorts as $x_1, \ldots, x_n$ and let $r \in$ Obs such that $c[t/x, u_1/x_1, \ldots, u_n/x_n] \approx r$.

Then $c[x, u_1/x_1, \ldots, u_n/x_n] \in W(\Sigma, x)|_{S_P}$ and therefore $c[t'/x, u_1/x_1, \ldots, u_n/x_n] \langle\sim\rangle_{\approx} r$.

By (sc1) there exist $t'', u_1'', \ldots, u_n''$ such that $t''\langle\sim\rangle_{\approx} t', u_1''\langle\sim\rangle_{\approx} u_1, \ldots, u_n''\langle\sim\rangle_{\approx} u_n$ and $c[t''/x, u_1''/x_1, \ldots, u_n''/x_n] \approx r$ holds. Because of $t'\langle\sim\rangle_{\approx} t''$, we have $S_{\approx}(\langle\sim\rangle_{\approx}, t', t'')$. Hence Lemma 1 implies $c[t'/x, u_1'/x_1, \ldots, u_n'/x_n] \approx r$ for some appropriately chosen $u_1', \ldots, u_n'$ with $u_1'\langle\sim\rangle_{\approx} u_1'', \ldots, u_n'\langle\sim\rangle_{\approx} u_n''$. Because of the transitivity of $\langle\sim\rangle_{\approx}$ we can apply Lemma 1. Hence $S_{\approx}(\langle\sim\rangle_{\approx}, t, t')$ holds and therefore $t\langle\sim\rangle_{\approx} t'$.

□

4. The Lattice of Simulation Congruences

If a simulation relation is included into another, then the associated simulation congruences are ordered by $\leq$.

Fact 6. Assume that $\sim_1, \sim_2 \in S(\approx)$ hold for some hierarchical partial congruence $\approx$ with $\sim_1 \subseteq \sim_2$. Then $\langle\sim_1\rangle_{\approx} \leq \langle\sim_2\rangle_{\approx}$ holds for the associated simulation congruences.

Proof. Because of $D(t)\langle\sim_1\rangle_{\approx} true$ iff $D(t) \approx true$ iff $D(t)\langle\sim_2\rangle_{\approx} true$ for all $t \in G(\Sigma)$, ($\leq$1) and ($\leq$2) hold obviously for all $t, t' \in G(\Sigma)$.

Because of property (sc2), the inclusion of $\langle\sim_1\rangle_{\approx}$ in $\langle\sim_2\rangle_{\approx}$ implies ($\leq$1) also for all $t \in W(\Sigma) - G(\Sigma)$. Then, by a straightforward structural induction, ($\leq$2) can be proven using (sc1). □

Proposition 2. *Let $\approx$ be a given partial Σ-congruence (hierarchical w.r.t. Σ_P). Then the following holds:*

(1) *$SC(\approx)$ forms a complete lower semilattice;*

(2) *the smallest element of $SC(\approx)$ is induced by the following simulation relation $\sim_I$: for all $u, v \in G(\Sigma)|_{S^+}$: $u \sim_I v$ iff for all $c \in OC_{\Sigma}(x)$ and $r \in Obs$: $c[u/x] \approx r$ iff $c[v/x] \approx r$.*

(3) *$SC(\approx)$ contains a greatest element which is induced by the simulation relation $\sim_Z =_{def} \bigcap_{i<\omega_1} \sim_i$ where, for all $u, v \in G(\Sigma)|_{S^+}$ and $i < \omega_1$, $u \sim_i v$ is defined as follows:*
 $u \sim_o v$ iff u and v are of the same sort;
 $u \sim_{i+1} v$ iff $S_{\approx}(\sim_i, u, v)$;
 $u \sim_{\alpha} v$ iff $u \sim_i v$ for all $i < \alpha$, α limit ordinal.

(4) *$SC(\approx)$ forms a complete lattice.*

(Hence the smallest element $\sim_I$ is fully abstract w.r.t. $\approx$ and the greatest element $\sim_Z$ is the "most abstract" simulation congruence, the socalled "bisimulation congruence").

Proof of (1). Let a family Con $\subseteq SC(\approx)$ of simulation congruences be given. We define a $\subseteq$-decreasing chain $\{\sim_i^{\text{Con}}\}$ of relations over $G(\Sigma)|_{S+}$ which are contained in all simulation relations generating the elements of Con such that the limit of these relations is a simulation relation generating the greatest lower bound of Con.

Let for all $u, v \in G(\Sigma)|_{S+}$ $u \sim_i^{\text{Con}} v$ $(i < \omega_1)$ be defined as follows:

$u \sim_o^{\text{Con}} v$ iff $u \sim v$ for all $\sim\in$ Con;
$u \sim_{i+1}^{\text{Con}} v$ iff $S_\approx(\sim_i^{\text{Con}}, u, v)$;
$u \sim_\alpha^{\text{Con}} v$ iff $u \sim_i^{\text{Con}} v$ for all $i < \alpha$, if α is a limit ordinal.

Then we have to prove that (1.1) $\sim_i^{\text{Con}}$ is a decreasing chain, (1.2) $\sim_{\text{Con}} =_{def} \lim_{i<\omega_1} \sim_i^{\text{Con}}$ is a simulation relation and (1.3) $\sim_{\text{Con}}$ is the greatest lower bound of the restriction of Con to $G(\Sigma)|_{S+}$. Then by Fact 6 the simulation congruence generated by $\sim_{\text{Con}}$ is the greatest lower bound of Con.

Proof of (1.1). $\{\sim_i^{\text{Con}}\}_{i<\omega_1}$ is a decreasing chain. (Proof by transfinite induction)

(a) $i = 0$.
Let $u, v \in G(\Sigma)|_{S+}$ and $u \sim_1 v$ hold. Then $S_\approx(\sim_o, u, v)$ holds. Hence by Lemma 1, for all $c \in C[x, x_1, \ldots, x_n]$, $u_1, \ldots, u_n \in G(\Sigma)|_{S+}$ of appropriate sorts and $r \in$ Obs there exist $v_1, \ldots, v_n$ such that
$c[u/x, u_1/x_1, \ldots, u_n/x_n] \approx r$ implies
  $u_i \sim v_i$ for all $\sim\in$ Con and $i = 1, \ldots, n$ and
  $c[v/x, v_i/x_i, \ldots, v_n/_x n] \approx r$.
Thus again by Lemma 1, $S_\approx(\sim, u, v)$ holds for all $\sim\in$ Con. Since any $\sim\in$ Con is a simulation relation, this means $u \sim v$ holds for all $\sim\in$ Con .
Therefore $\sim_1^{\text{Con}} \subseteq \sim_o^{\text{Con}}$.

(b) $i \Rightarrow i + 1$.
Let $u, v \in G(\Sigma)|_{S+}$ and $u \sim_{i+2}^{\text{Con}} v$ hold. As in (a) we apply twice Lemma 1.

Then for all $c \in C[x, x_1, \dots, x_n]$, $u_1, \dots, u_n \in G(\Sigma)|_{S^+}$ and $r \in \text{Obs}$ the validity of $c[u/x, u_1/x_1, \dots, u_n/x_n] \approx r$ implies the existence of $v_1, \dots, v_n$ with $v_j \sim_{i+1}^{\text{Con}} u_j$ for $j = 1, \dots, n$ and $c[v/x, v_1/x_1, \dots, v_n/x_n] \approx r$.
Because of the inductive hypothesis $v_j \sim_i^{\text{Con}} u_j$ holds for $j = 1, \dots, n$ and therefore $S_{\approx}(\sim_i^{\text{Con}}, u, v)$ which is equivalent to $u \sim_{i+1}^{\text{Con}} v$.

(c) α limit ordinal.
Obviously $u \sim_\alpha^{\text{Con}} v$ implies $u \sim_i^{\text{Con}} v$ for $i < \alpha$ by definition. □ 1.1

Proof of (1.2). $\sim_{\text{Con}}$ is a simulation relation.

Since $G(\Sigma)|_{S^+}{}^2$ is denumerable, every subset of $G(\Sigma)|_{S^+}{}^2$ is denumerable and hence every decreasing chain of subsets is denumerable. Hence every decreasing chain of relations is stationary for some $i_o < \omega_1$, i.e. for all $j \geq i_o$ (and $j \leq \omega_1$) $\sim_j^{\text{Con}} = \sim_{i_o}^{\text{Con}}$ holds.

Hence we conclude: $u \sim_{\text{Con}} v$ iff $u \sim_{i_o+1}^{\text{Con}} v$ iff $S_{\approx}(\sim_{i_o}^{\text{Con}}, u, v)$ iff $S_{\approx}(\sim_{\text{Con}}, u, v)$. □ 1.2.

Proof of (1.3). Every lower bound $lb \in S(\approx)$ of Con (w.r.t. $\subseteq$) is a lower bound of $\sim_{\text{Con}}$. Proof by transfinite induction that for all $i < \omega_1$, $\sim_{lb} \subseteq \sim_i^{\text{Con}}$ holds:

(a) $i = 0$ is obvious since $\sim_{lb}$ is a lower bound of Con;

(b) $i \Rightarrow i+1$: let for $u, v \in G(\Sigma)|_{S^+}{}^2$, $u \sim_{lb} v$ hold. Because of $\sim_{lb} \in S(\approx)$, $S_{\approx}(\sim_{lb}, u, v)$ holds. By Lemma 1 for all $c \in C[x, x_1, \dots, x_n]$, $u_1, \dots, u_n \in G(\Sigma)$ and $r \in \text{Obs}$, the validity of $c[u/x, u_1/x_1, \dots, u_n/x_n] \approx r$ implies the existence of $v_1, \dots, v_n$ with $u_j \sim_{lb} v_j$ for $j = 1, \dots, n$ and $c[v/x, v_1/x_1, \dots, v_n/x_n] \approx r$. By inductive hypothesis we get $u_j \sim_i^{\text{Con}} v_j$ for $j = 1, \dots, n$. Hence $S(\sim_i^{\text{Con}}, u, v)$ holds. By definition of $\sim_{i+1}^{\text{Con}}$ we get $u \sim_{i+1}^{\text{Con}} v$.

(c) α limit ordinal: if $\sim_{lb} \subseteq \sim_i^{\text{Con}}$ for all $i < \alpha$, then obviously $\sim_{lb} \subseteq \bigcap_{i<\alpha} \sim_i^{\text{Con}}$. □ 1.3. □ 1.

Proof of (2). We have to show that (2.1) $\sim_I \in S(\approx)$ and that (2.2) $\sim_I$ is a lower bound of $S(\approx)$. Then by Fact 6 the induced simulation congruence $\sim_I$ is the smallest element of $SC(\approx)$.

Proof of (2.1). $\sim_I \in S(\approx)$. Let $u, v \in G(\Sigma)|_{S^+}$. If $u \sim_I v$ and $c[u/x, u_1/x_1, \dots, u_n/x_n] \approx r$ for $c \in C[x, x_1, \dots, x_n]$, $u_1, \dots, u_n \in$

$G(\Sigma)$, $r \in$ Obs, then one may just choose $v_j \equiv u_j$ for $j = 1, \ldots, n$. Then by definition of $\sim_I$, $c[v/x, v_1/x_1, \ldots, v_n/x_n] \approx r$ and hence by Lemma 1 $S_{\approx}(\sim_I, u, v)$ holds.

On the other hand, let $S_{\approx}(\sim_I, u, v)$ hold. We have to show that for all $c \in OC_{\Sigma}(x)$, u and v are $\sim_I$-equivalent. Now any $c \in OC_{\Sigma}(x)$ can be seen as $c'[u_1/x_1, \ldots, u_n/x_n]$ for some $n \in N, c' \in C[x, x_1, \ldots, x_n]$ and $u_1, \ldots, u_n \in G(\Sigma)$.

Because of $S_{\approx}(\sim_I, u, v)$, for all $r \in Obs$ with $c[u/x] = c'[u/x, u_1/x_1, \ldots, u_n/x_n] \approx r$ there exist $v_1, \ldots, v_n \in G(\Sigma)$ such that $u_j \sim_I v_j$ for $j = 1, \ldots, n$ and $c'[v/x, v_1/x_1, \ldots, v_n/x_n] \approx r$ hold. Now, $c'[v/x, ./x_1, \ldots, v_n/x_n]$ can be considered as x_1-context in $OC_{\Sigma}(x_1)$. Hence by definition of $\sim_I, c'[v/x, u_1/x_1, v_2/x_2, \ldots, v_n/x_n] \approx r$ holds.

By induction on i we get for all $i = 1, \ldots, n$ that

$$c'[v/x, u_1/x_1, \ldots, u_i/x_i, v_{i+1}/x_{i+1}, \ldots, v_n/x_n] \approx r$$

is valid; hence $c[v/x] \equiv c'[v/x, u_1/x_1, \ldots, u_n/x_n] \approx r$. □ 2.1

Proof of (2.2). $\sim_I$ is a lower bound of $S(\approx)$ (w.r.t. $\subseteq$).

Let $\sim \in S(\approx)$ and $u \sim_I v$ for $u, v \in G(\Sigma)|_{S^+}$. Then $S_{\approx}(\sim, u, v)$ holds by choosing v_j identical to u_j for $j = 1, \ldots, n$ which implies $u \sim v$. □ 2.2 □ 2

Proof of (3). We have to prove that (3.1) $\{\sim_i\}$ is a decreasing chain, that (3.2) $\sim_Z$ is a simulation relation and that (3.3) is an upper bound of all simulation relations in $S(\approx)$.

Proof of (3.1). $\{\sim_i\}_{i<\omega_1}$ is a decreasing chain.

The proof is done by transfinite induction where the case $i = 0$ is obvious since $\sim_o$ is the largest relation on $G(\Sigma)|_{S^+}$ and where the induction is exactly as in the Proof 1.1. □ 3.1

Proof of (3.2). $\sim_Z$ is a simulation relation.

The proof is exactly as the one for 1.2. □ 3.2

Proof of (3.3). $\sim_Z$ is an upper bound for $S(\approx)$.

The proof (by transfinite induction) that for all $i < \omega_1$ and all $\sim \in S(\approx)$ holds is completely analogous to the proof of 1.3. □ 3.3. □ 3.

5. Models of Hierarchical Specifications

We consider hierarchical specifications with non-primitive axioms being first-order equivalent to the following special form:

(∗) $\forall x_1{:}s_1, \ldots, x_m{:}s_m.\ (D(r_1) \wedge u_1 = r_1) \wedge \ldots \wedge (D(r_n) \wedge u_n = r_n) \Rightarrow e$ where $m, n > 0$, and for some $j < m$ $s_1, \ldots, s_j \in S^+$, $s_{j+1}, \ldots, s_m \in S_P$, $u_1, \ldots, u_n \in OC_\Sigma(x_1, \ldots, x_m)$, $r_1, \ldots, r_n \in G(\Sigma, x_{j+1}, \ldots, x_m)|_{S_P}$ and where e has of one of the following forms:

(i) $D(u)$ with u of non-primitive sort;

(ii) $u = v$ such that either $u, v \in G(\Sigma, x_1, \ldots, x_m)$ or, if u and v are of primitive sort , then $D(u)$ follows from the premises of the axiom and the primitive axioms.

Note that the above conditions on the form of the axioms imply almost the partial completeness of the hierarchical specification T. If T satisfies the following stronger version of (i) and (ii):

(i') $D(u)$ with $u \in G(\Sigma, x_1, \ldots, x_m)|_{S^+}$ or

(ii') $u = v$ such that either $u, v \in G(\Sigma, x_1, \ldots, x_m)$ or $[u \in OC_\Sigma(x_1, \ldots, x_m)$ and $v \in G(\Sigma, x_1, \ldots, x_m)|_{S_P}]$,

then T is partially complete.

Example CCSO. The non-primitive axioms of CCSO are first-order equivalent to axioms of the form (∗): since $D(tr)$ holds in the primitive specification BOOL, the definedness of all left-hand sides of premises and conclusion of the axioms (▪1)–(C3) can be deduced. Moreover, all these left-hand-sides are in $OC_{\text{Sig(CCSO)}}(a, p, p', q, q')$. The terms of axiom (D) are in $G(\text{Sig(CCSO)}, p, q)$. Hence CCSO satisfies also the condition (i') and (ii'); therefore it is partially complete.

The form (∗) of the axioms is a reasonable restriction. For example also the axioms of the algebraic specification of CSP in Broy and Wirsing (1982) are first-order equivalent to this form.

The following proposition shows in particular that for such specifications all simulation congruences of a specification T (that are the simulation congruences w.r.t. the ground atomic theory $=_T$ of T) are hierarchical models of T.

Proposition 3. *Let T be a partially complete and hierarchy-consistent hierarchical specification such that the primitive specification is monomorphic and all non-primitive axioms of T are first-order equivalent to formulas of the form $(*)$.*

Then the set $SC(=_T)$ of all simulation congruences w.r.t. the ground atomic theory $=_T$ of T satisfies the following properties:

(1) *$SC(=_T)$ is a complete lattice the smallest element of which is the congruence associated with the weakly terminal algebra of T.*
(2) *$SC(=_T) \subseteq \mathrm{Mod}(T)$.*

Proof of (1). By Proposition 1, $SC(=_T)$ forms a complete lattice with a smallest element $\sim_I$ which, according to Broy (1982), is just the weakly terminal algebra in T.

Proof of (2). We have to show that every $\sim \in SC(=_T)$ satisfies the non-primitive axioms of T. Let such an axiom be given, which is first-order equivalent to the form $(*)$.

Let $t_1, \ldots, t_j \in G(\Sigma)|_{S^+}$ and $t_{j+1}, \ldots, t_m \in G(\Sigma)|_{S_P}$ be given such that $D^{\sim}(t_i)$ holds for $i = 1, \ldots, m$. Assume that the premises

$$\begin{aligned}
&D(r_1[t_{j+1}/x_{j+1}, \ldots, t_m/x_m]),\\
&\qquad u_1[t_1/x_1, \ldots, t_m/x_m] = r_1[t_{j+1}/x_{j+1}, \ldots, t_m/x_m], \ldots\\
&\ldots\\
&D(r_n[t_{j+1}/x_{j+1}, \ldots, t_m/x_m]),\\
&\qquad u_n[t_1/x_1, \ldots, t_m/x_m] = r_n[t_{j+1}/x_{j+1}, \ldots, t_m/x_m]
\end{aligned}$$

hold in $\sim$. According to Fact 5(1), the definedness of the r_i does not depend on the particular congruence; thus all terms in the premises are defined in $=_T$.

By the property (sc1) of $\sim \in SC(=_T)$ there exist $t'_1, \ldots, t'_j$ with $t_1 \sim t'_1, \ldots, t_j \sim t'_j$ such that $u_i[t'_1/x_1, \ldots, t'_j/x_j, t_{j+1}/x_{j+1}, \ldots, t_m/x_m] =_T r_i[t_{j+1}/x_{j+1}, \ldots, t_m/x_m]$ holds for $i = 1, \ldots, n$. Therefore the conclusion

$$e' =_{def} e[t'_1/x_1, \ldots, t'_j/x_j, t_{j+1}/x_{j+1}, \ldots, t_m/x_m]$$

of the axiom holds in $=_T$.

Case 1. e' has the form $D(t)$ or $u = v$ with u, v defined in $=_T$. Then $D(t)$ and $u = v$, resp., hold in the weakly terminal congruence of T which is the smallest congruence $\sim_I$ of $SC(=_T)$. Because of $\sim_I \subseteq \sim$, e' holds in $\sim$. The equivalences $t_i \sim t'_i$ for $i = 1, \ldots, j$ then imply that $e[t_1/x_1, \ldots, t_m/x_m]$ holds in $\sim$.

Case 2. If e' contains undefined terms w.r.t. $=_T$, then e' is of form $u = v$, with $u, v \in G(\Sigma)$ and of non-primitive sort. Since u and v are undefined w.r.t. $=_T$, Fact 5(1) implies that u and v are both undefined w.r.t. $\sim$. Thus $u \sim v$ holds. □

Every partial congruence corresponds to an isomorphism class of term generated algebras (cf. e.g. (Broy, 1984)). Hence, if we consider models instead of congruences, we obtain the following corollary:

Corollary 2. *Let T be a specification which satisfies the same assumptions as in Proposition 3. Then T possesses a complete lattice of (isomorphism classes) of fully abstract models the smallest of which is the weakly terminal algebra of T and the greatest element of which is the "bisimulation algebra" of T (which corresponds to the bisimulation congruence). All these fully abstract models are associated with simulation congruences of $=_T$.*

Example CCSO. The specification CCSO satisfies the assumptions of Proposition 3. As already stated above, the non-primitive axioms are first-order equivalent to a form which implies the partial completeness. The hierarchy-consistency can be proven by the confluence of the axioms when considered as a left-to-right term rewriting system.

Therefore, CCSO possesses a complete lattice (of isomorphism classes of) fully abstract models; in particular, CCSO has a weakly terminal model and (as expected) a bisimulation model which corresponds to strong bisimulation.

Reichel (1979) and Andréka (1980) use instead of strong equality the so-called "existential equality" $=^e$ for partial algebra semantics of computation structures. The existential equality $t =^e t'$ holds between two terms in an algebra A iff both terms are defined in A and their interpretation is equal. For specifications written with such positive conditional "existential" equations one obtains immediately the following corollary:

Corollary 3. *Let T be a partially complete and hierarchy-consistent specification such that the primitive specification is monomorphic and all non-primitive axioms of T are first-order equivalent to formulas of the form* $\forall x_1{:}s_1, \ldots, x_m{:}s_m.\ u_1 =^e r_1 \wedge \ldots \wedge u_n =^e r_n \Rightarrow u =^e v$ *where* $m, n > 0$ *and for some* $j < m$ $s_1, \ldots, s_j \in S^+$; $s_{j+1}, \ldots, s_m \in S_p$; $u_1, \ldots, u_n \in OC_\Sigma(x_1, \ldots, x_m)$; $r_1, \ldots, r_n \in G(\Sigma, x_{j+1}, \ldots, x_m)|_{S_P}$ *and either* ($u \in OC_\Sigma(x_1, \ldots, x_m)$ *and* $v \in G(\Sigma, x_{j+1}, \ldots, x_m)|_{S_P}$) *or* $u, v \in G(\Sigma, x_1, \ldots, x_m)$.

Then the set $SC(=_T)$ of all simulation congruences w.r.t. the ground atomic theory $=_T$ of T satisfies the following properties:

(1) $SC(=_T) \subseteq \mathrm{Mod}(T)$ *and*
(2) $SC(=_T)$ *is a complete lattice the smallest element of which is the congruence associated with the weakly terminal algebra of T.*

Proof. Because of $u =^e r \Leftrightarrow (D(u) \wedge u = r)$ the corollary is a special case of the previous proposition. □

Acknowledgment

Thanks to Gianna Reggio for reading drafts of the paper. We also gratefully acknowledge the careful typing of Margit Pockes, Thomas Ramke and Ombretta Arvigo. Thanks go to Bernhard Reus for converting the MacIntosh-version to Latex-format.

Appendix A

Basic Notions on Partial Algebras

Here we collect the essential definitions on heterogeneous partial algebras used throughout the paper. For a reference see (Broy, 1982). Let Σ be a signature; we denote by $Opns(\Sigma)$ and $Sorts(\Sigma)$ respectively the operations and the sorts of Σ. If Σ is a signature, then a partial Σ-algebra is defined as a total Σ-algebra, except that the interpretations of the function symbols can be partially defined functions (partial functions).

The class of partial Σ-algebras will be denoted by $PAlg(\Sigma)$. For $s \in Sorts(\Sigma)$ and $A \in PAlg(\Sigma)$ we denote the carrier set of A corresponding to s by $A|_s$. If $\Sigma' \subseteq \Sigma$, then we denote by $A|_{\Sigma'}$ the **Σ'-reduct** of A, i.e. the Σ'-algebra where the carriers and the functions are the same as the corresponding carriers and functions in A. The **Σ-algebra of terms** on a $Sorts(\Sigma)$-sorted set of variables X will be denoted by $W(\Sigma, X)$. $W(\Sigma, \emptyset)$, **the algebra of ground terms** (i.e. without variables), will usually be written $W(\Sigma)$; hence for any sort s, $W(\Sigma, X)|_s$ is the set of terms of sort s. We often write $x{:}s$ to denote a variable x of sort s, i.e. $x \in X_s$.

If $A \in PAlg(\Sigma), t \in W(\Sigma, X)$ and $v : X \to A$ is a total valuation function, i.e. a sort-respecting assignment to the variables, then the **interpretation** $t^{A,v}$ is defined as usual, but note that now the value may be undefined; if t is a ground term, then we use the notation t^A. For any partial algebra A, any assignment v of values to X in A, and any sort s, a total definedness predicate $D_s^{A,v}$ on $W(\Sigma, X)|_s$ is defined, such that $D_s^{A,v}(t)$ is true iff $t^{A,v}$ is defined. We will usually drop the subscript indicating the sort.

For a given signature Σ, an **axiom** is a many-sorted first-order formula on Σ with an equality predicate symbol $=$ and a **definedness predicate** symbol D as the only predicate symbols.

The interpretation of an axiom in an algebra A is defined as usual, with the following remarks:

- for terms t_1, t_2 of the same sort and the valuation function v, $t_1 = t_2$ is true (written $A, v \models t_1 = t_2$) iff either $t_1^{A,v}$ and $t_2^{A,v}$ are both defined and equal in A or both are undefined (we say that $=$ denotes the "**strong equality**")
- existentially and universally quantified variables range over *defined values.*

The rules for first-order logic with partial algebras are the same as for total algebras except that for substituting terms for variables one has to take the definedness into account : for all terms t and variables x of the same sort, one has the inference rule

$$\frac{\Phi}{D(t) \Rightarrow \Phi[t/x]} . \tag{A1}$$

Moreover, since partial functions are strict (i.e. if an argument is undefined, then the result is undefined) the following axiom holds

$$D(f(t_1, \ldots, t_n)) \Rightarrow D(t_i) \quad \text{for} \quad i = 1, \ldots, n. \tag{A2}$$

Finally, note that strong equality implies that implicitly we assume for terms t, t' of the same sort:

$$\neg D(t) \wedge \neg D(t') \Rightarrow t = t'. \tag{A3}$$

If E is a set of formulas and Φ a formula, then by the notation $E \vdash \Phi$ we indicate that Φ can be derived from E in the first-order partial calculus; we use the same notation $T \vdash E$ if $T = (\Sigma, E)$ where Σ is a signature and E a set of formulas.

If Φ is a formula, v is a valuation function for the variables in Φ and A is an algebra, then $A, v \models \Phi$ indicates that Φ is true in the interpretation under the valuation v and $A \models \Phi$ that Φ is true in A for any valuation; in particular if Φ is a **sentence**, i.e. a formula without free variables, then $A \models \Phi$ indicates that Φ is true in A (then we say that A satisfies Φ).

For a given set E of sentences on a signature Σ, $PAlg((\Sigma, E))$ denotes the class of **models** of (Σ, E), i.e. the class of partial algebras on Σ satisfying all sentences in E. By $PGen((\Sigma, E))$ we indicate the class of the term generated algebras of $PAlg((\Sigma, E))$, i.e. the algebras in $PAlg((\Sigma, E))$ for which every element is the valuation of a ground term in W_Σ.

For any class C of Σ-algebras we define by

$$\begin{aligned} Mdef(C) = \{A \in C : \text{ for any } t \in W(\Sigma), A \\ \models D(t) \text{ iff } [B \models D(t) \text{ for any } B \in C] \} \end{aligned} \tag{A4}$$

the class of **minimally defined algebras** in C. A **minimally defined model** of (Σ, E) is an algebra in $Mdef(PAlg((\Sigma, E)))$.

Let $A, B \in PAlg(\Sigma)$ be partial Σ-algebras; then a family $(\varphi_s : A|_s \rightarrow |B_s)_{s \in S}$ of possibly partial operations is called **Σ-homomorphism** from A into B (denoted by $\varphi : A \rightarrow B$) iff for all

$$\begin{aligned} & f : s_1 \times \ldots \times s_n \rightarrow s \in Opns(\Sigma) \text{ and all } x_1 \in A|_{s_1}, \ldots, x_n \in A|_{s_n} \\ & f^A(x_1, \ldots, x_n) \text{ defined } \Rightarrow \varphi_s(f^A(x_1, \ldots, x_n)) \\ & \qquad = f^B(\varphi_{s_1}(x_1), \ldots, \varphi_{s_n}(x_n)). \end{aligned} \tag{A5}$$

Therefore on total algebras this notion of homomorphism coincides with the usual one.

In particular, if a Σ-homomorphism $\varphi : A \rightarrow B$ is total, then for all $f : s_1 \times \ldots \times s_n \rightarrow s \in Opns(\Sigma)$ and all $x_1 \in A|_{s_1}, \ldots, x_n \in A|_{s_n}$: $f^A(x_1, \ldots, x_n)$ defined $\Rightarrow f^B(\varphi_{s_1}(x_1), \ldots, \varphi_{s_n}(x_n))$ defined.

A Σ-homomorphism $\varphi : A \to B$ is said to be **weak**, iff for all $f : s_1 \times \ldots \times s_n \to s \in Opns(\Sigma)$ and all $x_1 \in A|_{s_1}, \ldots, x_n \in A|_{s_n}$

$$\begin{aligned} f^B(\varphi_{s_1}(x_1), \ldots, \varphi_{s_n}(x_n)) &\text{ defined} \\ \Rightarrow f^A(x_1, \ldots, x_n) &\text{ defined.} \end{aligned} \tag{A6}$$

Then $\varphi : A \to B$ is a weak Σ-homomorphism iff for all $f : s_1 \times \ldots \times s_n \to s \in Opns(\Sigma)$ and all $x_1 \in A|_{s_1}, \ldots, x_n \in A|_{s_n}$

$$\varphi({f_s}^A(x_1, \ldots, x_n)) = {f_s}^B(\varphi_{s_1}(x_1), \ldots, \varphi_{s_n}(x_n)). \tag{A7}$$

Again " $=$ " denotes the strong equality. Between two term generated Σ-algebras there exists at most one total or one weak Σ-homomorphism. If there exist both, a total one and a weak one, then they are identical. A Σ-homomorphism both total and weak is called **strong Σ-homomorphism**.

For a class C of Σ-algebras we define particular elements of C:

a Σ-algebra $I \in C$ is called **initial** (**strongly initial** resp.) in C iff for all $A \in C$ there exists a unique total (strong resp.) Σ-homomorphism $\varphi : I \to A$;

a Σ-algebra $Z \in C$ is called **terminal** (**strongly terminal** resp.) in C iff for all $A \in C$ there exists (at least) a weak (strong resp.) Σ-homomorphism $\varphi : A \to Z$;

a Σ-algebra $Z \in C$ is called **weakly terminal** in C iff it is terminal in *Mdef*(C).

A formula is said to be **positive conditional** iff it is first-order equivalent to

$$\bigwedge_{1 \le j \le n} (D(u_j) \wedge u_j = r_j) \Rightarrow \Phi \tag{A8}$$

where Φ is either $D(t')$ or $t' = t$ and u_j, r_j, t', t are Σ-terms and $n \geq 0$.

A **specification** T is pair (Σ, E), where E is a set of positive conditional formulas. By $Sig(T)$ we denote the signature Σ of T. A Σ-algebra I is said to be **initial in the specification** T iff I is initial in *PAlg*(T). A Σ-algebra Z is said **weakly terminal in the specification** T iff it is weakly terminal in *PAlg*(T).

A specification is **monomorphic** iff all its term generated non-trivial models are isomorphic. (An algebra is trivial iff every carrier is the one-point set.)

Given a formula Φ with free variables $x_1, \ldots, x_n$. A **defined ground instantiation** $\Phi 1$ of Φ is the formula obtained by replacing $x_1, \ldots, x_n$ in Φ by ground terms $t_1, \ldots, t_n$ (of the same sorts as $x_1, \ldots, x_n$) such that $E \vdash D(t_i)$ for $i = 1, \ldots, n$. The **ground atomic theory**, often denoted by $=_T$, of T consists of all ground equations and ground definedness formulae which hold in T, i.e.

$$\begin{aligned} =_T \ &= \{t = t' : t, t' \in W(\Sigma) \text{ and } PAlg(T) \models t = t'\} \\ &\cup \{D(t) : t \in W(\Sigma) \text{ and } PAlg(T) \models D(t)\}. \end{aligned} \tag{A9}$$

Let $T = (\Sigma, E, P)$ with $P = (\Sigma', E')$ be a hierarchical specification. A term $t \in W(\Sigma', X)$ is called **primitive**. If $t = f(t_1, \ldots, t_n)$, where $f : s_1 \times \ldots \times s_n \to s'$ and $s' \in Sorts(\Sigma')$, then t is called a term of **primitive sort**. The only primitive terms of sort bool are assumed to be "*tr*" and "*fa*."

A hierarchical specification $T = (\Sigma, E, P)$ with $P = (\Sigma', E')$ is called **partially complete** iff for every term $t \in W(\Sigma)$ of primitive sort for which the definedness is provable (i.e. $T \vdash D(t)$) there exists a primitive term $t' \in W(\Sigma')$ with $T \vdash t = t'$. Analogously we define: the specification T is **hierarchy-consistent** iff for all primitive terms $t', t'' \in W(\Sigma')$ we have $T \vdash t' = t''$ only if $P \vdash t' = t''$ and $T \vdash D(t')$ only if $P \vdash D(t')$.

It is **consistent** iff not $T \vdash$ true = false.

References

Andréka, H., Burmeister, B., and Németi, I., (1980). *Quasivarieties of partial algebras—a unifying approach towards a two-valued model theory for partial algebras.* TH Darmstadt, FB Mathematik, Preprint No. 557.

Astesiano, E., Reggio, G., and Wirsing, M., (1986). *Relational specifications and observational semantics.* In *Proc. MFCS'86 (Bratislava)*, (J. Gruska, B. Rovan, J. Wiedemann, eds.) LNCS 233, Berlin, Springer Verlag.

Astesiano, E., and Reggio, G., (1987). "Direct semantics of concurrent languages in the SMoLCS approach." IBM Journal of Research and Development, **31**, 5, 512–534.

Astesiano, E., and Reggio, G., (1987). An outline of the SMoLCS methodology. In *Mathematical Models for the Semantics of Parallelism*, (M. Venturini-Zilli, ed.) LNCS 280, 81–113, Berlin, Springer Verlag.

Astesiano, E., Giovini, A., Mazzanti, F., Reggio, G., and Zucca, E., (1986). The Ada challenge for New Formal Semantic Techniques. In *Ada: Managing the Transition—Proc. of the Ada—Europe International Conference 1986.* Cambridge University Press.

Astesiano, E., Bendix Nielsen, C., Botta, N., Fantechi, A., Giovini, A., Inverardi, P., Karlsen, E., Mazzanti, F., Storbank Pedersen, J., Reggio, G., and Zucca, E., (1987). Deliverable 7 of the CEC MAP project: The DraftFormal Definition of ANSI/MIL-STD 1815A Ada.

Bauer, F. L., Berghammer, R., Broy, M., Dosch, W., Geiselbrechtinger, F., Gnatz, R., Hangel, E., Hesse, W., Kriegs-Brückner, B., Laut, A., Matzner, T., Möller, B., Nickl, F., Partsch, H., Pepper, P., Samelson, K., Wössner, H., Wirsing, M., (1985). *The Munich project CIP*, Vol.1: The wide spectrum language CIP-L, LNCS 183, Berlin, Springer Verlag.

Bergstra, J. A., and Tucker, J. V., (1984). Top-down design and the algebra of communicating processes. Centrum for Wiskunde en Informatica, Amsterdam, report CS-R8401.

Broy, M., and Wirsing, M., (1980). Programming languages as abstract data types. In *Proc. CAAP 80* (M. Dauchet, ed.), Lille, University of Lille, 160–177.

Broy, M., and Wirsing, M., (1982). "Partial abstract types," *Acta Informatica* **18**, 47–64.

Broy, M., and Wirsing, M., (1982). On the algebraic specification of finitary infinite communicating sequential processes. In *Proc. IFIP TC2 Working Conference on "Formal Description of Programming Concepts II,"* D. Bjorner, ed.), Garmisch 1982, Amsterdam, North-Holland.

Broy, M., and Wirsing, M., (1983). "Algebraic definition of a functional programming language and its semantic models," *R.A.I.R.O. Theoretical Informatics* **17**, 137–161.

Broy, M., Pair, C., and Wirsing, M., (1984). "A systematic study of models of abstract data types," *Theoretical Computer Science* **33**, 139-174.

Hußmann, H., (1985). Rapid prototyping for algebraic specifications—RAP-System User's Manual, Universität Passau, Technical Report, MIP 8504.

Milner, R., (1977). "Fully abstract semantics of typed λ-calculi," *Theoretical Computer Science* **4**, 1–22.

Milner, R., (1980). A calculus of communicating systems. LNCS 92, Berlin, Springer Verlag.

Milner, R., (1983). "Calculi for synchrony and asynchrony," *Theoretical Computer Science* **25**, 267–310.

Mosses, P., (1982). *Abstract semantic algebra!* In *Proc. IFIP TC2 Working Conference on "Formal Description of Programming Concepts II,"* (D. Bjorner, ed.), Garmisch 1982, Amsterdam, North-Holland.

Nivat, M., (1975). *On the Interpretation of Recursive Polyadic Program Schemes.* Istituto Nazionale di Alta Matematica, Symposia Mathematica XV, London, Academic Press, 255–281.

Park, D., (1981). Concurrency and automata on infinite sequences. In *Proc. 5th GI Conference on Theoretical Computer Science*, (P. Deussen, ed.), LNCS 104, Berlin, Springer Verlag.

Reichel, H., (1979). *Theorie der Aequoide. Dissertation* B, Humbold-Universität Berlin.

Wagner, E. G., Thatcher, J. W., and Wright, J. B., (1978). Programming languages as mathematical objects. In *Proc. MFCS 78*, (J. Winkowski, ed.), Zakopane, LNCS 64, 84–101, Berlin, Springer Verlag.

2 Characteristic Sets and Gröbner Bases in Geometry Theorem Proving

SHANG-CHING CHOU, WILLIAM F. SCHELTER, and JIN-GEN YANG

University of Texas, Austin, Texas[1]

1. An Introduction to the Methods

1.1. Historical Background

In the area of mechanical theorem proving, A. Tarski first gave a decision procedure for what he called elementary geometry, based on an algebraic method (Tarski, 1951). In spite of the subsequent improvements by A. Seidenberg (Seidenberg, 1954), G. Collins (Collins, 1975), and others along this line, the variants of Tarski's method still seem far away from mechanically proving difficult geometry theorems in practice. It was really a surprise when Wu Wen-tsün introduced an algebraic method which can be used to prove many geometry theorems whose traditional proofs need enormous amounts of human intelligence.

One of the key observations of Wu was that theorems in Euclidean geometry not involving the order relation (betweenness) can be relatively easily dealt with by other methods, completely different from

[1]Supported by NSF Grants DCR-8503498 and CCR-8702108. The first author wishes to thank his former thesis advisors R. S. Boyer and J. S. Moore for their guidance and help in his contributions to the work reported here. He also thanks Dr H. P. Ko for helpful discussions.

ISBN 0-12-046370-9

Tarski's. The power of the method can be illustrated by experiments on computers (Wen-tsün (1984), Wen-tsün (1982), Chou (1984), Chou (1986), Chou (1988)) in which hundreds of geometry theorems were proved, including Morley's trisector theorem, a conjecture of V. Thébault, and several possibly new theorems.

The success of Wu's method stimulated researchers to apply the Gröbner basis method to the same class of geometry theorems which can be addressed by Wu's method. Recently, applications of the Gröbner basis method to geometry theorem proving have been reported (Chou and Schelter (1986), Kapur (1986), Kutzler and Stifter (1986)).

1.2. Three Examples

Wu's method was introduced as a mechanical method for proving those statements in elementary geometries for which, in their algebraic forms,[2] the hypotheses and the conclusion can be expressed by *polynomial equations.*[3]

For such a geometry statement, after adopting an appropriate coordinate system, the hypotheses can be expressed as a set of polynomial equations:

$$\begin{aligned} h_1(u_1, \ldots, u_d, x_1, \ldots, x_t) &= 0 \\ h_2(u_1, \ldots, u_d, x_1, \ldots, x_t) &= 0 \\ &\vdots \\ h_n(u_1, \ldots, u_d, x_1, \ldots, x_t) &= 0, \end{aligned} \tag{1}$$

and the conclusion is also a polynomial equation $g = g(u_1, \ldots, u_d, x_1, \ldots, x_t) = 0$, where $h_1, \ldots, h_n$ and g are polynomials in $\mathbf{Q}[u_1, \ldots, u_d, x_1, \ldots, x_t]$. Variables $u_1, \ldots, u_d$ are parameters or independent variables and variables $x_1, \ldots, x_t$ are algebraically dependent on u's, being restricted by (1) and under normal conditions. The selection of the parameters u's comes from the assumption that the figure in the geometry statement should be in a general (generic) position, an assumption that is often implicit in geometry statements. The use

[2]For the relationship between axiomatic geometries and number systems, the reader is strongly urged to read Hilbert (1971).

[3]Dealing with inequality is generally beyond the method.

of two different kinds of variables goes back to the Hilbert mechanization theorem for affine geometry (see theorem 62 in Hilbert, 1971).

Now let us use three examples to illustrate how Wu's method works. The proofs of the three examples were actually produced by the prover, including the selection of coordinates.

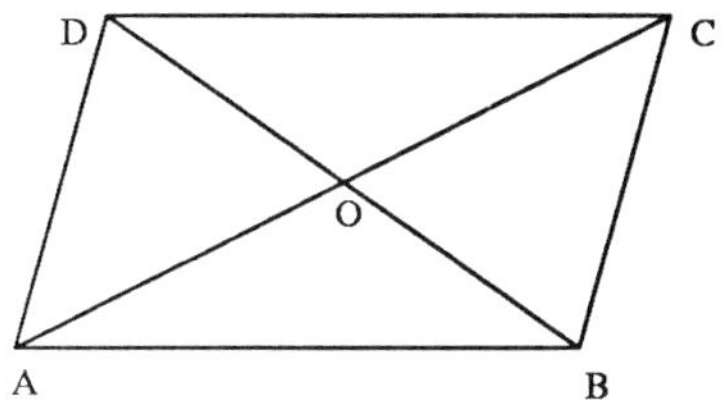

Fig. 1.

Example 1. Let $ABCD$ be a parallelogram, O be the intersection of the diagonals AC and BD. Show $AO \equiv OC$ (Fig. 1).

The implicit assumption that the parallelogram is in a general position means that any three points among the four points A, B, C, and D can be arbitrarily chosen. Thus we can let $A = (0,0)$, $B = (u_1, 0)$, $C = (u_2, u_3)$, $D = (x_2, x_1)$, and $O = (x_4, x_3)$. The hypothesis equations are:

$h_1 = u_1x_1 - u_1u_3 = 0$	AB is parallel to DC
$h_2 = u_3x_2 - (u_2 - u_1)x_1 = 0$	DA is parallel to CB
$h_3 = x_1x_4 - (x_2 - u_1)x_3 - u_1x_1 = 0$	O is on BD
$h_4 = u_3x_4 - u_2x_3 = 0$	O is on AC.

The conclusion $AO \equiv OC$ is: $g = 2u_2x_4 + 2u_3x_3 - u_3^2 - u_2^2 = 0$. Now one might think that our algebra formulation of the problem would be

$$h_1 = 0 \wedge h_2 = 0 \wedge h_3 = 0 \wedge h_4 = 0) \Rightarrow g = 0. \qquad (2)$$

However, (2) is not a valid statement. The reason is that the important nondegenerate conditions were overlooked. Wu's method can produce a sufficient number of nondegenerate conditions to make (2) valid.

The next step is to triangulate h_1, h_2, h_3, h_4 so that each polynomial only introduces one new (dependent) variable x_i. Thus h_1, h_2,

h_3, h_4 are not in triangular form because h_3 introduces two new variables, x_3 and x_4, at the same time. We can use a simple elimination procedure to obtain a triangular form: let $f_1 = h_1$, $f_2 = h_2$, $f_3 = prem(h_4, h_3, x_4)$,[4] $f_4 = h_4$. Then we have a triangular form:

$$\begin{aligned} f_1 &= u_1x_1 - u_1u_3 = 0 \\ f_2 &= u_3x_2 - (u_2 - u_1)x_1 = 0 \\ f_3 &= (u_3x_2 - u_2x_1 - u_1u_3)x_3 + u_1u_3x_1 = 0 \\ f_4 &= u_3x_4 - u_2x_3 = 0. \end{aligned}$$

Now we do successive pseudo divisions:

$$\begin{aligned} R_3 &= prem(g, f_4, x_4) = (2u_3^2 + 2u_2^2)x_3 - u_3^3 - u_2^2u_3 \\ R_2 &= prem(R_3, f_3, x_3) = (-u_3^4 - u_2^2u_3^2)x_2 \\ &\quad + ((u_2 - 2u_1)u_3^3 + (u_2^3 - 2u_1u_2^2)u_3)x_1 + u_1u_3^4 + u_1u_2^2u_3^2 \\ R_1 &= prem(R_2, f_2, x_2) = (-u_1u_3^4 - u_1u_2^2u_3^2)x_1 + u_1u_3^5 + u_1u_2^2u_3^3 \\ R_0 &= prem(R_1, f_1, x_1) = 0. \end{aligned}$$

Because the *final remainder* R_0 is zero, the theorem follows from appropriate subsidiary conditions. To see this, let us recall the simple and important *remainder formula* for successive pseudo divisions of g with respect to a triangular form $f_1, \ldots, f_r$:

$$I_1^{s_1} \ldots I_r^{s_r} g = Q_1 f_1 + \cdots + Q_r f_r + R_0$$

where I_k are the leading coefficients of f_k in x_k.

Since $R_0 = 0$, then $g = 0$ under $I_k \neq 0$ $(k = 1, \ldots, r)$. The subsidiary conditions $I_k \neq 0$ are usually connected with non-degeneracy and are also called nondegenerate conditions. For this particular case, (2) is valid under $I_1 = u_1 \neq 0$, $I_2 = u_3 \neq 0$, $I_3 = u_3x_2 - u_2x_1 - u_1u_3 \neq 0$, and $I_4 = u_3 \neq 0$.

The conditions $u_1 \neq 0$ and $u_3 \neq 0$ mean that A, B, and C are not collinear. The condition $u_3x_2 - u_2x_1 - u_1u_3 \neq 0$ means that line AC and line BD should have a normal intersection.

[4] $prem(h_4, h_3, x_4)$ denotes the pseudo remainder obtained by dividing h_4 by h_3 in variable x_4.

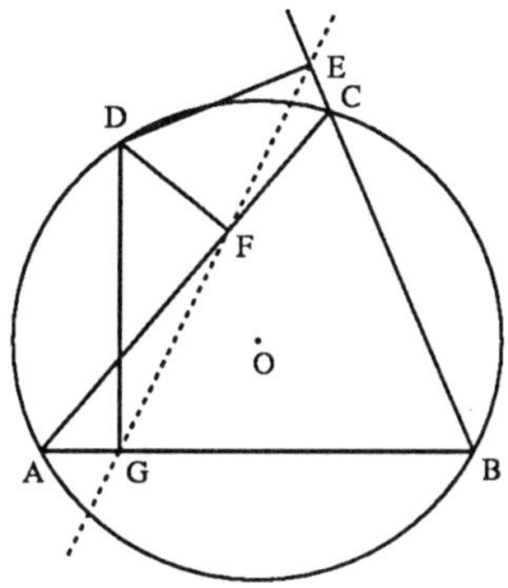

Fig. 2.

Example 2. (Simson's theorem) Let D be a point on the circumscribed circle (O) of triangle ABC. From D three perpendiculars are drawn to the three sides BC, CA and AB of $\triangle ABC$. Let E, F and G be the three feet respectively. Show that E, F and G are collinear (Fig. 2).

Let $A = (0,0)$, $B = (u_1, 0)$, $C = (u_2, u_3)$, $O = (x_2, x_1)$, $D = (x_3, u_4)$, $E = (x_5, x_4)$, $F = (x_7, x_6)$, and $G = (x_3, 0)$. Then the hypothesis equations are:

$h_1 = 2u_2x_2 + 2u_3x_1 - u_3^2 - u_2^2 = 0$	$OA \equiv OC$
$h_2 = 2u_1x_2 - u_1^2 = 0$	$OA \equiv OB$
$h_3 = -x_3^2 + 2x_2x_3 + 2u_4x_1 - u_4^2 = 0$	$OA \equiv OD$
$h_4 = u_3x_5 + (-u_2 + u_1)x_4 - u_1u_3 = 0$	Points E, B and C are collinear
$h_5 = (u_2 - u_1)x_5 + u_3x_4 + (-u_2 + u_1)x_3 - u_3u_4 = 0$	DE is perpendicular to BC
$h_6 = u_3x_7 - u_2x_6 = 0$	Points F, A and C are collinear
$h_7 = u_2x_7 + u_3x_6 - u_2x_3 - u_3u_4 = 0$	DF is perpendicular to AC.

The conclusion that points E, F and G are collinear is equivalent to $g = x_4x_7 + (-x_5 + x_3)x_6 - x_3x_4 = 0$. Now we can triangulate $h_1, \ldots, h_7$ by letting $f_1 = prem(h_1, h_2, x_2)$, $f_2 = h_2$, $f_3 = h_3$, $f_4 =$

$prem(h_4, h_5, x_5)$, $f_5 = h_4$, $f_6 = prem(h_6, h_7, x_7)$, $f_7 = h_7$:

$$
\begin{aligned}
f_1 &= 4u_1u_3x_1 - 2u_1u_3^2 - 2u_1u_2^2 + 2u_1^2u_2 = 0\\
f_2 &= 2u_1x_2 - u_1^2 = 0\\
f_3 &= -x_3^2 + 2x_2x_3 + 2u_4x_1 - u_4^2 = 0\\
f_4 &= (-u_3^2 - u_2^2 + 2u_1u_2 - u_1^2)x_4\\
&\quad + (u_2 - u_1)u_3x_3 + u_3^2u_4 + (-u_1u_2 + u_1^2)u_3 = 0\\
f_5 &= u_3x_5 + (-u_2 + u_1)x_4 - u_1u_3 = 0\\
f_6 &= (-u_3^2 - u_2^2)x_6 + u_2u_3x_3 + u_3^2u_4 = 0\\
f_7 &= u_2x_7 + u_3x_6 - u_2x_3 - u_3u_4 = 0.
\end{aligned}
$$

Here we see that $deg(f_3, x_3) = 2$. Thus successive pseudo division is more convenient.

$$
\begin{aligned}
R_6 &= prem(g, f_7, x_7) = (-u_2x_5 - u_3x_4 + u_2x_3)x_6 + u_3u_4x_4\\
R_5 &= prem(R_6, f_6, x_6) = (u_2^2u_3x_3 + u_2u_3^2u_4)x_5 + (u_2u_3^2x_3 - u_2^2u_3u_4)\\
&\quad x_4 - u_2^2u_3x_3^2 - u_2u_3^2u_4x_3\\
R_4 &= prem(R_5, f_5, x_5) = ((u_2u_3^3 + (u_2^3 - u_1u_2^2)u_3)x_3 - u_1u_2u_3^2u_4)\\
&\quad x_4 - u_2^2u_3^2x_3^2 + (-u_2u_3^3u_4 + u_1u_2^2u_3^2)x_3 + u_1u_2u_3^3u_4\\
R_3 &= prem(R_4, f_4, x_4) = u_1u_2u_3^4x_3^2 - u_1^2u_2u_3^4x_3 + u_1u_2u_3^4u_4^2\\
&\quad + (-u_1u_2u_3^5 + (-u_1u_2^3 + u_1^2u_2^2)u_3^3)u_4\\
R_2 &= prem(R_3, f_3, x_3) = -2u_1u_2u_3^4x_3x_2 + u_1^2u_2u_3^4x_3\\
&\quad - 2u_1u_2u_3^4u_4x_1 + (u_1u_2u_3^5 + (u_1u_2^3 - u_1^2u_2^2)u_3^3)u_4\\
R_1 &= prem(R_2, f_2, x_2) = -4u_1^2u_2u_3^4u_4x_1\\
&\quad + (2u_1^2u_2u_3^5 + (2u_1^2u_2^3 - 2u_1^3u_2^2)u_3^3)u_4\\
R_0 &= prem(R_1, f_1, x_1) = 0.
\end{aligned}
$$

Since the final remainder R_0 is 0, we have proved Simson's theorem under the following subsidiary conditions: $I_1 = 4u_1u_3 \neq 0$, $I_2 = 2u_1 \neq 0$, $I_4 = -u_3^2 - u_2^2 + 2u_1u_2 - u_1^2 \neq 0$, $I_5 = u_3 \neq 0$, $I_6 = -u_3^2 - u_2^2 \neq 0$, $I_7 = u_2 \neq 0$.

Example 3. (The Butterfly Theorem). A, B, C and D are four points on circle (O). E is the intersection of AC and BD. Through E draw a line perpendicular to OE, meeting AD at F and BC at G. Show $FE \equiv GE$ (Fig. 3).

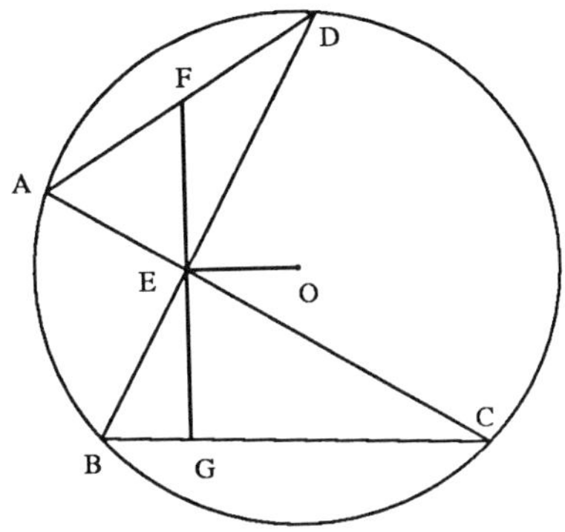

Fig. 3.

Let $E = (0,0)$, $O = (u_1, 0)$, $A = (u_2, u_3)$, $B = (x_1, u_4)$, $C = (x_3, x_2)$, $D = (x_5, x_4)$, $F = (0, x_6)$, and $G = (0, x_7)$. Then the hypothesis equations are:

$$\begin{aligned}
h_1 = &-x_1^2 + 2u_1x_1 - u_4^2 \\
&+ u_3^2 + u_2^2 - 2u_1u_2 = 0 && OA \equiv OB \\
h_2 = &-x_3^2 + 2u_1x_3 - x_2^2 \\
&+ u_3^2 + u_2^2 - 2u_1u_2 = 0 && OA \equiv OC \\
h_3 = &-u_3x_3 + u_2x_2 = 0 && C, A \text{ and } E \text{ are collinear} \\
h_4 = &-x_5^2 + 2u_1x_5 - x_4^2 \\
&+ u_3^2 + u_2^2 - 2u_1u_2 = 0 && OA \equiv OD \\
h_5 = &-u_4x_5 + x_1x_4 = 0 && \text{Points } D, B \text{ and } E \text{ are collinear} \\
h_6 = &(-x_5 + u_2)x_6 \\
&+ u_3x_5 - u_2x_4 = 0 && \text{Points } F, A \text{ and } D \text{ are collinear} \\
h_7 = &(-x_3 + x_1)x_7 \\
&+ u_4x_3 - x_1x_2 = 0 && \text{Points } G, B \text{ and } C \text{ are collinear.}
\end{aligned}$$

The conclusion is $g = x_7 + x_6 = 0$.

We can let $f_1 = h_1$, $f_2 = prem(h_2, h_3, x_3)$, $f_3 = h_3$, $f_4 = prem(h_4,$

$h_5, x_5)$, $h_5 = f_5$, $f_6 = h_6$, $f_7 = h_7$ to obtain a triangular form:

$$
\begin{aligned}
f_1 &= -x_1^2 + 2u_1x_1 - u_4^2 + u_3^2 + u_2^2 - 2u_1u_2 = 0\\
f_2 &= (-u_3^2 - u_2^2)x_2^2 + 2u_1u_2u_3x_2 + u_3^4 + (u_2^2 - 2u_1u_2)u_3^2 = 0\\
f_3 &= -u_3x_3 + u_2x_2 = 0\\
f_4 &= (-x_1^2 - u_4^2)x_4^2 + 2u_1u_4x_1x_4 + (u_3^2 + u_2^2 - 2u_1u_2)u_4^2 = 0\\
f_5 &= -u_4x_5 + x_1x_4 = 0\\
f_6 &= (-x_5 + u_2)x_6 + u_3x_5 - u_2x_4 = 0\\
f_7 &= (-x_3 + x_1)x_7 + u_4x_3 - x_1x_2 = 0.
\end{aligned}
$$

If we do successive pseudo divisions of g with respect to the above triangular form, then the final remainder R_0 is not zero. The reason is that the above triangular form is *reducible* (see Section 2.2). This reducibility comes from a special kind of degeneracy or *ambiguity* when using algebraic equations to encode certain geometric conditions. Equations $h_2 = 0$ and $h_3 = 0$ specify point C: C is on circle (O) and on line AE. However, there are two points satisfying the equations: one is C, which we really want; the other is point A. Because A has been constructed before C, reducibility arises. We might use the complete method of Wu to deal with reducibility. However, for such special reducibility we can use an elementary method, which is more satisfactory in geometry (see the last footnote of Chou (1984) or (1985)).

Because $f_2 = 0$ and $f_3 = 0$ have two solutions (one is for $A = (u_2, u_3)$ and the other is for $C = (x_3, x_2)$), $x_2 - u_3$ is a factor of f_2 under the previous geometric conditions: $f_2 = (x_2 - u_3)((-u_3^2 - u_2^2)x_2 - u_3^3 + (-u_2^2 + 2u_1u_2)u_3)$.

Thus we can use the division to obtain $f_2' = (-u_3^2 - u_2^2)x_2 - u_3^3 + (-u_2^2 + 2u_1u_2)u_3$ and replace f_2 by f_2'. In the same way we have:

$$f_4 = (x_4 - u_4)f_4' + r$$

where $f_4' = (-x_1^2 - u_4^2)x_4 - u_4x_1^2 + 2u_1u_4x_1 - u_4^3$, and $r = -u_4^2x_1^2 + 2u_1u_4^2x_1 - u_4^4 + (u_3^2 + u_2^2 - 2u_1u_2)u_4^2$.

We have $prem(r, f_1, x_1) = 0$. Hence under the previous conditions $f_i = 0$ and $I_i \neq 0$ $(i = 1, 2, 3)$: $f_4 = (x_4 - u_4)f_4'$. We can replace f_4 by f_4' to obtain the nondegenerate triangular form $f_1, f_2', f_3, f_4', f_5, f_6, f_7$.

Now we can do the successive divisions of g with respect to the new triangular form. The *final remainder* $R_0 = 0$. Thus the theorem follows from the following subsidiary conditions: $I_2 = u_3^2 + u_2^2 \neq 0$, $I_3 = u_3 \neq 0$, $I_4 = x_1^2 + u_4^2 \neq 0$, $I_5 = u_4 \neq 0$, $I_6 = x_5 - u_2 \neq 0$, $I_7 = x_3 - x_1 \neq 0$. In addition, the following nondegenerate conditions, which come from reducibility and have been detected by our prover, should be also added: $D \neq B$ (i.e., $x_2 - u_3 \neq 0$); $C \neq A$ (i.e., $x_4 - u_4 \neq 0$).

So far, we have been talking about the elementary part of Wu's method. A number of questions arise.

(1) After successive pseudo divisions, *if the final remainder R_0 is not zero, what conclusion can we infer?*
(2) *In what theory does the method prove theorems?*

These are quite complicated problems. We will discuss them in Sections 3 and 4.

1.3. The Three Examples Using the Gröbner Basis Method

Now we are using the three examples in Section 1.2 to illustrate how the Gröbner basis method can be used to prove geometry theorems. As we know, one of the important applications of the Gröbner basis method is to decide ideal membership for polynomial ideals.

Example 4. Example 1 of Parallelogram.

Let the u, the x, $h_1, \ldots, h_4$ and g be the same as in Example 1. Thus the question seems to show $g \in \text{Ideal}(h_1, h_2, h_3, h_4)$, or more precisely $g \in \text{Radical}(h_1, h_2, h_3, h_4)$. However, neither of the above assertions is true if we consider h_1, h_2, h_3, h_4 as polynomials in $\mathbf{Q}[u_1, u_2, u_3, x_1, \ldots, x_4]$. The reason is the same as we explained in Section 1.2: nondegenerate conditions are missing. Nondegenerate conditions can be produced if we work on polynomial ring $\mathbf{Q}(u)[x]$ instead of $\mathbf{Q}[u, x]$, where $\mathbf{Q}(u)$ is the quotient field of polynomial ring $\mathbf{Q}[u]$.

Thus we can ask whether $g \in \text{Radical}(h_1, \ldots, h_4)$ in $\mathbf{Q}(u)[x]$. By Hilbert's Nullstellensatz, this is equivalent to showing whether Radical $(h_1, \ldots, h_4, gz - 1) = (1)$ in $\mathbf{Q}(u)[x]$. We can check this by computing the Gröbner basis of $\{h_1, \ldots, h_4, gz - 1\}$[5] to see whether it

[5]When we talk about a Gröbner basis of an ideal, we always mean the unique

is $\{1\}$. This time, our GB program confirmed the theorem of parallelogram. The subsidiary (nondegenerate) conditions were implicitly used in the computation: the denominators in the computation should be nonzero. It is easy to collect these denominators and to remove the repeated factors. For this example, we need "$u_3 \neq 0$, $u_1 \neq 0$" in the computation.

As a first approximation, we can check whether $g \in$ Ideal $(h_1, \ldots, h_4)$ in $\mathbf{Q}(u)[x]$ first. Thus, we can compute the Gröbner basis of $h_1, \ldots, h_4$ in the pure lexicographic ordering $x_1 < x_2 < x_3 < x_4$; then to check whether it reduces the conclusion polynomial g to zero. $\mathrm{GB}(h_1, \ldots, h_4) = \{x_1 - u_3,\ x_2 - u_2 + u_1,\ x_3 - u_3/2,\ x_4 - u_2/2\}$, which reduces the conclusion polynomial $g = 2u_2x_4 + 2u_3x_3 - u_3^2 - u_2^2$ to zero. Thus the statement was confirmed. Again, we need subsidiary conditions, $u_3 \neq 0$ and $u_1 \neq 0$, in the computation.

For all theorems we found in practice this first approximation is enough. It is more informative and faster than using radical ideals.

Example 5. Example 2 (Simson's Theorem). $\mathrm{GB}(h_1, \ldots, h_7)$ (in $\mathbf{Q}(u)[x]$) =

$$\begin{aligned}
&x_1 - (u_3^2 + u_2^2 - u_1u_2)/2u_3\\
&x_2 - u_1/2\\
&x_3^2 - (u_1u_3x_3 - u_3u_4^2 + (u_3^2 + u_2^2 - u_1u_2)u_4)/u_3\\
&x_4 - ((u_2 - u_1)u_3x_3 + u_3^2u_4 + (-u_1u_2 + u_1^2)u_3)/(u_3^2 + u_2^2\\
&\qquad - 2u_1u_2 + u_1^2)\\
&x_5 - ((u_2^2 - 2u_1u_2 + u_1^2)x_3 + (u_2 - u_1)u_3u_4 + u_1u_3^2)/(u_3^2 + u_2^2\\
&\qquad - 2u_1u_2 + u_1^2)\\
&x_6 - (u_2u_3x_3 + u_3^2u_4)/(u_3^2 + u_2^2)\\
&x_7 - (u_2^2x_3 + u_2u_3u_4)/(u_3^2 + u_2^2).
\end{aligned}$$

The Gröbner basis reduces the conclusion polynomial g to zero. Thus we proved Simson's theorem to be generically true. The subsidiary conditions needed for the computation are: $u_2 \neq 0$, $u_1 \neq 0$, $u_3 \neq 0$, $u_3^2 + u_2^2 - 2u_1u_2 + u_1^2 \neq 0$, $u_3^2 + u_2^2 \neq 0$.

Example 6. The Butterfly theorem: Example 3.

---

reduced Gröbner basis with respect to certain compatible ordering among monomials and denote it by $\mathrm{GB}(h_1, \ldots, h_n)$, where $h_1, \ldots, h_n$ are generators of the ideal.

We choose slightly different coordinates: $E = (0,0)$, $O = (u_1, 0)$, $A = (u_2, u_3)$, $B = (u_4, x_1)$, $C = (x_2, x_3)$, $D = (x_4, x_5)$, $F = (0, x_6)$, and $G = (0, x_7)$. Then the hypotheses and the conclusion are:

$$
\begin{aligned}
h_1 &= x_1^2 + u_4^2 - 2u_1u_4 - u_3^2 - u_2^2 + 2u_1u_2 = 0\\
h_2 &= x_3^2 + x_2^2 - 2u_1x_2 - u_3^2 - u_2^2 + 2u_1u_2 = 0\\
h_3 &= u_2x_3 - u_3x_2 = 0\\
h_4 &= x_5^2 + x_4^2 - 2u_1x_4 - u_3^2 - u_2^2 + 2u_1u_2 = 0\\
h_5 &= u_4x_5 - x_1x_4 = 0\\
h_6 &= (x_4 - u_2)x_6 + u_2x_5 - u_3x_4 = 0\\
h_7 &= (x_2 - u_4)x_7 + u_4x_3 - x_1x_2 = 0\\
g &= x_7 + x_6 = 0
\end{aligned}
$$

However, neither $\text{GB}(h_1, \ldots, h_7, zg - 1) = \{1\}$, nor $\text{GB}(h_1, \ldots, h_7)$ reduces g to zero. The reason is the same as we explained in Example 3: in specifying of point C and B we have ambiguities. These ambiguities can be easily detected by the program and, in the case of using the Gröbner basis method, can be removed by adding two new equations $s_1 = (x_2 - u_2)v_1 - 1 = 0$ $(A \neq C)$, $s_2 = (x_4 - u_4)v_2 - 1 = 0$ $(B \neq D)$. Thus $\text{GB}(s_1, s_2, h_1, \ldots, h_7) =$

$$
\begin{aligned}
&v_1 + (2u_1u_4 + u_3^2 + u_2^2 - 2u_1u_2)/(2u_1u_4^2 + (2u_3^2 + 2u_2^2 - 4u_1u_2)u_4)\\
&v_2 + (u_3^2 + u_2^2)/(2u_2u_3^2 + 2u_2^3 - 2u_1u_2^2)\\
&x_1^2 + u_4^2 - 2u_1u_4 - u_3^2 - u_2^2 + 2u_1u_2\\
&x_2 + (u_2u_3^2 + u_2^3 - 2u_1u_2^2)/(u_3^2 + u_2^2)\\
&x_3 + (u_3^3 + (u_2^2 - 2u_1u_2)u_3)/(u_3^2 + u_2^2)\\
&x_4 + ((u_3^2 + u_2^2 - 2u_1u_2)u_4)/(2u_1u_4 + u_3^2 + u_2^2 - 2u_1u_2)\\
&x_5 + ((u_3^2 + u_2^2 - 2u_1u_2)x_1)/(2u_1u_4 + u_3^2 + u_2^2 - 2u_1u_2)\\
&x_6 + ((u_2u_3^2 + u_2^3 - 2u_1u_2^2)x_1 + (-u_3^3 + (-u_2^2 + 2u_1u_2)u_3)u_4)/\\
&\qquad ((u_3^2 + u_2^2)u_4 + u_2u_3^2 + u_2^3 - 2u_1u_2^2)\\
&x_7 - ((u_2u_3^2 + u_2^3 - 2u_1u_2^2)x_1 + (-u_3^3 + (-u_2^2 + 2u_1u_2)u_3)u_4)/\\
&\qquad ((u_3^2 + u_2^2)u_4 + u_2u_3^2 + u_2^3 - 2u_1u_2^2).
\end{aligned}
$$

The Gröbner basis reduces g to zero. Thus, we have proved the Butterfly theorem. In the computation we needed the following subsidiary conditions: $(u_3^2 + u_2^2)u_4 + u_2u_3^2 + u_2^3 - 2u_1u_2^2 \neq 0$, $u_4 - u_2 \neq 0$,

$2u_1u_4 + u_3^2 + u_2^2 - 2u_1u_2 \neq 0$, $u_4 \neq 0$, $u_1u_4 + u_3^2 + u_2^2 - 2u_1u_2 \neq 0$, $u_3^2 + u_2^2 \neq 0$, $u_2 \neq 0$, $u_3^2 + u_2^2 - u_1u_2 \neq 0$.

Note that the computation of the Gröbner basis is very sensitive to the variable ordering. If we choose the same coordinates that are in Example 3, then the computation would take much longer. For the present choice of coordinates, the computation took about 24 seconds on a Symbolics 3600.

2. Ritt's Characteristic Set Method

Wu introduced his method in 1977. Later on, he found that the algebraic tools and algorithms he needed were already begun in the work of J. F. Ritt (Ritt, 1938, 1950). In (Wen-tsün, 1986) and (Wen-tsün, 1984), Wu revised Ritt's work for his own need of mechanically proving geometry theorems of certain types. Our presentation here is based on the work of Ritt and Wu. It is essential for understanding the complete method of Wu.

2.1. Ascending Chains and Characteristic Sets

Let K be a field, and $y = y_1, y_2, \ldots, y_m$ be indeterminates. Unless stated otherwise, all polynomials mentioned in this section are in $A = K[y_1, \ldots, y_m] = K[y]$. We fix the order of the indeterminates as $y_1 < y_2 < \cdots < y_m$, which is essential for the subsequent discussion. Unless stated otherwise, we assume this order among variables $y_1, \ldots, y_m$. When talking about an algorithm about A, we always assume that the field K is computable.

Let f be a polynomial. Denote the degree of f in the variable y_i by $deg(f, y_i)$. The *class* of f is the smallest integer c such that f is in $K[y_1, \ldots, y_c]$. We denote it by $class(f)$. If f is in K we define $class(f) = 0$. Let $c = class(f)$ be non-zero and $lv(f)$ denote the *leading variable* y_c of f. Considering f as a polynomial in y_c, let $lc(f)$ denote the *leading coefficient* of f which is also called the *initial* of f. Let $ld(f)$ denote the *leading degree* of f, i.e., $deg(f, y_c)$. A polynomial g is *reduced with respect to f* if $deg(g, y_c) < deg(f, y_c)$, where $c = class(f) > 0$.

Let f and g be in $K[y]$. Suppose that y_i really occurs in f. Considering f and g as polynomials in y_i, we can do pseudo division in a standard way. Let $c = class(f) > 0$, then $prem(g, f, y_c)$ is reduced with respect to f; we denote $prem(g, f, y_c)$ simply by $prem(g, f)$.

Definition 1. Let $C = f_1, f_2, \ldots, f_r$ be a sequence of polynomials in $K[y]$. We call it a *quasi ascending chain* or a triangular form if either $r = 1$ and $f_1 \neq 0$, or $r > 1$ and $0 < class(f_1)$, $class(f_i) < class(f_j)$ for $i < j$.

Let $f_1, \ldots, f_r$ be a quasi ascending chain with $class(f_1) > 0$. We define $prem(g, f_1, \ldots, f_r)$ inductively to be $prem((prem(g, f_2, \ldots, f_r), f_1)$.

(i) A quasi ascending chain is called an *ascending chain* if f_j is reduced with respect to f_i for $i < j$.

(ii) A quasi ascending chain is called an *ascending chain in Wu's sense* if the initials I_j of the f_j are reduced with respect to f_i for $i < j$.

(iii) A quasi ascending chain is called an *ascending chain in weak sense* if $prem(I_i, f_1, \ldots, f_r) \neq 0$, for $i = 1, \ldots, r$.

Obviously, an ascending chain is an ascending chain in Wu's sense; an ascending chain in Wu's sense is an ascending chain in weak sense.

We define a partial order $<$ in $K[y]$: $f < g$ (g is of *higher rank* or *higher* than f) if $class(f) < class(g)$ or $class(f) = class(g) > 0$ and $ld(f) < ld(g)$. If neither $f < g$ nor $g < f$, then we say f and g are of the same rank.

Proposition 1. *The partial order $<$ in A is well-founded, i.e., there are no infinite, strictly decreasing sequences of polynomials $p_1 > p_2 > \cdots > p_n > \cdots$.*

Proof. As we know, the well-founded condition is equivalent to that every nonempty set S of polynomials has a minimal element in that partial order $<$; i.e., an element which is not higher than any other element in S. If S contains an element in K, any such element is minimal. Otherwise, a polynomial in S with the least leading degree among the subset of all elements with the least class in S is minimal. In the case of finite S, this provides an algorithm to obtain a minimal element in S. □

Definition 2. Let $C = f_1, \ldots, f_r$ and $C_1 = g_1, \ldots, g_m$ be two ascending chains. We define $C < C_1$ if there is an s such that $s \leq \min(r, m)$ and f_i and g_i are of the same rank for $i < s$ and that $f_s < g_s$, or $m < r$ and f_i and g_i are of the same rank for $i \leq m$.

Proposition 2. *The partial order $<$ among the set of all ascending chains is well-founded. In other words, every nonempty set Σ of ascending chains has a minimal element.*

Proof. Let Σ_1 be all elements of Σ whose first polynomials are minimal among the first polynomials of all elements in Σ. If all elements (ascending chains) in Σ_1 consist of one polynomial, then any element in Σ_1 is minimal in Σ. Suppose that there are elements in Σ_1 which consist of more than one polynomial. We can let Σ_2 be the subset of all elements of Σ_1, whose second polynomials are minimal. If the elements in Σ_2 all have just two polynomials, any of those elements serves our purpose. If not, we continue, reaching a minimal ascending chain in no more than m steps. □

Proposition 2 can be rephrased: there are no infinite, strictly decreasing sequences of ascending chains $C_1 > C_2 > \cdots > C_k > \cdots$. This is important for the termination of Ritt's algorithms (see Sections 3 and 4).

Definition 3. Let S be a nonempty polynomial set. A minimal ascending chain in the set of all chains formed from polynomials in S is called a *characteristic set* of S.

Unless stated otherwise, whenever we talk about a finite polynomial set S, we assume S does not contain zero. By Proposition 2, every nonempty polynomial set S has a characteristic set.

Proposition 3. *Let $C = f_1, \ldots, f_r$ be a characteristic set of polynomial set S with $0 < class(f_1)$. Let g be nonzero and reduced with respect to C. Then the set $S_1 = S \cup \{g\}$ has characteristic sets lower than C.*

Proof. If $class(g) \leq class(f_1)$, then g alone is an ascending chain lower than C. Suppose $class(g) > class(f_1)$, let $j = max\{i | class(f_i) < class(g)\}$, then $f_1, \ldots, f_j, g$ form an ascending chain lower than C. □

Proposition 4. *Let $C = f_1, \ldots, f_r$ be an ascending chain in the polynomial set S with $0 < class(f_1)$. C is a characteristic set of S if and only if S contains no nonzero polynomial reduced with respect to C.*

Proof. If there is a polynomial g in S reduced with respect to C, then from the proof of Proposition 3 we have an ascending chain in S lower than C. Thus C is not a characteristic set of S. If C is not a characteristic set of S, then there is an ascending chain $C_1 = g_1, \ldots, g_m$ in S lower than C. From the definition 2, we can find a g_i reduced with respect to C. □

Theorem 1. *Every nonempty polynomial set S has a characteristic set. In the case of finite set S, there is an algorithm to construct a characteristic set of S.*

Proof. Let f_1 be a polynomial with minimal rank in S. If f_1 is of class zero, then it is a characteristic set of S. Now let f_1 be of positive class. Let S_1 be the set of all polynomials in S, reduced with respect to f_1. If S_1 is empty, then f_1 is a characteristic set of S. Now suppose S_1 is nonempty, then every polynomial in S_1 is of higher class than f_1. Let f_2 be a polynomial with minimal rank in S_1. Let S_2 be the set of all polynomials of in S_1, reduced with respect to f_1, f_2. If S_2 is empty, then f_1, f_2 is a characteristic set. Otherwise continuing, we arrive at a characteristic set of S in no more than m steps. □

Proposition 5. *Let $f_1, \ldots, f_r$ be a characteristic set of the ideal I.*

(i) *If $g \in I$, then $prem(g, f_1, \ldots, f_r) = 0$.*
(ii) *If I is prime, then $prem(g, f_1, \ldots, f_r) = 0 \Rightarrow g \in I$.*

Proof. See Chou (1988).

2.2. Irreducible Ascending Chains

Let $C = f_1, \ldots, f_r$ be an ascending chain, not consisting of a constant. After a **suitable renaming** of the y_j, we may assume that $class(f_1) = d + 1$ and $m = d + r = class(f_r)$, where $d \geq 0$. We distinguish the y_i for $i \le d$ by calling them u_i and use x_i to denote $lv(f_i)$.

Thus C has the following "triangular" form:

$$\begin{array}{l} f_1(u_1, \ldots, u_d, x_1) \\ f_2(u_1, \ldots, u_d, x_1, x_2) \\ \quad \ldots \\ f_r(u_1, \ldots, u_d, x_1, \ldots, x_r). \end{array} \tag{3}$$

Definition 4. An ascending chain $f_1, \ldots, f_r$ of the form (3) is called *irreducible* if each f_i is irreducible in the polynomial ring $K(u)[x_1, \ldots, x_i]/(f_1, \ldots, f_{i-1})$. Thus the sequence $F_0 = K(u)$, $F_1 = F_0[x_1]/(f_1)$, ..., $F_r = F_{r-1}[x_r]/(f_r) = F_0[x]/(f_1, \ldots, f_r)$ is a tower of field extensions.[6]

Example 7. Let C be the ascending chain $f_1 = x_1^2 - u_1$, $f_2 = x_2^2 - 2x_1x_2 + u_1$. f_1 is irreducible over $F_0 = \mathbf{Q}[u_1]$; but f_2 is reducible over $F_1 = F_0[x_1]/(f_1)$ because $f_2 = (x_2 - x_1)^2$ under $x_1^2 - u_1 = 0$. Thus C is reducible.

Theorem 2. *Let ascending chain $f_1, \ldots, f_r$ of the form (3) be irreducible,[7] g be a polynomial in $K[u, x]$ and F_r be the same as in Definition 4. Let $prem(g, f_1, \ldots, f_r)$ be the remainder obtained from successive pseudo division of g by $f_1, \ldots, f_r$. Then the following conditions are equivalent:*

(i) *$prem(g, f_1, \ldots, f_r) = 0$.*
(ii) *Let E be an extension field of K. If $\mu = (\tilde{u}_1, \ldots, \tilde{u}_d, \tilde{x}_1, \ldots, \tilde{x}_r)$ in E^{d+r} is a common zero of $f_1, \ldots, f_r$ with $\tilde{u}_1, \ldots, \tilde{u}_d$ transcendental over K, then μ is also a zero of g, i.e., $g(\mu) = 0$.*
(iii) *The polynomial g, considered as an element in F_r, is 0; i.e., the image of g in F_r under the canonical mapping is 0.*
(iv) *There are finite non-zero polynomials $c_1, \ldots, c_s$ in $K[u_1, \ldots, u_d]$ such that $c_1 \cdots c_s g \in \text{Ideal}(f_1, \ldots, f_r)$ (in $K[u, x]$).*

Proof. See Chou (1988).

---

[6]In contrast to the previous notation, here $(f_1, \ldots, f_r)$ etc. denotes the polynomial ideal of $K(u)[x]$ (not of $A = K[u, x]$), generated by $f_1, \ldots, f_r$.

[7]Here the ascending chain can be in weak sense.

We call $\mu = (\tilde{u}_1, \ldots, \tilde{u}_d, \tilde{x}_1, \ldots, \tilde{x}_r)$ in (ii) a *generic point* of that irreducible ascending chain in field E.

The theorem is no longer true if $f_1, \ldots, f_r$ is reducible. We can find such an example by letting f_1, f_2 be the same as in Example 7 and $g = x_2 - x_1$.

Proposition 6. *Let $f_1, \ldots, f_r$ be irreducible, g any polynomial. If $prem(g, f_1, \ldots, f_r) \neq 0$, then there are polynomial q and r such that $qg - r \in$ the ideal $(f_1, \ldots, f_r)$ of A and $r \in K[u]$.*

Proof. This can be seen from the proof of (i) $\Rightarrow$ (iv) of the above theorem. □

Theorem 3. *Let $f_1, \ldots, f_r$ be an ascending chain. Suppose that $f_1, \ldots, f_{k-1}$ $(0 < k \leq r)$ is irreducible, but $f_1, \ldots, f_k$ is reducible. Then there are polynomials g and h in $K[u, x]$ reduced with respect to $f_1, \ldots, f_r$ such that $class(g) = class(h) = class(f_k)$ and $gh \in$ the ideal $(f_1, \ldots, f_k)$ of A.*

Proof. By assumption, f_k is reducible in the field $L = F_{k-1}[x_k]$. Thus there are h'' and g'' in L such that $f_k - g''h'' = 0$ with $0 < deg(g'', x_k) < deg(f_k, x_k)$ and $0 < deg(h'', x_k) < deg(f_k, x_k)$.[8] Clearing denominators, we have a polynomial $p = Qf_k - g'h'$, where g' and h' are in $K[u, x_1, \ldots, x_k]$ obtained from g'' and h'' by clearing denominators and $Q \in K[u]$. Polynomial $p = 0$ in L, thus by (iii) of Theorem 2 $prem(p, f_1, \ldots, f_{k-1}) = 0$. Thus by the remainder formula, there are integers $s_i \geq 0$ such that $q = I_1^{s_1} \ldots I_{k-1}^{s_{k-1}} g'h'$ is in the ideal $(f_1, \ldots, f_k)$ of A, where the I_i are the initials of the f_i. Let $g = prem(I_1^{s_1} \ldots I_{k-1}^{s_{k-1}} g', f_1, \ldots, f_{k-1})$ and $h = prem(h', f_1, \ldots, f_{k-1})$, then g and h have the specifications in the theorem. □

Theorem 4. *Let $f_1, \ldots, f_r$ be an irreducible ascending chain and $P =$*

$$\{g \mid g \in K[u, x] \text{ and } \mathrm{prem}(g, f_1, \ldots, f_r) = 0\},$$

then

(i) *P is a prime ideal, with $f_1, \ldots, f_r$ as its characteristic set.*

---

[8] This is one of the key steps in Ritt's decomposition algorithm (4.1). Factorization over algebraic extensions is needed.

(ii) *A generic point of* $f_1, \ldots, f_r$ *is a generic zero of prime ideal* P.

(iii) *If* E *is algebraically closed, then* $dim(V(P)) = d$, *where* $V(P)$ *denotes the set of common zeros of all polynomials of* P. *A polynomial* g *vanishes on* $V(P)$ *if and only* $prem(g, f_1, \ldots, f_r) = 0$.

(iv) *For any field* E, *if* $prem(g, f_1, \ldots, f_r) = 0$, *then polynomial* g *vanishes on* $V(P)$.

Proof. Let μ be a generic point of $f_1, \ldots, f_r$. By Theorem 2, $P = \{g \mid g(\mu) = 0\}$. It is now obvious that P is a prime ideal with the generic zero μ. Since there are no nonzero polynomials in P reduced with respect to $f_1, \ldots, f_r$, $f_1, \ldots, f_r$ is a characteristic set of P. If E is algebraically closed, then $I(V(P)) = P$, where $I(V)$ denotes the set of all polynomials vanishing on V. Therefore, a polynomial g vanishes on $V(P)$ if and only $g \in P$, i.e., $prem(g, f_1, \ldots, f_r) = 0$. The dimension of $V(P)$ is that of P, i.e., d. Since every field E has an algebraically closed extension, (iv) follows from (iii). □

Remark 1. If E is not algebraically closed, then $V(P)$ can be irreducible or reducible. However, if $V(P)$ is of dimension d, then $V(P)$ is irreducible. Indeed, let V_1 be one of its components with dimension d. Then $P_1 = I(V_1)$ is a prime ideal containing P and having dimension d. Thus $P = P_1$, $V(P) = V(P_1) = V_1$. $V(P)$ is irreducible. In that case, therefore, a polynomial g vanishes on $V(P)$ if and only if $prem(g, f_1, \ldots, f_r) = 0$.

Theorem 5. *Let* P *be a nontrivial prime ideal of* $K[u, x]$, *and* $f_1, \ldots, f_r$ *be a characteristic set of* P. *Then* $f_1, \ldots, f_r$ *is irreducible.*

Proof. As we know from Proposition 5, $P = \{g \mid prem(g, f_1, \ldots, f_r) = 0\}$. Suppose $f_1, \ldots, f_r$ is reducible. Then there is a $k > 0$ such that $f_1, \ldots, f_{k-1}$ is irreducible, but $f_1, \ldots, f_k$ is reducible. By Theorem 3 we can find two polynomials g and h reduced with respect to $f_1, \ldots, f_r$ such that $gh \in (f_1, \ldots, f_k) \subset P$, $0 < deg(f, x_k)$ and $0 < deg(g, x_k)$. Since g and h are reduced with respect to $f_1, \ldots, f_r$, they are not in P. Therefore, P is not a prime ideal. It contradicts the hypothesis. □

2.3. A Complete Triangulation Procedure: Ritt's Principle

Now we shall give a complete triangulation procedure, which was implicitly in Ritt's work (Ritt 1938, 1950) and was rewritten by Wu in detail (Wen-tsün 1984, 1986). This triangulation procedure is called *Ritt's Principle* by Wu. Wu considers it the basis of his method.

Theorem 6. **(Ritt's Principle).** *Let* $S = \{h_1, \ldots, h_n\}$ *be a finite non-empty polynomial set in* $A = K[y_1, \ldots, y_m]$*, and* I *be the ideal* $(h_1, \ldots, h_n)$ *of* A*. There is an algorithm to obtain an ascending chain* C *such that either*

1. C *consists of a polynomial in* $K \cap I$*, or*
2. $C = f_1, \ldots, f_r$ *with* $class(f_1) > 0$ *and such that* $f_i \in I$ *and* $prem(h_j, f_1, \ldots, f_r) = 0$ *for all* $i = 1, \ldots, r$ *and* $j = 1, \ldots, n$.

Ascending chain C *is called an extended characteristic set of* S.

Proof. By Theorem 1, we can construct a characteristic set C_1 of $S_1 = S$. If C_1 consists of only one nonzero constant, then we have (1). Otherwise, we can expand S_1 to S_2 by adding all nonzero remainders of elements of S_1 with respect to C_1. If $S_2 = S_1$, then we have (2). Otherwise, we can construct a characteristic set C_2 of S_2. If C_2 does not consist of one nonzero constant, then $C_1 > C_2$ by Proposition 3 and we can expand S_2 to S_3 using the same procedure. Thus we have a strictly increasing sequence of polynomial sets:

$$S_1 \subset S_2 \subset \cdots,$$

with the corresponding strictly decreasing sequence of characteristic sets

$$C_1 > C_2 > \cdots$$

By Proposition 2, this decreasing sequence can be only finite. Thus, there is an integer $k \geq 1$ such that either C_k consists of a nonzero constant or $S_k = S_{k+1}$; then we have either (1) or (2), respectively. □

Corollary 1. *Let* S *be the same as in Theorem 6. Suppose that the case (2) happens and* $f_1, \ldots, f_r$ *is an extended characteristic set of*

S. Let I_k be the initials of the f_k. Let $S_k = S \cup \{I_k\}$ and $P = \{g \mid g \in A$ and $prem(g, f_1, .., f_r) = 0\}$. For any extension E of the field K, we have

(i) $V(f_1, \ldots, f_r) - (V(I_1) \cup V(I_2) \cup \cdots \cup V(I_r)) \subset V(P) \subset V(S) \subset V(f_1, \ldots, f_r)$.

(ii) $V(S) = V(P) \cup V(S_1) \cup \cdots \cup V(S_r)$.

Proof. (i) is an immediate consequence of (2) of Theorem 6 and the remainder formula. Since $V(S) = [V(S) - (V(I_1) \cup \cdots \cup V(I_r))] \cup [V(S) \cap (V(I_1) \cup \cdots \cup V(I_r))]$, (ii) follows from (i) and the standard set manipulations. □

2.4. Ritt's Decomposition Algorithm

Let E be any extension of the based field K. As before, when we mention the algebraic set $V(S)$, we mean the algebraic set $V(S)$ in E^m.

Theorem 7. (Ritt's Decomposition Algorithm). *For any finite non-empty polynomial set S of A, there is an algorithm to decide whether Ideal(S) = (1), or (in the opposite case) to decompose*

$$V(S) = V(P_1) \cup V(P_2) \cup \cdots \cup V(P_s),$$

where the P_i are prime ideals given by their irreducible characteristic sets.

Proof. Let D be a temporary variable, representing a set of characteristic sets and initialized to be empty. From the polynomial set S, using Ritt's Principle, we can construct an extended characteristic set C of S and the corresponding extended set S' of the set S (i.e., the set S_k in the proof of Theorem 6). There are three cases:

Case 1. If C consists of a constant, then $V(S)$ is empty. We stop processing S.

Case 2. The ascending chain $C = f_1, \ldots, f_r$ is irreducible. Let I_k be the initials of the f_k and $S_k = S' \cup \{I_k\}$, by (ii) of Corollary 1, we have

$$V(S) = V(S') = V(P_1) \cup V(S_1) \cup \cdots \cup V(S_r),$$

where P_1 is the prime ideal having C as its characteristic set. By Proposition 3, each S_k has a characteristic set strictly lower than C. We add C to the set D. We can take each S_k as a new S, and repeat the process.

Case 3. The ascending chain $C = f_1, \ldots, f_r$ is reducible. Then there is an integer $k > 0$ such that $f_1, \ldots, f_{k-1}$ is irreducible, but $f_1, \ldots, f_k$ is reducible. By Theorem 3, we can find[9] two polynomials g and h with $class(f_k) = class(g) = class(h)$, reduced with respect to $f_1, \ldots, f_r$ and $gh \in (f_1, \ldots, f_k)$. We have decomposition: $V(S) = V(S') = V(S_1) \cup V(S_2)$, where $S_1 = S' \cup \{g\}$ and $S_2 = S' \cup \{h\}$ have characteristic sets strictly lower than C by Proposition 3. We can take each of S_1 and S_2 as new S, and repeat the process.

The above process will eventually terminate, because otherwise we would have an infinite, strictly decreasing sequence of ascending chains, which contradicts Proposition 2. Upon termination, we have two cases:

(1) D is empty. This means that S does not have common zeros in any extensions E of K, in particular in any algebraically closed extensions. By Hilbert's Nullstellensatz, Ideal$(S) = (1) = A$.

(2) $D = \{C_1, \ldots, C_s\}$, then we have

$$V(S) = V(P_1) \cup V(P_2) \cup \cdots \cup V(P_s),$$

where the P_k are the prime ideals having C_k as their characteristic sets. □

Remark 2. If E is *algebraically closed*, then by Hilbert's Nullstellensatz, each $V(P_k)$ is irreducible. In particular, the dimension of $V(P_k)$ in $E^m = m - r$, where r is the number of elements of C_k. The above decomposition gives a decomposition of an algebraic set into the union of irreducible algebraic sets (not necessarily irredundant). Hence, Radical$(S) = P_1 \cap \cdots \cap P_s$.

Remark 3. If E is not an algebraically closed field, then it is possible that $V(P_k)$ is reducible or empty. In particular, $V(S)$ can

[9]Here we need the factorization of polynomials over algebraic extensions.

be empty even if $1 \notin \text{Ideal}(S)$. We still call $V(P_i)$ a *component* of of $V(S)$.

Remark 4. In case $1 \notin \text{Idea}(S)$, we can write the decomposition in a slightly different form:

$$V(S) = V(C_1/J_1) \cup \cdots \cup V(C_s/J_s),$$

where the J_i are products of the initials of all elements in C_i and $V(C_i/J_i) = V(C_i) - V(J_i)$. This formula provides a complete solution of a system of polynomial equations up to solving systems of polynomial equations in the form of irreducible ascending chains, assuming the corresponding initials are nonzero. For details see Wentsün (1986).

In an actual implementation of Ritt's algorithms, many details should be taken care of to save the computer time and space.

3. Algebra and Geometry

Before presenting the two methods in proving geometry theorems in more detail, it is important to understand the relationship between algebra and geometry. There are few doubts that geometry statements are confirmed by the methods. However, if the method fails to confirm a geometry statement, does it disprove the statement? If so, in what sense? This is quite a complicated problem. There are two major issues behind this problem.

(1) There is a variety of geometries. For example, proving or disproving geometry statements in affine geometry and Euclidean geometry is different since the field associated with Euclidean geometry is the field of real numbers, whereas almost any fields can be the fields associated with affine geometry; particularly, they can be algebraically closed fields. To understand this issue fully, we need to discuss the relationship between geometry and algebra (Section 3.1).

(2) In proving a geometry statement involving equalities (with certain unspecified nondegenerate conditions in inequation forms), Wu's method can produce these additional conditions

in inequation forms. From the point of view of logic, this complicates our understanding of proving and disproving a geometry statement. An ideal geometry machine only needs to answer "true" or "false" to a fully specified geometry statement (including nondegenerate conditions). As we will see, that approach, though perfect from the point of view of logic, is not satisfactory from a geometric point of view, especially from the practical point of view of mechanical theorem proving in geometry (Section 2).

3.1. Axiomatic Geometries and Number Systems

We shall discuss briefly the relationship between geometry and algebra in order to show that the Cartesian product of the field of real numbers is not the *only* realistic (plane) geometry. There are two approaches to defining geometry.

(A) **The Algebraic Approach.** Starting from some number systems E (usually fields), we can define geometry objects and relations between those objects in the Cartesian product E^n (or E^n/E^* in projective geometry). In modern geometry, especially in algebraic geometry, this approach indisputably prevails. If we take this approach, then there are only a few differences between algebra and geometry; geometry can be regarded as a part of algebra.

However, the second approach suggested by Artin is more attractive from the point of view of axiomatic geometry and the point of view of traditional proofs of geometry theorems in Euclidean style.

(B) **The Axiomatic Geometry Approach.** By this approach we mean the one that was approximately used by Hilbert in his classic book (Hilbert, 1971). We shall restrict ourselves to first order logic. Depending on the context, the term "a geometry" has two meanings. When we mention *the theory of a geometry* we meaning *a first-order theory* defined in Shoenfield (1967), i.e., a formal system in first-order logic. Here "a geometry" means a class of geometry, e.g., affine geometry. On the other hand, when ordinary geometry textbook mention "a geometry", it usually means *a model* of a certain theory of geometry.

Thus Euclidean geometry (i.e., $\mathbf{R}^2$) is an affine geometry; i.e., Euclidean geometry is a model of the theory of affine geometry. Of course, this model has *much richer structures* than an ordinary affine geometry.

A very attractive feature of this approach is that algebra can be introduced in an elegant way (see Hilbert (1971)). For each model of a theory of geometry, we can prove the existence of a number system (usually a field) inherent to that model. This field is called the *field associated* E with that model. That model then can be represented as the Cartesian product E^2 of its associated field.

3.1.1. Affine Geometry

The only basic relation in the theory of affine geometry is "*incidence*", which satisfies three groups of axioms: *axioms of incidence*, *axiom of parallels*, and *Pappus' axiom*. Each model of the theory can be represented as E^2, where E is a field with certain restriction on its characteristic.

3.1.2. Metric Geometry

Metric geometry is a class of affine geometry in which there are two new basic relations "*perpendicularity*" and "*Congruence*" (Wen-tsün, 1984). These new relations satisfy two groups of new axioms (*axioms of perpendicularity* and *axioms of congruence*, see Wen-tsün (1984) or Chou (1985)). These two groups of axioms, together with axioms of affine geometry, makeup *the theory of metric geometry.* Each model of the theory can be represented as E, where E is Hilbert field of characteristic zero (a *Hilbert field* is a field in which the sum of the squares of two elements has a square root).

3.1.3. Hilbert Geometry

The theory of Hilbert geometry consists of the first four groups of axioms in Hilbert (1971). As we know, each model in this theory can be represented as E^2, where E is an *ordered* Hilbert field.

Hilbert geometry is *ordered* metric geometry. We can define an angle as two rays from one point. In general (unordered) metric geometry, we cannot define rays. However, we can define an angle as a pair of two lines which have a common point. Many theorems involving angle congruence in Euclidean geometry are still valid in unordered geometry as our prover and the prover of Wu verified. Feuerbach's theorem, Morley's trisector theorem, V. Thébault theorem are among such theorems.

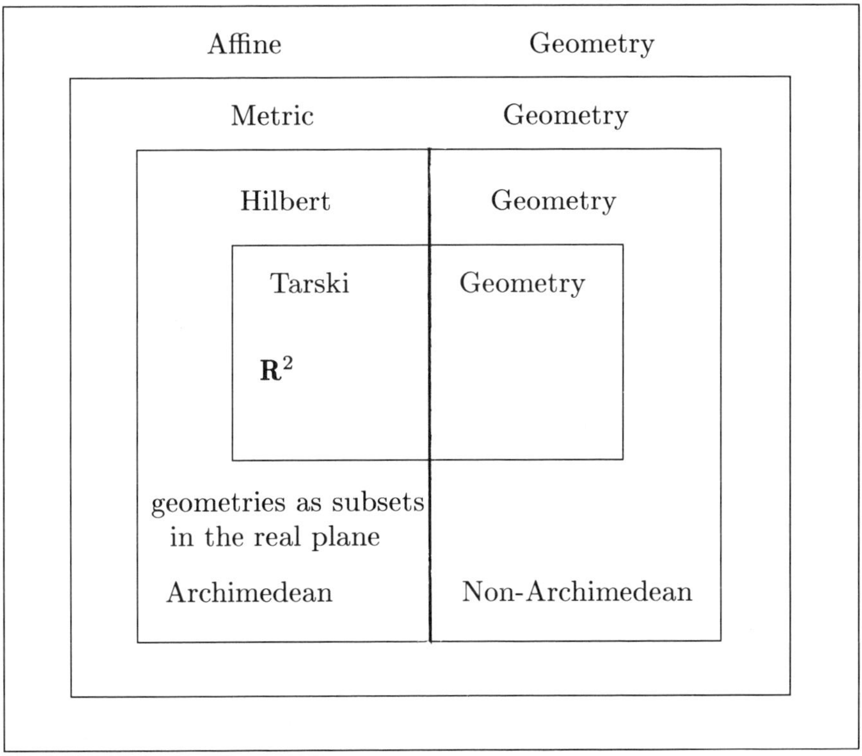

3.1.4. Tarski Geometry

The theory of Tarski geometry was given fully in Tarski (1959). Each model of the theory of Tarski geometry can be represented as R^2, where R is a real closed field, Tarski (1959). Due to a fundamental contribution by Tarski (1951), the theory of Tarski geometry is decidable. Note that the decision procedure is carried out in number systems (real closed fields), i.e., at the model level. This geometry is what Tarski called elementary geometry. As we know, there are Non-Archimedean Tarski geometries. The meaning of the term "elementary geometry" depends on who uses it. Following Wu, we prefer elementary geometry to include all geometries in which no differentiation is involved. Elementary geometry in this sense includes not only all geometries discussed so far, but also those "advanced" geometries, such as Bolyai-Lobachevskian geometry and Riemannian geometry. The inclusion relationship among the models of the four geometries discussed above can be illustrated by the preceding diagram.

The four geometries above are compatible with Euclidean geometry in the sense that Euclidean geometry is a model of the theories of the four geometries. Among geometries incompatible with Euclidean geometry, we mention projective geometry, Bolyai-Lobachevskian geometry, Riemannian geometry, etc. For those geometries, algebra and fields can be also introduced.

3.2. On the Algebraic Formulation of Certain Geometry Statements

We shall list several possible formulations and discuss related issues. For simplicity, we confine our discussion to a metric geometry whose associated field E is algebraically closed. Furthermore, we only discuss those geometric statements whose hypotheses and conclusion can be expressed by the basic relations in metric geometry – incidence, parallelism, perpendicularity, and congruence (with universal quantifier); i.e., those geometry statements for which, after adopting appropriate coordinate systems, the hypotheses can be expressed as a set of polynomial equations:

$$
\begin{aligned}
h_1(y_1, y_2, \ldots, y_m) &= 0 \\
h_2(y_1, y_2, \ldots, y_m) &= 0 \\
&\cdots \\
h_n(y_1, y_2, \ldots, y_m) &= 0,
\end{aligned}
$$

and the conclusion can also be expressed as an equation $g(y_1, \ldots, y_m) = 0$. Note that the h_i and g are in $\mathbf{Q}[y]$.

As we know from many examples before, the formula

$$\forall y_i \in E(h_1 = 0 \wedge \cdots \wedge h_n = 0 \Rightarrow g = 0)$$

is not valid for most geometry theorems because some nondegenerate conditions are missing. Note that all nondegenerate conditions are in inequation forms $s_1 \neq 0, \ldots, s_k \neq 0$, where the s_i are polynomials in $\mathbf{Q}[y]$.

3.2.1. Formulation F1

Together with $h_1 = 0, \ldots, h_n = 0$, all nondegenerate conditions necessary for the validity of the statement are specified in the forms $s_1 \neq 0, \ldots, s_k \neq 0$.

Then the algebraic formulation is:

$$\forall y_i \in E[(h_1 = 0 \wedge \cdots \wedge h_n = 0 \wedge s_1 \neq 0 \wedge \cdots \wedge s_k \neq 0) \Rightarrow g = 0]. \quad (4)$$

Theoretically, this formulation seems perfect and standard. In our opinion, however, finding all nondegenerate conditions essential for the validity of the statements is not easy.

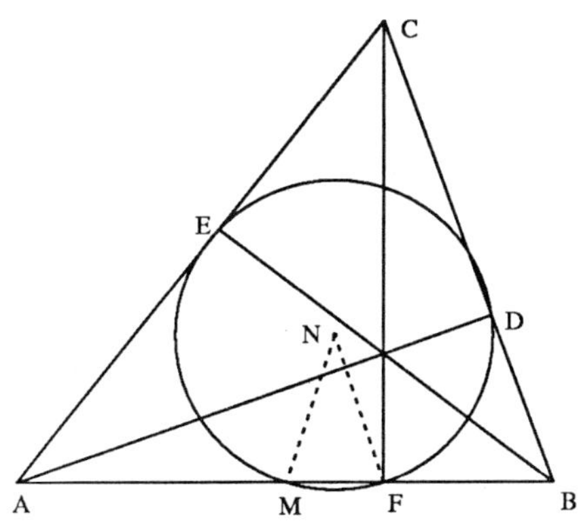

Fig. 4.

Example 8. (One form of the nine-point circle theorem) Let points D, E and F be the three feet of the altitudes of triangle ABC. Let

N be such that $NF \equiv NE$ and $NF \equiv ND$, and M be the midpoint of AB. Then $NM \equiv NF$ (Fig. 4).

In the above precise specification of the equality part, besides the condition that A, B and C are not collinear, we need at least the nondegenerate condition that triangle ABC is not a right triangle, which is not easy to detect.

Example 9. Example 2 in Section 1: Simson's Theorem in metric geometry. If we want to investigate whether Simson's theorems is a *logic consequence* of axioms related to incidence, perpendicularity, and congruence, then the "obvious" nondegenerate condition "point A, B and C are not collinear" is not enough.

In more complicated geometry statements are hidden nondegenerate conditions even harder to find.

In this formulation there are no differences between the following two cases: (a) a geometry statement is false because a condition that "a triangle in the statement is a right (or isosceles) triangle" is missing; (b) a geometry statement is false because a condition that "a triangle in the statement is *not* a right (or isosceles) triangle" is missing. However, from the point of view of mathematics or geometry, the two cases are essentially different. For the first case, the geometry statement is *generally untrue*, whereas the geometry statement is *generally true* for the second case.

Thus, this theoretically perfect formulation seems to be unsatisfactory from the practical point of view of mechanical theorem proving. The issues we raised here are important for implementing a practical prover based on this formulation.

In Kapur (1986), there is an approach based on this formulation and it was claimed: "subsidiary conditions, which are an integral part of a geometry statement, are handled in a natural way in proposed approach." This approach was independently experimented with by the authors Chou and Schelter (1986). We will give two examples in Section 5.3 to see whether this approach is natural.

We would prefer those formulations and methods which can mechanically produce nondegenerate conditions. Since the term "nondegenerate conditions" hasn't been formalized, one should take extremely precautions. To emphasize this important issue, we first give a formulation appearing in current literature (Kutzler and Stifter

(1986), Kapur (1986)), which (with its variants) can cause serious problems.

3.2.2. Formulation F2

In this formulation, to prove a geometry statement is to find a polynomial s in $\mathbf{Q}[y]$, such that

$$\forall y_i \in E[(h_1 = 0 \wedge \cdots \wedge h_n = 0 \wedge s \neq 0) \Rightarrow g = 0] \tag{5}$$

and

$$\neg(\forall y_i \in E[(h_1 = 0 \wedge \cdots \wedge h_n = 0) \Rightarrow s = 0]). \tag{6}$$

The algebraic problem in this formulation is well defined. However, the polynomial s sometimes has nothing to do with nondegenerate conditions in geometry. To make things worse, this formulation is *unsound* from a geometric point of view, as illustrated by the following examples.

Example 10. Let $ABCD$ be a parallelogram, O be the intersection of diagonals AC and BD. Show A, B and O are collinear (Fig. 1).

Let the u, the x and the h be the same as in Example 1. Now the conclusion that A, B and O are collinear is $g = u_1x_3 = 0$.

Let $s = u_3x_2 - (u_2 - u_1)u_3$. Then s and g satisfy both (5) and (6). Thus we have proved a statement which is generally not true in geometry by adding $s \neq 0$. It is interesting to note that for the above s, "$h_1 = 0, \ldots, h_4 = 0, s \neq 0 \Rightarrow g_1 = 0$" is not valid, where $g_1 = 2u_2x_4 + 2u_3x_3 - u_3^2 - u_2^2$ is the condition that $AO = OC$, which is *generally true* in geometry!

3.3. F3: Formulation Based on General Positions

This formulation is based on the notion of general positions. Our method is based on this formulation. In an elementary geometry statement there is an implicit assumption that the geometric figure to be considered should be in a general position. This means that among all points involved in a geometry statement,[10] certain points

[10]We only consider those geometry statements which can be expressed by relations between points only. As Tarski pointed out (Tarski, 1959), a geometric figure of an elementary geometry statement is determined by a fixed, finite number of points.

can be more or less arbitrarily chosen. The choice of these points, though not unique, is determined by the meaning of a geometry statement. As we have seen in many examples, we can use parameters to express the choice of arbitrarily chosen points (see Hilbert's proof of his theorem 62 in Hilbert (1971)). Thus, we can choose some variables among $y_1, \ldots, y_m$ to be independent variables or parameters $u_1, \ldots, u_d$. The remaining variables $x_1, \ldots, x_r$ can be considered dependent variables. Once these parameters $u_1, \ldots, u_d$ are fixed, the remaining variables $x_1, \ldots, x_r$ are (algebraically) determined under normal (nondegenerate) conditions. Based on the notion of general positions, our formulation consists of two parts:

(1) To define the exact meaning of whether a geometry statement is true for general values of $u_1, \ldots, u_d$, or a geometry statement is *generally* (*generically*) *true*.

(2) To identify those special values of the u, for which the geometry statement might be invalid, i.e., to identify degenerate cases.

3.3.1. The Generic Validity of a Geometry Statement

Let $V = V(h_1, \ldots, h_n)$ be the zeroes of $\{h_1, \ldots, h_n\}$ in E^m. Using Ritt's decomposition algorithm Theorem 7 of Section 2, V can be decomposed into irreducible components, represented by their irreducible characteristic sets:

$$V = V(P_1^*) \cup \cdots \cup V(P_c^*) \cup V(P_1^\dagger) \cup \cdots \cup V(P_e^\dagger) \cup V(P_1) \cup \cdots \cup V(P_l), \tag{7}$$

where P_i^*, $P_i^\dagger$ and P_i are prime ideals, $V(P_i^*)$ and $V(P_i^\dagger)$ are all those components on which $u_1, \ldots, u_d$ are algebraically independent, and $dim(V_i^*) = d$, $dim(V_i^\dagger) > d$. We call any $V_i^* = V(P_i^*)$ a component *general* for u or a *general component*. The condition $c > 0$ serves as a necessary criterion to check whether the parameters have been selected correctly.

Remark 5. Even if we use the notion of general positions, it is still possible that some components general for u correspond to what people usually consider degenerate cases. This is due to ambiguities in specifying points using algebraic equations, similar to that mentioned in Example 3 of Section 1 (the Butterfly theorem). However, for practical problems it is easy to detect such kinds of ambiguities and to exclude them mechanically. Thus we can assume that equations $h_1 = 0, \ldots, h_n = 0$ have already been *properly modified* so that there are no such components general for u.[11]

Remark 6. It is extremely rare that $e > 0$. In more than 400 nontrivial geometry theorems proved by our prover, none fit this case. If $e > 0$, then there might be some hidden property (theorem) of the geometric figure, and the problem is possibly beyond the original intention of the statement. In our current implementation we discard such cases, considering them inherently degenerate. However, the program also outputs the related information to let the user take a closer look at the problem. Thus in what follows we also assume $e = 0$. This assumption is related to Dimensionality Restriction in Section 4.

Remark 7. Under the assumptions made in Remarks 5 and 6, it is clear that, given a set of equations and a set of parameters, the $V(P_i^*)$ correspond to nondegenerate cases or generic cases, and the $V(P_j)$ correspond to degenerate cases because on those components the u_i are algebraically dependent. Thus the $V(P_i^*)$ are also called *nondegenerate components.*

Now we are in a position to define what is meant by "generically true." There are three possibilities:

(F3.1) The polynomial g vanishes on all components general for u; then we say the geometry statement is *generically or generally true.*

(F3.2) The polynomial g vanishes on none of the components general for u, then the statement is *generically or generally untrue.*

For more than 95 percent of the problems we encountered in practice, $c = 1$. In such case, we have a clear-cut answer to the problem: either it is generically true, or it is *generically untrue.* However, we also found a few problems belonging to the following case:

---

[11] In a recent experiment proving more than 360 theorems, such an approach was totally successful. See Chou (1986).

(F3.3) The polynomial g vanishes on some components, but not on all.

The answer to such problems is not clear cut. Some people may say that the geometry statement is not generically true because it is not true in some nondegenerate cases. However, others could counter that under some subsidiary conditions, which exclude those *nondegenerate* cases, the geometry statement is true. In our opinion, the problem should be investigated further. If the original problem comes from Euclidean geometry, then it is usually connected with order relation. We should use the order relation to obtain a complete answer. We will give examples in Section 6. Here we only give an example to illustrate (F3.1) and (F3.2).

Example 11. Example of parallelogram. The same as example 10. Under these geometric conditions, we want to decide whether (i) $AO \equiv CO$; (ii) point A, O and B are collinear.

The choice of coordinates are the same as in example 10. Since points A, B and C can be arbitrarily chosen, u_1, u_2 and u_3 can be chosen as parameters. Once these points are normally fixed, the remaining points D and O can be determined. Thus their coordinates x_1, x_2, x_3 and x_4 are algebraically dependent on u_1, u_2 and u_3 under normal (nondegenerate) conditions, being restricted by equations $h_1 = 0, \ldots, h_4 = 0$. We have two conjectures: $g_1 = 2u_2x_4 + 2u_3x_3 - u_3^2 - u_2^2 = 0$ ($AO \equiv CO$); $g_2 = u_1x_3 = 0$ (A, O and B are collinear). According to Ritt's decomposition algorithm, we can decompose $V(h_1, \ldots, h_4)$ into four irreducible components:[12]

$$V = V(P_1^*) \cup V(P_1) \cup V(P_2) \cup V(P_3),$$

where P_1^*, P_1, P_2, and P_3 have

$C_1^* = \{f_1, f_2, f_3, f_4\}$, ($f_1, \ldots, f_4$ are the same as in Example 1 of Section 1.)

$C_1 = \{u_3, x_1, x_3\}$,

$C_2 = \{u_1, u_3x_2 - u_2x_1, u_2x_3 - u_3x_4\}$, and

$C_3 = \{u_2, u_3, x_1, x_2 - u_1\}$

for their characteristic sets, respectively. Thus we have a unique nondegenerate component $V(P_1^*)$. To check whether $g_1 = 0$ or $g_2 = 0$

[12] Actually, we don't have to get all irreducible components. See Section 4.2.

are generally true, we only need to check whether g_1 or g_2 vanishes on $V(P_1^*)$. According to Theorem 4 of Section 2, it is equivalent to checking whether $prem(g_1, C_1^*) = 0$ or $prem(g_2, C_1^*) = 0$. Now $prem(g_1, C_1^*) = 0$ but $prem(g_2, C_1^*) \neq 0$. Thus, (i) is generally true, but (ii) is generally untrue.

Note that $dim(V(P_1)) = dim(V(P_2)) = 4 > dim(V(P_1^*)) = 3$. *Dimensions of degenerate components can be even higher than that of general components.* Thus, we cannot distinguish degenerate cases from general cases based only on dimensionality consideration.

3.3.2. Identifying Nondegenerate Conditions

The second part of our formulation is to identify those special values of the u for which the geometry statement might be invalid, i.e., to identify degenerate cases. We have already seen a practical method for giving sufficient subsidiary conditions so that the conclusion $g = 0$ is valid under $h_1 = 0, \ldots, h_n = 0$ and under conditions $I_1 \neq 0, \ldots, I_r \neq 0$ (where the I_i are initial of the f_i obtained from triangulation procedure). In general, we have the following theorem:

Theorem 8. *Let E be an algebraically closed field. $g = 0$ is generally true under $h_1 = 0, \ldots, h_n = 0$ and with respect to a parameter set $u_1, \ldots, u_d$, if and only if there is a non-zero polynomial s containing the u (called a u–polynomial) only such that $sg \in$* Radical$(h_1, \ldots, h_n)$.

Proof. Let

$$V(h_1, .., h_n) = V_1^* \cup \cdots \cup V_c^* \cup V_1 \cup \cdots \cup V_l, \tag{8}$$

where the V_i^* are all components general for u, and the u are algebraically dependent on the V_i.

(i) Suppose $g = 0$ is generally true, i.e., g vanishes on each V_i^*. Since u are algebraically dependent on each V_j, there is a u–polynomial s_j such that s_j vanishes on V_j. Let $s = s_1, ..s_l$, then

$$V(h_1, \ldots, h_n) - V(s) \subset V_1^* \cup \cdots \cup V_c^*. \tag{9}$$

Since g vanishes on $V_1^* \cup \cdots \cup V_c^*$, sg vanishes on $V(h_1, \ldots, h_n)$. Thus, $sg \in$ Radical$(h_1, \ldots, h_n)$ by Hilbert's Nullstellensatz.

(ii) Conversely, if $sg \in \text{Radical}(h_1, \ldots, h_n)$ for some u–polynomial s, then sg vanishes on $V(h_1, \ldots, h_n)$. Since s contains the u only, s does not vanishes on each V_k^*. Since each V_k^* is irreducible, g vanishes on each V_k^*. □

3.3.3. The Generic Validity of a Geometry Statement in an Arbitrary Field

The above formulation is for geometries with associated fields algebraically closed. Now we extend the formulation to geometries with any associated fields. Our starting point is the decomposition equation (7). Since the associated field E now is not necessarily algebraically closed, this decomposition is generally not an irreducible decomposition. However, the u are algebraically dependent on any components of the algebraic set $V(P_i)$, and thus we can regard $V(P_i)$ as degenerate cases. We will still call any $V(P_i^*)$ a component general for u or a nondegenerate component (note that $V(P_i^*)$ can be reducible or even empty).

(F3.1) If the polynomial g vanishes on all $V(P_i^*)$, then we say the geometry statement is **generically true**.

(F3.2) If the polynomial g vanishes on none of the $V(P_i^*)$, then the statement is **generically untrue**.

(F3.3) If the polynomial g vanishes on some components, but not on all, this situation should be investigated further.

4. The Complete Method of Wu

Now we shall present the complete method of Wu in detail. Generally, this method works for any field but is complete only for those geometries whose associate fields are algebraically closed.

4.1. Ritt's Principle Revised

For a geometry statement (S), suppose its hypotheses (with nondegenerate conditions unspecified) can algebraically be expressed by a set of polynomial equations

$$
\begin{aligned}
h_1(u_1, \ldots, u_d, x_1, \ldots, x_t) &= 0 \\
h_2(u_1, \ldots, u_d, x_1, \ldots, x_t) &= 0 \\
&\ldots \\
h_n(u_1, \ldots, u_d, x_1, \ldots, x_t) &= 0,
\end{aligned}
$$

and the conclusion is also a polynomial equation $g = g(u_1, \ldots, u_d, x_1, \ldots, x_t) = 0$, where $h_1, \ldots, h_n$ and g are polynomials in $\mathbf{Q}[u_1, \ldots, u_d, x_1, \ldots, x_t]$. Here $u_1, \ldots, u_d$ are parameters and $x_1, \ldots, x_t$ are dependent variables.

Now we present Ritt's Principle in the context of geometry theorem proving: we can let $K = \mathbf{Q}[u_1, \ldots, u_d]$, $x_i = y_i$, $t = m$, and $I = (h_1, \ldots, h_n)$. Note that Ritt's principle also works when K is the ring $\mathbf{Q}[u]$. Thus, applying Ritt's principle to $h_1, \ldots, h_n$, we can obtain an ascending chain C such that either

Case 1. C consists of a polynomial in $K \cap I$, or

Case 2. $C = f_1, \ldots, f_r$ with $class(f_1) > 0$ and such that $f_i \in I$ and $prem(h_j, f_1, \ldots, f_r) = 0$ for all $i = 1, \ldots, r$ and $j = 1, \ldots, n$.

In the case of 1, we have a polynomial in the parameters u only. This cannot happen if the parameters are properly chosen. Whenever this happens the program stops and outputs the related information. The user should reformulate the problem in a appropriate way. If the user chooses the parameters, then he should check his choice carefully. If the statement is of constructive type and the choice of parameters is done by the program, this means that there possibly exist certain unexpected properties for the geometric figure in the statement. In Case 2, we add the following restriction:

Dimensionality Restriction. *Every dependent variable x_i appears as a leading variable of an f_i. This means that $t = r$, and $x_i = lv(f_i)$. Thus the ascending chain $f_1, \ldots, f_r$ has the following form:*

$$
\begin{aligned}
&f_1(u_1, \ldots, u_d, x_1) \\
&f_2(u_1, \ldots, u_d, x_1, x_2) \\
&\qquad \ldots \\
&f_r(u_1, \ldots, u_d, x_1, \ldots, x_r),
\end{aligned}
\tag{10}
$$

Suppose some x_i does not appear as a leading variable in any of the f_k, then we can rename x_i to be a new parameter as we did in Section 2.2. This can be done by programs. But in view of some possible problems, we prefer the user to take the responsibility of renaming variables if the parameters were specified by the user.

4.2. Ritt's Decomposition Algorithm Revised

Let $u_1, \ldots, u_d, x_1, \ldots x_r$ and $H = \{h_1, \ldots, h_n\}$ be the same as in the previous section. First we present a simplified Ritt's decomposition algorithm in the context of geometry theorem proving to obtain all general (nondegenerate) components. The presentation here is our clarification of Wu's work (Wen-tsün 1986, 1984).

Algorithm 1. (General Component Decomposition). *First set S to be H.*

Main Step. Apply Ritt's principle as revised in the last section to the polynomial set S to obtain an ascending chain C.

Case 1. C consists of a polynomial $p \in \text{Ideal}(S) \cap \mathbf{Q}[u]$. If S is the initial polynomial set H, then we stop the program and output the related information. Thus we assume this case does not happen. If If S is not the initial polynomial set H, then we record this polynomial p.

Case 2. C is reducible. Then we have case 3 in Ritt's decomposition algorithm (2.4). Thus, we have at least two polynomial sets S_1, S_2 such that $V(S) = V(S_1) \cup V(S_2)$. Take S_1 and S_2 as new S and for each new S, repeat the main step.

Case 3. C is irreducible, but does not satisfy the dimensionality restriction. This means the geometric configuration possibly has richer properties than we expected. Whenever this happens, the problem should be investigated further by the user. The program stops and outputs related information. Thus we assume this case does not happen (see Remark 6 of Section 3.3).

Case 4. C is irreducible and satisfies the dimensionality restriction. Then record C.

As in Ritt's original decomposition algorithm, each new S obtained from Case 2 (if any) has an extended characteristic set strictly lower than that of its predecessor. Thus the procedure will terminate. Upon termination, we have a set of polynomial sets $S_1, \ldots, S_c$ with extended irreducible characteristic sets $C_1, \ldots, C_c$ (from Case 4) in the form of equation (10) and a set of polynomials $p_1, \ldots, p_s$ (from Case 1). It is easy to see that $V(H) - V(p_1 \cdots p_s) \subset V(S_1) \cup \cdots \cup V(S_c)$. Let P_i be the prime ideal with characteristic set C_i. Then

Theorem 9. *$V(P_1), \ldots, V(P_c)$ are all components of $V(H)$, general for u.*

Proof. Obviously, $V(P_1), \ldots, V(P_c)$ are components general for u. There are no any other components general for u. The reason is as follows. Further decomposition in Ritt's algorithm will start from Case 1 or Case 4. In case 1, since there is a u–polynomial $p \in \text{Ideal}(S)$, the u are algebraically dependent on all components of $V(S)$. In Case 4, the further decomposition is to deal with polynomial sets $S \cup \{I_i\}$, where the I_i are the initials of the ascending chain C. By Proposition 6 of Section 2, Ideal$(S \cup \{I_i\})$ contain a u–polynomial. Thus, the u are algebraically dependent on all components of $V(S \cup \{I_i\})$. □

The complete method of Wu is based on this version of decomposition algorithm of Ritt. The main difference between this version and the original version is in the efficiency. 219 of 366 theorems in the *Collection* (Chou, 1986) are linear, i.e., $f_1, \ldots, f_r$ obtained from $h_1, \ldots, h_n$ are linear in its leading variables. Thus one main step (Ritt's principle) is sufficient (the irreducibility is trivial). It is very fast, usually within 1–2 seconds. However, if we use the original algorithm to obtain all components, then it will take much longer.

4.3. Complete Method of Wu – Irreducible Cases

Let all notations, (S), $h_1, \ldots, h_n$, g and $f_1, \ldots, f_r$, be the same as in Section 1.1. As we have seen from Section 1, if $prem(g, f_1, \ldots, f_r) = 0$, then (S) is valid under $I_k \neq 0$, where the I_k are initials of the f_k. This is elementary use of the method to confirm a geometry statement. But an implicit problem exists.

Example 12. Let $h_1 = x_1^2, h_2 = x_1x_2,\ g = x_2 + x_1$. Then $f_1 = h_1, f_2 = h_2$ and $prem(g, f_1, f_2) = 0$. Thus, $(f_1 = 0, f_2 = 0, x_1 \neq 0) \Rightarrow g = 0$. However, by adding $x_1 \neq 0$, the hypotheses become inconsistent.

Now we suppose the ascending chain $f_1, \ldots, f_r$ obtained from $h_1, \ldots, h_n$ by Ritt's principle as revised in the last section or its variants is irreducible.[13] Such a geometry statement (S) is said to be an *irreducible statement* or an *irreducible problem.* This definition depends on the choice of parameters.

Theorem 10. *Let* (S), $h_1, \ldots, h_n$, g, *and* $f_1, \ldots, f_r$ *be the same as above. Suppose* $f_1, \ldots, f_r$ *is irreducible. If* $prem(g, f_1, \ldots, f_r) = 0$, *then*

(i) (S) *is generally true in all geometries (fields), and*
(ii) *For all fields,* $(h_1 = 0, \ldots, h_n = 0,\ I_1 \neq 0, \ldots, I_r \neq 0) \Rightarrow g = 0$, *where the* I_k *are the leading coefficients of the* f_k.

Proof. (i) Let P be the prime ideal with the characteristic set $f_1, \ldots, f_r$. Then by Theorem 9, $V(P)$ is the only component general for u. By Theorem 4 of Section 2, $prem(g, f_1, \ldots, f_r) = 0$ implies g vanishes on $V(P)$. Thus (S) is generally true. (ii) follows from the remainder formula. □

Theorem 11. *Let* (S), $h_1, \ldots, h_n$, g, *and* $f_1, \ldots, f_r$ *be the same as above. Suppose ascending chain* $f_1, \ldots, f_r$ *is irreducible. Let* P *be the prime ideal with the characteristic set* $f_1, \ldots, f_r$. *Let* E *be the field associated with the geometry* G. $prem(g, f_1, \ldots, f_r) = 0$ *is a necessary condition for* (S) *to be generally true if one of the following conditions is satisfied:*

(i) *For* E, $V(P)$ *is of degree* d.
(ii) E *is algebraically closed field.*
(iii) $f_1, \ldots, f_r$ *has a generic point in* E.

Proof. (i) Suppose $dim(V(P)) = d$. By remark 1 in Section 2, $V(P)$ is irreducible and $I(V(P)) = P$. Thus g vanishes on $V(P)$ if

[13] It can be an ascending chain in weak sense.

and only if $g \in P$, i.e., $prem(g, f_1, \ldots, f_r) = 0$. (ii) and (iii) follow from (i) (see Theorem 4). □

Theorem 12. *Let (S) be an irreducible geometry statement. (S) is a generically valid statement in metric geometry, i.e., in all models of the theory of metric geometry, only if* $prem(g, f_1, \ldots, f_r) = 0$.

Proof. Since $\mathbf{C}^2$ is a metric geometry, where $\mathbf{C}$ is the field of complex numbers, the theorem follows from (ii) of Theorem 11. □

Theorem 13. *Let (S) be an irreducible statement in Euclidean geometry* $\mathbf{R}^2$, *where* $\mathbf{R}$ *is the field of real numbers. Suppose* $f_1, \ldots, f_r$ *has a generic point in* $\mathbf{R}$. *(S) is generically true in Euclidean geometry only if* $prem(g, f_1, \ldots, f_r) = 0$.

Proof. This is an immediate consequence of (iii) of Theorem 11. □

By Theorem 13, to disprove a statement in Euclidean geometry requires proving the existence of generic points in **R**. The notion of generic points seems "non-elementary". However, the following theorem establishes an elementary understanding of "generic points" in **R**.

Theorem 14. *Let* $f_1, \ldots, f_r$ *be an irreducible ascending set. The following two conditions are equivalent:*

(i) *There are intervals* $U_1, \ldots, U_d$ *in* $\mathbf{R}$ *such that for all* $u_i \in U_i$, $f_1 = 0, \ldots, f_r = 0$ *has a solution in* $\mathbf{R}$.
(ii) *There exists a real generic point* $(e_1, \ldots, e_d, \tilde{x}_1, \ldots, \tilde{x}_r)$ *of* $f_1, \ldots, f_r$.

Proof. See (Chou, 1985). As we mentioned earlier, 219 among the 366 theorems in Chou (1986) can be expressed as linear cases (i.e., $deg(f_i, x_i) = 1$ for all f_i), including many statements involving the tritangent centers (the incenter and the three excenters) of a triangle. Those problems are *trivially* irreducible and corresponding real generic points *trivially* exist. Thus we can prove or disprove any conjectures about the configurations of such problems mechanically in all geometries, including Euclidean geometry. There is a generic nature in ordinary theorems in Euclidean geometry, i.e., the corresponding real generic points always exit. This is related to the fact that we can always draw the corresponding diagrams. For example, in Simson's theorem, all f's are linear but f_3. Note that f_3 is related to choosing an arbitrary point D on a circle. From Theorem 14 it is evident that corresponding real generic points exist. Thus we can prove and *disprove* any statement on Simson's configuration.

Example 13. Let A and B be two distinct points, let M be the midpoint of A and B. Let C be one of the intersections of the circle $(A, (AM))$ with the line L passing through B and perpendicular to AB, then $AC \equiv AB$.

This statement is not valid in $\mathbf{C}^2$, but is generically valid in Euclidean geometry. However, the method cannot prove this statement in Euclidean geometry.

4.4. Complete Method of Wu – General Cases

Now it is easy to see how to handle general cases. First we use the General Component Decomposition Algorithm 1 to obtain all general components $V(P_1), \ldots, V(P_c)$ of $V(h_1, \ldots, V_n)$, where each prime ideal P_i has a characteristic set C_i of the form of equation (10). Irreducible problems correspond to cases when $c = 1$.

Theorem 15. *If $prem(g, C_i) = 0$ $(i = 1, \ldots, c)$, then (S) is generally true in all geometries.*

Proof. This is immediate from the definition (F3.1) of Sections 3.3.1 or 3.3.3 and Theorem 4 of Section 2. □

Theorem 16. *Let E be the field associated with the geometry considered. $prem(g, C_i) = 0$ for all i are necessary conditions for (S) to be generally true if one of the following conditions is satisfied:*

(i) *For E, all $V(P_i)$ is of degree d $(i = 1, \ldots, c)$.*
(ii) *E is an algebraically closed field.*
(iii) *Each ascending chain C_i has a generic point in E.*

In each of cases (i)–(iii), if $prem(g, C_i) \neq 0$ for all i, then (S) is generally untrue.

Proof. It is similar to the proof of Theorem 11. □

If $prem(g, C_i) \neq 0$ for some i but not for all i, then the statement should be under further investigation. The geometric configuration in consideration becomes more interesting in such case. A notable example is V. Thèbault's conjecture (see Example 20). For more examples, see Chapter 5 of Chou (1988).

Theorem 17. *If (S) is generally true for an algebraically closed field E, then it is generally true for all fields E.*

Proof. Suppose (S) is generally true for an algebraically closed field E. Then by Theorem 16, $prem(g, C_i) = 0$ for all $i = 1, \ldots, c$. By (iv) of Theorem 4, g vanishes on all $V(P_i)$ for any fields E, i.e., (S) is generally true. □

5. Geometry Theorem Proving Using the Gröbner Basis Method

In Section 1, we give a brief review of the basic notions in the Gröbner basis method necessary for our purpose. Buchberger's paper (Buchberger, 1985) is an excellent and comprehensive review.

5.1. A Review of the Gröbner Basis Method

Let K be a (computable) field and $A = K[y_1, \ldots, y_n]$ be the polynomial ring over K with indeterminates $y_1, \ldots, y_n$.[14]

Definition 5. A total order $<$ among the set M of all monomials (the power products of the y) in $K[y]$ is called *compatible* (with multiplication), if

(i) $1 < m$ for all $m \neq 1$.
(ii) $m_1 < m_2 \Rightarrow sm_1 < sm_2$, for all s, m_1, $m_2 \in M$.

[14] We are only interested in Gröbner bases in polynomial rings over fields. There are many generalizations of fields to certain commutative rings.

Example 14. The following two compatible orderings are mostly in common use in the Gröbner basis method.

(i) **The pure lexicographic ordering.** $y_1^{s_1} \ldots y_n^{s_n} < y_1^{t_1} \ldots y_n^{t_n}$ if there is $k \leq n$ such that $s_i = t_i$ for $i < k$ and $s_k < t_k$.
(ii) **The total degree ordering.** $y_1^{s_1} \ldots y_n^{s_n} < y_1^{t_1} \ldots y_n^{t_n}$ if $s = s_1 + \cdots + s_n < t = t_1 + \cdots + t_n$, or $s = t$ and there is $k \leq n$ such that $s_i = t_i$ for $i < k$, $s_k < t_k$.

Theorem 18. *Any compatible order is well-ordered.*

Proof. Suppose $m_1 > m_2 > \cdots > m_k > \ldots$ is an infinite, strictly decreasing sequence of monomials. We assert that such a sequence does not exist. One can use induction on n to prove the assertion. Here we give an indirect proof using the Hilbert basis theorem. Let $J_k = (m_1, \ldots, m_k)$ be the polynomial ideals in $K[y_1, \ldots, y_n]$, generated by monomials $m_1, \ldots, m_k$. Then $J_1 \subset J_2 \subset \ldots$ is an ascending sequence of ideals. By the Hilbert basis theorem, there is an integer t such that $J_t = J_{t+k}$ for all $k > 0$. Thus $m_{t+1} = p_1 m_1 + \cdots + p_t m_t$. Hence $m_{t+1} \geq m_s$ for some $s < t + 1$ by (ii) of Definition 5. This contradicts our assumption. □

In what follows, we fix a compatible ordering $<$; all discussions are relative to this ordering. Thus a polynomial p can be expressed in the form:

$$p = c_1 m_1 + \cdots + c_k m_k,$$

where the c_i are nonzeros in K and the m_i are decreasing monomials: $m_1 > m_2 > \ldots$. This representation is unique. The monomial m_1 is called the leading monomial of p, $c_1 m_1$ the leading term of p. We also say that $m_1, \ldots, m_k$ are monomials of p.

Let S be a polynomial set and g be a polynomial. Polynomial g is *b-reducible* modulo S if there is an $f \in S$ such that the leading monomial m_1 of f is a factor of a monomial m of g. In that case, letting $f = c_1 m_1 + f_1$ and $m = s m_1$, we can replace m in g by $-s f_1 / c_1$ to obtain a polynomial g_1. We say g is reduced to g_1 modulo S via one step. Obviously, $g - g_1 \in \text{Ideal}(S)$. If g_1 is b-reducible, then we can repeat the process to obtain a new polynomial g_2. We call this process *reduction*. From Theorem 18, it is not hard to prove that this process terminates. Thus, g is eventually reduced to a polynomial g_k, b-irreducible modulo S. This polynomial is called a *normal form* of g modulo S. Note that a normal form of polynomial g modulo S is not necessarily unique. Different reductions of g modulo S can lead to different normal forms.

Definition 6. A set of generators G of an ideal $I \subset K[y]$ is called a Gröbner basis of I if any reduction of any polynomial g modulo G leads to the same normal form of g.

Theorem 19. *If G is a Gröbner basis of ideal I, then $g \in I$ if and only if the normal form of g modulo G is zero.*

Proof. See (Bachmair and Buchberger, 1980).

This theorem means that once we have a finite Gröbner basis of an ideal I, then we can decide the ideal membership (one of the most important problems in polynomial ideal theory) easily.

Definition 7. Let $p_1 = c_1 m_1 + f_1$ and $p_2 = c_2 m_2 + f_2$ be two nonconstant polynomials with leading terms $c_1 m_1$ and $c_2 m_2$, respectively. Let m be the (monic) least common multiplier of m_1 and m_2, s_1 and s_2 be monomials such that $m = s_1 m_1 = s_2 m_2$. Then polynomial $p = s_1 f_1 - (c_1/c_2) s_2 f_2$ is called the S-polynomial of p_1 and p_2.

Note that if $\mathrm{GCD}(m_1, m_2) = 1$, then the S-polynomial p of p_1 and p_2 can be reduced to zero modulo $\{p_1, p_2\}$.

Algorithm 2. *(Buchberger's algorithm for constructing Gröbner bases). Given a finite set S of generators of an ideal I, we can use the following procedure to construct a Gröbner basis of I.*

Form all S-polynomials of all pairs of polynomials in $S_0 = S$. Reduce these S-polynomials to their normal forms modulo S_0 and add any nonzero normal forms to S to get a new polynomial set S_1. Repeat the process for S_1 to obtain a new polynomial set S_2, and so on.

Theorem 20. *The procedure in Algorithm 2 always terminates, i.e., for the sequence $S_0 \subset S_1 \subset S_2 \subset \ldots$, there is k such that $S_k = S_{k+1}$ and S_k is a Gröbner basis of I.*

Proof. Considering the leading monomials of polynomials of the set S_i in the order in which new polynomials are added, we can use similar reasoning as in the proof of Theorem 18 to prove the termination of the procedure; i.e., there is k such that $S_k = S_{k+1}$. For the proof of the fact that S_k is a Gröbner basis of I, see (Bachmair and Buchberger, 1980).

Our actual implementation is that, whenever a new S-polynomial is added to the polynomial set, we reduce each polynomials modulo others in certain order until every polynomial is b-reduced modulo the others; then we form new S-polynomials again. In this way we can obtain a *reduced* Gröbner basis of I, i.e., a Gröbner basis of I, each polynomial of which is monic and reduced modulo the remaining polynomials. Note that the implementation details can affect the efficiency greatly (see Section 6.2). For reduced Gröbner bases, we have the following uniqueness theorem:

Theorem 21. *For a given compatible order, the reduced Gröbner basis of an ideal I is unique regardless of the starting set S of generators $p_1, \ldots, p_m$.*

Proof. See Buchberger's paper (Buchberger, 1976).

Unless stated otherwise, whenever we mention a Gröbner basis of an ideal $(p_1, \ldots, p_m)$, we mean this unique reduced Gröbner basis and denote it by $\mathrm{GB}(p_1, \ldots, p_m)$.

5.2. A Proof Method for Formulation F3

Basically, the Gröbner basis method can be used to prove the same class of geometry theorems that can be addressed by Wu's method. Our approachto proving geometry theorems using the Gröbner basis method is based on the following theorems.

Let (S), $u = u_1, \ldots, u_d$, $x = x_1, \ldots, x_r$, $h_1, \ldots, h_n$, and g be the same as in IV. We assume that the field E associated with the geometry in consideration is algebraically closed. As we already knew (Theorem 17), if we can confirm (S) to be generally true in E, then (S) is generally true in any field. Let I be the ideal $(h_1, \ldots, h_n)$ in $\mathbf{Q}[u, x]$, J be the ideal $(h_1, \ldots, h_n)$ in $\mathbf{Q}(u)[x]$. Obviously, $I \subset J$ and $g \in J$ if and only if there is a nonzero polynomial $p \in \mathbf{Q}[u]$ such that $pg \in I$. Let $L = \mathrm{Radical}(h_1, \ldots, h_n)$ in $\mathbf{Q}(u)[x]$; i.e., $\mathrm{Radical}(J)$.

Theorem 22. *Let all notations be the same as in the previous paragraph. The geometry statement (S) is generically true (in E) if and only if $g \in L$.*

Proof. Suppose (S) is generically true, then by Theorem 8, there is a u-polynomial s such that $sg \in \text{Radical}(I)$ of $\mathbf{Q}[u, x]$, i.e, there is an integer $t > 0$ such that $(sg)^t \in I$. Hence $g^t \in J$, i.e., $g \in L$.

Conversely, suppose that $g \in L$, then there is an integer $t > 0$, $g^t \in J$. Thus there is a u–polynomial s such that $sg^t \in I$. By Theorem 8, g is generically true. □

Theorem 23. *If $g \in J$, then (S) is generically true.*

Proof. It is obvious from Theorem 22 since $g \in J \subset L$. □

Theorem 24. *Let all notations be the same as in (2.1) and z be a new variable other than the u and the x. $GB(h_1, \ldots, h_n, zg-1) = \{1\}$ (in $\mathbf{Q}(u)[x, z]$) if and only if $g \in L$. Thus, (S) is generally true if and only if $GB(h_1, \ldots, h_n, zg - 1) = \{1\}$.*

Proof. If $g \in L$, then $h_1, \ldots, h_n, zg - 1$, considered as polynomials in $\mathbf{Q}(u)[x, z]$, do not have common zeros in any extension of $\mathbf{Q}(u)$, in particular in algebraically closed extensions of $\mathbf{Q}(u)$. Thus, by Hilbert's Nullstellensatz, the ideal $(h_1, \ldots, h_n, zg - 1)$ of $\mathbf{Q}(u)[x, z]$ is the unit ideal (1), i.e., $\text{GB}(h_1, \ldots, h_n, zg - 1) = \{1\}$. If $\text{GB}(h_1, \ldots, h_n, zg - 1) = \{1\}$, then there are q_i in $\mathbf{Q}(u)[x, z]$ such that

$$q_1 h_1 + \cdots + q_n h_n + q_{n+1}(zg - 1) = 1.$$

Let $z = 1/g$, and clearing denominators, we have

$$a_1 h_1 + \cdots + a_n h_n = g^t,$$

for some $t \geq 0$ and polynomials $a_i \in \mathbf{Q}(u)[x]$. Thus $g \in L$. □

The main method we use in our geometry theorem prover is based on Theorem 22, 23, and 24:

Method 1. *We assume $x_1 < \cdots < x_r$ and the pure lexicographic ordering among monomials (see Example 14 of the last section). Note that variables $u_1, \ldots, u_d$ are now in the based field $\mathbf{Q}(u)$. Thus there are no orderings among the u.*

Step 1. Compute the Gröbner basis of $h_1, \ldots, h_n$ in $\mathbf{Q}(u)[x]$ to see whether $g \in J$. If it is, then (S) is generically true. If it is not, then

Step 2. Compute the Gröbner basis GB of $h_1, \ldots, h_n, zg-1$ in $\mathbf{Q}(u)[x]$ to see whether it is $\{1\}$. (S) is generically true if and only if $\text{GB} = \{1\}$. (Note that E is algebraically closed.)

We can also find subsidiary conditions by collecting all denominators (removing repeated factors) during the computation of the Gröbner basis.

Step 1 can be considered as a first approximation. However, for all theorems we have found in practice, $J = L$. Thus, Step 1 is usually sufficient (see Examples in Section 1.3). The reader can find many other examples in Chou and Schelter (1986) and in Chou (1986).

The following theorem gives another method for finding subsidiary conditions.

Theorem 25. *Let all notations be the same as in Theorem 22 and z be a new variable other than the u and the x. (S) is generally true if and only if the Gröbner basis of $I_z = (h_1, \ldots, h_n, gz-1)$ in $\mathbf{Q}[u, x, z]$ (in purely lexicographic orderings $u < x$ and $u < z$) contains at least a u–polynomial.*

Proof. Suppose (S) is generally true. Let J_z be the ideal generated by $h_1, \ldots, h_n$, $gz - 1$ in $\mathbf{Q}(u)[x, z]$. Then by Theorem 24, $J_z = (1)$. A polynomial $a \in J_z$ if and only if there is a u–polynomial s such that $sa \in I_z$. Since $1 \in J_z$, there is a u–polynomial s such that $s = s \cdot 1 \in I_z$. Since we are using purely lexicographic orderings with $u < x$ and $u < z$, $\text{GB}(h_1, \ldots, h_n, gz-1)$ of I_z contains at least a u–polynomial. Conversely, suppose $\text{GB}(h_1, \ldots, h_n, gz-1)$ of I_z contains a u–polynomial s. Then there are polynomials q_i in $\mathbf{Q}[u, x, z]$ such that

$$q_1 h_1 + \cdots + q_n h_n + q_{n+1}(zg - 1) = s.$$

Let $z = 1/g$, and clearing denominators, we have

$$a_1h_1 + \cdots + a_nh_n = sg^t,$$

for some $t \geq 0$ and $a_i \in \mathbf{Q}[u, x]$. Thus (S) is generally true. $\square$

Method 2. *Compute the Gröbner basis GB of ideal* $(h_1, \ldots, h_n, gz - 1)$ *in* $\mathbf{Q}[u, x, z]$ *in a purely lexicographic ordering* $u < x$, $u < z$ *(where z is a new variable) to see whether it contains a u–polynomial. (S) is generally true if and only if GB contains a u–polynomial. In that case, let* $s_1, \ldots, s_k$ *be all u–polynomials in GB, then*

$$\forall y_i \in E[(h_1 = 0 \wedge \cdots \wedge h_n = 0) \wedge (s_1 \neq 0 \vee \cdots \vee s_k \neq 0) \Rightarrow g = 0].$$

The idea of this method is due to Kapur (1986). However, Kapur's original method causes problems, as shown in Example 10. This method can produce subsidiary conditions weaker than Method 1 or Wu's method. But it is extremely time-consuming. Therefore, we are using Method 1 in our prover.

5.3. A Proof Method for Formulation F1

Formulation F1 in Section 3.2 requires that nondegenerate conditions of a geometry statement be fully specified by the user. Thus the equivalent algebraic formula of a geometry statement is

$$\forall y_i(h_1 = 0 \wedge \cdots \wedge h_n = 0 \wedge s_1 \neq 0 \wedge \cdots \wedge s_k \neq 0 \Rightarrow g = 0), \quad (11)$$

where $h_1, \ldots, h_n$, $s_1, \ldots, s_k$, g are polynomials in $\mathbf{Q}[y_1, \ldots, y_m]$. Obviously, equation (11) is equivalent to

$$\begin{aligned}\forall y_i[\exists z_j(h_1 &= 0 \wedge \cdots \wedge h_n = 0 \wedge s_1z_1 - 1 \\ &= 0 \wedge \cdots \wedge s_kz_k - 1 = 0) \Rightarrow g = 0],\end{aligned} \quad (12)$$

where $z_1, \ldots, z_k$ are new variables. Since $z_1, \ldots, z_k$ do not occur in g, equation (12) is equivalent to

$$\begin{aligned}\forall y_i \forall z_j[(h_1 &= 0 \wedge \cdots \wedge h_n = 0 \wedge s_1z_1 - 1 \\ &= 0 \wedge \cdots \wedge s_kz_k - 1 = 0) \Rightarrow g = 0].\end{aligned} \quad (13)$$

Our method (Chou and Schelter, 1986) is based on the following theorem.

Theorem 26. *For an algebraically closed field E, equation (13) is valid if and only if the ideal $I = (h_1, \ldots, h_n, s_1z_1 - 1, \ldots, s_kz_k - 1, gv - 1)$ of $\mathbf{Q}[y_1, \ldots, y_m]$ is the unit ideal, where v is a new variable.*

Proof. Suppose $I = (1)$. Similar to the proof of Theorem 24, we can infer that there is $t > 0$ such that $g^t \in (h_1, \ldots, h_n, s_1z_1 - 1, \ldots, s_kz_k - 1)$; thus equation (13) is valid. If equation (13) is valid, then polynomials $h_1, \ldots, h_n, s_1z_1 - 1, \ldots, s_kz_k - 1, gv - 1$ do not have common zeros in E. Because E is algebraically closed, $I = (1)$ by Hilbert's Nullstellensatz. □

Method 3. *(A proof method for formulation F1). Check whether $GB(h_1, \ldots, h_n, s_1z_1 - 1, \ldots, s_kz_k - 1, gv - 1)$ of $\mathbf{Q}[y_1, \ldots, y_m]$ is $\{1\}$. If it is, then the statement is confirmed in any field. If it is not, then the statement is not valid in any algebraically closed field E, but not necessarily in other fields E.*

Example 15. Example 1: Parallelogram again. It seems plausible that "A, B and C are non-collinear" (i.e., $u_1u_3 \neq 0$) is the only nondegenerate condition. Thus we can ask whether or not

$$(h_1 = 0, h_2 = 0, h_3 = 0, h_4 = 0, u_1u_3 \neq 0) \Rightarrow g = 0.$$

In an algebraically closed field, this is equivalent to checking whether $\mathrm{GB}(h_1, \ldots, h_4, u_1u_3z_1 - 1, gz - 1) = \{1\}$. Our prover showed this is the case. Thus we proved the theorem.

As the first approximation, we can compute $\mathrm{GB} = \mathrm{GB}(h_1, \ldots, h_4, u_1u_3z_1 - 1)$, then check whether GB reduces g to zero. $\mathrm{GB}(h_1, \ldots, h_4, u_1u_3z_1 - 1) = \{x_1 - u_3, x_2 - (u_2 - u_1), x_3 - u_3/2, x_4 - u_2/2, z_1u_1u_3 - 1\}$.

Remark 8. The reader might think the nondegenerate condition "line AC and line BD intersect" (have only one common point) is redundant because this is always the case for a nondegenerate parallelogram. However, it is not easy to find a traditional proof of this trivial fact. (Please try it!) Note that if AC and BD do not intersect normally, then the problem is meaningless (degenerate). Again, this extremely simple example reveals the fact that a nondegenerate condition, obvious to one person, might not be obvious to another and difficult to a third. In the real plane, success of a traditional proof depends heavily on many "trivial facts." People accept these "trivial facts" simply by looking at the corresponding diagrams.

Example 16. Example 2: Simson's theorem. By looking at diagrams (Fig. 2), we are sure that the "obvious" nondegenerate condition "A, B and C are non-collinear", i.e., $u_1u_3 \neq 0$, is enough. We can think the algebraic equivalence of Simson's theorem (fully specified) is: $[(h_1=0,\ldots,h_7=0,u_1u_3\neq 0)\Rightarrow g=0]$.

However, $\mathrm{GB}(h_1,\ldots,h_7,u_1u_3z_1-1,gz-1)\neq\{1\}$. Thus the theorem cannot be confirmed. The real situation is that if the Gröbner basis method confirms a statement, then the statement is valid in all fields. Thus we have to add other nondegenerate conditions $c_1=u_3^2+u_2^2-2u_1u_2+u_1^2\neq 0$, $c_2=u_3^2+u_2^2\neq 0$.

Our prover confirmed that $\mathrm{GB}(h_1,\ldots,h_7,u_1u_3z_1-1,(u_3^2+u_2^2-2u_1u_2+u_1^2)z_2-1,(u_3^2+u_2^2)z_3-1,gz-1)=\{1\}$. Thus the theorem has been confirmed.

We can also compute $\mathrm{GB}(h_1,\ldots,h_7,u_1u_3z_1-1,(u_3^2+u_2^2-2u_1u_2+u_1^2)z_2-1,(u_3^2+u_2^2)z_3-1)$ first. Then check whether the GB rewrites g to zero. Our prover also confirmed the theorem in this way.

Remark 9. As we have seen from this example, one has to add some seemingly "unnatural" nondegenerate conditions ($c_1\neq 0$, $c_2\neq 0$) in order to confirm the theorem. Ordinary users will find it difficult to understand these conditions, and it is almost impossible for them to find such conditions. This is the drawback of this method and any method based on formulation F1 in general (see the discussion in Section 3.2). In comparison with Method 1, this method is slow, at least in our current implementation. For example, it took about 30 seconds to prove Simson's theorem using this method, whereas it only took 2.5 seconds using Method 1 in the previous section. For this method, the time for computing Gröbner bases is more sensitive to variable ordering. For example, it would take much longer time if we choose z_1, z_2, and z_3 to be the highest variables.

6. A Comparison of the Two Methods

Among Methods 1, 2, and 3, 1 is most efficient. In a recent experiment, 358 among 365 theorems in Chou (1986) were proved using Method 1. Now we compare Wu's method mainly with Method 1.

6.1. Some Theoretical Connections

First we shall establish some theoretical connections between the characteristic set method and the Gröbner basis method in the context of geometry theorem proving.

Let $u_1, \ldots, u_d$, $x_1, \ldots, x_r$, $h_1, \ldots, h_n$ be the same as in Section 4.1 with $\mathbf{Q}$ replaced by any (computable) filed K. Furthermore, suppose that $\{h_1, \ldots, h_n\}$ satisfies the Dimensionality Restriction (Section 4). Let $f_1, \ldots, f_r$ be obtained from $h_1, \ldots, h_n$ using Ritt's principle as revised in Section 4.1 let I and J be the ideals of $K[u, x]$ and $K(u)[x]$ respectively, and generated by $\{h_1, \ldots, h_n\}$. Let $P = K[u, x] \cap J$. Obviously, $I \subset P$, and $g \in P$ iff there is a $c \in K[u]$ such that $cg \in I$. In this section I, J and P *always* denote the above ideals. In this section, a Gröbner basis of an ideal always means the unique reduced Gröbner basis. First we have:

Proposition 7. *Suppose $\{h_1, \ldots, h_n\}$ satisfies the Dimensionality Restriction. For the following three conditions*

(i) *$f_1, \ldots, f_r$ is an irreducible ascending chain,*
(ii) *J is maximal,*
(iii) *$V = V(h_1, \ldots, h_n)$ has only one component general for u,* *have (i) $\Longrightarrow$ (ii) $\Longrightarrow$ (iii).*

Proof. If (i) holds then the sequence of ring extensions

$$K(u)[x_1]/(f_1) \subset K(u)[x_1, x_2]/(f_1, f_2) \subset \ldots K(u)[x]/(f)$$

is actually a sequence of field extensions. Thus the (f) generate a maximal ideal, but it is contained in J. Suppose (ii) holds and suppose there were two components of V general for u. Then these would correspond to two maximal ideals of $K(u)[x]$ both containing J. This is a contradiction. Clearly (iii) does not imply (ii) but only that Rad(J) be maximal. Also if (ii) holds we do not necessarily have an irreducible ascending chain. □

Theorem 27. *Suppose $\{h_1, \ldots, h_n\}$ satisfies the Dimensionality Restriction. Let $G = \{b_1, \ldots, b_j\}$ be the Gröbner Basis of J. If J is a prime ideal (or more strongly, $f_1, \ldots, f_r$ is irreducible) then*

(i) $j = r$.

(ii) b_i *is a monic polynomial in* x_i *with coefficients in* $K(u)[x_1, \ldots, x_{i-1}]$.

(iii) *Let* $b'_i = c_i b_i \in I$ *where* $c_i \in K[u, x]$ *and* $\notin P$, *then* $b'_1, \ldots, b'_r$ *is an irreducible characteristic set of prime ideal* P.

Proof. (i) and (ii). Since J is prime, we know by the condition on the dimensions that in fact J is maximal in $K(u)[x]$. Let $F = K(u)[x]/J$. We induct on r. The induction step is based on the fact that if $E \subset E[t]/A$ is a field extension where t is an indeterminate, then the ideal A is generated by a single monic polynomial in $E[t]$. Thus if we let F_i be the subfields of F generated by the first i of the $\bar{x}_k$ where $\bar{x}_k$ is canonical image of x_k in F, and let A_i be their defining ideal then A_i is generated by adding a monic polynomial d_i in $F_{i-1}[x_i]/A_{i-1}$ to A_{i-1}. Thus we can choose inverse images $b_1, \ldots, b_r$ of $d_1, \ldots, d_r$ in $K(u)[x]$ such that b_i are monic and $deg(b_j, x_i) < deg(b_i, x_i)$ for $i < j$. It is easy to see that $b_1, \ldots, b_r$ are generators of J. But such a set of *reduced* generators has no overlaps among the leading monomials of the generators, hence is the Gröbner basis by the uniqueness of reduced Gröbner bases.

(iii) Let $C = b'_1, \ldots, b'_r$. Because we have field extensions, the b'_i are irreducible in $K(u)[x_1, \ldots, x_i]/(b_1, \ldots, b_{i-1}) = K(u)[x_1, \ldots, x_i]/(b'_1, \ldots, b'_{i-1})$. Thus C is an irreducible ascending chain. Let b be a polynomial in P reduced with respect to C. Since the canonical image $\bar{b}$ of b in the field $F = K(u)[x]/(b_1, \ldots, b_r) = K(u)[x]/(b'_1, \ldots, b'_r)$ is zero, it follows that b is zero. Thus C is a characteristic set of P. Since $P = K[u, x] \cap J$ and J is prime, P is prime. □

In practice, for 95% theorems from geometry, $f_1, \ldots, f_r$ is irreducible; thus computing $\text{GB}(h_1, \ldots, h_n)$ in $K(u)[x]$ in Method 1 is actually computing a special kind of characteristic sets of P. It usually takes more time than using Ritt's principle (or its variants) to compute $f_1, \ldots, f_r$.

Theorem 28. *Assume that* $\{h_1, \ldots, h_n\}$ *satisfies the Dimensionality Restriction. Let* $G = \{g_1, \ldots, g_h\}$ *be the Gröbner basis of* I. *If* J *is prime, then there will be a polynomial* B_i *in* G *for each variable* x_i *whose leading term is* $p_i x_i^{n_i}$ *with* $p_i \in k[u]$.

(i) $\{B_1, \ldots, B_r\}$ *is an irreducible ascending chain in Wu's sense with the property that* $prem(g, B_1, \ldots, B_r) = 0$ *for all* $g \in P$.

(ii) $\{B_1, \ldots, B_r\}$ *is an unreduced Gröbner basis for* J *in* $k(u)[x]$.

Proof. Let $b_1, \ldots, b_r$ be the Gröbner basis of J. For each b_i, there is a nonzero $d_i \in k[u]$ such that $d_i b_i$ is in I. Since $d_i b_i$ can be reduced to zero by G, there must be a B_i in G whose head term (monomial) is $mx_i^{n_i}$, where m is a monomial containing u's only and $n_i \leq ld(b_i)$. By the order of the monomials, the leading term of B_i should be $p_i x_i^{n_i}$ with $p_i \in k[u]$.

Let $b'_i = B_i/p_i$ and $F_k = k(u)[x_1, \ldots, x_k]/(b_1, \ldots, b_k)$ for $k = 1, \ldots, r$. Since b'_i is in J, the canonical image of b'_i in F_i is zero. It follows that $\bar{b}'_i = a\bar{b}_i$ for some nonzero $a \in F_{i-1}[x_i]$, where $\bar{b}_i$ and $\bar{b}'_i$ are canonical images of b_i and b'_i in $F_{i-1}[x_i]$ respectively. Because $n_i = ld(b'_i) \leq ld(b_i)$, $a \in F_{i-1}$. Since $\bar{b}_i$ and $\bar{b}'_i$ are monic, $a = 1$. It follows, by induction on i, that $F_i = k(u)[x]/(b'_1, \ldots, b'_i) = k(u)[x]/(B_1, \ldots, B_i)$. Thus the $B_1, \ldots, B_r$ generate ideal J. They form an irreducible ascending chain in Wu's sense. Let $g \in P$, then the canonical image $\bar{g}$ of g in the field $F = k(u)[x]/(b_1, \ldots, b_r) = k(u)[x]/(B_1, \ldots, B_r)$ is zero. Thus, $prem(g, B_1, \ldots, B_r) = 0$, so (i) holds. We also have that $B_1, \ldots, B_r$ form an unreduced Gröbner basis, since they generate J, and there can't be any overlaps since the leading terms are all powers of different x_i. □

Remark 10. Repeating the same arguments as in the proof, we can see that B_i is the first element of G whose leading variable is x_i and whose leading coefficient is in $k[u]$. Furthermore, we can prove that, if there is a g_k in G *before* B_i such that $lv(g_k) = x_i$, then the canonical image of the leading coefficient of g_k in F is zero.

6.2. The Efficiency

So far, there are no theoretical studies comparing the efficiency of these two methods. In our current implementation, the Gröbner basis method is generally slower than Wu's method for most problems. For timing comparisons in experiments see Chou and Schelter (1986), Chou (1986), and especially Chou (1988). Our observation is based on at least 600 nontrivial geometry theorems.

Here we would emphasize that comparisons are often implementation dependent. For example, we proved Pascal's theorem in about

1,900 seconds on a Symbolics 3600 using Method 1; the same example was used and proved in Kutzler and Stifter (1986) in 13,335 seconds on an IBM 4341 using a method similar to Method 1.[15] It took only 2.5 second using Wu's method.

In this review, we only discuss the basic algorithms, without going into implementation details in speeding up the algorithms. There are many ways to speed up the characteristic set method in the context of geometry theorem proving. For a brief review, see Chou (1988).

One reason that the CS method is generally faster than the GB method is that computing a GB usually takes more time than computing a CS. Once we have the f or the b, we can use them to reduce the conclusion polynomial g to see if it is zero. In this step, the situation is not clear-cut. There are many problems which favor GB. However, there are also many other problems which favor the CS version.

One reason that some problems favor GB is that there is a cancellation of common factors containing the u's only, when working over $\mathbf{Q}(u)$. Since at least for irreducible problems, GB is a special kind of CS. This tells us we should probably do a similar cancellation for CS (removing common factors of coefficients of the f) for whose problems for which the reduction (successive division) produce large polynomials. We call such a CS a simplified CS. For timing comparisons between the simplified CS version and the default CS version used in our prover, see Chou (1988).

For linear cases (i.e., $deg(f_i, x_i) = 1$ for all i), a simplifiedCS is essentially the GB. It is faster to produce simplified CS than to produce GB in our current implementations. In the reduction step, it is general faster to use simplified CS than to use GB. Again this observation is based on our implementations.

6.3. The Scope

Basically the two methods have the same scope: they can address those geometry statements, for which the hypotheses and the conclusion can be expressed as polynomial equations with certain (unspecified) nondegenerate (subsidiary) conditions in inequation forms.

[15]The method in Kutzler and Stifter (1986) includes only Step 1 of our Method 1; thus it is incomplete. It is unable to prove, e.g., "$x^2 = 0 \Rightarrow x = 0$".

The two methods can confirm whether a geometry statement is generally valid and, in the case of general validity, produce subsidiary conditions in a natural way.

Wu's method was originally aimed at proving geometry theorems. Thus, it can address more delicate problems in geometry. Here we would discuss a remarkable fact in elementary geometry which is a consequence of the theory of irreducibility and reducibility developed in Wu's method.

Example 17. (Feuerbach's theorem). *The nine-point circle* (N) *of a triangle is tangent to the incircle and to each of the excircles of the triangle* (Fig. 5).[16]

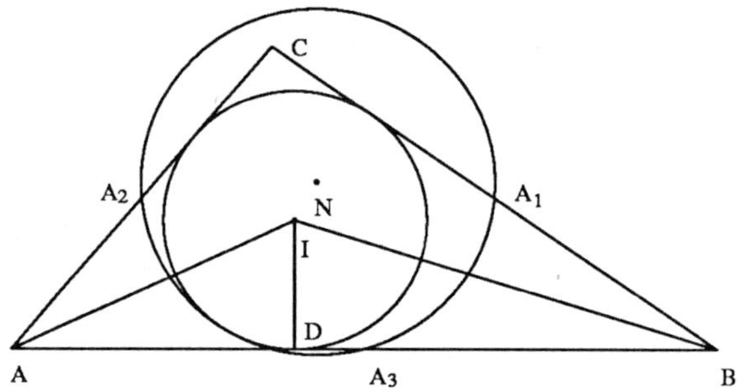

Fig. 5.

Superficially, the fact that (N) touches the incircle and *at the same time* also touches the three excircles seems a coincidence. However, from the algebraic formulation of geometry problems, this fact is natural once the irreducibility has been checked.

Suppose that we try to prove that circle (N) touches only the incircle (I). We can use the bisectors of angles to determine incenter I. However, if we use a polynomial equation to specify the bisector of an angle, we always have an ambiguity: we have an irreducible polynomial of degree two, which means that there are *two* bisectors of an angle. One is internal; the other is external.[17] Because of the

[16]This statement is for Euclidean geometry. In metric geometry, we cannot even distinguish the incenter from the excenters. For this example and examples in this section, we assume real generic points exist, as geometry textbooks implicitly do.

[17]This is in ordered geometry. In unordered geometry we cannot talk about "internal" or "external".

irreducibility, we cannot distinguish one from the other. To make the discussion concrete, we use the following specification.

Let $A = (0,0)$, $B = (u_1, 0)$, $C = (u_2, u_3)$, $I = (x_1, x_2)$, $D = (x_1, 0)$, $A_1 = (x_3, x_4)$, $A_2 = (x_5, x_6)$, $A_3 = (x_7, 0)$, and $N = (x_8, x_9)$. Then the hypothesis polynomials are:

$$
\begin{aligned}
h_1 &= u_1u_3x_2^2 + ((2u_1u_2 - 2u_1^2)x_1 - 2u_1^2u_2 + 2u_1^3) \\
&\quad x_2 - u_1u_3x_1^2 + 2u_1^2u_3x_1 - u_1^3u_3 = 0 && \tan(CBI) = \tan(IBA) \\
h_2 &= u_1u_3x_2^2 + 2u_1u_2x_1x_2 - u_1u_3x_1^2 = 0 && \tan(CAI) = \tan(IAB) \\
h_3 &= 2x_3 - u_2 - u_1 = 0 \\
h_4 &= 2x_4 - u_3 = 0 && A_1 \text{ is the midpoint of } B \text{ and } C \\
h_5 &= 2x_5 - u_2 = 0 \\
h_6 &= 2x_6 - u_3 = 0 && A_2 \text{ is the midpoint of } A \text{ and } C \\
h_7 &= 2x_7 - u_1 = 0 && A_3 \text{ is the midpoint of } A \text{ and } B \\
h_8 &= 2x_6x_9 + (-2x_7 + 2x_5) \\
&\quad x_8 + x_7^2 - x_6^2 - x_5^2 = 0 && NA_3 \equiv NA_2 \\
h_9 &= 2x_4x_9 + (-2x_7 + 2x_3) \\
&\quad x_8 + x_7^2 - x_4^2 - x_3^2 = 0 && NA_3 \equiv NA_1.
\end{aligned}
$$

The conclusion that circle (N) is tangent to circle (I) is equivalent to $g = ((8x_2x_7 - 8x_1x_2)x_8 - 4x_2x_7^2 + 4x_1^2x_2)x_9 + (-4x_7^2 + 8x_1x_7 + 4x_2^2 - 4x_1^2)x_8^2 + (4x_7^3 - 4x_1x_7^2 + (-8x_2^2 - 4x_1^2)x_7 + 4x_1^3)x_8 - x_7^4 + (4x_2^2 + 2x_1^2)x_7^2 - x_1^4 = 0$. The triangulation procedure gives the following (simplified) CS:

$$
\begin{aligned}
f_1 &= 4x_1^4 - 8u_1x_1^3 + (-4u_3^2 - 4u_2^2 + 4u_1u_2 + 4u_1^2)x_1^2 \\
&\quad + (4u_1u_3^2 + 4u_1u_2^2 - 4u_1^2u_2)x_1 - u_1^2u_3^2 = 0 \\
f_2 &= (2x_1 + 2u_2 - 2u_1)x_2 - 2u_3x_1 + u_1u_3 = 0 \\
f_3 &= 2x_3 - u_2 - u_1 = 0 \\
f_4 &= 2x_4 - u_3 = 0 \\
f_5 &= 2x_5 - u_2 = 0 \\
f_6 &= 2x_6 - u_3 = 0 \\
f_7 &= 2x_7 - u_1 = 0 \\
f_8 &= 4x_8 - 2u_2 - u_1 = 0 \\
f_9 &= 4u_3x_9 - u_3^2 + u_2^2 - u_1u_2 = 0.
\end{aligned}
$$

The fact that $deg(f_1, x_1) = 4$ means there are four solutions for I: one is the incenter and the other three are the excenters. Now f_1 is irreducible in x_1,[18] which means that we cannot distinguish the four centers by the above equations. If one subcase (e.g., the case of incenter) is true. Thus the following assertion that

If the nine point circle touches the incircle
it also touches the other three excircles (14)

is a consequence of our theory. Of course, Feuerbach's theorem can be easily proved by our prover in 2.7 seconds on a Symbolics 3600.

As a consequence of our general theory, for almost all theorems about inscribed circle, there are three counterparts about exscribed circles. We list the following two theorems to illustrate our general observation:

Example 18. The triangle formed by the points of contact of the sides of given triangle with excircles corresponding to these sides is equivalent (the same area) to the triangle formed by the points of contact of the sides of the triangle with the inscribed circle (Fig. 6).

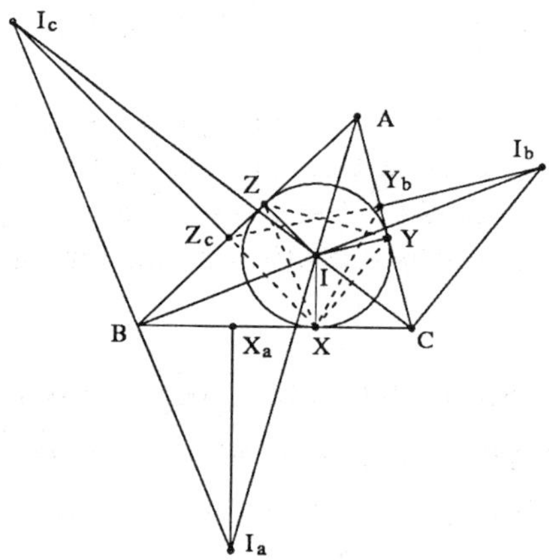

Fig. 6.

Example 19. (Nagel Point) The lines joining the vertices of a triangle to the points of contact of the opposite sides with the excircles relative to those sides are concurrent.

[18]Our prover can factor multivariate polynomials over **Q** as well as quadratic polynomials of successive quadratic extensions of fields of rational functions.

As a consequence of our theory, there are four pairs of equivalent triangles and four Nagel points. Examples 18 and 19 mention only one of the four. The proofs produced by our prover are for the four.

This general observation is not only for tritangent circles, but also for many other problems. Before giving further examples, Let us sum up the implications in Example 17.

For a given geometry statement (S) involving equality only, let $h_1 = 0, \ldots, h_n = 0$ be hypothesis polynomials, and $f_1, \ldots, f_r$ the ascending chain obtained from the h_i using Ritt's Principle as revised in Section 1.1. Assuming that the product of the leading degree of the f_i is greater than one (so there is more than one subcase), then

- (3.A) If $f_1, \ldots, f_r$ is irreducible and (S) is valid in one subcase, then $prem(g, f_1, \ldots, f_r) = 0$. Hence (S) is valid in all other subcases.
- (3.B) If $f_1, \ldots, f_r$ is irreducible and (S) is not valid in one subcase then (S) is not valid in all other subcases.
- (3.C) If (S) is valid in one subcase and $prem(g, f_1, \ldots, f_r) \neq 0$, then $f_1, \ldots, f_r$ is *reducible.*

Let us use the following example to illustrate our general observations (3.A)–(3.C).

Example 20. Through the vertex A of a triangle ABC a straight line AM is drawn, cutting the side BC in M. Let O and I be the centers of the circumscribed circle (O) and the *inscribed circle* (I) of ABC. The circle (w_1) and (w_2) with centers w_1 and w_2 are each tangent to (O) and the first is tangent also to two sides of angle AMB, while the second is tangent to the two sides of angle AMC. Prove that the straight line joining w_1 and w_2 passes through I (Fig. 7).

This was a conjecture proposed by V. Thébault of France in 1938 and proved by K.B. Taylor of England in 1983. We state the problem as follows (see Taylor, 1983).

There are four centers w_1 on one bisector of $\angle AMB$ and four w_2 on the other. For an appropriate algebraic specification of the problem (Chou, 1988) , if the problem were irreducible, we would have sixteen lines of collinearity instead of one as problem original proposed. However, $prem(g, f_1, \ldots, f_r) \neq 0$. By (3.C), this means that $f_1, \ldots, f_r$ is reducible if the conjecture (or the proof of Taylor)

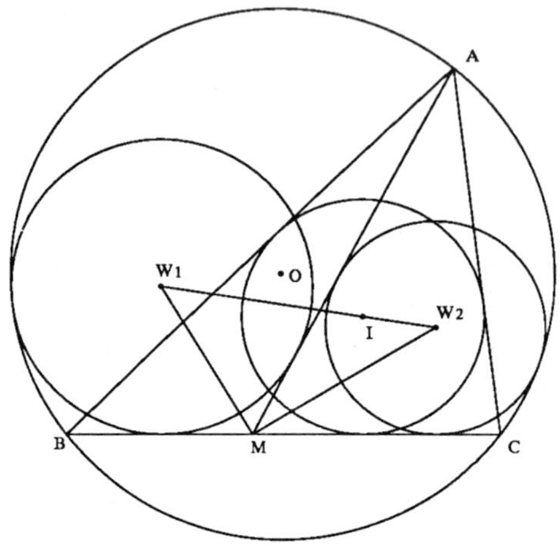

Fig. 7.

is correct. Our prover found that there are four components and the conclusion is valid for only one component. With a more careful interpretation of our result (Chou, 1986), we come to a *generalization* of Thébault-Taylor's theorem:

For each of the four centers I (the incenter and the three excenters) there exist a w_1 and a w_2 such that w_1, I and w_2 are collinear.

The following is another problem of this kind:

Example 21. If a cyclic quadrilateral Q has sides a, b, c d and semi-perimeter s, its area K is given by $K^2 = (s-a)(s-b)(s-c)(s-d)$.

Superficially, this problem is totally beyond our method: we cannot express the convexity without using order relation and we cannot express the fact $a > 0$, etc. If the formula is true for convex quadrilateral, then we know there are real generic points (i.e., when quadrilateral is convex, and $a > 0$, $b > 0$, $c > 0$, $d > 0$) to make $K^2 - (s-a)(s-b)(s-c)(s-d) = 0$ true. Thus by (3.C), the algebraic specification of the problem is reducible. We found two components of our algebraic specification, and confirmed the equation $K^2-(s-a)(s-b)(s-c)(s-d) = 0$ for one component. Actually the results we got are richer than the original formula. Q is not necessarily convex and K is the signed area of an oriented quadrilateral. When Q is not convex, the signs of a, b, c, d will be changed.

In solving Example 20 and 21, we have to use factorization over algebraic extensions. Some people think using factorization is a drawback of Wu's method in comparing with the Gröbner basis method. Actually, factorization is a case analysis approach, as shown by Examples 20 and 21. It is a useful approach. The Gröbner basis method is unable on its own to solve such problems unless decomposition (factorization) is introduced in certain way. So far, no work in this direction has been reported. We would like to see the further development of the Gröbner basis method to solve Examples 20 and 21 and many other similar problems.

References

Bachmair, L., and Buchberger, B., (1980). "A Simplifed Proof of the Characterization Theorem for Gröbner Bases," *ACM SIGSAM Bull* **14(4)**, 29–34.

Bergman, G., (1978)."Diamond Lemma," *Advances in Math.* **29**, 178–218.

Buchberger, B., (1976). "Some Properties of Gröbner bases for Polynomial ideals," *JACM SIGSAM Bull.* **10(4)**, 19–24.

Buchberger, B., (1985). "Gröbner Bases: An Algorithmic Method in Polynomial Ideal Theory," Chapter 6 in *Recent Trends in Multidimensional Systems Theory* (N.K. Bose, ed.). D. Reidel Publ. Comp.

Chou, S.C., (1984). "Proving Elementary Geometry Theorems Using Wu's Algorithm," In *Automated Theorem Proving: After 25 years* (W.W. Bledsoe and D. Loveland, eds.), AMS Contemporary Mathematics Series **29**, 243–286.

Chou, S.C., (1985). "Proving and Discovering Theorems in Elementary Geometries Using Wu's Method," PhD Thesis, Department of Mathematics, University of Texas, Austin.

Chou, S.C., (1986). " Proving Geometry Theorems Using Wu's Method: A Collection of Geometry Theorems Proved Mechanically," Technical Report 50. Institute for Computing Science, University of Texas at Austin, July 1986.

Chou, S.C., (1986). "An Introduction to Wu's Method for Mechanical Theorem Proving in Geometry." Accepted to *Journal of Automated Reasoning*, September 1986.

Chou, S.C., (1988). *Mechanical Geometry Theorem Proving.* D. Reidel Publishing Company, Netherlands.

Chou, S.C., and Schelter, W.F., (1986). "Proving Geometry Theorems with Rewrite Rules," *Journal of Automated Reasoning* **2(4)**, 253–273.

Collins, G.E., (1975). "Quantifier Elimination for Real Closed Fields by Cylindrical Algebraic Decomposition," *Lecture Notes In Computer Science* **33**, Springer-Verlag, Berlin, 134–183.

Hilbert, D., (1971). *Foundations of Geometry.* Open Court Publishing Company, La Salla, Illinois.

Kapur, D., (1986). "Geometry Theorem Proving Using Hilbert's Nullstellensatz," In *Proceedings of the 1986 Symposium on Symbolic and Algebraic Computation*, Waterloo, Ontario, 202–208.

Kutzler, B., and Stifter, S., (1986). "Automated Geometry Theorem Proving Using Buchberger's Algorithm," In *Proceedings of the 1986 Symposium on Symbolic and Algebraic Computation*, Waterloo, Ontario, 209–214.

Newman, M. H. A., (1942). "On theories with a combinatorial definition of 'equivalence'," *Ann. of Math.* **43**, 223–243.

Ritt, R. F., (1938). *Differential Equation from Algebraic Standpoint.* AMS Colloquium Publications **14**, New York.

Ritt, R. F., (1950). *Differential Algebra.* AMS Colloquium Publications, New York.

Seidenberg, A., (1954). "A New Decision Method for Elementary Algebra," *Annals of Math.* **60**, 365–371.

Shoenfield, J. R., (1967). *Mathematical Logic.* Addison-Wesley Publishing Company.

Tarski, A., (1951). *A Decision Method for Elementary Algebra and Geometry, Second Edition.* University of California Press, Berkeley and Los Angeles.

Tarski, A., (1959). "What is Elementary Geometry," In *The Axiomatic Method* (Tarski et al., eds.). North-Holland Publishing Company, Amsterdam.

Taylor, K.B., (1983). "Three Circles with Collinear Centers," *American Mathematical Monthly* **90**, 486–487.

Wen-tsün, Wu, (1978). "On the Decision Problem and the Mechanization of Theorem Proving in Elementary Geometry," *Scientia Sinica* **21**, 157–179.

Wen-tsün, Wu, (1982). "Mechanical Theorem Proving in Elementary Geometry and Differential Geometry," *Proc. 1980 Beijing Symposium on Differential Geometry and Differential Equations* **2**. Science Press, 125–138.

Wen-tsün, Wu, (1984). "Basic Principles of Mechanical Theorem Proving in Geometries," *J. of Sys. Sci. and Math. Sci.* **4(3)**, 207–235. Republished in *Journal of Automated Reasoning* **2(4)**, 221–252, (1986).

Wen-tsün, Wu, (1984). "Some Recent Advances in Mechanical Theorem-Proving of Geometries," In *Automated Theorem Proving: After 25 years.* American Mathematical Society, Contemporary Mathematics **29**, 235–242.

Wen-tsün, Wu, (1984). *Basic Principles of Mechanical Theorem Proving in Geometries.* (in Chinese) Peking.

Wen-tsün, Wu, (1986). "On Zeros of Algebraic Equations—An Application of Ritt Principle," *Kexue Tongbao* **31(1)**, 1–5.

3 On Recognizable Sets and Tree Automata

B. COURCELLE
Laboratoire d'Informatique
Bordeaux–I University
Talence, France[1]

Une science bien traitée n'est qu'une langue bien faite.

—Condillac (1780–1814)

1. Introduction

This paper reviews a few aspects of recognizability in general algebraic structures and examines from this point of view several notions of trees. In each case an appropriate class of tree-automata is defined.

Before describing the content of the paper we compare the two usual meanings of the term "recognizable." The first one is algebraic, and the second one is algorithmic.

Let M be a set, the subsets of which one wants to describe. Typical examples are the sets of words, of terms, of finite trees (of various types), of finite graphs, of k-tuples of words, and of infinite words. A subset L of M is **recognizable** in the algebraic sense iff it is a union of congruence classes for some congruence having finitely many classes. This notion is meant with respect to a given algebraic structure on M. This definition arose from the characterizations of regular

[1]Formation associée au CNRS. Reprints can be obtained from the author by electronic mail at mcvax!inria!geocub!courcell on UUCP network (or in the new syntax: Courcell @ geocub.greco-prog.fr). This work has been supported by the "Programme de Recherches Coordonnées: Mathématiques et Informatique."

Resolution of Equations in
Algebraic Structures
Volume 1

ISBN 0-12-046370-9

languages given by Nerode and Myhill and was formulated for arbitrary algebraic structures by Mezei and Wright in a fundamental paper that we shall quote frequently in this work (Mezei and Wright, 1967).

The second sense of recognizability, the algorithmic one, is not as precise as the first. It refers to the existence of a membership algorithm testing whether an element m of M belongs to L, that can be formulated by means of a "finite-state automaton." This notion is very vague. Whereas it is well-established in the cases of words and terms (and it is because it defines in these two cases the family of recognizable sets in the algebraic sense), it is not in many of the above cases.

Let us consider, for instance, the case of pairs of words, $M = X^* \times Y^*$. Deterministic finite state automata (called 2-tape automata) have been defined by Rabin and Scott (1959) and nondeterministic ones by Elgot and Mezei (1965). The family of subsets of $X^* \times Y^*$ defined by the former class of automata is closed under complementation but not under union, while the family defined by the latter is closed by union but is not by complementation. As in every algebraic structure, the family of recognizable subsets of $X^* \times Y^*$ is a Boolean algebra. Hence neither of these two notions of automata formalizes the algebraic notion of recognizability. (See Example 2 below.)

In the case of graphs, an algebraic structure has been introduced by Bauderon and Courcelle (1987), from which follows a notion of recognizability (in the algebraic sense) investigated by Courcelle (1986b). But no notion of graph-automaton (extending the special types of graph automata defined, for instance, by Kamimura et al. (1982), Bossut et al. (1987); further references can be found in Petrov (1978)) represents algorithmically the notion of recognizability.

There is in addition a third use of the term "recognizable" in theories of sets of infinite words, and of sets of infinite trees. (See the volume edited by Nivat and Perrin (1985).) This use corresponds to an extension of the "algorithmic" sense of recognizability: a word (or a tree) is "accepted" by an automaton if there exists an infinite computation of the automaton satisfying certain conditions concerning the states occurring infinitely many times. This notion of acceptance is not effective since it concerns infinite objects. (Note that these infinite computations cannot be "approximated" as in the theory of domains issued from Scott (1976).) It is not algebraic either since the corresponding sets of infinite trees or words are not recognizable with respect to any (known) algebraic structure.

No firmly grounded general theory of recognizability can be built from the algorithmic meaning of the term due to the variety of notions of automata that can be defined and to the lack of criteria for preferring one notion to another. On the contrary, an algebraic theory of recognizability can be developed for arbitrary algebraic structures, as we shall see below. "Arbitrary" means that the notion is not restricted to "discrete objects" like words, trees, or graphs although in this paper we shall only consider these cases.

This discussion of the term "recognizable" may sound scholarly but it reflects the necessity of establishing the fundamental concepts of the theory. And as wrote Condillac in his "Traité des Systèmes" (chap. XVIII):

A well-grounded science is nothing but a well-designed language.

This paper consists of three parts. The first one reviews the general properties of recognizability (in the algebraic sense), and compares this notion with that of **equationality** (i.e. the definability of a set as a component of the least solution of system of equations) that is equally fundamental.

The second part examines the recognizable subsets of free T-algebras, where T is an equational theory like the theory of monoids in the case of words. This case (or rather this family of cases) includes the cases of words, of terms, of finite trees of various types, and of finite graphs by the results of Bauderon and Courcelle (1987), Courcelle (1986b), Courcelle (1987a), and Courcelle (1987b). For every theory T presented by a signature F and a set of axioms E, some interesting properties of the family of recognizable subsets of the free T-algebras can be formulated in terms of E considered as a term rewriting system and of the recognizable sets of terms written over F. In particular, we give syntactic conditions on E ensuring that in the associated free algebras, the equational sets are recognizable, or that the family of recognizable sets is closed under special homomorphisms called **relabellings**.

The last result involves the extension of the classical determinization algorithm for finite state (word or tree) automata to theories presented by bilinear equations. It happens that all the theories that are useful for formalizing words, trees, and graphs are of this type, but the deep reasons of this fact are not immediately clear.

In the third part, we define several theories corresponding to several notions of trees. Trees can be ranked or unranked, ordered or unordered. This gives four general notions (with auxiliary "subnotions"). In each case, we define an appropriate theory (having a bilinear presentation) and we examine some properties of the corresponding families of recognizable sets (closure properties and comparison with the family of equational sets). We also define tree-automata thus making the connection between the algebraic and the algorithmic sense of recognizability.

2. Recognizable and Equational Sets: General Properties

Definition 1. Magmas and theories. As in many previous works, we call F-**magma** what is more usually called an F-algebra. The reason is that the term "algebra" is used in many different situations with different meanings, so that we prefer to avoid it and use a fresh word. The same remark applies to the term "algebraic." For a set, we shall use the term "equational," introduced by Mezei and Wright (1967), rather than the term "algebraic," introduced by Eilenberg and Wright (1967).

Let $\mathcal{S}$ be a set called the set of **sorts**. An $\mathcal{S}$-signature F is a set of symbols given with two mappings $\alpha : F \to \mathcal{S}^*$ (the **arity** mapping) and $\sigma : F \to \mathcal{S}$ (the **sort** mapping). The **profile** of f in F is the pair $(\alpha(f), \sigma(f))$ written $s_1 \times s_2 \times \cdots \times s_n \to \sigma(f)$ where $\alpha(f) = (s_1, \ldots, s_n)$. The **rank** $\rho(f)$ of f is the length of $\alpha(f)$.

A **many-sorted** F-**magma** is an object $\mathbf{M} = \langle (M_s)_{s\in\mathcal{S}}, (f_\mathbf{M})_{f\in F}\rangle$ where M_s is a nonempty set for each s called the **domain** (or **carrier**) of sort s and $f_\mathbf{M}$ is a total mapping: $M_{s_1} \times \cdots \times M_{s_k} \to M_s$ for f of profile $s_1 \times \cdots \times s_k \to s$.

Let U be an $\mathcal{S}$-**sorted** set of nullary symbols called here **variables** and disjoint from F. Every u in U has a sort $\sigma(u)$ in $\mathcal{S}$. We denote by $M(F,U)_s$ the set of well-formed terms of sort s (i.e. with a leading symbol of sort s) built with $F \cup U$. If $t \in M(F, \{u_1, \ldots, u_k\})_s$ then $t_\mathbf{M}$ denotes a mapping $M_{s_1} \times \cdots \times M_{s_k} \to M_s$ also called a **derived operator** of $\mathbf{M}$ (where $s_i = \sigma(u_i)$ for $i = 1, \ldots, k$). If $t \in M(F) = M(F, \emptyset)$ then $t_\mathbf{M} \in M_{\sigma(t)}$. The mapping $t \mapsto t_\mathbf{M}$ is the unique F-homomorphism of $\mathbf{M}(F)$ (the initial F-magma) into $\mathbf{M}$. We shall denote it by $h_{\mathbf{M}}$. We say that F **generates** $\mathbf{M}$ if $h_{\mathbf{M}}$ is onto, i.e., if every element of M_s is $h_{\mathbf{M}}(t)(= t_{\mathbf{M}})$ for some $t \in \mathbf{M}(F)_s$.

The set of sorts and the signature may be infinite. This is the case for graphs in Bauderon and Courcelle (1987), Courcelle (1986b), (1987a), (1987b), (1987c) but we shall not consider them here. In this paper, we shall use only "small" sets of sorts $\mathcal{S}$, enumerated in a fixed way. Typically, $\mathcal{S} = \{s_1, s_2, s_3\}$. Then an $\mathcal{S}$-sorted set U is an ordered triple (U_1, U_2, U_3) of pairwise disjoint sets where U_i is the set of elements of U of sort s_i. An F-magma $\mathbf{M}$ where F is an $\mathcal{S}$-signature is then defined as an object $\langle (M_1, M_2, M_3), (f_{\mathbf{M}})_{f \in F}\rangle$ where M_i is the domain of sort s_i.

A **theory** is a pair $\mathcal{T} = (F, E)$ consisting of an $\mathcal{S}$-signature F (for some $\mathcal{S}$) and a set E of (**equational**) **axioms**, i.e., a set of pairs of elements of $M(F, U)$ where two elements in a pair are of the same sort.

A $\mathcal{T}$**-magma** is an F-magma $\mathbf{M}$ such that $t_{\mathbf{M}} = t'_{\mathbf{M}}$ for every pair (t, t') in E. We denote by $\mathbf{M}(\mathcal{T}, V)$ the **free** $\mathcal{T}$**-magma** generated by V, where V is an $\mathcal{S}$- sorted set (of nullary symbols) called here the set of **generators**. It is well-known that $\mathbf{M}(\mathcal{T}, V) = \mathbf{M}(F, V))/ \overset{*}{\longleftrightarrow}_E$ where $\overset{*}{\longleftrightarrow}_E$ is the congruence on $\mathbf{M}(F, V)$ generated by E considered as a term rewriting system. (See the survey of Huet and Oppen (1980) on term rewriting systems.)

Example 1. Words. The set of all words written over an alphabet X is denoted by X^*. The empty word is denoted by ε. The result of the concatenation of two words u and v is denoted by uv or by $\cdot(u, v)$ for emphasis. It is well-known that $\mathbf{X}^* = \langle X^*, \cdot, \varepsilon\rangle$ is the free monoid generated by X, i.e., is $\mathbf{M}(\mathcal{T}_m, X)$ where $\mathcal{T}_m = (\{\cdot, \varepsilon\}, E_m)$ is the one-sort theory with the set of axioms:

$$\{\cdot(u, \cdot(v, w)) = \cdot(\cdot(u, v), w), \cdot(u, \varepsilon) = u, \cdot(\varepsilon, u) = u\}.$$

There is another useful theory descriptive of X^*. Its basic operation is not the concatenation of two words but the concatenation of a word with one letter. One can describe this theory by using one unary function symbol x for each $x \in X$, such that $x(u) = xu$ for every $u \in X^*$. But in this way the letters are no longer the generators. Changing the alphabet necessitates changing the set of function symbols (hence the theory). In order to avoid this, we introduce a signature with two sorts, a sort **letter** and a sort **word**.

So we let $\mathcal{S}_u$ be {**letter,word**}, ε be a constant of sort **word**, q be a function symbol of profile : **letter**$\times$**word**$\rightarrow$**word**, and F_u be $\{\varepsilon, q\}$. We define an F_u-magma $\mathbf{T}_u(X) = \langle(X, X^*), q_\mathbf{T}, \varepsilon\rangle$ where ε is the empty word and $q_\mathbf{T}(x, u) = xu$ for all $x \in X, u \in X^*$. It is not hard to see that $\mathbf{T}_u(X) = \mathbf{M}(\mathcal{T}_u, (X, \emptyset))$.

Definition 2. Polynomial systems and equational sets. Let $\mathcal{S}, F$ and $\mathbf{M}$ be as above. We augment F into F_+ by adding for every sort s in $\mathcal{S}$ a new symbol $+_s$ of profile $s \times s \rightarrow s$ and a new constant $\boldsymbol{\Omega}_s$ of sort s. With $\mathbf{M}$ we associate its **power-set magma** $\mathcal{P}(\mathbf{M}) = \langle(\mathcal{P}(M_s))_{s\in\mathcal{S}}, (f_{\mathcal{P}(\mathbf{M})})_{f\in F_+}\rangle$ where for $A_1, \ldots, A_k \subseteq M_{s_1}, \ldots, M_{s_k}$

$$\begin{aligned}
A_1 +_{s\mathcal{P}(\mathbf{M})} A_2 &= A_1 \cup A_2 \text{ (with } s = s_1 = s_2) \\
f_{\mathcal{P}(\mathbf{M})}(A_1, \ldots, A_k) &= \{f_\mathbf{M}(a_1, \ldots, a_k)/a_1 \in A_1, \ldots, a_k \in A_k\} \\
&\quad \text{(where } \alpha(f) = s_1, \ldots, s_k) \\
\boldsymbol{\Omega}_{\mathbf{s}\mathcal{P}(\mathbf{M})} &= \emptyset.
\end{aligned}$$

Hence $\mathcal{P}(\mathbf{M})$ is an F_+-magma.

A **polynomial system** over F is a sequence of equations $S = \langle u_1 = p_1, \ldots, u_n = p_n\rangle$ where $U = \{u_1, \ldots, u_n\}$ is the $\mathcal{S}$-sorted set of **unknowns**. Each p_i is a **polynomial**, i.e., a term of the form $\boldsymbol{\Omega}_s$ or $t_1 +_s t_2 +_s \cdots +_s t_m$ where the t_j's belong to $M(F \cup U)_s$ and $s = \sigma(u_i)$. The terms $t_1, \ldots, t_m$ are called **monomials**. The subscript s is usually omitted in $+_s$ and $\boldsymbol{\Omega}_s$.

A mapping $S_{\mathcal{P}(\mathbf{M})} : \mathcal{P}(\mathbf{M}_{\sigma(u_1)}) \times \cdots \times \mathcal{P}(\mathbf{M}_{\sigma(u_n)})$ into itself is associated with S and $\mathbf{M}$ as follows: for $A_1 \subseteq M_{\sigma(u_1)}, \ldots, A_n \subseteq M_{\sigma(u_n)}$, $S_{\mathcal{P}(\mathbf{M})}(A_1, \ldots, A_n) = (A'_1, \ldots, A'_n)$ where $A'_i = p_{i\mathcal{P}(\mathbf{M})}(A_1, \ldots, A_n)$ for $i = 1, \ldots, n$. A **solution** of S in $\mathcal{P}(\mathbf{M})$ is an n-tuple $(A_1, \ldots, A_n)$ such that $(A_1, \ldots, A_n) = S_{\mathcal{P}(\mathbf{M})}(A_1, \ldots, A_n)$. Such a system S has at least a solution in $\mathcal{P}(\mathbf{M})$ w.r.t. set inclusion. Its component corresponding to u_i is denoted by $L((S, \mathbf{M}), u_i)$. A systematic study of polynomial systems can be found in Courcelle (1986a). The components of the least solution in $\mathcal{P}(\mathbf{M})$ of a polynomial system are said to be **equational** (w.r.t. $\mathbf{M}$). They are the **M-equational** sets. We denote by **Equat**($\mathbf{M}$) the family of $\mathbf{M}$-equational sets.

It is well known that for every context-free grammar one can construct a polynomial system over F_m the unknowns of which are the non-terminals of the grammar. The least solution of this system is the tuple of languages generated by the non-terminals of the grammar (Ginsburg and Rice (1962), Schützenberger (1963), Mezei and Wright (1967)).The same construction works for certain tree-grammars (called regular tree-grammars) (see Gecseg and Steinby (1984)) and for graph-grammars of various types (Bauderon and Courcelle (1987), Habel and Kreowski (1987), Courcelle (1987b)). All these cases show the importance of the notion of an equational set.

The equations of polynomial systems are considered as *defining* (*recursive*) equations. The least solution of a polynomial system is considered as the solution of interest. It associates a value (possibly the empty set) with each unknown of the system.

Let us note that in this paper we use nullary symbols (from sets $X, U, V, W, \dots$) in the following different ways:

- as universally quantified variables in axioms of theories,
- as generators in free $\mathcal{T}$-magmas,
- as unknowns in polynomial systems.

The signatures can also have nullary symbols, usually called **constants**. (An example is the symbol ε denoting the empty word.)

We recall an important result due to Mezei and Wright (1967, Lemma 5.3).

Proposition 1. *Let S be a polynomial system over F, let $\mathbf{M}$ and $\mathbf{M}'$ be two F-magmas and $h : \mathbf{M} \to \mathbf{M}'$ be a homomorphism. The least solution of S in $\mathcal{P}(\mathbf{M}')$ is the image under h of the least solution of S in $\mathcal{P}(\mathbf{M})$.*

We now define recognizable sets.

Definition 3. Recognizable Sets. Let F and $\mathcal{S}$ be as above. (They may be infinite.) An F-magma $\mathbf{A}$ is **locally-finite** if every domain $A_s, s \in \mathcal{S}$, is finite.

Let $\mathbf{M}$ be an F-magma and $s \in \mathcal{S}$. A subset B of M_s is **M-recognizable** if there exists a locally-finite F-magma, a homomorphism $h : \mathbf{M} \to \mathbf{A}$ and a (finite) subset C of A_s such that$B = h^{-1}(C)$. The pair $(h, \mathbf{A})$ is called a **semi-automaton** and the triple $(h, \mathbf{A}, C)$ is called an **automaton**.

We denote by $\mathbf{Rec}(\mathbf{M})$ the family of $\mathbf{M}$-recognizable sets. Recognizable subsets of M_s can also be characterized as unions of congruence classes for a congruence $\sim = (\sim_{s'})_{s' \in \mathcal{S}}$ on $\mathbf{M}$ such that each equivalence relation $\sim_{s'}$ (on $M_{s'}$) has finitely many classes. The equivalence of the two definitions is easy to establish.

It is well known that $\mathbf{Rec}(\mathbf{X}^*)$ is the family of languages that can be defined by finite-state automata. Recognizable subsets of $M(F)$ can be characterized by tree-automata of various types (top-down or bottom-up, deterministic or not; see Gecseg and Steinby (1984)). The classical identification of terms with finite, ordered, ranked trees explains the qualification of "tree"-automaton.

But there are several other notions of trees. More precisely, there are several **theories of trees** as we shall see below. Appropriate notions of tree automata can be defined, and the recognizable sets do not have the same properties in all cases.

Recognizable sets of graphs have been investigated in Courcelle (1986b, 1987b, 1987c). In these works **infinite signatures** have been introduced to deal with graphs. But no notion of automaton has been obtained for characterizing recognizable sets of graphs. Since there are uncountably many sets of graphs (Courcelle 1986b, section 5) there cannot exist such a notion, at least in the usual sense where an automaton is a finite object. This finiteness condition implies that there exist only countably many automata.

Let us point out that the notions of a recognizable set and of an equational set depend on the underlying algebraic structure. This is especially clear in the case of words.

Proposition 2. (1) $\mathbf{Rec}(\mathbf{X}^*) = \mathbf{Rec}(\mathbf{T}_u(X)) = \mathbf{Equat}(\mathbf{T}_u(X))$. *This is the family of rational languages on* X.

(2) $\mathbf{Equat}(\mathbf{X}^*)$ *is the family of context-free languages. If* $\mathbf{Card}(X) \geq 2$, *then* $\mathbf{Rec}(\mathbf{X}^*)$ *is properly contained in* $\mathbf{Equat}(\mathbf{X}^*)$.

Proof. (1) If $\mathbf{M}$ is an F-magma and $\mathbf{M}'$ is a G-magma with the same sorts and domains and such that every operation $g_{\mathbf{M}'}$ is a derived operator of $\mathbf{M}$, then one says that $\mathbf{M}'$ is a **derived magma** of $\mathbf{M}$, and then $\mathbf{Rec}(\mathbf{M}) \subseteq \mathbf{Rec}(\mathbf{M}')$. The inclusion $\mathbf{Rec}(\mathbf{X}^*) \subseteq \mathbf{Rec}(\mathbf{T}_u(X))$ is immediate since $\mathbf{T}_u(X)$ is a derived magma of $\mathbf{X}^*$.

The inclusion $\mathbf{Rec}(\mathbf{T}_u(X)) \subseteq \mathbf{Rec}(\mathbf{X}^*)$ is a particular property of $\mathbf{T}_u(X)$ and $\mathbf{X}^*$. Its proof consists in constructing from a homomorphism $h : \mathbf{T}_u(X) \to \mathbf{Q}$ a homomorphism $k : \mathbf{X}^* \to \mathbf{S}(Q)$ where $\mathbf{S}(Q)$ is the monoid of total mappings: $Q \to Q$, in such a way that for every $Q' \subseteq Q$, $h^{-1}(Q') = k^{-1}(S')$ where S' is the set of mappings $\beta : Q \to Q$ such that $\beta(h(\varepsilon)) \in Q'$. (See Eilenberg (1974) for technical details.) These facts have also been observed in Steinby (1981).

The equality $\mathbf{Rec}(\mathbf{T}_u(X)) = \mathbf{Equat}(\mathbf{T}_u(X))$ follows from the **freeness** of $\mathbf{T}_u$. We mean by this that $E_u = \emptyset$; see Proposition 6 below.

From a polynomial system over F_u defining $L \subseteq X^*$ one can construct a rational expression defining L. (This technique is due to Arden.) And it is known that the language defined by a rational expression is recognizable.

(2) It is well known that $\mathbf{Equat}(\mathbf{X}^*)$ is the family of context-free languages, hence, that it contains properly $\mathbf{Rec}(\mathbf{X}^*)$ if $\mathbf{Card}(X) \geq 2$. □

We shall now state a few general properties of equational and recognizable sets (Propositions 3 and 4 have been stated in the paper by Steinby (1981)).

In order to simplify the notation, the statements of the results and their proofs will be given for a one-sort (possibly infinite) signature F. The extension to several sorts is straightforward.

Proposition 3. (1) $\mathbf{Equat}(\mathbf{M})$ *contains* $\emptyset$ *and is closed under* $\cup$ *and the operations*$f_{(\mathcal{P}\mathbf{M})}$ *for all* $f \in F$.

(2) *If* $h : \mathbf{M} \to \mathbf{M}'$ *is a homomorphism and* $L \in \mathbf{Equat}(\mathbf{M})$ *then* $h(L) \in \mathbf{Equat}(\mathbf{M}')$.

(3) *If* F *generates* $\mathbf{M}$ *then every finite subset of* M *is equational. Furthermore if* F *is finite, then* M *is equational.*

Proof. (1) and (3) can be proved by easy extensions of the constructions classically done for context-free grammars.

(2) is nothing but the result of Mezei and Wright recalled in Proposition 1. □

Proposition 4. (1) $\mathbf{Rec}(\mathbf{M})$ *contains* $\emptyset, M$ *and is closed under union, intersection, and difference.*

(2) *If* $h : \mathbf{M} \to \mathbf{M}'$ *is a homomorphism and* $L' \in \mathbf{Rec}(\mathbf{M}')$ *then* $h^{-1}(L) \in \mathbf{Rec}(\mathbf{M})$.

Proof. If L_i is recognized by $(h_i, \mathbf{A}_i, C_i), i = 1, 2$, then L_1 and L_2 are both recognized by the same semi-automaton, namely $(h_1 \times h_2, \mathbf{A}_1 \times \mathbf{A}_2)$, with respective sets of "final states" $C_1 \times A_2$ and $A_1 \times C_2$. The closure under the Boolean operations follows immediately. The other assertions are easy to verify. □

Proposition 5. *If* $L \in \mathbf{Equat}(\mathbf{M})$ *and* $K \in \mathbf{Rec}(\mathbf{M})$*, then* $L \cap K \in \mathbf{Equat}(\mathbf{M})$.

Proof. Without loss of generality we can assume that L is defined by a uniform system $S = \langle u_i = p_i, i \in [n] \rangle$such that $L = \cup\{L((S,\mathbf{M}),u_i)/i \in I\}$ for some subset I of $[n]$. (**Uniform** means that each monomial is of the form $f(w_1, w_2, \ldots, w_k)$ for some $f \in F$, some $w_1, \ldots, w_k \in U$. See Courcelle (1986a, p. 63) for more details.)

Let $K = \mu^{-1}(B')$ for some homomorphism $\mu : \mathbf{M} \to \mathbf{B}$ where $\mathbf{B}$ is finite and $B' \subseteq B$.For every $i \in [n]$ and $b \in B$ we introduce a new unknown $[u_i, b]$. We shall define a system S' such that:

$$L((S',\mathbf{M}),[u_i,b]) = \{m \in L((S,\mathbf{M}),u_i)/\mu(m) = b\}.$$

It follows that $L((S,\mathbf{M}),u_i)$ is partitioned into a family $\{L((S',\mathbf{M}), [u_i,b])/b \in B\}$ (some elements of which may be empty).

Then S' is defined as $\langle [u_i,b] = q_{i,b}/i \in [n], b \in B\rangle$ where $q_{i,b}$ is the polynomial constructed as follows from p_i. Every monomial of p_i is of the form $f(u_{i_1}, u_{i_2}, \ldots, u_{i_k})$. We associate with it the polynomial:

$$\sum\{f([u_{i_1},b_1],\ldots,[u_{i_k},b_k])/b_1,\ldots,b_k \in B, b = f_{\mathbf{B}}(b_1,\ldots,b_k)\}.$$

We let $q_{i,b}$ be the sum of the polynomials associated in this way with the monomials of p_i. It has been proved by Courcelle (1986b, Prop. 2.3) that S' satisfies the required properties. Hence, $L \cap K = \cup\{L((S',\mathbf{M}),[u_i,b])/i \in I, b \in B\}$ and is equational. □

This proof is effective even if F is infinite: since a polynomial system is finite, finitely many symbols of F occur in it. Hence, only a finite part of F is concerned in the above construction. The case of an infinite signature is used by Courcelle (1986b, 1987c) for proving closure properties of certain context-free sets of graphs. □

A special case of this proposition is the well-known result saying that the intersection of a context-free language and a regular one is context-free. (See Proposition 2.)

The following result is due to Mezei and Wright (1967, Theorem 4.6). We shall recall its proof in Proposition 14 below.

Proposition 6. *If* F *is finite, then* $\mathbf{Rec}(\mathbf{M}(F)) = \mathbf{Equat}(\mathbf{M}(F))$.

From this result and the following two characterizations of $\mathbf{Rec}(\mathbf{M})$ and $\mathbf{Equat}(\mathbf{M})$, it is clear that the recognizable subsets of $\mathbf{M}(F)$

play a central role. They can be characterized by various types of tree-automata. We shall use bottom-up deterministic and non-deterministic tree-automata in Section 3.

Proposition 7. *A subset L of M is* $\mathbf{M}$*-equational iff* $L = h_{\mathbf{M}}(T)$ *for some* $T \in \mathbf{Rec}(\mathbf{M}(F'))$ *where* F' *is a finite subset of* F. *(Mezei and Wright, 1967, theorem 5.5.)*

Let us recall that $h_{\mathbf{M}}$ denotes the unique homomorphism: $\mathbf{M}(F') \rightarrow \mathbf{M}$.

Proposition 8. *If F generates* $\mathbf{M}$, *then a subset K of M is* $\mathbf{M}$*-recognizable iff* $h_{\mathbf{M}}^{-1}(K)$ *is* $\mathbf{M}(F)$*-recognizable.*

Proof. The verification is easy. □

Corollary 1. *If F is finite and generates* $\mathbf{M}$, *then* $\mathbf{Rec}(\mathbf{M}) \subseteq \mathbf{Equat}(\mathbf{M})$. *(Mezei and Wright, 1967, theorem 4.8.)*

3. Recognizable Subsets of Free $\mathcal{T}$-Magmas

In this part we shall concentrate our attention on the recognizable subsets of $\mathbf{M}(\mathcal{T}, V)$ where $\mathcal{T} = (F, E)$ is a theory (or rather the *presentation* of a theory) and V is a finite set called the set of generators.

We consider $\mathcal{T}$ as fixed: the operations of F define "construction schemes" and their properties are fully described by E. The elements of V are the finitely many "atomic objects" with which the objects of interest (like words, trees, graphs) are built.

In order to simplify the definitions, the statements of the results, and their proofs they will be given for one-sort signatures and theories. The extension to several sorts is straightforward. We shall use many-sorted theories in Section 4.

Definition 4. Free $\mathcal{T}$-magmas. Let $\mathcal{T} = (F, E)$ be a theory. We let $\mathbf{Equat}(\mathcal{T}) = \cup\{\mathbf{Equat}(\mathbf{M}(\mathcal{T}, V))/V$ is a finite set of generators$\}$ and similarly $\mathbf{Rec}(\mathcal{T}) = \cup\{\mathbf{Rec}(\mathbf{M}(\mathcal{T}, V))/V$ is a finite set of generators$\}$.

Let $j : U \to M(F,V)$ be a sort preserving mapping (where U is another $\mathcal{S}$-sorted set of generators). It extends in a unique way into a homomorphism: $\mathbf{M}(F,U) \to \mathbf{M}(F,V)$ also denoted by j and called a **first-order substitution**. Another notation for $j(t)$ is $t[j(u_1)/u_1, \ldots, j(u_k)/u_k]$ (if $U = \{u_1, \ldots, u_k\}$), or $t[j(u_1), \ldots, j(u_k)]$ (if the ordering of U is known by the context).

In the context of a fixed set of equational axioms E we denote by $[t]$ the equivalence class of t modulo $\stackrel{*}{\longleftrightarrow}_E$ (for $t \in M(F,V)$). For $T \subseteq M(F,V)$ we let $[T]$ denote $\cup\{[t]/t \in T\}$.

A homomorphism $h : \mathbf{M}(\mathcal{T},U) \to \mathbf{M}(\mathcal{T},V)$ is completely determined by its restriction to $\{[u]/u \in U\}$, i.e., by a mapping $j : U \to M(F,V)$ so that $h([u]) = [j(u)]$ for all $u \in U$. We denote this homomorphism h by $j_{\mathcal{T}}$.

A **relabelling** is a homomorphism $j_{\mathcal{T}} : \mathbf{M}(\mathcal{T},V) \to \mathbf{M}(\mathcal{T},W)$ such that $j(V) \subseteq W$. It is said to be **surjective** if $j(V) = W$. It is **injective** if j is injective as a mapping : $V \to W$.

Given a theory $\mathcal{T}$, we shall consider the following properties:

P1: $\mathbf{Equat}(\mathcal{T}) = \mathbf{Rec}(\mathcal{T})$.

P2: $\mathbf{Rec}(\mathcal{T})$ is F-closed (i.e., is closed under the operations of F).

P3: $\mathbf{Rec}(\mathcal{T})$ is closed under homomorphisms.

P4: $\mathbf{Rec}(\mathcal{T})$ is closed under relabellings.

From Proposition 3, it is clear that P1 implies P2, P3, P4. Property P3 implies P4 trivially. Examples will prove that the reverse implications do not hold.

We shall try to characterize properties P1 to P4 in terms of the axioms of the theory.

For finitely presented theories, one obtains four decision problems. The first one, namely the problem of deciding whether $\mathbf{Equat}(\mathcal{T}) = \mathbf{Rec}(\mathcal{T})$ for a finitely presented theory $\mathcal{T}$ is undecidable (Tison, 1987). The other ones seem to be open.

Proposition 9 and Corollary 2 below reduce these problems to questions concerning recognizable subsets of $M(F,V)$ and term rewriting systems. Sufficient conditions ensuring P1 and P4 follow then.

A subset T of $M(F,V)$ is **saturated** (implicitly w.r.t. $\stackrel{*}{\longleftrightarrow}$) if $[T] = T$.

Proposition 9. *If F is finite, then* $\mathbf{Equat}(\mathbf{M}(\mathcal{T},V)) = \mathbf{Rec}(\mathbf{M}(\mathcal{T},V))$ *iff for every recognizable subset T of $M(F,V)$, the set $[T]$ is a recognizable subset of $M(F,V)$.*

Proof. Immediate consequence of Propositions 7, 8 and Corollary 1. □

A term is **linear** if each variable occurs there at most once. An equational axiom is **bilinear** if its left-hand and right-hand sides are both linear. It is **separated** if no variable occurs on both sides. It is **balanced** if the same variables occur on both sides.

Corollary 2. *If all axioms of $\mathcal{T}$ are bilinear and separated, then* $\mathbf{Rec}(\mathcal{T}) = \mathbf{Equat}(\mathcal{T})$. *Hence,* $\mathbf{Rec}(\mathcal{T})$ *is closed under homomorphisms, relabellings, and under the operations of F.*

Proof. We prove that the condition of Proposition 9 is satisfied. Since V can be considered as a set of nullary symbols in F, we shall do the proof for $V = \emptyset$. This will simplify the notation.

A **ground tree transducer** as defined in Dauchet et al. (1987) is a finite set K of pairs (L,R) where L and R are two nonempty recognizable subsets of $M(F)$.

The relation $\rightarrow_K$ on $M(F)$ associated with K is defined by:

$t \rightarrow_K t'$ iff there exist $w, t_1, \ldots, t_n, t'_1, \ldots, t'_n$, such that $t = w[t_1, \ldots, t_n], t' = w[t'_1, \ldots, t'_n]$, and for all $i = 1, \ldots, n$, there exists (L,R) in K such that $t_i \in L, t'_i \in R$.

It is proved in Dauchet et al. (1987) that $\overset{*}{\longleftrightarrow}_K$ is defined by another ground tree transducer K' that one can construct from K.

We claim that $\overset{*}{\longleftrightarrow}_E$ is defined by a ground tree transducer, and this will achieve the proof since for every recognizable subset of $M(F)$ and every ground tree transducer K, $\{t' \in M(F)/t \rightarrow_K t'$ for some $t \in T\}$ is recognizable. (This can be proved directly by standard manipulations of tree-automata).

For every $t \in M(F,X)$ that is linear w.r.t. X, the set $t[M(F)]$ defined as $\{t[t_1/x_1, \ldots, t_n/x_n]/t_1, \ldots, t_n \in M(F)\}$ (where $X = \{x_1, \ldots, x_n\}$) is recognizable.

With E assumed to be bilinear and separated, we associate:

$$K = \{(t[M(F)], t'[M(F)])/t = t' \text{ or } t' = t \text{ belongs to } E\}.$$

It is clear that $\rightarrow_K \subseteq \overset{*}{\longleftrightarrow}_E$ and that $\rightarrow_E \subseteq \rightarrow_K$. It follows that $\overset{*}{\longleftrightarrow}_K = \overset{*}{\longleftrightarrow}_E$, hence, that $\overset{*}{\longleftrightarrow}_E$ is defined by a ground tree transducer. This concludes the proof. □

The axiom $f(x_1, x_2, \ldots, \Omega, \ldots, x_n) = \Omega$ expressing that a function f is strict in one of its arguments is of the type requested by Corollary 2. (The constant Ω stands for "undefined.")

There are other cases where E satisfies Proposition 9 without being bilinear and separated. See Claim 2.

A problem similar to P1 is considered in the theory of partially commutative monoids: given a finite set of commutation rules on X, the problem is to find conditions on a recognizable language $L \subseteq X^*$ ensuring that $[L]$ is recognizable. Such conditions can be found in Métivier (1986) and Ochmanski (1985). (The decision problem is not solvable, so the best one can hope is to have decidable sufficient conditions.)

Proposition 10. (1) *Let $f \in F$ be of rank n. The family* $\mathbf{Rec}(\mathbf{M}(\mathcal{T}, V))$ *is closed under f iff for all recognizable saturated subsets $T_1, \ldots, T_n$ of $M(F, V)$, the set $[f(T_1, \ldots, T_n)]$ is recognizable.*

(2) *Let $j : V \rightarrow M(F, W)$ be a first-order substitution and let $j_{\mathcal{T}} : \mathbf{M}(\mathcal{T}, V) \rightarrow \mathbf{M}(\mathcal{T}, W)$ be the associated homomorphism. Then $j_{\mathcal{T}}(\mathbf{Rec}(\mathbf{M}(\mathcal{T}, V))) \subseteq \mathbf{Rec}(\mathbf{M}(\mathcal{T}, W))$ iff for every recognizable saturated subset T of $M(F, V)$, the set $[j(T)] \subseteq M(F, W)$ is recognizable.*

Proof. Immediate verifications follow from the definitions. □

Corollary 3. *If $\mathcal{T} = (F, E)$ and E is a set of bilinear and balanced axioms, then* $\mathbf{Rec}(\mathcal{T})$ *is closed under relabellings. If the axioms of E are only assumed to be bilinear, the closure holds w.r.t. surjective relabellings.*

Proofs. Let us first prove that if $t \in M(F, V)$ and $s \in M(F, W)$ are such that $j(t) \longleftrightarrow_E s$ then $s = j(t')$ for some $t' \in M(F, V)$ such that $t \longleftrightarrow_E t'$.

We assume that E is bilinear.

Let $j(t) \longleftrightarrow_E s$ by an axiom of E of the form $z = z'$ where $t = z[t_1, \dots, t_k]$; then $s = z'[j(t_1), \dots, j(t_k), s_1, \dots, s_h]$. (The terms $s_1, \dots, s_h$ correspond to the variables in z' that are not in z). Since $j : V \to W$ is surjective, there exist $s'_1, \dots, s'_h$ in $M(F, V)$ such that $j(s'_i) = s_i$. Hence, $t' = z'[t_1, \dots, t_k, s'_1, \dots, s'_h]$ is the desired term.

If E is bilinear and balanced, then $h = 0$ and there is no need to assume that j is surjective. We omit the remaining details.

Now let $T \subseteq M(F, V)$ be recognizable and saturated. Then $[j(T)] \subseteq j(T)$ by the fact proved above. Hence, $[j(T)] = j(T)$ and $[j(T)]$ is recognizable as a subset of $M(F, V)$.

The result follows then from Proposition 10, Assertion (2). □

We shall give below another proof using a generalization to $\mathcal{T}$-automata of the classical construction of a deterministic automaton equivalent to a nondeterministic one.

Examples and counterexamples. The theory $\mathcal{T}_u$ introduced in Example 1 to formalize words has no equations. It satisfies P1 to P4 by Propositions 6 and 9. The polynomial systems over F_u correspond to finite-state automata (Mezei and Wright, 1967). Hence they define again the regular languages and $\mathbf{Rec}(\mathbf{M}(\mathcal{T}_u, X)) = \mathbf{Equat}(\mathbf{M}(\mathcal{T}_u, X))$. (See Proposition 2.) The homomorphisms reduce to relabellings: one cannot transform a letter into a word since they are of different sorts.

The theory of monoids, $\mathcal{T}_m$, also introduced in Example 1, satisfies P2 (the family of regular languages is closed under concatenation), P3 and P4 (this family of languages is closed under homomorphisms) but not P1 since $\mathbf{Equat}(\mathbf{X}^*)$ is the family of context-free languages and since it includes properly the family of regular languages.

We now exhibit a theory for which P4 holds (by Corollary 3) but P3 does not.

Let $F = \{f, g\}$ and E be the axiom $f(x) = g(f(g(x)))$. Let $V = \{v\}$ and $\mathcal{T} = (F, E)$. Consider $\mathbf{M} = \mathbf{M}(\mathcal{T}, V)$ and $h_{\mathbf{M}} : M(F, V) \to \mathbf{M}$ (the canonical homomorphism). Then $h_{\mathbf{M}}^{-1}([v]) = \{v\}$ and, hence, is a recognizable subset of $M(F, V)$. Hence, $\{[v]\}$ is a recognizable subset of $\mathbf{M}(\mathcal{T}, V)$. Now let j map v to $f(v)$. Then $h_{\mathbf{M}}^{-1}([f(v)]) = \{t \in M(F, V)/t \overset{*}{\longleftrightarrow}_E f(v)\}$. It is not hard to see that this set is $\{g^n f g^n(v)/n \geq 0\}$. Hence, that it is not a recognizable subset of $M(F, V)$. (Otherwise, $\{g^n f g^n/n \geq 0\}$ would be a regular set of words.)

We finally define a theory $\mathcal{T}'$ having a single axiom, that is bilinear but not balanced, and such that $\mathbf{Rec}(\mathcal{T}')$ is not closed under all relabellings. It is similar to the previous one. We let F' be $\{f, g\}$, E' be $\{f(x) = g(y, f(g(z, x)))\}$, and $\mathcal{T}'$ be (F', E'). Then it is not hard to establish that, if V is strictly included in W, then $\{t' \in M(F', W)/t' \overset{*}{\longleftrightarrow}_{E'} t$, for some $t \in M(F', V)\}$ is not recognizable as a subset of $M(F', W)$. This proves that, if we denote by j the inclusion mapping: $V \to W$, then $j_{\mathcal{T}}(M(\mathcal{T}', V))$ is *not* in $\mathbf{Rec}(\mathbf{M}(\mathcal{T}', W))$. This example shows that for an arbitrary theory $\mathcal{T}$, an element of $\mathbf{Rec}(\mathcal{T})$ must be considered as a pair (L, V) where $L \in \mathbf{Rec}(\mathbf{M}(\mathcal{T}, V))$. In other words, one cannot identify an element of $\mathbf{Rec}(\mathbf{M}(\mathcal{T}, V))$ with its images under injective relabellings although every injective relabelling is actually injective as a mapping $: M(\mathcal{T}, V) \to M(\mathcal{T}, W)$. Such an identification can be done if the theory $\mathcal{T}$ under consideration is bilinear and balanced.

For the elements of $\mathbf{Equat}(\mathcal{T})$, the identification can be made without any restriction on $\mathcal{T}$, for the following reason. Let $j : V \to W$ be injective. If $L \in \mathbf{Equat}(\mathbf{M}(\mathcal{T}, V))$, then $j_{\mathcal{T}}(L) \in \mathbf{Equat}(\mathbf{M}(\mathcal{T}, W))$. Conversely, if $L \subseteq M(\mathcal{T}, V)$ and $j_{\mathcal{T}}(L) \in \mathbf{Equat}(\mathbf{M}(\mathcal{T}, W))$ then $L \in \mathbf{Equat}(\mathbf{M}(\mathcal{T}, V))$.

Example 2. The theory of pairs of words. Let A and B be two disjoint alphabets. Let $F_p = A \cup B \cup \{e\}$ with $\rho(a) = 1$ for $a \in A \cup B$ and $\rho(e) = 0$. Let $\mathbf{M}$ be the F_p-magma such that $M = A^* \times B^*, e_{\mathbf{M}} = (\varepsilon, \varepsilon)$, $a_{\mathbf{M}}((w_1, w_2)) = (w_1 a, w_2)$, $b_{\mathbf{M}}((w_1, w_2)) = (w_1, w_2 b)$ for all $a \in A, b \in B, w_1 \in A^*$ and $w_2 \in B^*$. Let E_p be the set of equations $\{a(b(x)) = b(a(x))/a \in A, b \in B\}$ and $\mathcal{T}_p = (F_p, E_p)$. It is clear that $\mathbf{M} = \mathbf{M}(\mathcal{T}_p, \emptyset)$.

Two notions of two-tape (one-way, finite-state) automata have been introduced to define subsets of $A^* \times B^*$: deterministic automata by Rabin and Scott (1959), and nondeterministic ones by Elgot and Mezei (1965).

Let $\mathbf{Det}(A^* \times B^*)$ and $\mathbf{Ndet}(A^* \times B^*)$ denote the two corresponding families of subsets of $A^* \times B^*$. Neither of them is equal to $\mathbf{Rec}(\mathbf{M})$ since the former is not closed under union (Rabin and Scott, 1959) and the latter is not closed under intersection. These automata actually define equational subsets of $\mathbf{M}$ as shown in Courcelle (1986a, p.103). Hence, to summarize:

$$\mathbf{Rec}(\mathbf{M}) \subset \mathbf{Det}(A^* \times B^*) \subset \mathbf{Ndet}(A^* \times B^*) = \mathbf{Equat}(\mathbf{M}).$$

The inclusions are strict.

From a homomorphism $h : \mathbf{M} \to \mathbf{Q}$ where $\mathbf{Q}$ is finite, it is not hard to construct a deterministic 2-tape automaton that reads first all symbols of the first tape and then all symbols of the second one. This establishes the first inclusion. The remaining details are left to the reader. □

Our purpose now is to present tree-automata. They offer an appropriate algebraic framework for comparing recognizable and equational sets.

Definition 5. $\mathcal{T}$-automata. We now assume that F is finite. The definition of a homomorphism $h : \mathbf{M}(F, V) \to \mathbf{Q}$ where $\mathbf{Q}$ is a finite F-magma can be concretely given in the form of a 4-tuple $A = \langle F, V, Q, \delta \rangle$ called a **semi-F-automaton** such that:

> Q is a finite set, called the set of **states**,
> δ is a mapping associating:
> with $v \in V$ an element $\delta(v)$ of Q,
> with $f \in F$ a mapping $\delta(f) : Q^{\rho(f)} \to Q$ (where $\rho(f)$ denotes the rank of f).

With h, one associates the **semi-F-automaton** $A_h = \langle F, V, Q, \delta \rangle$ such that Q is the domain of $\mathbf{Q}$, $\delta(v)$ is $h(v)$ for all $v \in V$ and $\delta(f)$ is the mapping $f_{\mathbf{Q}}$. Conversely, every semi-F-automaton $\langle F, V, Q, \delta \rangle$ is A_h for some F-magma $\mathbf{Q}$ and some homomorphism $h : \mathbf{M}(F, V) \to \mathbf{Q}$.

A 5-tuple $A = \langle F, V, Q, \delta, Q' \rangle$ where $\langle F, V, Q, \delta \rangle$ is as above and Q' is a subset of Q (called the set of **final states**), is called an **F-automaton**. The subset of $M(F, V)$ defined by A is $T(A) = h_A^{-1}(Q')$ where h_A is the homomorphism defined by the first four components of A. It is recognizable (w.r.t. $\mathbf{M}(F, V)$) and every $\mathbf{M}(F, V)$-recognizable set is of this form.

Now let $\mathcal{T} = (F, E)$ be a theory. A **$\mathcal{T}$-automaton** is an F-automaton A such that the associated F-magma $\mathbf{Q}$ satisfies E, i.e., such that the functions $\delta(f)$ for $f \in F$, satisfy the properties defined by E. One can decide whether a given 5-tuple is indeed a $\mathcal{T}$-automaton if E is finite.

Let A be a $\mathcal{T}$-automaton $\langle F, V, Q, \delta, Q' \rangle$. There is one and only one homomorphism $k : \mathbf{M}(\mathcal{T}, V) \to \mathbf{Q}$ such that the diagram of Fig. 1 is commutative (where A is a $\mathcal{T}$-automaton):

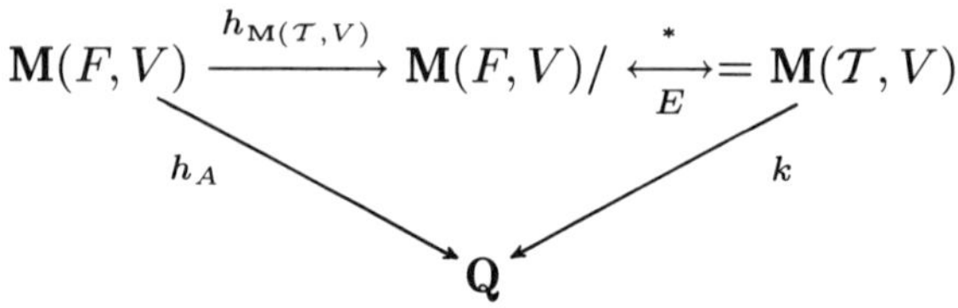

Fig. 1

The subset of $M(\mathcal{T}, V)$ **defined by** A is then:

$$L(A) := h_{\mathbf{M}(\mathcal{T},V)}(T(A)).$$

It is clear that $L(A) = k^{-1}(Q')$ (where Q' is the set of final states of A). Hence, $L(A) \in \mathbf{Rec}(\mathbf{M}(\mathcal{T}, V))$. Note also that $T(A) = h^{-1}_{\mathbf{M}(\mathcal{T},V)}(L(A))$. Of course $T(A) = L(A)$ if $\mathcal{T}$ is a **free** theory, i.e., has an empty set of axioms.

Conversely, let L be a recognizable subset of $\mathbf{M}(\mathcal{T}, V)$, defined as $h^{-1}(Q')$ for some $Q' \subseteq Q$ and some homomorphism $h : \mathbf{M}(\mathcal{T}, V) \to \mathbf{Q}$ where $\mathbf{Q}$ is finite. Let $Q_1 = h(M(\mathcal{T}, V))$, and $Q'_1 = Q' \cap Q_1$. The set Q_1 is finite, closed under the operations $f_{\mathbf{Q}}, f \in F$. Hence, it is the domain of a sub-F-magma $\mathbf{Q}_1$ of $\mathbf{Q}$, h is a homomorphism: $\mathbf{M}(\mathcal{T}, V) \to \mathbf{Q}_1$ and $L = h^{-1}(Q'_1)$. It is easy to establish that $\mathbf{Q}_1$ satisfies the equations of E. Hence, $A = \langle F, V, Q_1, \delta, Q'_1 \rangle$ (where $\delta(f)$ is the restriction of $f_{\mathbf{Q}}$ to Q_1) is a $\mathcal{T}$-automaton. Hence, we have established the following result.

Proposition 11. $\mathbf{Rec}(\mathbf{M}(\mathcal{T}, V))$ *is the set of subsets of* $M(\mathcal{T}, V)$ *defined by* $\mathcal{T}$*-automata.*

This proposition actually gives a characterization of $\mathbf{Rec}(\mathbf{M})$ where $\mathbf{M}$ is an F-magma that is generated by F (assumed to be finite). In this case, $\mathbf{M}$ is isomorphic to $\mathbf{M}(F)/ \overset{*}{\longleftrightarrow}_E$ for some finite or infinite set of axioms, i.e., to $\mathbf{M}(\mathcal{T}_{\mathbf{M}}, \emptyset)$for a certain theory $\mathcal{T}_{\mathbf{M}}$.

Hence, a set is $\mathbf{M}$-recognizable iff it is defined by a $\mathcal{T}_{\mathbf{M}}$-automaton of the form $\langle F, \emptyset, Q, \delta, Q' \rangle$.

We now characterize the equational sets in terms of nondeterministic automata.

Definition 6. Nondeterministic $\mathcal{T}$***-automata.*** Nondeterministic finite automata are nothing but uniform polynomial systems in disguise. Introducing them is useful in order to make the comparison with (deterministic) $\mathcal{T}$-automata.

A **nondeterministic semi-F-automaton** is a 4-tuple$A = \langle F, V, Q, \eta \rangle$ such that F is a finite ranked alphabet, V is a finite set of nullary symbols, Q is a finite set (the set of **states**), η is the transition relation defined as a finite set of tuples of the form (v, q) with $v \in V$ and $q \in Q$, or $(q_1, \ldots, q_k, f, q)$ with $k = \rho(f), q_1, \ldots, q_k, q \in Q$. Note that η associates a (possibly empty) set of states with every v in V and a multivalued function : $Q_1 \times \cdots \times Q_k \to Q$ with every f in F of rank k.

A **nondeterministic F-automaton** is a 5-tuple $A = \langle F, V, Q, \eta, Q' \rangle$ where $\langle F, V, Q, \eta \rangle$ is a nondeterministic semi-F-automaton and Q' is a subset of Q called the set of **final states**.

Let A be a semi-F-automaton as above. Let $t \in M(F, V)$. A **computation of** A on t is a mapping ν from the set of nodes N of t (considered as usual as a finite tree) into Q such that for every node μ of N, if f labels μ and $(\mu_1, \ldots, \mu_k)$ is the sequence of its successors (with $k = \rho(f)$) then $(\nu(\mu_1), \ldots, \nu(\mu_k), f, \nu(\mu)) \in \eta$ and if $v \in V$ labels μ then $(v, \nu(\mu)) \in \eta$.

Then we define $T(A, q)$ as the set of elements of $M(F, V)$ having at least one computation for which the value of the root is q. If A has in addition a set of final states Q', then $T(A)$ is defined as $\cup\{T(A, q)/q \in Q'\}$.

Let now $\mathbf{M}$ be an F-magma. We define $L(A)$, the subset of M defined by A, as $\{t_{\mathbf{M}}/t \in T(A)\}$ (and similarly $L(A, q) = \{t_{\mathbf{M}}/t \in T(A, q)\}$).

Proposition 12. *A subset of M is $\mathbf{M}$-equational iff it is defined by a nondeterministic F-automaton of the form $\langle F, \emptyset, Q, \eta, Q' \rangle$.*

Proof. Let $A = \langle F, \emptyset, Q, \eta, Q' \rangle$. We associate with A a polynomial system $S = \langle u_q = p_q; q \in Q \rangle$. Its set of unknowns is $\{u_q/q \in Q\}$, constructed from Q. The polynomial p_q is the sum of all monomials of the form $f(u_{q_1}, \ldots, u_{q_k})$ such that $(q_1, \ldots, q_k, f, q) \in \eta$. It can then be proved (for instance with the help of Courcelle (1986a, Prop. 13.3)) that $L(A, q) = L((S, \mathbf{M}), u_q)$ for every $q \in Q$. Hence, $L(A) = \cup\{L((S, \mathbf{M}), u_q)/q \in Q'\}$, and is $\mathbf{M}$-equational.

From a uniform polynomial system defining $L \subseteq M$, one can obtain by reversing the above construction a nondeterministic automaton A that defines L. $\square$

The above result actually holds in the more general case where **M** has multi-valued (nondeterministic) operations. (The notion of an equational set can also be defined. See Dubinsky (1975) where such an object is called a **nondeterminate algebra**.)

Let us now consider a theory $\mathcal{T} = (F, E)$. We cannot define the notion of a nondeterministic $\mathcal{T}$-automaton because we do not know what it can mean to say that multivalued functions (the ones associated with a transition relation) satisfy a set of equational axioms. Dubinsky (1975) introduces nondeterminate algebras, but he does not introduce equational axioms defining them.

Definition 7. ***Semi-deterministic $\mathcal{T}$-automata.*** We introduce a new notion that generalizes the notion of a $\mathcal{T}$-automaton.

A **semi-deterministic $\mathcal{T}$-automaton** automaton is a nondeterministic F-automaton $A = \langle F, V, Q, \eta, Q' \rangle$ such that:

(1) the multi-valued functions associated with η are total single-valued mappings (i.e., A is deterministic w.r.t. F),
(2) these mappings satisfy the axioms of E (i.e., A is a (deterministic) $\mathcal{T}$-automaton except that sets of states, as opposed to states, are associated with the elements of V).

This notion is not introduced for the mere pleasure of generalization as shown by the following proposition. Let **Relab**(**Rec**($\mathcal{T}$)) be the set of relabellings of sets in **Rec**($\mathcal{T}$).

Proposition 13. **Relab**(**Rec**($\mathcal{T}$)) *is the family of sets defined by semi-deterministic $\mathcal{T}$-automata.*

Proof. Let $j : V \to W$ be a relabelling and $L \in \mathbf{Rec}(M(\mathcal{T}, V))$ be defined by a $\mathcal{T}$-automata $A = \langle F, V, Q, \delta, Q' \rangle$.

Then $j_{\mathcal{T}}(L) = L(B)$ where B is the semi-deterministic $\mathcal{T}$-automaton $\langle F, W, Q, \eta', Q' \rangle$ where:

$$(q_1, \ldots, q_k, f, q) \in \eta \text{ iff } \delta(f)(q_1, \ldots, q_k) = q$$
$$(w, q) \in \eta \text{ iff } \delta(v) = q \text{ for some } v \in V \text{ such that } j(v) = w.$$

Proving that every set defined by a semi-deterministic $\mathcal{T}$-automaton is in **Relab**(**Rec**($\mathcal{T}$)) is not difficult. □

We now recall a well-known construction. With every nondeterministic F-automaton $A = \langle F, V, Q, \eta, Q' \rangle$, one associates the (deterministic) F-automaton $\mathbf{det}(A) := \langle F, V, \mathcal{P}(Q), \delta, P \rangle$ where:

$$\begin{aligned} \delta(v) &:= \{q \in Q / (v, q) \in \eta\} \\ \delta(f)(Q_1, \ldots, Q_k) &:= \{q \in Q / (q_1, \ldots, q_k, f, q) \in \eta \\ &\qquad \text{for some } q_1 \in Q_1, \ldots, q_k \in Q_k\} \\ &\qquad \text{where } Q_1, \ldots, Q_k \subseteq Q, \\ P &:= \{Q_1 \subseteq Q / Q_k \cap Q' \neq \emptyset\}. \end{aligned}$$

Proposition 14. (1) $T(\mathbf{det}(A)) = T(A) (\subseteq M(F, V))$.

(2) *If A is a semi-deterministic $\mathcal{T}$-automaton and if the axioms of $\mathcal{T}$ are bilinear and balanced, then* $\mathbf{det}(A)$ *is a $\mathcal{T}$-automaton. Hence, the subset of $M(\mathcal{T}, V)$ defined by A is $\mathbf{M}(\mathcal{T}, V)$-recognizable.*

Proof. (1) This result is well known. It is presented in terms of systems of equations in Mezei and Wright (1967, Theorem 5.3) and as above in Gecseg and Steinby (1984).

(2) This follows from a theorem of Gautam (1957) (recalled and used by Lescanne (1982) and by Courcelle (1986a, Proposition 10.8)) saying that the validity of a bilinear and balanced equational axiom is preserved under the powerset construction (together with its natural extension to functions). □

Assertion (1) establishes that $\mathbf{Equat}(\mathbf{M}(F, V)) \subseteq \mathbf{Rec}(\mathbf{M}(F, V))$ and, hence, proves Proposition 6.

Assertion (2) proves that $\mathbf{Relab}(\mathbf{Rec}(\mathcal{T})) \subseteq \mathbf{Rec}(\mathcal{T})$ and, hence, gives another proof of Corollary 3.

If E is a set of bilinear axioms (not necessarily balanced) and $A = \langle F, V, Q, \eta, Q' \rangle$ is the semi-deterministic $\mathcal{T}$-automaton defining $j_{\mathcal{T}}(L)$ where $L \in \mathbf{Rec}(\mathbf{M}(\mathcal{T}, W))$ and j is a surjective mapping : $W \to V$ (A is obtained by the construction of Proposition 13) then an automaton $\mathbf{det}'(A)$ can be constructed as $\mathbf{det}(A)$ except that its states are the nonempty subsets of Q. This is possible since j is surjective so that $\delta(v) \neq \emptyset$ for all $v \in V$.

And then $\mathbf{det}'(A)$ satisfies the axioms of E because the validity of a bilinear (though not necessarily balanced) equation is preserved under the extension of an F-magma consisting in taking its nonempty subsets (with the canonical extension of functions to sets). This gives another proof of the second assertion of Corollary 3.

The validity of the determinization construction for $\mathcal{T}$-automata where $\mathcal{T}$ is a bilinear theory is mentioned by Eilenberg and Wright (1967) in their abstract but the idea is not developed in the paper that deals only with free theories.

Let us finally observe that all the above results hold in the case of a many-sorted signature provided they are appropriately reformulated. Hence, if $\mathcal{S} = \{s_1, \ldots, s_n\}$, if F is an $\mathcal{S}$-signature, an F-automaton is a 5-tuple : $A = \langle F, (V_1, \ldots, V_n), (Q_1, \ldots, Q_n), \delta, Q' \rangle$ where $Q_i \cap Q_j = \emptyset$ for $i \neq j$ and $Q' \subseteq Q_i$ for some i. This automaton defines a recognizable subset of $M(F, V)_{s_i}$ denoted by $T(A)$. If A is a $\mathcal{T}$-automaton then $L(A) \in \mathbf{Rec}(\mathbf{M}(\mathcal{T}, V))_{s_i}$.

4. Various Notions of Trees

There are many notions of **trees** in current use in computer science. Roughly speaking, trees can be either ranked or unranked and either ordered or unordered. This gives four notions of trees. (**Ranked** means that each node has a fixed number of successors, i.e., of sons; this number is determined by the label of the node. **Ordered** means that the successors of a node are ordered in a relevant way.)

For each of these families, we define an appropriate theory $\mathcal{T}$ such that the set of all finite trees of the family under consideration, built with the set of labels V, is $\mathbf{M}(\mathcal{T}, V)$. (This means that appropriate operations on these trees have been defined). We examine the corresponding $\mathcal{T}$-automata, and we check in some cases the validity of properties P1 to P4 introduced in Definition 4.

As in the theory $\mathcal{T}_u$ introduced in Example 1 in order to deal with words (and for the same reasons), we shall use several sorts.

Definition 8. The theory of unordered binary trees. Let V and W be two disjoint sets of labels. Let U be the set of finite unordered binary trees, the leaves of which are labeled in V and the internal nodes of which are labeled in W.

Let $p : W \times U \times U \to U$ be the mapping that associates with $a \in W$, $t_1, t_2 \in U$ the tree in U, the root of which is labeled by a, and that has subtrees t_1 and t_2.

Let $\mathcal{S}_{ub}$ consist of two sorts, **intlab** and **tree**. Let $F_{ub} = \{p\}$ where p is of profile: **intlab** $\times$ **tree** $\times$ **tree** $\to$ **tree**. Let E_{ub} consist of the single axiom:

$$p(x, y, z) = p(x, z, y).$$

Then $\mathcal{T}_{ub} = (F_{ub}, E_{ub})$ is called the **theory of unordered binary trees.**

Let $\mathbf{U}(W, V)$ be the F_{ub}-magma $\langle (W, U), p \rangle$ where W is the domain of sort **intlab** and U is that of sort **tree**. The elements of V are of the sort **tree**.

Claim 1. $\mathbf{U}(W, V) = \mathbf{M}(\mathcal{T}_{ub}, (W, V))$.

Easy manipulations of tree-automata (in the sense of Gecseg and Steinby (1984) or Engelfriet (1975)) allow one to establish the validity for $\mathcal{T}_{ub}$ of the condition of Proposition 9. Hence,

Claim 2. $\mathbf{Equat}(\mathcal{T}_{ub}) = \mathbf{Rec}(\mathcal{T}_{ub})$. *This family of sets of trees is closed under* p*, under homomorphisms, and under inverse homomorphisms.*

The extension to unordered, ranked trees is straightforward: for each $k > 0$, one introduces a sort **intlab-**k for labels of rank k, i.e., for those which label nodes with k successors, and one keeps the sort **tree** for trees and for labels of rank 0. Claims 1 and 2 hold as well when appropriately reformulated.

We now consider $\mathcal{T}_{ub}$-automata. The sets U, V, W are as above. From Definition 5 and our conventions concerning the many-sorted case, a $\mathcal{T}_{ub}$-automaton intended to define a subset of U is a 5-tuple $A = \langle F_{ub}, (W, V), (Q_1, Q_2), \delta, Q' \rangle$ where $Q' \subseteq Q_2$ and such that:

$$\delta(p)(q_1, q_2, q_2') = \delta(p)(q_1, q_2', q_2)$$

for all $q_1 \in Q_1$, all $q_2, q_2' \in Q_2$. We shall simplify this definition, in particular, by eliminating the set of states Q_1.

A **simplified** $\mathcal{T}_{ub}$**-automaton** is a 6-tuple $B = \langle W, V, Q, \delta_1, \delta_2, Q' \rangle$ where Q is a finite set, $Q' \subseteq Q$, δ_1 and δ_2 are mappings $\delta_1 : V \to Q$ and $\delta_2 : W \times Q \times Q \to Q$ such that:

$$\delta_2(w, q, q') = \delta_2(w, q', q)$$

for all $w \in W, q, q' \in Q$.

The behavior of B is characterized by a mapping $\delta^* : U \to Q$ such that

$$\delta^*(v) = \delta_1(v) \text{ if } v \in V,$$

$$\delta^*(t) = \delta_2(w, q, q') \text{ if } t \text{ is}$$

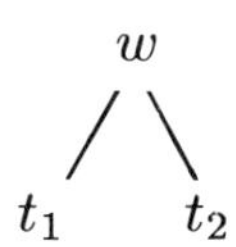

and $\delta^*(t_1) = q, \delta^*(t_2) = q'$.

We let $L(B) = \{t \in U/\delta^*(t) \in Q'\}$ be the set of trees **defined by** B.

With any $\mathcal{T}_{ub}$-automaton $A = \langle F_{ub}, (W,V), (Q_1, Q_2), \delta, Q' \rangle$ such that $Q' \subseteq Q_2$, we associate the simplified automaton $\mathbf{simp}(A) = \langle W, V, Q_2, \delta_1, \delta_2, Q' \rangle$ where: $\delta_1(v) = \delta(v)$ and $\delta_2(w, q, q') = \delta(p)(\delta(w), q, q')$.

It is clear that $L(\mathbf{simp}(A)) = L(A)$.

Conversely, for any simplified automaton B, one can construct an automaton A such that $\mathbf{simp}(A) = B$. (For a set of states Q_1 of A, one can take W and one then lets $\delta(w) = w$ for $w \in W$.)

This shows that the notion of a $\mathcal{T}_{ub}$-automaton is not so far from the concrete ones used classically for tree-automata (see Gecseg and Steinby, 1984).

Definition 9. The theory of ordered binary trees. The theory of **ordered binary trees** is almost the same except that the commutativity of p (or of δ in the corresponding automata) is no longer required. Hence, it is a **free theory** (i.e., without axioms) and the results corresponding to Claims 1 and 2 hold trivially.

Incomplete binary (ordered) trees are sometimes considered. (See for instance the book by Horowitz and Sahni (1976) on data structures.) They can be "coded" as ordered binary trees in the above sense by means of a special label $\perp$ of type **tree** standing for a missing subtree. Hence, for instance, the "incomplete" tree over $X = \{a, b, c, d\}$

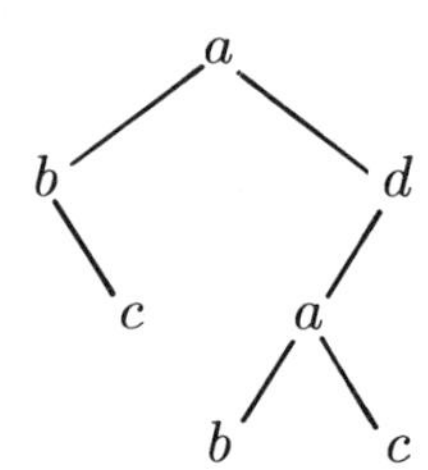

can be coded as the ordered binary tree

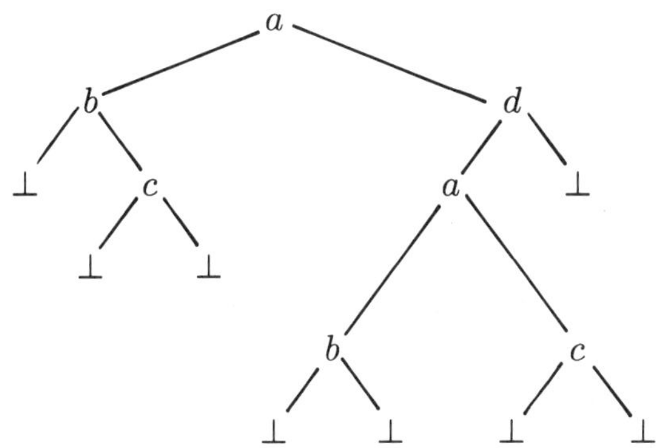

constructed with X as set of internal labels and $\{\perp\}$ as set of labels of type **tree**. Hence, the binary trees of Horowitz and Sahni (1976) can be investigated in this framework.

With more sorts, one can also deal with ordered ranked trees of larger rank. The reader may consider that this is a fairly complicated way to deal with a very well-known object, namely, the initial F-magma $\mathbf{M}(F)$. But as in the case of words (see in Example 1 the discussion introducing $\mathbf{T}_u$) it is useful to consider the set of node labels of trees as a parameter and not to "freeze" it in a signature F.

Definition 10. The theory of unordered unranked trees. The three basic notions are those of a **label**, of a **tree**, and of a **forest** (i.e., a finite multiset of trees). A tree is formed by means of a label (labeling its root) and a forest (the multiset of the subtrees issued from the root).

So we let $\mathcal{S}_{uu} = \{\mathbf{lab}, \mathbf{tree}, \mathbf{forest}\}$, we let $F_{uu} = \{\phi, p, r, \uplus\}$where:

> ϕ is a nullary symbol of sort **forest** denoting the empty forest,
> $p : \mathbf{lab} \times \mathbf{forest} \to \mathbf{tree}$ forms a tree from a label and a forest.
> $r : \mathbf{tree} \to \mathbf{forest}$ makes any tree into the forest reduced to that tree.
> $\uplus : \mathbf{forest} \times \mathbf{forest} \to \mathbf{forest}$ makes the union of two forests (considered as multisets).

We let E_{uu} express that $\uplus$ is associative, commutative, and has unit ϕ, and we let $\mathcal{T}_{uu} = (F_{uu}, E_{uu})$.

We denote by $\mathbf{Z}(W)$ the F_{uu}-magma of labels (from W), trees, and forests (over W) with the above defined operations. Hence, $Z(W)_{\mathbf{tree}}$

denotes the set of trees over W and $Z(W)_{\mathbf{forest}}$ denotes the set of forests over W.

Claim 3. $Z(W) = \mathbf{M}(\mathcal{T}_{uu}, (W, \emptyset, \emptyset))$.

Since E_{uu} is bilinear and balanced, we have the following result.

Claim 4. $\mathbf{Rec}(\mathcal{T}_{uu})$ *is closed under relabellings.*

Proposition 15. $\mathbf{Rec}(\mathbf{Z}(W))$ *is strictly included in* $\mathbf{Equat}(\mathbf{Z}(W))$ *if* W *has at least two elements.*

The proof can be found in the appendix.

Proposition 16. $\mathbf{Rec}(\mathbf{Z}(W))$ *is* F_{uu}*-closed.*

Proof. The condition of Proposition 10 is easily verified for p and r because these operations do not appear in E_{uu}. The proof for ϕ is easy. The case of $\uplus$ is much more complicated, and the proof can be found in the appendix. □

We now look at $\mathcal{T}_{uu}$-automata. From Definition 5, a $\mathcal{T}_{uu}$-automaton is a 5-tuple $A = \langle F_{uu}, (W, W', W''), (Q_1, Q_2, Q_3), \delta, Q' \rangle$. If $Q' \subseteq Q_2$ then L(A) is a set of trees, and if $Q' \subseteq Q_3$, then L(A) is a set of forests. (The case $Q' \subseteq Q_1$ is not interesting.)

As for unordered binary trees, we can simplify this definition. We consider only the case where a set of trees is defined.

A **simplified** $\mathcal{T}_{uu}$**-automaton** is a 6-tuple $B = \langle W, Q, \delta_1, \delta_2, \delta_3, Q' \rangle$ where $Q' \subseteq Q$ and $\delta_1, \delta_2, \delta_3$ are total mappings:

$$\delta_1 : W \to Q$$
$$\delta_2 : W \times Q \to Q$$
$$\delta_3 : Q \times Q \to Q$$

such that δ_3 is associative and commutative.

The behaviour of B is defined by a mapping $\delta^* : Z(W)_{\mathbf{tree}} \to Q$ defined as follows:

$\delta^*(t) = \delta_1(w)$ if t is reduced to one node labelled by w,

$\delta^*(t) = \delta_2(w, \delta^*(t'))$ if t is of the form  ,

and $\delta^*(t) = \delta_2(w, \delta_3(q_1, \delta_3(q_2, \dots, \delta_3(q_{k-1}, q_k)) \dots))$,

if t is of the form

and $q_i = \delta^*(t_i)$ for $i = 1, \dots, k$

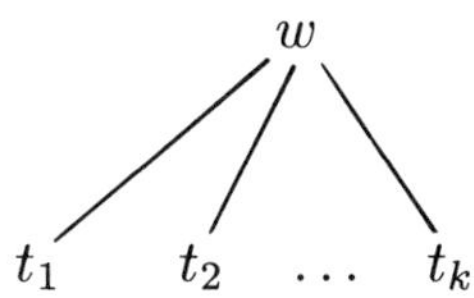

We then let $L(B, q)_{\mathbf{tree}} = \{t \in Z(W)_{\mathbf{tree}} / \delta^*(t) = q\}$ and $L(B) = \cup\{L(B, q)/q \in Q'\}$.

Let A be a $\mathcal{T}_{uu}$-automaton. One can transform it so as to ensure that $\delta(r) : Q_2 \to Q_3$ is injective. (One replaces, if necessary, Q_3 by $Q_2 \times Q_3$ and $\delta(r)$ by $\delta'(r)$ such that $\delta'(r)(q) = (q, \delta(r)(q)))$. Hence, one can assume that A satisfies this condition.

Then we let $\mathbf{simp}(A) = \langle W, Q_3, \delta_1, \delta_2, \delta_3, Q'' \rangle$ where:

$$\begin{aligned} \delta_1(w) &= \delta(r)(\delta(p)(w, \delta(\phi))), \\ \delta_2(w, q) &= \delta(r)(\delta(p)(w, q)), \\ \delta_3(q, q') &= \delta(\uplus)(q, q'), \\ Q'' &= \delta(r)(Q'), \end{aligned}$$

and it is not hard to prove that the simplified $\mathcal{T}_{uu}$-automaton $\mathbf{simp}(A)$ defines $L(A)$.

Conversely, from a simplified automaton B one can construct a $\mathcal{T}_{uu}$-automaton A such that $\mathbf{simp}(A) = (B)$. (One takes $Q_3 = Q$ augmented if necessary with a new element q_0 serving as a unit for δ_3; one takes for Q_2 an isomorphic copy of Q, and for Q_1, one takes W.) Again this shows that the notion of a $\mathcal{T}_{uu}$-automaton is not as far from what is usually considered as an automaton, as one could think on first sight.

Definition 11. Two theories for dealing with ordered unranked trees. The first theory is very close to the theory of forests dealt with above. We only indicate a few modifications. We introduce $\mathcal{S}_{ou} = \{\mathbf{lab}, \mathbf{tree}, \mathbf{hedge}\}$ where a **hedge** is a finite sequence of ordered unranked trees. Then $F_{ou} = \{\varepsilon, r, \bullet, p\}$ where:

ε denotes the empty hedge,
r : **tree** $\to$ **hedge** makes a tree into a hedge reduced to it,

p : **lab** $\times$ **hedge** $\to$ **tree** makes a tree by connecting the trees of a hedge with a new root labeled by the first argument,

$\bullet$: **hedge** $\times$ **hedge** $\to$ **hedge** concatenates two hedges.

We let $\mathcal{T}_{ou} = (F_{ou}, E_{ou})$ where E_{ou} expresses that $\bullet$ is associative with unit ε. The F_{ou}-magma of set labels W, of trees, and of hedges over W is denoted by $\mathbf{H}(W)$.

Claim 5. $\mathbf{H}(W) = \mathbf{M}(\mathcal{T}_{ou}, (W, \emptyset, \emptyset))$.

As in Proposition 15 for $\mathbf{Z}(W)$ one can prove:

Claim 6. $\mathbf{Rec}(\mathbf{H}(W))$ *is strictly included in* $\mathbf{Equat}(\mathbf{H}(W))$ *provided* W *has at least two symbols.*

Claim 7. $\mathbf{Rec}(\mathcal{T}_{ou})$ *is closed under the homomorphisms and the operations of* F_{ou}.

The closure under relabellings is an immediate consequence of the form of E_{ou}. The other assertions must be established independently. The $\mathcal{T}_{ou}$-automata and their simplified forms can be easily defined as for $\mathcal{T}_{uu}$, but we leave this up to the reader.

This way of defining ordered unranked trees and the related one we shall now describe are adapted from Pair (1970).

Here is the second theory for dealing with ordered unranked trees. It can be defined as a derived theory of $\mathcal{T}_{ou}$. We let $\mathcal{S}'_{ou} = \{$**lab**, **hedge**$\}$, and $F'_{ou} = \{\varepsilon, s\}$. The mapping s : **lab** $\times$ **hedge** $\times$ **hedge** $\to$ **hedge** is defined by:

$$s(w, h_1, h_2) = \bullet\ (r(p(w, h_1)), h_2).$$

This makes $(W, \mathbf{H}(W)_{\mathbf{hedge}})$ into an F'_{ou}-magma $\mathbf{H}'(W)$. Let $\mathcal{T}'_{ou} = (F'_{ou}, \emptyset)$. Then:

Claim 8. $\mathbf{H}'(W) = \mathbf{M}(\mathcal{T}'_{ou}, (W, \emptyset))$. *(Pair, 1970).*

Let us take an example in order to illustrate the theory $\mathcal{T}'_{ou}$.

The tree t shown below is the value of a unique term t' of $M(F'_{ou}, (W, \emptyset, \emptyset))$ also shown in the same figure:

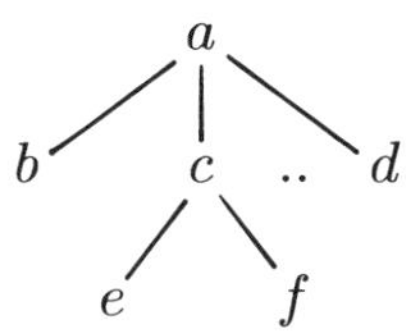

tree t

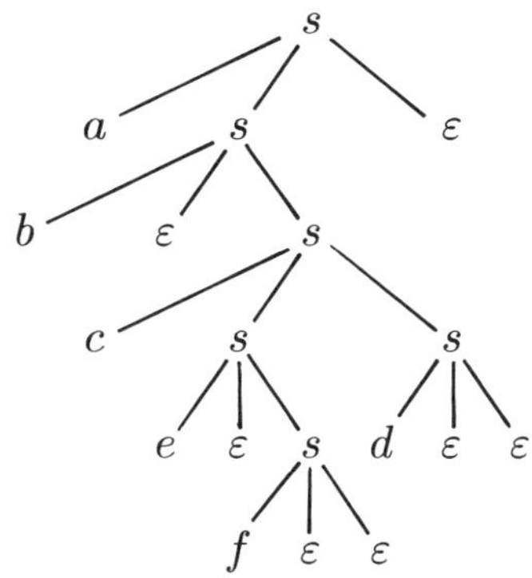

term t'

This shows that this correspondence is something very well known, the representation of an ordered tree by means of a binary tree with two functions: first son and right brother.

Going back to the properties of $\mathbf{Rec}(\mathcal{T}'_{ou})$ we have, since $E'_{ou} = \emptyset$:

Claim 9. $\mathbf{Rec}(\mathcal{T}'_{ou}) = \mathbf{Equat}(\mathcal{T}'_{ou})$ *and this family is closed under the homomorphisms and the operations of* F'_{ou}.

An immediate consequence is then:

Claim 10. $H(W)_{\mathbf{tree}} = s(W, H(W)_{\mathbf{hedge}}, \{\varepsilon\})$. *Hence, this set is both equational and recognizable w.r.t.* $\mathbf{H}'(W)$.

Proposition 17. *A subset* K *of* $H(W)_{\mathbf{hedge}}$ *is recognizable w.r.t.* $\mathbf{H}(W)$ *iff it is recognizable w.r.t.* $\mathbf{H}'(W)$.

Proof. The "only if" part is immediate because $\mathbf{H}'(W)$ is a derived magma of $\mathbf{H}(W)$. The "if" part can be proved by a generalization of the proof of the first equality of Proposition 2. □

The $\mathcal{T}'_{ou}$-automata define recognizable sets of trees, but they behave in a way that is not very intuitive in terms of trees.

5. Conclusion

This paper has presented the main properties of the algebraic notion of recognizability.

It leaves open some problems on term rewriting systems listed in Definition 4.

An important continuation would be an algebraic characterization of "ω-recognizability." This notion should coincide with the definability of sets of infinite words by Büchi-automata, of sets of infinite trees by Rabin-automata, and it should also work for arbitrary ω-complete magmas.

Appendix: Technical Proofs

Proof of Proposition 15. We shall prove that the inclusion of $\mathbf{Rec}(\mathbf{M}(\mathcal{T}_{uu}, (W, \emptyset, \emptyset)))$ in $\mathbf{Equat}(\mathbf{M}(\mathcal{T}_{uu}, (W, \emptyset, \emptyset)))$ is strict if W contains at least two symbols.

Let $W = \{a, b\}$. Let T be the set of trees of the form

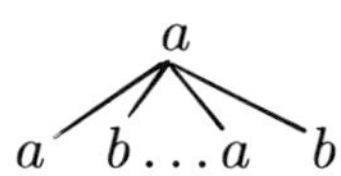

with as many leaves labeled by a and by b. It is equational. (It is not hard to find a recognizable subset of $M(F_{uu}, (W, \emptyset, \emptyset))$ of which it is the image by the canonical homomorphism $h : M(F_{uu}, (W, \emptyset, \emptyset)) \to \mathbf{M}(\mathcal{T}_{uu}, (W, \emptyset, \emptyset))$.)

Let $T' = h^{-1}(T) \subseteq M(F_{uu}, (W, \emptyset, \emptyset))$. We shall prove that T' is not recognizable. For every word $w = a_1 \ldots a_n$ in W^+(with $a_1, \ldots,$ $a_n \in W$), we let t_w be the term

$$r(p(a_1, \phi)) \uplus r(p(a_2, \phi)) \uplus \ldots (p(a_n, \phi)).$$

It denotes the forest $\{a_1, a_2, \ldots, a_n\}$. (Since $\uplus$ is associative, we use infix notation without parentheses.)

The set $R = \{p(a, (t_{a^n} \uplus t_{b^m}))/n \geq 1, m \geq 1\}$ is recognizable. The set $R \cap T'$ is equal to $\{p(a, (t_{a^n} \uplus t_{b^n}))/n \geq 1\}$ and hence, is not recognizable. Hence, T' is not recognizable. $\square$

Proof of Proposition 16. It says that if L and L' are two recognizable sets of forests, then $L \uplus L' = \{f \uplus f'/f \in L, f' \in L'\}$ is also recognizable. This proof is complicated. It will use $\mathcal{T}_{uu}$-automata and needs a few technical lemmas.

We let $T = Z(W)_{\mathbf{tree}}$ and $F = Z(W)_{\mathbf{forest}}$. A forest $f \in F$ can be considered as a mapping $f : T \to \mathbf{N}$ such that $f(t) \neq 0$ for finitely many trees t. In particular, $f \uplus f'$ is the function $f + f'$ such that $(f + f')(t) = f(t) + f'(t)$.

Let A be a $\mathcal{T}_{uu}$-semi-automaton of the form $\langle F_{uu}, (W, \emptyset, \emptyset), (Q_1, Q_2, Q_3), \delta\rangle$. We denote by δ^* the associated homomorphism of $\mathbf{Z}(W)$ into the F_{uu}-magma with domains (Q_1, Q_2, Q_3) canonically associated with A.

With a forest $f \in F$, we associate the mapping $\tilde{f} : Q_2 \to \mathbf{N}$ such that:

$$\tilde{f}(q) = \sum\{f(t)/\delta^*(t) = q\}.$$

Since $\delta(\uplus) : Q_3 \times Q_3 \to Q_3$ is associative and commutative, it will be denoted as an addition, and its extension to a finite set will be denoted by $\sum$. The notation $n.q$ for $n \in \mathbf{N}, q \in Q$ will be used as an abbreviation of $\delta(\uplus)(q, \delta(\uplus)(q, \ldots, \delta(\uplus)(q, q)) \ldots))$ with n times q. We use $0.q$ for $\delta(\phi)$ which is a unit for $\delta(\uplus)$ since A is a $\mathcal{T}_{uu}$-automaton.

For $g \in \mathbf{N}^{Q_2}$ (the set of mappings : $Q_2 \to \mathbf{N}$), we let:

$$\tilde{\delta}^*(g) = \sum\{g(q).\delta(r)(q)/q \in Q_2\} \in Q_3.$$

Fact 1. *For all* $f \in F, \delta^*(f) = \tilde{\delta}^*(\tilde{f})$.

Since $\tilde{\delta}^*$ is a homomorphism $\mathbf{N}^{Q_2} \to Q_3$ (w.r.t. the additive structures of $\mathbf{N}^{Q_2}$ and Q_3), the set $D(q) := \{g \in \mathbf{N}^{Q_2}/\tilde{\delta}^*(g) = q\}$ for $q \in Q_3$ is a recognizable subset of $\mathbf{N}^{Q_2}$. Hence, if $Q' \subseteq Q_3$ is a set of final states for A, then:

$$\{f \in F/\delta^*(f) \in Q'\} = \{f \in F/\tilde{f} \in K\}$$

where K is the $\mathbf{N}^{Q_2}$-recognizable set $\cup\{D(q)/q \in Q'\}$.

Conversely, it is not difficult to establish:

Fact 2. *For every* $K \in \mathbf{Rec}(\mathbf{N}^{Q_2})$, *the set* $\{f \in F/\tilde{f} \in K\}$ *is* $\mathbf{Z}(W)$-*recognizable.*

The last technical result we shall need is

Fact 3. *If K and $K' \in \mathbf{Rec}(\mathbf{N}^{Q_2})$, then $K + K' \in \mathbf{Rec}(\mathbf{N}^{Q_2})$.*

Proof. Observe that the recognizable subsets of $\mathbf{N}$ are the ultimately periodic sets. It follows that $K + K' = \{n + n'/n \in K, n' \in K'\} \in \mathbf{Rec}(\mathbf{N})$ if $K, K' \in \mathbf{Rec}(\mathbf{N})$.

For every product monoid $\mathbf{M} \times \mathbf{M}'$, $\mathbf{Rec}(\mathbf{M} \times \mathbf{M}')$ is the set of finite unions of sets of the form $L \times L'$ where $L \in \mathbf{Rec}(\mathbf{M}), L' \in \mathbf{Rec}(\mathbf{M}')$. (See Eilenberg (1974); this result does not hold for arbitrary magmas $\mathbf{M}$ and $\mathbf{M}'$.) It follows that $\mathbf{Rec}(\mathbf{N}^{Q_2})$ is the set of finite unions of products of ultimately periodic sets.

So consider $K \in \mathbf{Rec}(\mathbf{N}^{Q_2})$. Let $q = \mathbf{Card}(Q_2)$. There are $n \in N$, and sets $A_{i,j} \in \mathbf{Rec}(\mathbf{N})$ for $i \in [n], j \in [q]$ such that $K = \cup\{A_{i,1} \times \cdots \times A_{i,q}/i = 1, \ldots, n\}$. Similarly $K' = \cup\{A'_{j,1} \times \cdots \times A'_{j,q}/j = 1, \ldots, n'\}$. Hence $K + K' = \cup\{(A_{i,1} \times \cdots \times A_{i,q}) + (A'_{j,1} \times \cdots \times A'_{j,q})/i = 1, \ldots, n, j = 1, \ldots, n'\}$. But for each pair (i, j):

$$\begin{aligned}(A_{i,1} \times \cdots \times A_{i,q}) &+ (A'_{j,1} \times \cdots \times A'_{j,q}) \\ &= (A_{i,1} + A'_{j,1}) \times (A_{i,2} + A'_{j,2}) \times \cdots \times (A_{i,q} + A'_{j,q}).\end{aligned}$$

Each component of this product is in $\mathbf{Rec}(\mathbf{N})$ by the initial remark. Hence, the result follows from the characterization of $\mathbf{Rec}(\mathbf{N}^{Q_2})$. □

We can now complete the proof of the proposition.

Let L and L' be two recognizable sets of forests over W. By the proof of Proposition 4, there exists a semi-automaton A as above and two subsets Q and Q' of Q_3 such that $L = \{f \in F/\delta^*(f) \in Q\}$ and $L' = \{f \in F/\delta^*(f) \in Q'\}$. There exist two recognizable subsets of $\mathbf{N}^{Q_2}$, K and K' such that $L = \{f/\tilde{f} \in K\}$ and $L' = \{f/\tilde{f} \in K'\}$. Hence, $L \uplus L' = \{f \uplus f'/f \in L, f' \in L'\} = \{f \in F/\tilde{f} \in K + K'\}$. This set is recognizable by Facts 3 and 2. □

Acknowledgements

I thank A. Arnold, A. Podelski, and J. C. Raoult for their helpful comments on the first draft of this paper. I thank K. Callaway who helped me with the English style.

References

Bauderon, M., and Courcelle, B., (1987). "Graph expressions and graph rewritings," *Mathematical Systems Theory* **20**, 83-127.

Bossut, F., Dauchet, M., and Warin, B., (1987). Rationality and recognizability on planar dags, Report IT 108-87, Lille-1 University.

Condillac, E., (1798). "Traité des Systèmes," Paris.

Courcelle, B., (1986a). "Equivalences and transformations of regular systems. Applications to recursive program schemes and grammars," *Theor. Comp. Sci.* **42**, 1–122.

Courcelle, B., (1986b). Recognizability and second order definability for sets of finite graphs, Research report 8634, Bordeaux I University. Submitted for publication.

Courcelle, B., (1987a). "A representation of graphs by algebraic expressions and its use for graph rewriting systems," *Proceedings of the 3rd International Workshop on Graph Grammars* (H. Ehrig and G. Rozenberg, eds.), L.N.C.S. **291**, Springer, 112–132.

Courcelle, B., (1987b). "On context-free sets of graphs and their monadic second-order theory," *Proceedings of the 3rd International Workshop on Graph Grammars* (H. Ehrig and G. Rozenberg, eds.), L.N.C.S. **291**, Springer, 133–146.

Courcelle, B., (1987c). "An axiomatic definition of context-free rewriting and its application to NLC graph grammars," *Theoretical Computer Science*, **55**, 141–181.

Dauchet, M., Tison, S., Heuillard, T., and Lescanne, P., (1987). "Decidability of the confluence of ground term rewriting systems," Second conference on Logic in Computer Science, Ithaca, 1987 Proceeding published by IEEE Computer Society, 353–360.

Dubinski, A., (1975). Computation on arbitrary algebras. In "λ-calculus and Computer Science theory," *Lec. Notes Comp. Sci.* **37**, Springer, 319–341.

Eilenberg, S., (1974). *Automata, languages and machines* **A**, Academic Press.

Eilenberg, S., and Wright, J., (1967). "Automata in general algebras," *Information and Control* **11**, 52–70.

Elgot, C., and Mezei, J., (1965). "On relations defined by generalized finite automata," *IBM J. Res. Dev.* **9**, 47–68.

Engelfriet, J., (1975). "Bottom-up and top-down tree transformations, a comparison," *Math. Systems Theory* **9**, 198–231.

Gautam, N., (1957). "The validity of equations in complex algebras," *Arch. Math. Logik Grundlag.* **3**, 117–124.

Gecseg, F., and Steinby, M., (1984). "Tree automata," Akademiai Kiado, Budapest.

Ginsburg, S., and Rice, H., (1962). "Two families of languages related to ALGOL," *JACM* **9**, 350–371.

Habel, A., and Kreowski, H.J., (1987). "Some structural aspects of hypergraph languages generated by hyperedge replacements," *Proceedings of STACS 1987*, Lec. Notes Comp. Sci. **247**, Springer, 207–219.

Horowitz, E., and Sahni, S., (1976). *Fundamentals of data structures.* Pitman, London, G.B.

Huet, G., and Oppen, D., (1980). "Equations and rewrite rules, a survey," In *Formal Languages, Perspectives and Open Problems* (R. Book, ed.), Academic Press.

Kamimura, T., and Slutzki, G., (1982). "Transductions of dags and trees," *Math. Systems Theory.* **15**, 225–249.

Lescanne, P., (1982). "Modèles non déterministes des types abstraits," *RAIRO Inform. Théor.* **16**, 225–244.

Métivier Y., (1986). "On Recognizable subsets of free partially commutative monoids," 13th *ICALP, Rennes, Lec. Notes Comp. Sci.* **226**, Springer, 254–264.

Mezei, J., and Wright, J., (1967). "Algebraic automata and context-free sets," *Information and Control* **11**, 3–29.

Nivat, M., Perrin, D. (eds.), (1985). "Automata on infinite words," *Lec. Notes Comp. Sci.* **192**, Springer, 56–67.

Ochmanski, E., (1985), "Regular behaviour of concurrent systems," *Bull. of EATCS*, no. **27**, October, 56–67.

Pair, C., (1970). "Sur des notions algébriques liées à l'analyse syntaxique," *R.I.R.O.*, $4^{\text{ième}}$ année, R-3, 3–29.

Petrov, S., (1978). "Graph grammars and automata (survey)," *Automation and Remote Control* **39**, 1034–1050.

Rabin, M., Scott, D., (1959). "Finite automata and their decision problems," *IBM Journal of Res. and Dev.*, April 114–125.

Schützenberger, M., (1963). "On context-free languages and push-down automata," *Inf. and Control* **6**, 246–264.

Scott, D., (1976). "Data types as lattices," *SIAM Jour. on Computing* **5**, 522–587.

Steinby, M., (1981). Some algebraic aspects of recognizability and rationality, In "Fundamentals of Computation Theory," *Lec. Notes Comp. Sci.* **117**, Springer, 360–372.

Tison, S., (1987). Unpublished communication to the author (November).

4 The Idea of a Diagram

DESMOND FEARNLEY-SANDER

Department of Mathematics
University of Tasmania
Hobart, Australia

> ... although [mathematicians] make use of the visible forms and reason about them, they are thinking not of these, but of the ideals which they resemble; not of the figures which they draw, but of the absolute square and the absolute diameter, and so on—the forms which they draw or make are converted by them into images, but they are really seeking to behold the things in themselves, which can only be seen with the eye of the mind.
>
> —Plato, *The Republic*, Book VI

A reasonable programming language for geometry would

- implement some kind of uniform representation of geometric entities;
- transparently organize geometric entities into natural hierarchical structures;
- be capable of deducing properties of geometric entities from their specification alone;
- be able to compute geometric quantities such as volume;
- embody a formalism which is terse, natural and preferably resembles traditional mathematical usage;
- transparently avoid coordinates whenever possible, but use them when necessary;
- make use of the natural geometric typing that, for example, distinguishes points from vectors;
- handle transparently the symmetries inherent in typical geometric propositions;

ISBN 0-12-046370-9

- be powered by a strong theorem prover, which in the course of attempting a proof can uncover more useful information than merely whether the attempt succeeds or fails;
- cope painlessly with the assumptions of genericity that are often implicit in geometric statements;
- be suited to an interactive implementation (in accordance with the natural way of applying geometry, for example, in CAAD);
- interface smoothly with graphic and logic languages.

The critical problem in the development of such a language is the first of these items—how to represent geometric knowledge, and, in particular, capture the mathematician's notion of a diagram. This paper focuses on a solution of this problem. It is presented in terms of a particular formalism, but will be seen to be meaningful for other algebraic formalisms. Examples are given of how theorems of affine geometry—properties of diagrams—may be proved by equational reasoning. The ideas have been partially implemented (see Fearnley-Sander (1986)).

1. Grassmann Algebras

The basic theory of Grassmann algebras, on which the formalism is based, is given (with minor differences in terminology from the present paper) in Fearnley-Sander (1980).

A *Grassmann algebra* is a ring which is generated by (the union of) disjoint subsets $\mathcal{K}$ and $\mathcal{P}$ such that

GA1 $\mathcal{K}$ is a field (under the ring operations);
GA2 $\mathcal{P}$ is an affine space over $\mathcal{K}$ (under the ring operations);
GA3 $aA = Aa$ for every $a \in \mathcal{K}$, $A \in \mathcal{P}$;
GA4 $BA = -AB$ for every $A, B \in \mathcal{P}$.

Here, **GA2** means that

$$A, B \in \mathcal{P}, a, b \in \mathcal{K} \text{ and } a + b = 1 \Rightarrow aA + bB \in \mathcal{P}. \qquad (1)$$

An important consequence of **GA4**, provided the characteristic of $\mathcal{K}$ is not 2, is

GA5 $A^2 = 0$ for every $A \in \mathcal{P}$.

It is easy to show that elements of $\mathcal{V} = \mathcal{P} - \mathcal{P}$ also have the properties **GA4** and **GA5**.

If there is a maximal finite set of elements of $\mathcal{P}$ which is linearly independent over $\mathcal{K}$, then the Grassmann algebra is said to be *finite-dimensional*, and the cardinality, less 1, of such a set is called its *dimension*. For a given field $\mathcal{K}$ and natural number n, there is, up to isomorphism, precisely one Grassmann algebra of dimension n.

In the present paper we take $\mathcal{K}$ to be the field of rational numbers.

2. Grassmann Geometry

Grassmann geometry is the language in which one speaks about Grassmann algebras. The *expressions* of Grassmann geometry are the terms generated by

point variables $A, B, C, \ldots$
vector variables $U, V, W, \ldots$
point constant O
vector constants $X = X_1, Y = X_2, Z = X_3, X_4, \ldots$
scalar constants $0, 1, -1, \frac{1}{2}, \ldots$ (rational numbers)
operation symbols $+, *, -$ (of arities 2,2,1)

We use infix notation for addition and multiplication, and usually indicate multiplication by juxtaposition. An *atom* is a variable or constant. A *ground* expression is one all of whose atoms are constants.

Rewrite rules corresponding to the Grassmann algebra axioms (together with "built-in" addition and multiplication tables for scalar constants) may be used to transform any expression into a sum of products of atoms. Indeed by ordering the generating atoms a unique *normal form* is obtainable. The set of normal expressions may be endowed with an algebraic structure having the Grassmann algebra signature in the natural way (by defining, for example, the sum of two normal expressions exp_1 and exp_2 to be the normal form of the expression $\mathsf{exp}_1 + \mathsf{exp}_2$). We call this the UNIVERSE. We refer to an expression in normal form as a *polynomial*, and to the equivalence class of all expressions having a common normal form as a *quantity*. A polynomial is a sum of products of atoms, and from now on we call these products the *terms* of the polynomial (and refer to elements of the term algebra as expressions).

The *degree* of a term is the number of point and vector atoms occurring in it. If all the terms in the normal form of an expression have the same degree, then the expression is said to be *homogeneous*.

An expression of degree zero is called a *scalar*. A first degree homogeneous expression is called a *point*, if the sum of the coefficients of all the point atoms in its normal form is 1, and a *vector*, if this sum is 0. We denote by NUMBERS, POINTS and VECTORS the sets, respectively, of all scalar constants, of all points and of all vectors. This typing may be extended in a natural way to higher degree terms.

Taking $\mathcal{K}$ to be NUMBERS, $\mathcal{P}$ to be POINTS and $\mathcal{V}$ to be VECTORS, the UNIVERSE is a Grassmann algebra. For each natural number n, the subalgebra WORLD_n generated by $O, X_1, X_2, \ldots, X_n$ is a Grassmann algebra of dimension n; we call it the *world of dimension n*. By convention, WORLD_0 is NUMBERS.

Although it is the UNIVERSE that we are concerned with, in practice it is handy to allow its elements to be represented in non-normal form since this makes interpretation easier—for example $(B-A)(D-C)$ is immediately recognizable as a product of two vectors, while its normal form $BD - BC - AD + AC$ is not.

A *configuration of type* (k, m) is a finite sequence of k points followed by m vectors—that is, an element of $\mathsf{POINTS}^k \times \mathsf{VECTORS}^m$. A configuration is called *atomic* (or *variable* or *constant*) if all its elements are atoms (or, respectively, variables or constants). This notion of configuration, which suffices for the current purpose, can usefully be generalized.

3. Equations

An *equation* is simply a monic homogeneous polynomial *poly*. It is convenient to allow it to be represented by any pair of expressions $(\mathsf{exp}_1, \mathsf{exp}_2)$ such that $\mathsf{exp}_1 - \mathsf{exp}_2$ represents the same quantity as some non-zero constant scalar multiple of poly; and to write such a pair in the familiar form

$$\mathsf{exp}_1 = \mathsf{exp}_2. \tag{2}$$

Particularly useful representations are the *polynomial representation*

$$\mathsf{poly} = 0 \tag{3}$$

and the *head-body representation*

$$\mathsf{head} = \mathsf{body}, \tag{4}$$

where head is the maximal product term of *poly*.

A *labelled object* is a homomorphism of the UNIVERSE into itself which maps finitely many atomic variables (say, $\mathsf{var}_1, \mathsf{var}_2, \ldots, \mathsf{var}_k$) to constants ($\mathsf{val}_1$, $\mathsf{val}_2, \ldots$, val_k) and leaves fixed all other variables; it is completely determined by the set of equations

$$\mathsf{var}_1 = \mathsf{val}_1, \mathsf{var}_2 = \mathsf{val}_2, \ldots , \mathsf{var}_k = \mathsf{val}_k. \tag{5}$$

We denote by LABELLINGS the set of all labelled objects.

If σ is a labelled object, we call σ (exp) the *value of the expression exp on the labelled object.* If σ (exp) $= 0$, we say that the *expression vanishes on the labelled object.*

Example 1. The equations

$$\begin{aligned} A &= O \\ B &= O + X \\ C &= O + X + Y \\ D &= O + Y \end{aligned}$$

describe a labelled object. The expression $ABC - ACD$ vanishes on this labelled object. The expressions $(B - A)(C - A)$ and $AB + BC + CA$ have the same normal form; and on this labelled object, they have the same value, namely XY. $\square$

In general, expressions exp_1 and exp_2 have the same normal form (that is, represent the same quantity) if and only if they have the same value on every labelled object. Thus quantities have well-defined values on labelled objects.

An object labelling σ is said to *satisfy* (or be a *solution* of) an equation

$$\mathsf{exp}_1 = \mathsf{exp}_2 \tag{6}$$

if exp_1 and exp_2 have the same value on σ .

Example 2. The labelled object of the previous example satisfies the equations

$$ABC = ACD \tag{7}$$

and

$$C - A = X + Y. \square \tag{8}$$

We shall often need to speak of sets of equations, and coin the term *description* as an abbreviation. We denote by DESCRIPTIONS the set of all descriptions.

A labelled object is said to *satisfy* a description if it satisfies every equation in it. For a description eqns, we denote by SOLUTIONS(eqns) the set of all labelled objects that satisfy it.

A description eqns is called

- *true* or *valid* if SOLUTIONS (eqns) = LABELLINGS (that is, if every labelled object satisfies it), and invalid otherwise;
- *unsatisfiable* or *false* if SOLUTIONS (eqns) = $\emptyset$ (that is, if there exists no labelled object that satisfies it), and *satisfiable* otherwise.

If eqns consists of ground equations, then eqns is necessarily either true or false.

Type-checking alone can establish that individual equations are unsatisfiable; hence an implementation can reject attempts to input equations like $C - B + A = 0$ (because 0 is a vector while $C - B + A$ is not) and $AB = XY$ (because AB and XY are of different types).

4. Theorems

An *elementary assertion* is a pair of descriptions (hyps,conc); it is called an *elementary theorem*, and we say that hyps *semantically implies* conc, and write

$$\mathsf{hyps} \models \mathsf{conc}, \tag{9}$$

if SOLUTIONS(hyps) $\subseteq$ SOLUTIONS(conc) (that is, if every labelled object that satisfies hyps satisfies conc). Clearly hyps semantically implies conc if and only if hyps semantically implies each equation of conc. We say that an equation eqn is a *consequence* of a description hyps if $\mathsf{hyps} \models \{\mathsf{eqn}\}$. We denote by $\mathsf{CON}(\mathsf{hyps})$ the set of all consequences of hyps. Then CON is a closure operation on the space of all descriptions.

Example 3. It will be helpful to represent a typical assertion like $(M = \frac{1}{2}(A+B), AM = MB)$ in the more expansive notation:

$$\begin{array}{ll} \mathsf{hyp} & M = \frac{1}{2}(A+B) \\ \mathsf{conc} & AM = MB \end{array} \tag{10}$$

This is actually a theorem, as we show below. □

Two descriptions eqns_1 and eqns_2 are called *semantically equivalent* if

$$\mathsf{eqns}_1 \models \mathsf{eqns}_2 \text{ and } \mathsf{eqns}_2 \models \mathsf{eqns}_1 \tag{11}$$

(that is, if the sets of labelled objects that satisfy them are identical). Note that $\{A - B + C - D = 0\}$ and $\{P - Q + R - S = 0\}$ are *not* semantically equivalent.

For a set of labelled objects objs, the set of quantities

$$\begin{aligned} \mathsf{IDEAL}(\mathsf{objs}) &= \{\mathsf{poly} : \sigma \in \mathsf{objs} \Rightarrow \sigma(\mathsf{poly}) = 0\} \\ &= \{\mathsf{poly} : \text{ every } \sigma \in \mathsf{objs} \text{ satisfies } \mathsf{poly} = 0\} \end{aligned} \tag{12}$$

is an ideal in the $\mathsf{UNIVERSE}$. In this ring-theoretic notation,

$$\mathsf{CON}(\mathsf{eqns}) = \mathsf{IDEAL}(\mathsf{SOLUTIONS}(\mathsf{eqns})), \tag{13}$$

and the idcal generated by eqns is a subset (usually proper) of this ideal.

5. Diagrams

The fundamental entities of geometry are *diagrams*. We now define this simple but abstract notion.

A *diagram definition* is a pair $(\mathsf{varconfig}, \mathsf{eqns})$, consisting of a variable configuration $\mathsf{varconfig} = (\mathsf{pt}_1, \ldots, \mathsf{pt}_k, \mathsf{vec}_1, \ldots, \mathsf{vec}_m)$, say, and a description eqns. The *diagram* it defines is the map

$$\mathsf{POINTS}^k \times \mathsf{VECTORS}^m \to \mathsf{DESCRIPTIONS}, \tag{14}$$

whose value at any configuration

$$(\mathsf{ptexp}_1, \ldots, \mathsf{ptexp}_k, \mathsf{vecexp}_1, \ldots, \mathsf{vecexp}_m) \\ \in \mathsf{POINTS}^k \times \mathsf{VECTORS}^m \tag{15}$$

is the set of all consequences of the description obtained by replacing each occurrence of pt_1 in eqns by ptexp_1, each occurrence of pt_2 by ptexp_2, and so on.

We use an obvious notation to denote the assignment of a name to a diagram. For example,

$$\mathsf{def} \qquad \underline{\mathsf{parallel}}(V, W) \longmapsto VW = 0, \tag{16}$$

means that $\underline{\mathsf{parallel}}$ is the name given to the diagram defined by the pair $((V, W), \{VW = 0\})$.

A *call* of the diagram $\underline{\mathsf{dgm}}$ is a pair consisting of the name $\underline{\mathsf{dgm}}$ and a configuration of the appropriate type; and, for example, the notation

$$\underline{\mathsf{parallel}}(U + V, W) \longmapsto (U + V)W = 0 \tag{17}$$

(lacking the annotation def) means that the value of the diagram $\underline{\mathsf{parallel}}$ (previously defined) at the configuration $(U + V, W)$ is the set $\mathsf{CON}\{(U + V)W = 0\}$. In a diagram call, the type of the configuration to which the diagram is applied must match the type that was (implicitly) declared in the definition.

We may also define a diagram in terms of other diagrams, as in the example

$$\mathsf{def} \qquad \underline{\mathsf{vertical}}(A, B) \longmapsto \underline{\mathsf{parallel}}(B - A, Y). \tag{18}$$

(The symbol $\longmapsto$ may be read as "rewrites to.") This allows economical storage of diagram definitions, and imposes a natural hierarchical structure on the set of diagrams. For example:

$$\begin{aligned}
&\mathsf{def} \quad \underline{\mathsf{parallelogram}}(A, B, C, D) \longmapsto A - B + C - D = 0 \\
&\mathsf{def} \quad \underline{\mathsf{wall}}(A, B, C, D) \longmapsto \underline{\mathsf{parallelogram}}(A, B, C, D), \\
&\qquad\qquad \underline{\mathsf{horizontal}}(A, B), \\
&\qquad\qquad \underline{\mathsf{vertical}}(B, C) \\
&\mathsf{def} \quad \underline{\mathsf{door}}(A, B, C, D) \longmapsto \underline{\mathsf{parallelogram}}(A, B, C, D), \\
&\qquad\qquad B = A + 3X, \\
&\qquad\qquad C = B + 7Y
\end{aligned} \tag{19}$$

$$\text{def} \quad \underline{\mathsf{house}}(A,B,C,D,E,F,G,H,K) \longmapsto \begin{aligned}[t] &\underline{\mathsf{wall}}(A,B,C,D),\\ &\underline{\mathsf{vertical}}(\tfrac{1}{2}(A+B),E),\\ &E-\tfrac{1}{2}(C+D)\\ &\quad =\tfrac{1}{2}(D-A),\\ &\underline{\mathsf{door}}(F,G,H,K),\\ &\tfrac{1}{2}(F+G)=\tfrac{1}{2}(A+B)\end{aligned}$$

While a diagram definition is a finite specification of a function, we may also conceive it in a procedural way—to show, for example, that a configuration (A, B) is vertical, show that the configuration $(B - A, Y)$ is parallel.

6. Realizations

We say that a configuration config *realizes* a diagram $\underline{\mathsf{dgm}}$ if $\underline{\mathsf{dgm}}$ (config) is true. If config is constant then $\underline{\mathsf{dgm}}$ (config) is necessarily either true or false, and so we may even view a diagram as a predicate.

A diagram, such as the one defined by

$$\text{def} \quad \underline{\mathsf{fred's\_front\_door}}(A,B,C,D) \longmapsto \begin{aligned}[t] &A=O,\\ &B=O+3X,\\ &C=O+3X+7Y,\\ &D=O+7Y,\end{aligned} \tag{20}$$

whose description is a labelled object, is called an *object*. There is an obvious one-to-one correspondence between objects and constant configurations, and it is sometimes convenient to slur the distinction between the two concepts. An object may be viewed as belonging to any world that contains all its constants. For example $\underline{\mathsf{fred's\_front\_door}}$ belongs to WORLD_2 (and to WORLD_3).

Example 4. While adding constraints to existing diagrams is a natural mode of definition it is not the only way to proceed. For example, $\underline{\mathsf{new\_door}}$ defined by

$$\begin{aligned} \text{def} \quad & \underline{\mathsf{new\_door}}(A,B,C,D,V)\\ &\longmapsto \underline{\mathsf{fred's\_front\_door}}(A-V,B-V,C-V,D-V)\\ &\longmapsto A=O+V,\\ &\qquad B=O+3X+V,\\ &\qquad C=O+3X+7Y+V,\\ &\qquad D=O+7Y+V,\end{aligned} \tag{21}$$

is not an object. The realizations of new_door are precisely the translates of fred's_front_door. □

7. Properties

We say a diagram call $(\underline{\mathsf{dgm}}_1, \mathsf{config}_1)$ *has the property* $(\underline{\mathsf{dgm}}_2, \mathsf{config}_2)$ if every labelled object that satisfies the description $\underline{\mathsf{dgm}}_1(\mathsf{config}_1)$ also satisfies the description $\underline{\mathsf{dgm}}_2(\mathsf{config}_2)$, or, in other words, if

$$\underline{\mathsf{dgm}}_1(\mathsf{config}_1) \models \underline{\mathsf{dgm}}_2(\mathsf{config}_2) \tag{22}$$

is a theorem. For example, with harmless abuse of language, we may say that the labelled object $\underline{\mathsf{fred's\_front\_door}}(B, C, D, E)$ has the property $\underline{\mathsf{parallel}}(C - B, E - D)$.

We say that $\underline{\mathsf{dgm}}_1$ is_a $\underline{\mathsf{dgm}}_2$ if every configuration that realizes $\underline{\mathsf{dgm}}_1$ also realizes $\underline{\mathsf{dgm}}_2$ (or, equivalently, if, for every configuration config of appropriate type, $(\underline{\mathsf{dgm}}_1, \mathsf{config})$ has the property $(\underline{\mathsf{dgm}}_2, \mathsf{config})$); for example, according to our definitions, a door is_a wall. This relation is a partial order on the set of all diagrams, and objects (being those diagrams that are realized by just one configuration) are precisely the minimal diagrams under this relation.

Example 5.

The interplay of the different concepts that we have defined is exemplified by the equivalence (given the relevant diagram definitions) of the following statements:

- the configuration $(O, O + 3X, O + 3X + 7Y, O + 7Y)$ realizes the diagram door;
- for every variable configuration (A, B, C, D) in POINTS^4, the labelled object given by $\{A = O, B = O + 3X, C = O + 3X + 7Y, D = O + 7Y\}$ is a solution of the description $\{A - B + C - D = 0, B = A + 3X, C = B + 7Y\}$;
- $\{A = O, B = O + 3X, C = O + 3X + 7Y, D = O + 7Y\} \models \{A - B + C - D = 0, B = A + 3X, C = B + 7Y\}$;
- the assertion $(\underline{\mathsf{fred's\_front\_door}}(A, B, C, D), \underline{\mathsf{door}}(A, B, C, D))$ is a theorem;
- for every configuration $(\mathsf{exp}_1, \mathsf{exp}_2, \mathsf{exp}_3, \mathsf{exp}_4) \in \mathsf{POINTS}^4$, $\underline{\mathsf{fred's\_front\_door}}(\mathsf{exp}_1, \mathsf{exp}_2, \mathsf{exp}_3, \mathsf{exp}_4)$ has the property $\underline{\mathsf{door}}(\mathsf{exp}_1, \mathsf{exp}_2, \mathsf{exp}_3, \mathsf{exp}_4)$;

- fred's_front_door is_a door. □

Quite generally, a configuration realizes a diagram dgm if and only if the associated object is_a dgm.

8. Proof

The statement of an elementary assertion may be abbreviated (and its geometric meaning made more obvious) by allowing some of the equations to be replaced by diagram calls that produce them. Then elementary assertions have the form of Horn clauses with diagrams as explicit predicates and equations as implicit ones.

Thus there may be three different types of sentence involving predicates like wall—definitions, hypotheses (definition calls) and conclusions (or queries); and we also permit un-named equations as hypotheses or conclusion. Equations may be used directly, on a once off basis, as hypotheses of a theorem, or, for repeated use, may be incorporated in a data-base of diagrams. Current diagram definitions provide a global environment while un-named equations used as hypotheses provide a local one.

The simplest theorems are those with no hypotheses—they assert that a particular configuration realizes a diagram. A very basic mechanism available for proof is rewriting to normal form—for a single equation is valid if and only if its polynomial form is $0 = 0$. Rewriting alone gives, for example,

$$\begin{aligned} \text{conc} \quad & \underline{\mathsf{parallelogram}}(O, O+X, O+X+Y, O+Y) \\ & \longmapsto (O+Y)-(O+X+Y)+(O+X)-O=0 \quad (23) \\ & \longmapsto 0=0; \end{aligned}$$

in other words, the points $O, O+X, O+X+Y, O+Y$ form a parallelogram. Variable configurations give more interesting theorems:

$$\begin{aligned} \text{conc} \quad & \underline{\mathsf{parallelogram}}(\tfrac{1}{2}(P+Q), Q, \tfrac{1}{2}(Q+S), \tfrac{1}{2}(P+S) \\ & \longmapsto \tfrac{1}{2}(P+Q)-Q+\tfrac{1}{2}(Q+S)-\tfrac{1}{2}(P+S)=0 \quad (24) \\ & \longmapsto 0=0; \end{aligned}$$

in other words, for any points P, Q, S, the points $\frac{1}{2}(P+Q), Q, \frac{1}{2}(Q+S), \frac{1}{2}(P+S)$ form a parallelogram.

Another example:

$$\begin{aligned} \mathsf{conc} \quad & \underline{\mathsf{vertical}}(A + 7V, A + 7V + 5Y) \\ & \longmapsto (A + 7V + 5Y)Y - (A + 7V)Y = 0 \\ & \longmapsto 0 = 0; \end{aligned} \tag{25}$$

in other words, for any point A and vector V, the vector from $A+7V$ to $A + 7V + 5Y$ is vertical.

For our purposes a *proof* of an elementary assertion assert_1 may be taken to be a finite tree of elementary assertions with the following properties:

(1) the root is assert_1;
(2) every leaf has either a valid conclusion or unsatisfiable hypotheses (and hence is a theorem);
(3) every node is a theorem provided all its children are theorems.

In what follows we shall present some *inference rules*—they express how to determine legitimate children of a node. We shall sketch some sample proofs, but must leave the presentation of an efficient proof strategy for another occasion.

Rewriting of equations to normal form is, so to speak, a tacit inference rule (that reconciles our desire to work with polynomials with the practical need to represent them as expressions); we assume that it is automatically performed whenever new equations are created.

9. Reduction

An equation conc is said to be *reducible* via a description hyps if the head of some element of hyps occurs as a subproduct of some term of the conc; and then we say that conc *reduces to* conc′ *via* hyps if conc′ is the equation obtained from conc when some such occurrence of the head of an equation is replaced by its body. In a proof, the node (hyps, conc) may be followed by (hyps,conc′); we call this the **reduction inference rule**. It is sound by virtue of the replacement property of equality.

Example 6.

$$
\begin{array}{ll}
\textsf{hyp} & \underline{\textsf{parallel}}(B-A, D-C) \longmapsto BD-BC-AD+AC=0 \\
\textsf{conc} & \underline{\textsf{coplanar}}(A,B,C,D) \longmapsto ABCD=0
\end{array}
\tag{26}
$$

Reduction replaces the conclusion by

$$-AC(BC+AD-AC)=0 \tag{27}$$

which rewrites to $0=0$, completing a proof. □

Note: in such Examples the rewrite symbol $\longmapsto$ and the equations that follow it are not part of the theorem statement (and do not represent user input) but are included simply to remind the reader of how the hypotheses and conclusion are rewritten when a proof attempt commences (and even, as with coplanar above, to implicitly supply a definition).

Example 7.

$$
\begin{array}{ll}
\textsf{hyp} & \underline{\textsf{door}}(A,B,C,D) \\
 & \longmapsto A-B+C-D=0, B=A+3X, C=B+7Y \\
\textsf{conc} & \underline{\textsf{wall}}(A,B,C,D) \\
 & \longmapsto A-B+C-D=0, (B-A)X=0, \\
 & \qquad\qquad (C-B)Y=0
\end{array}
\tag{28}
$$

This is the assertion that a door is_a wall, and is immediately provable by reduction. □

Example 8.

$$
\begin{array}{ll}
\textsf{hyp}_1 & \underline{\textsf{parallelogram}}(A,B,C,D) \longmapsto A-B+C-D=0 \\
\textsf{hyp}_2 & \underline{\textsf{midpoint}}(A,C,P) \longmapsto P=\frac{1}{2}(A+C) \longmapsto 2P-C-A=0 \\
\textsf{hyp}_3 & \underline{\textsf{midpoint}}(B,D,Q) \longmapsto Q=\frac{1}{2}(B+D) \longmapsto 2Q-D-B=0 \\
\textsf{conc} & \underline{\textsf{coincides}}(P,Q) \longmapsto P-Q=0
\end{array}
\tag{29}
$$

Proof by reduction is straightforward:

$$P-Q=\tfrac{1}{4}(A+C)(B+D)=\tfrac{1}{4}(A+C)(A+C)=0. \ \square$$

For any finite set of equations hyps iterated reduction of any equation conc terminates after finitely many steps; however more than one irreducible equation may be obtainable. This may be remedied by critical pair completion. A variant of Buchberger's algorithm (Buchberger, 1985) can be used to replace the hypotheses by a semantically equivalent description via which reduction of any equation terminates uniquely. We refer to this process as *compilation*; though time-consuming, it need only be applied once to each diagram stored in the data-base.

Even after compilation, there may remain equations that are semantically implied by the hypotheses but cannot be deduced from them by reduction alone. Further inference rules are needed.

Example 9.
Compiling the hypotheses

$$\begin{array}{ll} \mathsf{hyp}_1 & AB = OX \\ \mathsf{hyp}_2 & BC = OY \end{array} \tag{31}$$

generates a new hypothesis

$$\mathsf{hyp}_3 \qquad OXC = OYA \tag{32}$$

which may be used to deduce, by reduction alone, the equation

$$\mathsf{conc} \qquad OXYC = 0, \tag{33}$$

but not, on the other hand, the equation

$$\mathsf{conc} \qquad ABC = OXY, \tag{34}$$

which is also implied by the hypotheses. □

10. Quantities

The expressiveness of Grassmann geometry springs, in part, from the fact that not only diagrams, but the quantities themselves may have geometrical meaning. For example $B-A$ is to be interpreted as the vector from A to B. Availability of both the notions of point and vector allows elementary theorems like the following to be succinctly and naturally expressed.

Example 10.

$$
\begin{array}{ll}
\mathsf{def} & \underline{\mathsf{vector\_triangle}}(U,V,W) \longmapsto U+V+W=0 \\
\mathsf{hyp}_1 & \underline{\mathsf{midpoint}}(B,C,D) \longmapsto D=\tfrac{1}{2}(B+C) \\
\mathsf{hyp}_2 & \underline{\mathsf{midpoint}}(C,A,E) \longmapsto E=\tfrac{1}{2}(C+A) \\
\mathsf{hyp}_3 & \underline{\mathsf{midpoint}}(A,B,F) \longmapsto F=\tfrac{1}{2}(A+B) \\
\mathsf{conc} & \underline{\mathsf{vector\_triangle}}(D-A,E-B,C-F) \\
 & \longmapsto A+B+C-D-E-F=0
\end{array}
\tag{35}
$$

Proof by reduction is trivial. □

An attempt to prove an assertion with conclusion

$$\mathsf{target}=0 \tag{36}$$

may be regarded more generally as seeking to compute the value of the **target** expression relative to the hypotheses. From this point of view a negative result may well convey useful information.

Example 11.

$$
\begin{array}{ll}
\mathsf{hyp} & \underline{\mathsf{door}}(A,B,C,D) \\
 & \longmapsto A-B+C-D=0, B=A+3X, C=B+7Y \\
\mathsf{conc} & AB+BC+CD+DA=0
\end{array}
\tag{37}
$$

The proof fails, concluding with the useful target value $42XY$—this shows that any door has area 21 (times the area of the "unit square"). There is a simple rationale for such interpretations, which we will not go into in the present paper. □

Example 12.
If the diagonals of a quadrilateral (A,B,C,D) have midpoints M and N, and a pair of its opposite sides meet at P, then the triangle

(P, M, N) has one quarter the area of (A, B, C, D). This is easily formalized as an elementary theorem:

$$\begin{array}{ll} \mathsf{hyp1} & \underline{\mathsf{collinear}}(A, D, P) \longmapsto AD + DP + PA = 0 \\ \mathsf{hyp2} & \underline{\mathsf{collinear}}(B, C, P) \longmapsto BC + CP + PA = 0 \\ \mathsf{hyp3} & \underline{\mathsf{midpoint}}(A, C, M) \longmapsto M = \tfrac{1}{2}(A + C) \\ \mathsf{hyp4} & \underline{\mathsf{midpoint}}(B, D, N) \longmapsto N = \tfrac{1}{2}(B + D) \\ \mathsf{conc} & 4(PM + MN + NP) = AB + BC + CD + DA \end{array} \tag{38}$$

And it is provable automatically by reduction:

$$\begin{aligned} &4(PM + MN + NP) \\ &\quad = 2P(A + C) + (A + C)(B + D) + 2(B + D)P \\ &\quad = 2(AD + DP + PA) - 2(BC + CP + PB) \\ &\qquad + (AB + BC + CD + DA) \\ &\quad = AB + BC + CD + DA. \end{aligned} \tag{39}$$

This is to be compared with the proof given in (Coxeter and Greitzer, 1967, page 55) in which two auxiliary points are created and two lemmas invoked. □

11. Instantiation

An expression of the form

$$\mathsf{num}_0 O + \mathsf{num}_1 X_1 + \cdots + \mathsf{num}_n X_n \tag{40}$$

has normal form 0 if and only if all the coefficients num_k are 0; in other words, the constants $O, X_1, \ldots, X_n$ are (linearly) independent over NUMBERS and hence form a basis for WORLD_n. A straightforward argument shows that ground points $\mathsf{pt}_1, \mathsf{pt}_2, \ldots, \mathsf{pt}_m$ (which necessarily belong to some finite-dimensional WORLD_n) are dependent if and only if

$$\mathsf{pt}_1 \mathsf{pt}_2 \ldots \mathsf{pt}_m = 0 \tag{41}$$

A useful "incremental" variant of this is the following result.

Theorem 1. *For ground points* $\mathsf{pt}_1,\mathsf{pt}_2,\ldots,\mathsf{pt}_m$

$$\mathsf{pt}_1\mathsf{pt}_2\ldots\mathsf{pt}_m = 0 \tag{42}$$

if and only if either

$$\mathsf{pt}_1\mathsf{pt}_2\ldots\mathsf{pt}_{m-1} = 0 \tag{43}$$

or there exist scalars $\mathsf{num}_1,\mathsf{num}_2,\ldots,\mathsf{num}_{m-1}$ *such that*

$$\mathsf{pt}_m = \mathsf{num}_1\mathsf{pt}_1 + \mathsf{num}_2\mathsf{pt}_2 + \cdots + \mathsf{num}_{m-1}\mathsf{pt}_{m-1} \tag{44}$$

Proof 1. Suppose that $\mathsf{pt}_1\mathsf{pt}_2\ldots\ \mathsf{pt}_m = 0$. As observed above, there exist scalars $\mathsf{num}_1,\mathsf{num}_2,\ldots,\mathsf{num}_m$, not all zero, such that

$$\mathsf{num}_1\mathsf{pt}_1 + \mathsf{num}_2\mathsf{pt}_2 + \cdots + \mathsf{num}_m\mathsf{pt}_m = 0. \tag{45}$$

If num_m is non-zero then the second of the stated alternatives holds. Otherwise one of $\mathsf{pt}_1,\mathsf{pt}_2,\ldots,\mathsf{pt}_{m-1}$ is a linear combination of the rest and hence $\mathsf{pt}_1\,\mathsf{pt}_2\ldots\ \mathsf{pt}_{m-1} = 0$.

The converse is obvious. □

In the latter case of the theorem, the sum of the coefficients, $\mathsf{num}_1,\mathsf{num}_2,\ldots,\mathsf{num}_m$ is necessarily 1.

These facts concerning the WORLD (which are valid in all finite-dimensional Grassmann algebras) are not captured by the algebraic semantics presented so far, but can be incorporated by the device of introducing scalar variables and an appropriate new inference rule.

11.1. Instantiation Inference Rule

Suppose that an equation of the form

$$P_1P_2\ldots P_m = 0 \tag{46}$$

may be deduced from an hypothesis of an assertion. Then the assertion may be replaced by a pair of assertions of which the first (called the *generic case*) is obtained by adding

$$P_k = a_1P_1 + a_2P_2 + \cdots + (1 - a_1 - a_2 - \cdots - a_{k-2})P_{k-1} \tag{47}$$

(where $a_1, a_2, \ldots, a_{k-2}$ are new scalar variables) to the hypotheses, and the second (the *exceptional case*) is obtained by adding

$$P_1P_2 \ldots P_{k-1} = 0 \tag{48}$$

to the hypotheses. □

This inference rule could be used to eliminate all hypotheses of degree greater than 1, but it is more efficient to confine its use to situations in which no reductions are possible.

The work of the prover may be considerably reduced, if desired, by making an assumption of genericity when appropriate: special relationships which are not implied by the data are assumed not to hold. This is analogous to the "negation as failure" rule.

In similar fashion, the fact that NUMBERS is a field is manifested by the **scalar division inference rule**: if an hypothesis of an assertion has the form

$$\mathsf{num.exp} = 0. \tag{49}$$

where num is a scalar expression and exp is an arbitrary expression, then the assertion may be replaced by a pair of assertions, of which the first (called the *generic case*) is obtained by replacing that hypothesis by

$$\mathsf{num} = 0, \tag{50}$$

and the second (called the *exceptional case*) by replacing it by

$$\mathsf{exp} = 0. \tag{51}$$

In particular, if exp is a non-zero ground polynomial, only the former case occurs.

Example 13.

$$\begin{array}{ll} \mathsf{hyp} & AM = MB \\ \mathsf{conc} & \underline{\mathsf{midpoint}}(A, B, M) \longmapsto M = \frac{1}{2}(A + B) \end{array} \tag{52}$$

The hypothesis implies that $ABM = 0$. No reduction of the target being possible, M is instantiated as $aA+(1-a)B$. Then hyp reduces to

$$(1 - 2a)AB = 0. \tag{53}$$

The assumption of genericity (namely that AB is nonzero) gives $a = \frac{1}{2}$ and hence $M = \frac{1}{2}(A + B)$. Finally, reduction of the target produces 0.

In this example, the assertion is also valid in the non-generic case as is easily proved:

$$\begin{array}{ll} \mathsf{hyp}_1 & AB = 0 \\ \mathsf{hyp}_2 & AM = MB \\ \mathsf{conc} & M = \frac{1}{2}(A + B) \end{array} \tag{54}$$

Instantiation produces, first, $B = A$ and, then, $M = A$, from which the conclusion follows immediately by reduction.

Since the converse (provable by reduction only) is also valid, the interpretation of the equation $AM = MB$ is forced. □

Example 14.

$$\begin{array}{ll} \mathsf{hyp}_1 & \underline{\mathsf{vertical}}(A, B) \longmapsto BY - AY = 0 \\ \mathsf{hyp}_2 & \underline{\mathsf{parallel}}(B - A, D - C) \longmapsto BD - BC - AD + AC = 0 \\ \mathsf{conc} & \underline{\mathsf{vertical}}(C, D) \longmapsto CY - DY = 0 \end{array} \tag{55}$$

Instantiating B as $A + bY$ reduces hyp_2 to

$$bDY - bCY = 0 \tag{56}$$

from which the conclusion follows provided that b is not 0. If $b = 0$ the hypotheses are eliminated, and the assertion is not proved. □

12. Inconclusive Proofs

An attempt to prove an assertion with satisfiable hypotheses succeeds if the final value computed for the target expression is 0 and fails if this final value is a non-zero constant. Otherwise the attempt is inconclusive; and this also may be useful.

Example 15.

$$\begin{array}{ll} \mathsf{hyp} & \underline{\mathsf{parallelogram}}(A, B, C, D) \longmapsto A - B + C - D = 0 \\ \mathsf{conc} & \underline{\mathsf{collinear}}(A, C, D) \longmapsto AC + CD + DA = 0 \end{array} \tag{57}$$

Although the attempt to prove this is inconclusive it terminates with the useful target value $AB+BC+CA$. This reveals that the assertion would become valid either if

$$AB + BC + CA = 0 \tag{58}$$

was added to the hypotheses (that is if A, B and C were collinear), or if the conclusion was replaced by

$$AC + CD + DA = AB + BC + CA \tag{59}$$

(that is by the conclusion that the triangles (A, C, D) and (A, B, C) have equal areas). Note that if with the same hypothesis an attempt is made to prove that $AB + BC + CA = 0$ the returned target value is $AB+BC+CA$, which reveals nothing; the user is able to exercise some degree of control by the ordering he sets up (by the names he gives them) for the points of the assertion. □

Example 16.

$$\begin{array}{ll} \mathsf{hyp}_1 & Q = O + X \\ \mathsf{hyp}_2 & R = O + 3X + 2Y \\ \mathsf{hyp}_3 & P = O + xX + yY \\ \mathsf{conc} & \underline{\mathsf{collinear}}(P, Q, R) \longmapsto PQ + QR + RP = 0 \end{array} \tag{60}$$

We must compute the value of

$$\mathsf{target} = PQ + QR + RP. \tag{61}$$

Reduction gives

$$\mathsf{target} = 2(y - x + 1)XY. \tag{62}$$

In the generic case the conclusion is false. In the exceptional case, where

$$y - x + 1 = 0 \tag{63}$$

the conclusion holds. Thus the equation of the line through P an d R is obtained. □

13. Coordinate Proofs

The following inference rule may be used to replace any assertion by an equivalent assertion of standard polynomial algebra over the rationals.

13.1. Coordinatization Inference Rule

An equation of an assertion that has the normal form

$$c_1 \mathsf{exp}_1 + \cdots c_n \mathsf{exp}_n = 0 \tag{64}$$

where $\mathsf{exp}_1, \ldots, \mathsf{exp}_n$ are ground product terms and $c_1, \ldots, c_n$ are scalars may be replaced by the set of equations

$$c_1 = 0, \ldots, c_n = 0. \quad \square \tag{65}$$

The soundness of this rule follows from the fact that each exp_i must be a product of distinct elements of the sequence $O, X_1, X_2, \ldots$ and that any finite set of distinct such products is necessarily independent.

To prove a theorem using coordinates is effectively to prove it at the semantic level. For example the coordinate proof below shows directly that any object labelling that satisfies the hypotheses of the assertion satisfies its conclusion.

Example 17. The elementary assertion

$$\begin{array}{ll} \mathsf{hyp}_1 & D = \frac{1}{2}(A+B) \\ \mathsf{hyp}_2 & E = \frac{1}{2}(A+C) \\ \mathsf{conc} & 4ADE = ABC \end{array} \tag{66}$$

is easily proved by reduction:

$$\begin{aligned} 4ADE - ABC &= A(A+B)(A+C) - ABC \\ &= 0. \end{aligned} \tag{67}$$

We may produce a coordinate proof by incorporating the following equations as additional hypotheses:

$$\begin{aligned} A &= O + a_1X_1 + a_2X_2 + a_3X_3 + a_4X_4 \\ B &= O + b_1X_1 + b_2X_2 + b_3X_3 + b_4X_4 \\ C &= O + c_1X_1 + c_2X_2 + c_3X_3 + c_4X_4 \\ D &= O + d_1X_1 + d_2X_2 + d_3X_3 + d_4X_4 \\ E &= O + e_1X_1 + e_2X_2 + e_3X_3 + e_4X_4 \end{aligned} \tag{68}$$

Pre-processing of the hypotheses followed by application of the coordinatization inference rule to each of them yields the coordinate constraints:

$$\begin{aligned} 2d_i - a_i - b_i &= 0, \qquad 1 \le i \le 4, \\ 2e_i - a_i - c_i &= 0, \qquad 1 \le i \le 4. \end{aligned} \tag{69}$$

And the conclusion, as one easily calculates, is equivalent to the scalar equations

$$\begin{aligned} 4(d_ie_j - d_je_i - a_ie_j + a_je_i + a_id_j - a_jd_i) \\ - (b_ic_j - b_jc_i - a_ic_j + a_jc_i + a_ib_j - a_jb_i) = 0, \\ 1 \le j < i \le 4. \end{aligned} \tag{70}$$

We see that proving this theorem is equivalent to a standard problem of inferring polynomial equations (6 of them, each homogeneous of degree 2) from polynomial equations (8 of them, each of degree 1) in several variables (20 of them) and could be solved, though with much unnecessary labour, by the Gröbner basis method (Buchberger, 1985). In practice one may well be happy with proofs that are valid at least in three dimensions; in that case only three basis vectors are needed and the complexity of the equivalent set of scalar equations is reduced accordingly. As a matter of fact, this example is necessarily planar (because the hypotheses entail that $ABCD = 0$ and $ABCE = 0$) and hence, if one could first somehow realize this fact, a genuine (not dimension-restricted) proof could be obtained by deriving just one second degree equation from 4 first degree equations in 10 scalar variables. □

In two dimensions, coordinate proofs become highly feasible as we know from the remarkable China prover (see Wen-tsün (1986), Chou (1984), and other papers of these authors). The point to be noted is that by going to coordinates Grassmann geometry computations are reduced to standard polynomial algebra; hence, what is known about the power of such methods is pertinent to Grassmann geometry. At the same time, the comparison makes evident the relative efficiency, even for such a simple example as the one above, of a proof method that handles geometric constraints directly without going to coordinates.

14. Conclusion

The specification of a diagram, as described here, is geometrical in character. To produce a picture, topological information must also be provided. We shall present elsewhere an extension of the formalism which allows this to be done in an elegant way; moreover topological properties can be proved by equational reasoning.

A limitation of the language as outlined is that it is confined to affine geometry. This seemed a reasonable discipline to impose in a paper whose purpose is to introduce the idea of a diagram. But the point may also be made that affine geometry is fundamental to applications such as computer-aided architectural design, and that it is in the nature of things that affine geometry can be more efficiently implemented than its extensions. In fact, though there is much work to be done, we believe that the Grassmann geometry formalism can be generalized to take in convex geometry, on the one hand, and metric geometries, on the other. But the idea of a diagram is independent of particular formalisms, and indeed is pertinent to areas other than geometry.

The pencil sketch which a mathematician draws when he is trying to prove a theorem about parallelograms is not, of course, a generic parallelogram, only a helpful representation of one. What, one may ask, is it that is so represented? The ideal parallelogram does not share all the properties of any given parallelogram, yet has the properties that are shared by all particular parallelograms. It might be thought that the set of all configurations that "are" parallelograms adequately represents the idea of a parallelogram, but this is not enough for practical purposes; for in order to speak about properties, points must be given names. At the same time, the properties themselves must be independent of the names—there must be names, but what they are does not matter. It is this conception that the notion of a diagram attempts to capture.

Acknowledgements

The author thanks Alan Robinson and Tim Stokes for conversations pertinent to the paper, and José Meseguer for drawing attention to several items in the bibliography.

Note: The bibliography, though not comprehensive, includes a representative selection of recent papers relevant to geometry theorem proving.

References

Arbab, Farhad, and Wing, Jeanette M. (1985). "Geometric reasoning: a new paradigm for processing geometric information," 80–89. In *International Symposium on New Directions in Computing*, IEEE Computer Science Press.

Borning, Alan (1981). "The programming language aspects of Thinglab, a constraint-oriented simulation laboratory." *ACM Trans. Prog. Languages and Systems*, **3**, 353–387.

Buchberger, Bruno (1985). "Gröbner bases: an algorithmic method in polynomial ideal theory," 184–232. In *Multi-dimensional Systems Theory* (N. K. Bose, ed.). Reidel, Dordrecht.

Chou, Shang-Chin (1984). "Proving elementary geometry theorems using Wu's algorithm," 243–286. In *Automated Theorem Proving: After 25 Years* (W. W. Bledsoe and D. W. Loveland, eds.). American Mathematical Society.

Chou, Shang-Chin, and Schelter, William F. (1986). "Proving geometry theorems with rewrite rules." *Automated Reasoning*, **2**, 153–273.

Chou, Shang-Chin, Schelter, William F., and Yang, Jin-Gen (1989). "Characteristic sets and Gröbner bases in geometry theorem proving." This volume.

Coelho, Helder, and Pereira, Luiz Moniz (1986). "Automated reasoning in geometry theorem proving with prolog." *Automated Reasoning*, **2**, 329–390.

Coxeter, H.S.M., and Greitzer, S.L. (1967). *Geometry Revisited.* Mathematical Association of America.

Fearnley-Sander, Desmond (1980). "Affine geometry and exterior algebra," *Houston Journal of Math.*, **6**, 53–58.

Fearnley-Sander, Desmond (1986). *Diagrams in Grassmann geometry.* Technical Report, University of Tasmania, Department of Mathematics.

Fearnley-Sander, Desmond (1985). *Using and computing symmetry in geometry proofs.* Research Paper 257, University of Edinburgh, Department of Artificial Intelligence.

Goguen, J.A. "Modular algebraic specification of some basic geometrical constructions." To appear.

Kapur, Deepar (1986). "Using Gröbner bases to reason about geometry problems." *J. Symbolic Computation*, **2**, 399–208.

Kutzler, B., and Stifter, S. (1986). "On the application of Buchberger's algorithm to automated geometry theorem proving." *J. Symbolic Computation*, **2**, 389-397.

van Wyck, Christopher J. (1982). "A high-level language for specifying pictures." *ACM Trans. Graphics*, **1**, 163–182.

Wen-tsün, Wu (1986). "Basic principles of mechanical theorem-proving in geometries." *J. Automated Reasoning*, **2**, 221–252.

5 Rigid E-Unification and Its Applications to Equational Matings

JEAN GALLIER
WAYNE SNYDER

Department of Computer and Information Science
University of Pennsylvania
Philadelphia, Pennsylvania

STAN RAATZ

Rutgers University
Department of Computer Science
New Brunswick, New Jersey[1]

1. Introduction

In this paper, we show that the method of matings due to Andrews (1981) (and also investigated by Bibel (1979, 1981, 1982)) can be extended to (first-order) languages with equality, and prove that this extension is both sound and complete. This paper is an expanded version of Gallier *et al.* (1987a) where the results were stated without proof. Recall that the method of matings applies to formulae in negation normal form, and that it was introduced with two motivations in mind: to avoid breaking a formula into parts, which can result in loss of information about the global structure, and to avoid transforming it to clausal form, which can result in an exponential

[1]This research was partially supported by the National Science Foundation under Grant No. DCR-86-07156.

ISBN 0-12-046370-9

increase in the number of literals due to the repeated use of the distributive law $(P \lor (Q \land R)) \equiv ((P \lor Q) \land (P \lor R))$.

The method of matings exploits the fundamental property given by the Skolem-Herbrand-Gödel theorem (Andrews, 1981; Andrews, 1986; Gallier, 1986). In short, the unsatisfiability of a (universally) quantified sentence can be reduced to the unsatisfiability of a *quantifier-free formula*, modulo guessing a ground substitution. The crucial observation due to Andrews is that a quantifier-free formula (in nnf) is unsatisfiable iff certain sets of literals occurring in A (called *vertical paths*) are unsatisfiable. Matings come up as a convenient method for checking that vertical paths are unsatisfiable. Roughly speaking, a *mating* is a set of pairs of literals of opposite signs (mated pairs) such that all these (unsigned) pairs are globally unified by some substitution. The importance of matings stems from the fact that a quantifier-free formula A has a mating iff there is a ground substitution θ such that $\theta(A)$ is unsatisfiable.

The extension to *equational matings* is nontrivial, and requires proving a generalization of Andrews's version of the Skolem-Herbrand-Gödel theorem (Andrews, 1981, 1986). It also requires extending the concept of a mating so that an equational mating is a set of sets of literals (mated sets), where a mated set consists of several positive equations and a single negated equation (rather than pairs of literals as in Andrews's case), and a form of unification modulo equational theories, or E-unification, first studied by Plotkin (1972). A related extension is sketched (without proofs) in Bibel (1982, Section V.3, pp. 234–242). However, Bibel's method and ours differ significantly. This is because standard unification is used in Bibel's method, and so, it is usually necessary to include extra literals arising from instances of the equality axioms to the mated sets. On the other hand, our method uses a form of E-unification, and we *never* include any extra literals (arising from equality axioms) to our mated sets. For a detailed comparison of our method with others, see subsection 5.4.

The method of matings is an incremental proof (or refutation) procedure that interleaves two steps: quantifier-duplication steps, and search for matings. It is an *analytic* proof procedure in the sense of Smullyan (1968), and, even though Andrews did not present it in terms of Gentzen or Tableaux systems (Szabo, 1970; Gallier, 1986; Smullyan, 1968) it can easily be presented in any of these formalisms.

For example, Fitting's method of tableaux (Fitting, 1986) is close in spirit to the method of matings.

Checking that a family of mated sets is unsatisfiable, i.e. an equational mating, leads to an interesting and nontrivial problem. This problem, which is central to this method, is a restricted version of unification modulo equations.

Problem 1. Given $E = \{E_i \mid 1 \leq i \leq n\}$ a family of n finite sets of equations and $S = \{\langle u_i, v_i \rangle \mid 1 \leq i \leq n\}$ a set of n pairs of terms, is there a substitution θ such that, treating each set $\theta(E_i)$ as a set of *ground* equations (i.e. holding the variables in E_i "rigid"), $\theta(u_i)$ and $\theta(v_i)$ are provably equal from $\theta(E_i)$ for $i = 1, \ldots, n$?

Equivalently, is there a substitution θ such that $\theta(u_i)$ and $\theta(v_i)$ can be shown congruent from $\theta(E_i)$ by the congruence closure method for $i = 1, \ldots, n$ (Kozen (1976, 1977), Nelson and Oppen (1980), Downey, Sethi, and Tarjan (1980))?

A substitution θ solving problem 1 is called a *rigid E-unifier of S*, and a pair $\langle E, S \rangle$ such that S has some rigid E-unifier is called an *equational premating*. It will be shown in Section 12 that deciding whether a pair $\langle E, S \rangle$ is an equational premating is an NP-complete problem. Since the problem of deciding whether a family of mated sets forms an equational mating is precisely the problem of finding whether a pair $\langle E, S \rangle$ is an equational premating, the former problem is also NP-complete. Actually, this result is an easy extension of a simpler problem.

Problem 2. Given a finite set $E = \{u_1 \doteq v_1, \ldots, u_n \doteq v_n\}$ of equations and a pair $\langle u, v \rangle$ of terms, is there a substitution θ such that, treating $\theta(E)$ as a set of ground equations, $\theta(u) \overset{*}{\cong}_{\theta(E)} \theta(v)$, that is, $\theta(u)$ and $\theta(v)$ are congruent modulo $\theta(E)$ (by congruence closure)? The substitution θ is called a *rigid E-unifier of u and v*.

Example 1. Let $E = \{fa \doteq a,\ x \doteq fa\}$, and $\langle u, v \rangle = \langle gx, x \rangle$. Then, the substitution $\theta = [ga/x]$ is a rigid E-unifier of u and v. Indeed, $\theta(E) = \{fa \doteq a,\ ga \doteq fa\}$, and $\theta(gx)$ and $\theta(x)$ are congruent modulo $\theta(E)$, since

$$\begin{aligned} \theta(gx) = gga &\longrightarrow gfa && \text{using } ga \doteq fa \\ &\longrightarrow ga = \theta(x) && \text{using } fa \doteq a. \end{aligned} \tag{1}$$

Note that θ is not the only rigid E-unifier of u and v. For example, $[gfa/x]$ or more generally $[gf^na/x]$ is a rigid E-unifier of u and v. However, θ is more general than all of these rigid E-unifiers (in a sense to be made precise later). It will be shown in section 10 that rigid E-unification is NP-complete, and that there is always a finite set of most general rigid E-unifiers, called a complete set of rigid E-unifiers.

Note that any substitution θ satisfying the above problem is an E-unifier of u and v. However, the equations in E are used in a restricted fashion. Contrary to E-unification, in which there is no bound on the number of instances of the equations in E used to show that $\theta(u) \xleftrightarrow{*}_E \theta(v)$, in our situation, *only* the m instances in $\theta(E)$ can be used.

The solution to Problem (2) is a significant extension of a result of Kozen, who has shown that the problem is NP-complete when all equations in E are ground, (Kozen, 1981). We also show that even when u, v are ground, and all equations in E except one regular equation are ground, the problem is NP-complete.

Rigid E-unification is exciting because it eliminates one of the two aspects of undecidability associated with the method of equational matings, namely, that of E-unification. This is particularly important here, since even if E-unification is decidable for the set of all equations occurring in a formula in nnf, it is necessary to consider subsets of this set of equations, and the E-unification problem for any subset can be undecidable.

2. Preliminaries

This section contains a brief review of the main concepts used in this paper. As much as possible, we stick to the definitions used in the literature on the subject. More specifically, we will follow Huet and Oppen (1982), and Gallier (1986). The purpose of this section is mainly to establish the terminology and the notation, and it can be omitted by readers familiar with the literature. First, we review the basics of many-sorted languages.

Definition 1. A set S of *sorts* (or *types*) is any nonempty set. Typically, S consists of types in a programming language (such as *integer*, *real*, *boolean*, *character*, etc.). An *S-ranked alphabet* is a pair (Σ, ρ) consisting of a set Σ together with a function $\rho : \Sigma \to S^* \times S$

assigning a *rank* (u, s) to each symbol f in Σ. The string u in S^* is the *arity* of f and s is the *sort* (or *type*) of f. If $u = s_1 \dots s_n$, $(n \geq 1)$, a symbol f of rank (u, s) is to be interpreted as an operation taking arguments, the i-th argument being of type s_i and yielding a result of type s. A symbol of rank (e, s) (when u is the empty string) is called a *constant of sort* s. For simplicity, a ranked alphabet (Σ, ρ) is often denoted by Σ.

Next, we review the definition of tree domains and trees (or terms). Let $\mathbf{N}$ denote the set of natural numbers, and $\mathbf{N}_+$ the set of positive natural numbers.

Definition 2. A *tree domain* D is a nonempty subset of strings in $\mathbf{N}_+^*$ satisfying the conditions:

(1) For all $u, v \in \mathbf{N}_+^*$, if $uv \in D$ then $u \in D$.
(2) For all $u \in \mathbf{N}_+^*$, for every $i \in \mathbf{N}_+$, if $ui \in D$ then, for every j, $1 \leq j \leq i$, $uj \in D$.

For every $n \in \mathbf{N}$, let $[n] = \{1, 2, \dots, n\}$, and $[0] = \emptyset$.

Definition 3. Given an S-sorted ranked alphabet Σ, a Σ-*tree (or term) of sort* s is any function $t : D \to \Sigma$, where D is a tree domain denoted by $dom(t)$, and t satisfies the following conditions:

1) The root of t is labeled with a symbol $t(e)$ in Σ of sort s.
2) For every node $u \in dom(t)$, if $\{i \mid ui \in dom(t)\} = [n]$, then if $n > 0$, for each ui, $i \in [n]$, if $t(ui)$ is a symbol of sort v_i, then $t(u)$ has rank (v, s'), with $v = v_1 \dots v_n$, else if $n = 0$, then $t(u)$ has rank (e, s'), for some $s' \in S$.

Given a tree t and some tree address $u \in dom(t)$, the *subtree of* t *rooted at* u is the tree t/u whose domain is the set $\{v \mid uv \in dom(t)\}$ and such that $t/u(v) = t(uv)$ for all v in $dom(t/u)$.

Given two tree addresses $\alpha, \beta \in dom(t)$ in a tree t, α is an *ancestor* of β iff α is a prefix of β,[2] and α is a *proper ancestor* of β iff it is an ancestor of β and $\alpha \neq \beta$. Addresses α and β are *independent* iff neither one is an ancestor of the other.

[2]That is, $\beta = \alpha\gamma$, for some $\gamma \in \mathbf{N}_+^*$.

The set of all finite trees of sort s is denoted by T_Σ^s, and the set of all finite trees by T_Σ.

In this paper, it is assumed that for every S-sorted alphabet Σ, there is a distinguished sort *bool* $\in$S. Symbols of sort *bool* are called *predicate symbols.* Terms of sort *bool* will be interpreted as logical formulae.

Let $X = (X_s)_{s \in S}$ be an S-indexed family of countable set of variables. We can form the sets of trees $T_\Sigma^s(X)$ obtained by adjoining each set X_s to the set of constants of sort s. To prevent free algebras from having empty carriers (so that the Herbrand-Skolem-Gödel theorem holds), we assume that every sort is *nonvoid.* We say that a sort s is *nonvoid* iff either there is some constant of sort s, or there is some function symbol f of rank $\rho(f) = (s_1 \ldots s_n, s)$ such that $s_1, \ldots, s_n$ are nonvoid. Then, for every sort s, the set T_Σ^s is nonempty, and it is well known that for every set X, $T_\Sigma(X)$ is the free Σ-algebra generated by X (see Gallier, 1986). This allows us to define substitutions.

Definition 4. Given a term t, the set of variables occurring in t is the set $\{x \in X \mid \exists u \in dom(t),\ t(u) = x\}$, and it is denoted by $Var(t)$.

Definition 5. A *substitution* is any (S-sorted) function $\sigma : X \to T_\Sigma(X)$, such that, $\sigma(x) \neq x$ for only finitely many $x \in X$. Since $T_\Sigma(X)$ is the free Σ-algebra generated by X, every substitution $\sigma : X \to T_\Sigma(X)$ has a unique homomorphic extension $\hat{\sigma} : T_\Sigma(X) \to T_\Sigma(X)$. In the sequel, we will identify σ and its homomorphic extension $\hat{\sigma}$.

Definition 6. Given a substitution σ, the *support* (or *domain*) of σ is the set of variables $D(\sigma) = \{x \mid \sigma(x) \neq x\}$. The set of variables *introduced by* σ is the set of variables $I(\sigma) = \bigcup_{x \in D(\sigma)} Var(\sigma(x))$. Given a substitution σ, if its support is the set $\{x_1, \ldots, x_n\}$, and if $t_i = \sigma(x_i)$, $1 \leq i \leq n$, then σ is also denoted by $[t_1/x_1, \ldots, t_n/x_n]$. Given a term (or formula) r, we also denote $\sigma(r)$ as $r[t_1/x_1, \ldots, t_n/x_n]$.

Given a substitution σ and a set W of variables, the *restriction* of σ to W, denoted by $\sigma|_W$, is the substitution θ defined such that, $\theta(x) = \sigma(x)$ for all $x \in W$, and $\theta(x) = x$ for all $x \notin W$.

Definition 7. Given two substitutions σ and θ, their *composition* is the substitution denoted by $\sigma \circ \theta$, such that, for every variable x, $\sigma \circ \theta(x) = \hat{\theta}(\sigma(x))$ (the composition of the functions σ and $\hat{\theta}$).

A substitution σ is *idempotent* if $\sigma \circ \sigma = \sigma$. It is easily seen that σ is idempotent iff $D(\sigma) \cap I(\sigma) = \emptyset$.

The operation of tree replacement (or tree substitution) will also be needed.

Definition 8. Given two trees t_1 and t_2 and a tree address u in t_1, the *result of replacing* t_2 at u in t_1, denoted by $t_1[u \leftarrow t_2]$, is the function whose graph is the set of pairs

$$\begin{aligned} &\{(v, t_1(v)) \mid v \in \mathrm{dom}(t_1),\ u \text{ is not a prefix of } v\} \\ &\qquad \cup \{(uv, t_2(v)) \mid v \in \mathrm{dom}(t_2)\}, \end{aligned} \tag{2}$$

and it is only defined provided that the sort of the root of t_2 is equal to the sort of $t_1(u)$.

If $\{\alpha_1, \ldots, \alpha_k\}$ is a set of pairwise independent addresses in a tree t, given any k trees $t'_1, \ldots, t'_k$, the tree $t[\alpha_1 \leftarrow t'_1, \ldots, \alpha_k \leftarrow t'_k]$ is defined as $t[\alpha_1 \leftarrow t'_1] \ldots [\alpha_k \leftarrow t'_k]$, which is well defined, and does not depend on the order of the α_i.

We also quickly review formulae in negation normal form. For details, see Gallier (1986).

Definition 9. An atomic formula is a term of the form either $Pt_1 \ldots t_n$, where P is a predicate symbol of rank $(s_1 \ldots s_n, \mathit{bool})$ and each t_i is a term of sort s_i, or a term of the form $(t_1 \doteq t_2)$, where t_1 and t_2 are terms of some identical sort s. An atomic formula of the form $(t_1 \doteq t_2)$ is called an *equation of sort s*. A *literal* is either an atomic formula or the negation of an atomic formula.

Definition 10. Formulae in *negation normal form* (for short, formulae in *nnf*) are defined inductively as follows. A formula A is in nnf iff either

(1) A is a literal, or
(2) $A = (B \vee C)$, where B and C are in nnf, or
(3) $A = (B \wedge C)$, where B and C are in nnf, or
(4) $A = \forall x B$, where B is in nnf, or
(5) $A = \exists x B$, where B is in nnf.

A *quantifier-free* formula in nnf is obtained by applying only clauses (1)–(3), and a *universal* formula in nnf by applying only clauses (1)–(4).

Definition 11. Given a formula A (resp. a term t), the set of variables occurring free in A (resp. t) is denoted by $Var(A)$ (resp. $Var(t)$). A *ground term* t is a term such that $Var(t) = \emptyset$, and similarly a *ground formula* A is a quantifier-free formula such that $Var(A) = \emptyset$. A *ground substitution* σ is a substitution such that $\sigma(x)$ is a ground term for every variable x in the support of σ.

Finally, we review some concepts related to term rewriting.

Definition 12. Let $\Longrightarrow$ be a binary relation $\Longrightarrow\ \subseteq A \times A$ on a set A. The transitive closure of $\Longrightarrow$ is denoted by $\stackrel{+}{\Longrightarrow}$ and the reflexive and transitive closure of $\Longrightarrow$ by $\stackrel{*}{\Longrightarrow}$. The *converse* (or *inverse*) of the relation $\Longrightarrow$ is the relation denoted as $\Longrightarrow^{-1}$ or $\Longleftarrow$, defined such that $u \Longleftarrow v$ iff $v \Longrightarrow u$. The symmetric closure of $\Longrightarrow$, denoted by $\Longleftrightarrow$, is the relation $\Longrightarrow \cup \Longleftarrow$.

Definition 13. A relation $\succ$ on a set A is *Noetherian* or *well founded* iff there are no infinite sequences $\langle a_0, \ldots, a_n, a_{n+1}, \ldots\rangle$ of elements in A such that $a_n \succ a_{n+1}$ for all $n \geq 0$.[3]

Definition 14. A *preorder* $\preceq$ on a set A is a binary relation $\preceq\ \subseteq A \times A$ that is reflexive and transitive. A *partial order* $\preceq$ on a set A is a preorder that is also antisymmetric. The converse of a preorder (or partial order) $\preceq$ is denoted as $\succeq$. A *strict ordering* (or *strict order*) $\prec$ on a set A is a transitive and irreflexive relation. Given a preorder (or partial order) $\preceq$ on a set A, the strict ordering $\prec$ associated with $\preceq$ is defined such that $s \prec t$ iff $s \preceq t$ and $t \npreceq s$. Conversely, given a strict ordering $\prec$, the partial ordering $\preceq$ associated with $\prec$ is defined

[3]We warn the readers that this is not the usual way of defining a well founded relation in set theory, as for example in Levy (1979). In set theory, the condition is stated in the form $a_{n+1} \prec a_n$ for all $n \geq 0$, where $\prec = \succ^{-1}$. It is the dual of the condition we have used, but since $\prec = \succ^{-1}$, the two definitions are equivalent. When using well founded relations in the context of rewriting systems, we are usually interested in the reduction relation $\Longrightarrow$ and the fact that there are no infinite sequences $\langle a_0, \ldots, a_n, a_{n+1}, \ldots\rangle$ such that $a_n \Longrightarrow a_{n+1}$ for all $n \geq 0$. Thus, following other authors, including Dershowitz, we adopt the dual of the standard set theoretic definition.

such that $s \preceq t$ iff $s \prec t$ or $s = t$. The converse of a strict ordering $\prec$ is denoted as $\succ$. Given a preorder (or partial order) $\preceq$, we say that $\preceq$ is well founded iff $\succ$ is well founded.[4]

Definition 15. Let $\longrightarrow$ be a binary relation $\longrightarrow \subseteq T_\Sigma(X) \times T_\Sigma(X)$ on terms. The relation $\longrightarrow$ is *monotonic* iff for every two terms s, t and every function symbol f, if $s \longrightarrow t$ then $f(\ldots, s, \ldots) \longrightarrow f(\ldots, t, \ldots)$. The relation $\longrightarrow$ is *stable* (under substitution) if $s \longrightarrow t$ implies $\sigma(s) \longrightarrow \sigma(t)$ for every substitution σ.

Definition 16. A strict ordering $\prec$ has the *subterm property* iff $s \prec f(\ldots, s, \ldots)$ for every term $f(\ldots, s, \ldots)$ (since we are considering symbols having a fixed rank, the deletion property is superfluous, as noted in Dershowitz, 1987). A *simplification ordering* $\prec$ is a strict ordering that is monotonic and has the subterm property. A *reduction ordering* $\prec$ is a strict ordering that is monotonic, stable, and such that $\succ$ is well founded. With a slight abuse of language, we will also say that the converse $\succ$ of a strict ordering $\prec$ is a simplification ordering (or a reduction ordering). It is shown in Dershowitz (1987) that there are simplification orderings that are total on ground terms.

Definition 17. Let $E \subseteq T_\Sigma(X) \times T_\Sigma(X)$ be a binary relation on terms. We define the relation $\longleftrightarrow_E$ over $T_\Sigma(X)$ as the smallest symmetric, stable, and monotonic relation that contains E. This relation is defined explicitly as follows: Given any two terms $t_1, t_2 \in T_\Sigma(X)$, then $t_1 \longleftrightarrow_E t_2$ iff there is some variant[5] (s, t) of a pair in $E \cup E^{-1}$, some tree address α in t_1, and some substitution σ, such that

$$t_1/\alpha = \sigma(s), \quad \text{and} \quad t_2 = t_1[\alpha \leftarrow \sigma(t)]. \tag{3}$$

(In this case, we say that σ is a *matching substitution* of s onto t_1/α. The term t_1/α is called a *redex*.) Note that the pair (s, t) is used as a two-way rewrite rule (that is, non-oriented). In such a case, we denote the pair (s, t) as $s \doteq t$ and call it an *equation*. When

---

[4]Again, we caution our readers that in standard set theory it is $\prec$ that is well founded! However, our definition is equivalent to the standard set-theoretic definition of a well founded partial ordering.

[5]In what follows we shall assume that before a pair (i.e., an equation) is used it has been renamed apart from all variables in current use. This is essential to prevent clashes among the variables. Thus we shall always state that a *variant* of an equation is being used.

$t_1 \longleftrightarrow_E t_2$, we say that we have an *equality step*. It is well known that the reflexive and transitive closure $\overset{*}{\longleftrightarrow}_E$ of $\longleftrightarrow_E$ is the smallest stable congruence on $T_\Sigma(X)$ containing E. When we want to fully specify an equality step, we use the notation

$$t_1 \longleftrightarrow_{[\alpha, s \doteq t, \sigma]} t_2 \tag{4}$$

(where some of the arguments may be omitted). A sequence of equality steps

$$u = u_0 \longleftrightarrow_E u_1 \longleftrightarrow_E \ldots \longleftrightarrow_E u_{n-1} \longleftrightarrow_E u_n = v \tag{5}$$

is called a *proof* of $u \overset{*}{\longleftrightarrow}_E v$.

Definition 18. Given a finite set E of equations (ground or not) and any two terms u, v (ground or not), we use the notation $u \overset{*}{\cong}_E v$ to express the fact that, treating E as a set of *ground equations*, $u \overset{*}{\longleftrightarrow}_E v$. Equivalently, $u \overset{*}{\cong}_E v$ iff u and v can be shown congruent from E by *congruence closure* (Kozen (1976, 1977), Nelson and Oppen (1980), Downey, Sethi, and Tarjan (1980)) again, treating all variables as constants — they are considered *rigid.*

Definition 19. When a pair $(s, t) \in E$ is used as an oriented equation (from left to right), we call it a *rule* and denote it as $s \to t$. The *reduction relation* $\longrightarrow_E$ is the smallest stable and monotonic relation that contains E. We can define $t_1 \longrightarrow_E t_2$ explicitly as in Definition 17, the only difference being that (s, t) is a variant of a pair in E (and not in $E \cup E^{-1}$). When $t_1 \longrightarrow_E t_2$, we say that t_1 *rewrites* to t_2, or that we have a *rewrite step.* When we want to fully specify a rewrite step, we use the notation

$$t_1 \longrightarrow_{[\alpha, s \to t, \sigma]} t_2 \tag{6}$$

(where some of the arguments may be omitted).

When $Var(r) \subseteq Var(l)$, then a rule $l \to r$ is called a *rewrite rule*; a set of such rules is called a *rewrite system.* A *degenerate equation* is an equation of the form $x \doteq t$, where x is a variable and $x \notin Var(t)$, and a *nondegenerate equation* is an equation that is not degenerate.

Definition 20. Let $\longrightarrow \subseteq T_\Sigma(X) \times T_\Sigma(X)$ be a binary relation on $T_\Sigma(X)$. We say that $\longrightarrow$ is *Church Rosser* iff for all $t_1, t_2 \in T_\Sigma(X)$, if $t_1 \stackrel{*}{\longleftrightarrow} t_2$, then there is some $t_3 \in T_\Sigma(X)$ such that $t_1 \stackrel{*}{\longrightarrow} t_3$ and $t_2 \stackrel{*}{\longrightarrow} t_3$. We say that $\rightarrow$ is *confluent* iff for all $t, t_1, t_2 \in T_\Sigma(X)$, if $t \stackrel{*}{\longrightarrow} t_1$ and $t \stackrel{*}{\longrightarrow} t_2$, then there is some $t_3 \in T_\Sigma(X)$ such that $t_1 \stackrel{*}{\longrightarrow} t_3$ and $t_2 \stackrel{*}{\longrightarrow} t_3$. A term s is *irreducible w.r.t.* $\longrightarrow$ iff there is no term t such that $s \longrightarrow t$.

It is well known that a relation is confluent iff it is Church Rosser (Huet, 1980). We say that a rewrite system R is Noetherian, Church Rosser, or confluent, iff the relation $\longrightarrow_R$ associated with R given in Definition 19 has the corresponding property. We say that R is *canonical* iff it is Noetherian and confluent.

3. Preview of the Method

In this section, we present the method of equational matings informally. Given a rectified[6] universal sentence A_0 in nnf, the method works in an incremental fashion as follows. The formula A_0 will evolve in steps called *quantifier duplication steps*, and we denote this evolving formula by A. We will also need to refer to the quantifier-free formula $\hat{A}$ obtained from A by deleting the quantifiers, called an *amplification* of A_0.

Initially, $A := A_0$.

Step 1. Construct a set $vp(\hat{A})$ of sets of literals called vertical paths, associated with $\hat{A}$. The set $vp(\hat{A})$ is defined inductively as follows:

If $\hat{A}$ is a literal, then $vp(\hat{A}) = \{\{\hat{A}\}\}$;
If $\hat{A} = (B \wedge C)$, then $vp(\hat{A}) = \{\pi_1 \cup \pi_2 \mid \pi_1 \in vp(B), \pi_2 \in vp(C)\}$;
If $\hat{A} = (B \vee C)$, then $vp(\hat{A}) = vp(B) \cup vp(C)$.

Step 2. Find whether there is a substitution σ such that for every vertical path $\pi \in vp(\hat{A})$, $\sigma(\pi)$ is unsatisfiable. If step 2 succeeds, go to step 4. Otherwise, go to step 3.

---

[6]A formula A is *rectified* iff no variable occurs both free and bound in A, and distinct occurrences of quantifiers bind distinct variables. It is well known that every formula is equivalent to a rectified formula. It is also well known that for every formula A, one can construct a universal formula B, a *Skolem form* of A, such that A is unsatisfiable iff B is unsatisfiable (see Gallier, 1986).

Step 3. Choose some universal subformula $\forall xB$ of the current A, and replace it by $(\forall xB \wedge \forall xB)$. Then, rectify variables in this new formula, obtaining A'. Let $A := A'$. This step is called a *quantifier duplication step.* Go back to step 1.

Step 4. Stop, A_0 is unsatisfiable (and so are $\hat{A}$ and A).

If A_0 is unsatisfiable, this procedure will stop after a finite number of quantifier duplication steps when it succeeds in finding some substitution closing all vertical paths in step 2. Roughly speaking, a set consisting of certain subsets of vertical paths, such that these subsets are unsatisfiable under some substitution and span all vertical paths, is called an *equational mating.* The heart of the method of matings is to find such equational matings.

The difficult step in the presence of equality is step 2. What is difficult is not to check that a substitution closes all vertical paths — this can be done using the congruence closure algorithm — but to determine whether such a substitution exists at all. This problem is indeed decidable, but NP-complete. For languages without equality, the checking is reduced to the existence of a standard unifier, which is easy. Unfortunately, whether or not equality is present, the number of vertical paths to be checked may be exponential. The following example illustrates the method.

Example 2. Let

$$A = (\forall x(x^2 \doteq x) \wedge ((Pa^3 \wedge \neg Pa) \vee (Pb^2 \wedge \neg Pb))). \tag{7}$$

There are 2 vertical paths

$$\{(x^2 \doteq x), Pa^3, \neg Pa\}$$

and

$$\{(x^2 \doteq x), Pb^2, \neg Pb\} \tag{8}$$

depicted as follows:

$$\begin{array}{ccc} & (x^2 \doteq x) & \\ Pa^3 & & Pb^2 \\ \neg Pa & & \neg Pb \end{array} \tag{9}$$

It is clear that there is no substitution that closes both paths. However, the substitution $[a/x]$ closes the first path, and the substitution $[b/x]$ closes the second path. Hence, we perform an amplification step. We obtain

$$A' = ((\forall x_1(x_1^2 \doteq x_1) \wedge \forall x_2(x_2^2 \doteq x_2)) \wedge ((Pa^3 \wedge \neg Pa) \vee (Pb^2 \wedge \neg Pb))). \tag{10}$$

There are 2 vertical paths

$$\{(x_1^2 \doteq x_1), (x_2^2 \doteq x_2), Pa^3, \neg Pa\}$$

and

$$\{(x_1^2 \doteq x_1), (x_2^2 \doteq x_2), Pb^2, \neg Pb\} \tag{11}$$

depicted as follows:

$$\begin{array}{ccc} & (x_1^2 \doteq x_1) & \\ & (x_2^2 \doteq x_2) & \\ Pa^3 & & Pb^2 \\ \neg Pa & & \neg Pb \end{array} \tag{12}$$

This time, it is easy to see that the substitution $\sigma = [a/x_1, b/x_2]$ closes both vertical paths, using the fact that Pa^3 rewrites to Pa in two steps using the equation $a^2 \doteq a$, and that Pb^2 rewrites to Pb in one step using the equation $b^2 \doteq b$. Hence, A is unsatisfiable.

It should be noted that our method does not require the inclusion of extra literals corresponding to instances of equational axioms during the amplification process, contrary to Bibel's method (1982). In this sense, equality is "built in". In the following sections, we shall define the method precisely and prove its completeness rigorously.

4. Equational Matings

In order to generalize matings to equational languages, it is necessary to consider sets of literals rather than pairs, as in Andrews's case (1981). Let us first consider the case of *quantifier-free formulae* in negation normal form. The general case will be lifted from the quantifier-free case via the Skolem-Herbrand-Gödel theorem, and using rigid E-unification.

Let A be a quantifier-free formula, and let $\{x_1, \ldots, x_n\}$ be the set of variables occurring in A. The *universal closure of* A is the sentence $\forall x_1 \ldots \forall x_n A$. It is also denoted as $\forall A$. Testing the unsatisfiability of a quantifier-free formula A is much easier that testing the unsatisfiability of its universal closure $\forall A$. In the former case, the congruence closure method gives a decision procedure, whereas in the latter case, unsatisfiability is undecidable.

The crucial observation due to Andrews is that a quantifier-free formula in nnf is satisfiable iff some conjunction of literals occurring in A is satisfiable (see also Bibel 1979, 1981, 1982).

Definition 21. Given a quantifier-free formula A in nnf, the set $vp(A)$ of *vertical paths* in A is the set of sets of literals defined inductively as follows:

If A is a literal, then $vp(A) = \{\{A\}\}$;
If $A = (B \wedge C)$, then $vp(A) = \{\pi_1 \cup \pi_2 \mid \pi_1 \in vp(B), \pi_2 \in vp(C)\}$;
If $A = (B \vee C)$, then $vp(A) = vp(B) \cup vp(C)$.

Let us say that a vertical path π is satisfiable iff the conjunction of the literals in π is satisfiable. The following simple lemma shows the crucial role played by vertical paths.

Lemma 1. *Given a quantifier-free formula A in nnf, A is unsatisfiable iff every vertical path in A is unsatisfiable.*

Proof. Straightforward induction on the structure of A. □

A criterion for the unsatisfiability of a conjunction of literals based on the concept of congruence closure is known. In order to explain this criterion, it is convenient to represent every atomic formula as an equation. This can be done by adding to our language (which already contains the special sort *bool*) the constant $\top$ of sort *bool*, interpreted as **true**. Then, every atomic formula $Pt_1 \ldots t_n$ of sort *bool* can be expressed as the equation $(Pt_1 \ldots t_n \doteq \top)$. Hence, we can assume that all atomic formulae are equations. The notations $Pt_1 \ldots t_n$ and $(Pt_1 \ldots t_n \doteq \top)$ will be used interchangeably for atomic formulae of sort *bool*.

Given a vertical path π, let us arrange the literals in π by grouping positive and negative literals together, to form a conjunction C_π of the form

$$(s_1 \doteq t_1) \wedge \ldots \wedge (s_m \doteq t_m) \wedge \neg(s'_1 \doteq t'_1) \wedge \ldots \wedge \neg(s'_n \doteq t'_n). \quad (13)$$

Let $\overset{*}{\cong}_E$ be the congruence closure (Kozen (1976, 1977), Nelson and Oppen (1980)) on the graph $G(C_\pi)$ of the relation

$$E = \{(s_1, t_1), \ldots , (s_m, t_m)\}. \quad (14)$$

The following result is well known (see Kozen (1976), Nelson and Oppen (1980), or Gallier (1986)).

Lemma 2. *π is unsatisfiable iff for some j, $1 \leq j \leq n$, $s'_j \overset{*}{\cong}_E t'_j$.*

The concept of an E-unifier will be needed later.

Definition 22. Let E be a finite set of (universally quantified) equations. Given two terms u and v, we say that a substitution σ is a *unifier of u and v modulo E*, for short, an *E-unifier of u and v*, iff $\sigma(u) \overset{*}{\longleftrightarrow}_E \sigma(v)$.

The definition of an equational mating is motivated by Lemma 2, the Skolem-Herbrand-Gödel theorem (Theorem 1), and Lemma 4. Indeed, combining Theorem 1 and Lemma 4, we have that a universal sentence A in nnf is unsatisfiable iff there is some quantifier-free formula D (an amplification of A) and some substitution σ such that $\sigma(D)$ is unsatisfiable. The concept of an equational mating is designed so that we have a criterion expressed in terms of vertical paths for testing whether given a quantifier-free formula D, there is some substitution σ such that $\sigma(D)$ is unsatisfiable (see Lemma 3).

Definition 23. Let A be a quantifier-free formula in nnf. An *equational mating $\mathcal{M}$ for A* is a pair $\langle MS, \sigma\rangle$, where MS is a set of sets of literals called *mated sets* and σ is a substitution, such that, each mated set is a subset of some vertical path $\pi \in vp(A)$ and is of the form

$$\{(s_1 \doteq t_1), \ldots , (s_m \doteq t_m), \neg(s \doteq t)\} \subseteq \pi, \quad (15)$$

where $m \geq 0$,[7] and, for every mated set $\{(s_1 \doteq t_1), \ldots , (s_m \doteq t_m), \neg(s \doteq t)\} \in MS$, the set of literals $\{\sigma(s_1 \doteq t_1), \ldots , \sigma(s_m \doteq$

[7]The case $m = 0$ is indeed possible when $\sigma(s) = \sigma(t)$, i.e., when σ is a unifier of s and t.

$t_m), \neg\sigma(s \doteq t)\}$ is unsatisfiable. The substitution associated with the mating $\mathcal{M}$ is also denoted as $\sigma_{\mathcal{M}}$. We also commit a slight abuse of language (and notation) and say that a mated set belongs to $\mathcal{M}$.

An equational mating $\mathcal{M}$ is a *refutation mating* iff $\sigma_{\mathcal{M}}(A)$ is unsatisfiable.

An equational mating $\mathcal{M}$ is *path acceptable*[8] (for short, *p-acceptable*), iff, for every path $\pi \in vp(A)$, there is some mated set $\{(s_1 \doteq t_1), \ldots, (s_m \doteq t_m), \neg(s \doteq t)\} \in \mathcal{M}$, such that

$$\{(s_1 \doteq t_1), \ldots, (s_m \doteq t_m), \neg(s \doteq t)\} \subseteq \pi. \tag{16}$$

A number of remarks are in order:

(1) Given the substitution σ, the mating condition can be tested using the congruence closure method. As mentioned in the introduction, it is decidable whether a mating substitution exists, but this is an NP-complete problem.

(2) Given a family MS of mated sets, let $E = \{E_{\mathcal{S}} \mid \mathcal{S} \in MS\}$ be the family of sets of equations of the form $E_{\mathcal{S}} = \{(s_1 \doteq t_1), \ldots, (s_m \doteq t_m)\}$ and $S = \{\langle s, t\rangle \mid \mathcal{S} \in MS\}$ the family of pairs where $E_{\mathcal{S}}$ and $\langle s, t\rangle$ are associated with the mated sets $\mathcal{S} = \{(s_1 \doteq t_1), \ldots, (s_m \doteq t_m), \neg(s \doteq t)\} \in MS$. Observe that $\mathcal{M} = \langle MS, \sigma\rangle$ is a mating iff σ is a solution of Problem 1 (discussed in the introduction) for $\langle E, S\rangle$, iff $\langle E, S\rangle$ is an equational premating. This key observation will be used in searching for the substitutions associated with matings. They are the rigid E-unifiers of S.

(3) It is obviously desirable to choose p-acceptable matings as minimal as possible. One can define a preorder on matings as follows. Given two matings $\mathcal{M}_1$ and $\mathcal{M}_2$, $\mathcal{M}_1 \sqsubseteq \mathcal{M}_2$ iff for every mated set $S_1 \in \mathcal{M}_1$, there is some mated set $S_2 \in \mathcal{M}_2$, such that, $S_1 \subseteq S_2$. A mating $\mathcal{M}$ is *minimal* iff it is minimal with respect to the preorder $\sqsubseteq$, that is, for any mating $\mathcal{M}'$, if $\mathcal{M}' \sqsubseteq \mathcal{M}$, then $\mathcal{M}' = \mathcal{M}$. It is obvious that if a p-acceptable mating exists, then a minimal p-acceptable mating also exists, but it may be difficult to find it efficiently. In order to find matings as small as possible, one can look for overlapping vertical paths that are spanned by some common mated set. We

[8] A path acceptable mating is also called a *spanning mating* by Miller, 1984.

leave the problem of discovering strategies for finding minimal matings as a topic for further research.

(4) If A does not contain equations, each mated set contains some atom which unifies with the negated atom. Let $Pt_1 \ldots t_n$ be the negated atom in a mated set. Any mated set for a formula without equational atoms is of the form $\{(A_1 \doteq \top), \ldots, (A_m \doteq \top), \neg(Pt_1 \ldots t_n \doteq \top)\}$, where $A_1, \ldots, A_m$ are nonequational atoms. Since the set $\{\sigma(A_1 \doteq \top), \ldots, \sigma(A_m \doteq \top), \neg\sigma(Pt_1 \ldots t_n \doteq \top)\}$ is unsatisfiable, there is some atom $A_i = Ps_1 \ldots s_n$, such that, $\sigma(Pt_1 \ldots t_n) = \sigma(Ps_1 \ldots s_n)$. Hence, σ is a unifier of $Pt_1 \ldots t_n$ and $Ps_1 \ldots s_n$.

Hence, when A does not contain equality, a mating can be defined as a set of pairs $\langle L, \neg L' \rangle$ of literals of opposite signs, such that $\sigma(L) = \sigma(L')$, as in Andrews (1981).

Lemma 3. *Given a quantifier-free formula A in nnf, the following properties hold:*

(1) *Given a substitution θ, if $\theta(A)$ is unsatisfiable, then there is a p-acceptable mating $\mathcal{M}$ for A.*
(2) *A p-acceptable mating $\mathcal{M}$ for A is a refutation mating for A, i.e. $\sigma_{\mathcal{M}}(A)$ is unsatisfiable.*

Proof. (1) Assume that $\theta(A)$ is unsatisfiable. By Lemma 1, every vertical path in $vp(\theta(A))$ is unsatisfiable. Note that every vertical path $\pi' \in vp(\theta(A))$ is of the form $\theta(\pi)$, for some vertical path $\pi \in vp(A)$. Since every path $\pi' \in vp(\theta(A))$ is unsatisfiable, by Lemma 2, there is some subset $\{(s'_1 \doteq t'_1), \ldots, (s'_m \doteq t'_m), \neg(s' \doteq t')\} \subseteq \pi'$ of literals in π' which is unsatisfiable. For every vertical path $\pi \in vp(A)$, since $\pi' = \theta(\pi)$ is a vertical path in $vp(\theta(A))$, we can choose a set of literals $\{(s_1 \doteq t_1), \ldots, (s_n \doteq t_n), \neg(s \doteq t)\} \subseteq \pi$, such that,

$$\begin{aligned}&\{\theta(s_1 \doteq t_1), \ldots, \theta(s_n \doteq t_n), \neg\theta(s \doteq t)\}\\ &\quad = \{(s'_1 \doteq t'_1), \ldots, (s'_m \doteq t'_m), \neg(s' \doteq t')\}\end{aligned} \qquad (*)$$

is unsatisfiable. We form a mating $\mathcal{M} = \langle MS, \theta \rangle$ for A by choosing MS as the set of sets of literals defined in $(*)$. Clearly, $\mathcal{M}$ is a p-acceptable mating for A.

(2) Assume that $\mathcal{M} = \langle MS, \sigma \rangle$ is a p-acceptable mating for A. We prove that every vertical path $\pi' \in vp(\sigma(A))$ is unsatisfiable. Indeed, every vertical path $\pi' \in vp(\sigma(A))$ is of the form $\sigma(\pi)$, for some vertical path $\pi \in vp(A)$. Since $\mathcal{M}$ is p-acceptable, for every vertical path $\pi \in vp(A)$, there is some mated set of literals $\{(s_1 \doteq t_1), \ldots, (s_m \doteq t_m), \neg(s \doteq t)\} \in \mathcal{M}$, such that

$$\{(s_1 \doteq t_1), \ldots, (s_m \doteq t_m), \neg(s \doteq t)\} \subseteq \pi. \tag{17}$$

Since $\mathcal{M}$ is a mating, the set $S_{\pi'} = \{\sigma(s_1 \doteq t_1), \ldots, \sigma(s_m \doteq t_m), \neg\sigma(s \doteq t)\}$ is unsatisfiable. Since $S_{\pi'}$ is a subset of the vertical path $\pi' \in vp(\sigma(A))$, π' is unsatisfiable. But then, by Lemma 1, $\sigma(A)$ is unsatisfiable, which establishes the fact that $\mathcal{M}$ is a refutation mating. □

The previous lemma implies the following useful corollary.

Corollary 1. *Given a quantifier-free formula A in nnf, there is a substitution θ such that $\theta(A)$ is unsatisfiable iff there is a p-acceptable mating $\mathcal{M}$ for A.* □

Let us give an example illustrating the use of the previous lemma.

Example 3. Consider the following Horn formula A, where x, y, z denote variables:

$$\begin{array}{l}
(a \doteq b) \land \\
((f^3x \doteq x) \lor \neg(fx \doteq fb)) \land \\
(Qa \lor \neg(f^3a \doteq a)) \land \\
((f^5y \doteq y) \lor \neg Qy) \land \\
(Ra \lor \neg(fa \doteq a) \lor \neg Pfa) \land \\
\neg Rfz \land \\
Pa
\end{array} \tag{18}$$

There are 24 vertical paths in A. Let $\theta = [a/x, a/y, a/z]$. It is easy to see that 21 of the vertical paths in $\theta(A)$ are unsatisfiable. These are the vertical paths containing the sets of literals $\{(f^3a \doteq a), \neg(f^3a \doteq a)\}$, $\{Qa, \neg Qa\}$, and $\{(a \doteq b), \neg(fa \doteq fb)\}$. A p-acceptable mating for A is given by θ and the following set of 6 sets of literals:

$$\begin{gathered}\{\{(f^3x \doteq x), \neg(f^3a \doteq a)\},\\ \{Qa, \neg Qy\},\\ \{(a \doteq b), \neg(fx \doteq fb)\},\\ \{(f^5y \doteq y), (f^3x \doteq x), Ra, \neg Rfz\},\\ \{(f^5y \doteq y), (f^3x \doteq x), \neg(fa \doteq a)\},\\ \{(f^5y \doteq y), (f^3x \doteq x), Pa, \neg Pfa\}\}.\end{gathered} \tag{19}$$

The above set is a mating because $(fa \doteq a)$ is equationally provable from $(f^3a \doteq a)$ and $(f^5a \doteq a)$. Indeed, $(f^3a \doteq a)$ implies $(f^4a \doteq fa)$, which implies $(f^5a \doteq f^2a)$, which, by transitivity, implies $(f^2a \doteq a)$. In turn, $(f^2a \doteq a)$ implies $(f^3a \doteq fa)$, and by one more application of transitivity, this implies $(fa \doteq a)$. According to Lemma 4, $\theta(A)$ is unsatisfiable. Since $\models \forall x\forall y\forall z A \supset \theta(A)$, the universal closure $\forall A$ of A is also unsatisfiable.

Unfortunately, in general, a universal sentence $\forall A$ may be unsatisfiable, but there may not be any substitution θ such that $\theta(A)$ is unsatisfiable (see Example 2). However, a version of the Skolem-Herbrand-Gödel theorem for first-order languages with equality ensures that some substitution instance of an *amplification* of A (a formula obtained from A by duplicating some universal subformulae of A) is unsatisfiable. It is the notion of amplification (due to Andrews (1981)), that will allow us to apply the method of matings to arbitrary universal sentences in nnf.

5. A Skolem-Herbrand-Gödel Theorem

First, we need the definition of a compound instance, due to Andrews (1981).

5.1. Compound Instances

From now on, it is assumed that we are dealing with *rectified universal formulae in nnf.* The standard statement of the Skolem-Herbrand-Gödel theorem (as in Gallier (1986)) says that given a universal *prenex* sentence $A = \forall x_1 \ldots \forall x_n B$ (where B is quantifier-free), A is unsatisfiable iff there exist some ground substitutions $\sigma_1, \ldots, \sigma_k$ such that $\sigma_1(B) \wedge \ldots \wedge \sigma_k(B)$ is unsatisfiable.

It would be nice if we could relax the condition that A is in prenex form, and have a statement referring to a single substitution. This can be achieved by introducing the ingenious concepts of a compound instance and of an amplification due to Andrews (1981).

Definition 24. Let A be a rectified universal sentence in nnf (Every variable occurring in A is universally quantified). The set of *compound instances* (for short, *c-instances*) of A is defined inductively as follows:

(i) If A is either a ground atomic formula B or the negation $\neg B$ of a ground atomic formula, then A is its only c-instance;
(ii) If A is of the form $(B{*}C)$, where $* \in \{\vee, \wedge\}$, for any c-instance H of B and c-instance K of C, $(H * K)$ is a c-instance of A;
(iii) If A is of the form $\forall x B$, for any $k \geq 1$ ground terms $t_1, \ldots, t_k$, if H_i is a c-instance of $B[t_i/x]$ for $i = 1, \ldots, k$, then $H_1 \wedge \ldots \wedge H_k$ is a c-instance of A.

The importance of c-instances lies in the following version of the Skolem-Herbrand-Gödel theorem, which is a generalization of a theorem of Andrews to first-order languages with equality (Andrews, 1981, 1986). For stating this theorem, we assume (without loss of generality) that there is a least one constant symbol in the language.

Theorem 1. *(Skolem-Herbrand-Gödel theorem) Given a universal sentence A in nnf, A is unsatisfiable iff some c-instance C of A is unsatisfiable.*

Proof. It is nontrivial. A proof is given in Gallier (1986), theorem 7.6.1, page 364. Showing that if some compound instance C is unsatisfiable implies that A is unsatisfiable is straightforward, because it is easily shown that $\models A \supset C$ (Gallier (1986), theorem 7.6.1). The proof of the converse is much harder. In Gallier (1986), this is derived proof-theoretically as a consequence of a sharpened Gentzen-like Hauptsatz (Gallier (1986), theorem 7.4.1, page 334, theorem 7.4.2, page 337, and lemma 7.6.2, page 360). A semantic proof in the line of Andrews's proof is also given in Gallier, Raatz, and Snyder (1988). $\square$

The connection between matings and compound instances is established through the notion of amplification, also due to Andrews (1981).

5.2. Amplifications and Compound Instances

Let A, B, C, and D be universal sentences in nnf.

Definition 25. We say that a sentence C is obtained from a sentence B by *quantifier duplication* iff C results from B by replacing some subformula of B of the form $\forall x M$ by $(\forall x M \wedge \forall x M)$. If there is a sequence $\langle C_1, \ldots, C_n \rangle$, $n \geq 1$, of formulae, such that, $B = C_1$, $C = C_n$, and C_{i+1} is obtained from C_i by quantifier duplication, for every i, $1 \leq i < n$, we say that C is obtained from B by some *sequence of quantifier duplications.*

If B is obtained from A by some sequence of quantifier duplications, C is a rectified sentence equivalent to B, and D is obtained from C by deleting the quantifiers in C, we say that D is an *amplification* of A. The following result can be shown easily.

Lemma 4. *Given a universal sentence A in nnf, C is a c-instance of A iff there is some amplification D of A and some (ground) substitution θ such that $C = \theta(D)$.*

Proof. The proof is by induction on the structure of A. The only case worth mentioning is the case in which $A = \forall x B$. In this case, there are k ground terms $t_1, \ldots, t_k$ and k formulae $H_1, \ldots, H_k$, such that, each H_i is a c-instance of $B[t_i/x]$, and $C = H_1 \wedge \ldots \wedge H_k$. By the induction hypothesis, for each i, $1 \leq i \leq k$, there is some amplification D_i of $B[t_i/x]$ and a substitution θ_i, such that, $H_i = \theta_i(D_i)$. It can also be assumed (using renaming) that the sets of variables occurring in these amplifications are disjoint. It is not difficult to show by induction on the length of a quantifier duplication sequence that for each $B[t_i/x]$ and D_i, there is some renamed copy B_i of B, some amplification D_i' of B_i, and a substitution σ_i, such that, $H_i = \theta_i(\sigma_i(D_i'))$ (σ_i is a substitution that substitutes t_i for renamed occurrences of x). It can also be assumed (using renaming) that the sets of variables occurring in these amplifications are disjoint. Then, note that $D = D_1' \wedge \ldots \wedge D_k'$ is an amplification of A that can be obtained by first applying k quantifier duplications, obtaining $\forall xB \wedge \ldots \wedge \forall xB$ (with k copies of $\forall xB$), and then by amplifying each copy of $\forall xB$ to D'_i. Furthermore, the substitution $\theta = \sigma_1 \circ \theta_1 \circ \ldots \circ \sigma_k \circ \theta_k$ is such that $C = \theta(D)$. $\square$

We can now state one of the main theorems of this paper.

Theorem 2. *Given a universal sentence A in nnf, A is unsatisfiable iff some amplification of A has a p-acceptable mating.*

Proof. First, assume that some amplification D of A has a p-acceptable mating $\mathcal{M}$. Let B be obtained by some sequence of quantifier duplications from A, C the rectified formula equivalent to B, and D the result of deleting quantifiers from C. Let $\forall x_1, \ldots, \forall x_n$ be the quantifiers of C, in the left-to-right order in which they occur in C. It is easy to show that $\models A \equiv B$, $\models B \equiv C$, $\models C \equiv \forall x_1 \ldots \forall x_n D$, and $\models \forall x_1 \ldots \forall x_n D \supset \sigma_{\mathcal{M}}(D)$. Since $\mathcal{M}$ is a p-acceptable mating, by Lemma 3, it is a refutation mating, and so $\sigma_{\mathcal{M}}(D)$ is unsatisfiable. Hence, A is also unsatisfiable.

Now, suppose that A is unsatisfiable. By the Skolem-Herbrand-Gödel theorem (Theorem 1), there is some c-instance C of A which is unsatisfiable. By Lemma 4, there is some amplification D of A and a substitution θ such that $C = \theta(D)$. By Lemma 3, since $\theta(D)$ is unsatisfiable and D is quantifier free, there is some p-acceptable mating $\mathcal{M}$ for D. $\square$

The following example illustrates Theorem 2.

Example 4. Let A be the following (equational) sentence:

$$\forall x \forall y \forall z (*(x, *(y, z)) \doteq *(*(x, y), z))) \wedge \tag{20}$$

$$\forall u (*(u, 1) \doteq u) \wedge \tag{21}$$

$$\forall v (*(1, v) \doteq v) \wedge \tag{22}$$

$$\forall w (*(w, w) \doteq 1) \wedge \tag{23}$$

$$\neg(*(a, b) \doteq *(b, a)). \tag{24}$$

The first three equations are the axioms for monoids (a binary operation $*$ which is associative and has an identity element 1), the fourth equation asserts that the square of every element is the identity, and the fifth asserts the negation of the commutativity of $*$ (A is the result of a Skolemization). The unsatisfiability of A asserts that any monoid such that the square of every element is the identity is commutative.

Consider the following amplification D of A in the top column and the set MS consisting of one set of literals in the bottom column:

$$\begin{aligned} D = & (*(u_1, 1) \doteq u_1) \\ & \wedge\ (*(w_1, w_1) \doteq 1) \\ & \wedge\ (*(x_1, *(y_1, z_1)) \doteq *(*(x_1, y_1), z_1))) \\ & \wedge\ (*(x_2, *(y_2, z_2)) \doteq *(*(x_2, y_2), z_2))) \\ & \wedge\ (*(w_2, w_2) \doteq 1) \\ & \wedge\ (*(1, v_1) \doteq v_1) \\ & \wedge\ (*(x_3, *(y_3, z_3)) \doteq *(*(x_3, y_3), z_3))) \\ & \wedge\ (*(x_4, *(y_4, z_4)) \doteq *(*(x_4, y_4), z_4))) \\ & \wedge\ (*(w_3, w_3) \doteq 1) \\ & \wedge\ \neg(*(a, b) \doteq *(b, a)). \end{aligned} \tag{25}$$

$$\begin{aligned} MS = \{\{ & (*(u_1, 1) \doteq u_1), \\ & (*(w_1, w_1) \doteq 1), \\ & (*(x_1, *(y_1, z_1)) \doteq *(*(x_1, y_1), z_1))), \\ & (*(x_2, *(y_2, z_2)) \doteq *(*(x_2, y_2), z_2))), \\ & (*(w_2, w_2) \doteq 1), \\ & (*(1, v_1) \doteq v_1), \\ & (*(x_3, *(y_3, z_3)) \doteq *(*(x_3, y_3), z_3))), \\ & (*(x_4, *(y_4, z_4)) \doteq *(*(x_4, y_4), z_4))), \\ & (*(w_3, w_3) \doteq 1), \\ & \neg(*(a, b) \doteq *(b, a))\}\}. \end{aligned}$$

Let θ be the substitution

$$\begin{aligned} & [a/u_1, (a*b)/w_1, a/x_1, (a*b)/y_1, (a*b)/z_1, \\ & a/x_2, a/y_2, b/z_2, a/w_2, b/v_1, \\ & b/x_3, (a*b)/y_3, b/z_3, a/x_4, b/y_4, b/z_4, b/w_3]. \end{aligned} \tag{26}$$

We claim that $\langle MS, \theta \rangle$ is a mating for D. For simplicity of notation let us adopt infix notation, and denote $*(s,t)$ as $s*t$. Then, we have:

$$\begin{aligned}
a*b &= \{a*1\}*b && \text{by (21)}\\
&= \{a*[(a*b)*(a*b)]\}*b && \text{by (23)}\\
&= \{[a*(a*b)]*(a*b)\}*b && \text{by (20)}\\
&= \{[(a*a)*b]*(a*b)\}*b && \text{by (20)}\\
&= \{[1*b]*(a*b)\}*b && \text{by (23)}\\
&= \{b*(a*b)\}*b && \text{by (22)}\\
&= b*\{(a*b)*b\} && \text{by (20)}\\
&= b*\{a*(b*b)\} && \text{by (20)}\\
&= b*\{a*1\} && \text{by (23)}\\
&= b*a, && \text{by (21)}
\end{aligned}$$

which shows that $\langle MS, \theta\rangle$ is a p-acceptable mating for D (there is a single vertical path in D). Note that Equation (21) instantiated by the substitution $[a/u_1]$ is used twice.

Theorem 2 suggests a procedure for showing that a universal sentence in nnf is unsatisfiable: Compute incrementally amplifications of D, and at each stage, test whether such an amplification has a p-acceptable mating. Such a procedure is presented in a later section.

5.3. Outermost Quantifier Duplication

Since the complexity of the search for an acceptable mating grows exponentially as the number of occurrences of literals in the amplification D increases, it is important to keep this number small.

A systematic scheme for duplicating quantifiers that both minimizes the number of occurrences of literals to some extent, and guarantees completeness, is to duplicate *outermost* quantifiers.

Definition 26. Given a universal formula A in nnf, a subformula $\forall xB$ of A is a *maximal quantified subformula* of A iff there is no quantified subformula $\forall yC$ of A, such that $\forall xB$ is a proper subformula of $\forall yC$.[9] If $\forall xB$ is a maximal quantified subformula of A, the quantifier $\forall x$ is called an *outermost* quantifier.

[9]For an inductive definition of this concept, see Gallier (1986).

Lemma 5. *Let A be a universal sentence in nnf. Then, A is unsatisfiable iff there is a refutation mating for some amplification D of A, such that, in forming D from A, only outermost quantifier duplications are performed.*

Proof. It is enough to show that the Skolem-Herbrand-Gödel theorem holds for c-instances obtained as substitution instances of formulae obtained from A by outermost quantifier duplications only. This can be shown in at least two ways. The first proof is already essentially contained in the proof of lemma 7.6.2 of Gallier (1986, page 360). Indeed, this lemma is obtained from theorem 7.4.1 (and theorem 7.4.2 (Gallier, 1986) page 334, and page 337), a Gentzen-like Hauptsatz for a proof system in which quantifier rules apply only to outermost quantifiers (the system $G2^{nnf}$, (Gallier, 1986), page 327). One simply has to verify that the induction in lemma 7.6.2 yields the right kind of c-instances, and this is straightforward. The other proof is obtained by observing that the proofs of lemma 5 and theorem 2 in Andrews (1981, page 208) go through unchanged, as they do not depend on the fact that the language contains equality. □

Hence, in searching for a mating, there is no loss of generality in duplicating outermost quantifiers only. However, this is not always the best strategy, and it would be useful to develop heuristics for duplicating quantifiers.

5.4. Comparison with Other Methods

An extension of the method of matings to first-order languages with equality is sketched (without proofs) in Bibel (1982, Section V.3, pp. 234–242) (under the name *connection method with equality*). Bibel's method uses mated sets similar to ours, except that they are dual to ours, since Bibel's method shows the validity rather than the unsatisfiability of a (Skolemized) sentence. Hence, sets of literals are interpreted as disjunctions. A set of the form

$$\{\neg(s_1 \doteq t_1), \ldots, \neg(s_m \doteq t_m), (s \doteq t)\} \tag{27}$$

is called an *eq-literal*, and a set of the form

$$\{\neg(s_1 \doteq t_1), \ldots, \neg(s_m \doteq t_m), \neg L, L'\} \tag{28}$$

where L and L' are nonequational atoms, is called an *eq-connection*. In our presentation, the use of a many-sorted language with the special sort *bool* allows us to treat a nonequational literal as the special equation $L \doteq \top$, and we only need the first kind of mated set, but this is an inessential detail.

Bibel's method and ours differ significantly in the criterion used for testing the validity (equivalently, unsatisfiability) of a mated set. Bibel defines an eq-literal to be *valid* iff there is some substitution σ such that $\sigma(s_i) = \sigma(t_i)$ for all i, $1 \leq i \leq m$, and $\sigma(s) = \sigma(t)$. An eq-connection is said to be *complementary* iff there is some substitution σ such that $\sigma(s_i) = \sigma(t_i)$ for all i, $1 \leq i \leq m$, and $\sigma(L) = \sigma(L')$.

It should be noted that the notion of a substitution used by Bibel is highly nonstandard. Bibel [1982, Section III.1.6, page 66] defines a substitution σ as a set of pairs $\{s_1/t_1, \ldots, s_n/t_n\}$, where each t_i is a term to be substituted for s_i, but where s_i itself can be a nonvariable term! Of course, substitutions are applied in a homomorphic fashion, but with this definition, a substitution is *not* necessarily defined on all terms.

To be completely accurate, with Bibel's definition of a substitution, the substitution σ mentioned in the definition of a valid eq-literal is such that is consists of pairs of the form s_i/t_i or t_i/s_i. Then, theorem V.3.6.C (page 237) states (in our language) that a formula F is valid iff for some amplification D of F, there is a spanning set W of eq-literals and eq-connections and some substitution σ such that, for every eq-literal $w \in W$, $\sigma(w)$ is valid, and for every eq-connection $w \in W$, $\sigma(w)$ is complementary.

This theorem does hold, provided that we allows eq-literals and eq-connections in the set W to contain extra literals arising from instances of the equality axioms. Hence, Bibel's method uses standard unification, but the mated sets may have to include extra literals corresponding to instances of the equality axioms.

In our method, we require that there be some substitution σ such that

$$\{\sigma(s_1 \doteq t_1), \ldots, \sigma(s_m \doteq t_m), \neg\sigma(u \doteq v)\} \tag{29}$$

is unsatisfiable, or equivalently, treating the equations in $\sigma(E)$ as ground equations, that $\sigma(u) \stackrel{*}{\longleftrightarrow}_{\sigma(E)} \sigma(v)$ holds. Hence, σ is a special kind of E-unifier (a rigid E-unifier), but there is *no need* to include extra literals corresponding to instances of the equality axioms to our mated sets. The following example should illustrate this point clearly. Consider the eq-literal

$$\{\neg(f^3 a \doteq a), \neg(f^5 a \doteq a), (fa \doteq a)\}. \tag{30}$$

It is valid, but yet, there is no substitution in the sense of Bibel demonstrating that it is valid. The only way to show validity is to add additional equality axioms to show that fa and a are *congruent* modulo the set of equations $\{(f^3 a \doteq a), (f^5 a \doteq a)\}$. Hence, in Bibel's method, this mated set would have to be expanded before it is shown to be valid. In our method, it would be found valid immediately (actually, its negation would be found unsatisfiable).

Hence, Bibel's method and ours differ in the type of unification and the methods used to check the validity (or unsatisfiability) of mated sets.

In Chapter 4 of his Ph.D dissertation, Pfenning (1987) presents a method for dealing with equality in a system of expansion proofs that involves matings. Pfenning's system applies to higher-order logic, and equality is treated as a defined symbol ($(A \doteq B)$ is an abbreviation for $\forall Q(Q(A) \supset Q(B))$, where Q is a predicate variable). As pointed out by Pfenning, it is theoretically possible to derive the mated sets arising in our method from the mated sets used in his method via the translation mentioned above. In some sense, our way of checking mated sets is an optimization of Pfenning's method restricted to the first-order case. However, it does not seem possible to obtain the completeness of our method in this fashion. Our method is also different in a more radical sense, which is that Pfenning's method uses higher-order unification, whereas we use a special form of E-unification that is decidable. This suggests that there may be a form of rigid higher-order unification, but we have not explored this possibility.

6. Complete Sets of Rigid E-Unifiers

We have already noted in remark (2) after Definition 23 that $\mathcal{M} = \langle MS, \sigma \rangle$ is an equational mating iff σ is a rigid E-unifier of S, where $E = \{E_{\mathcal{S}} \mid \mathcal{S} \in MS\}$ and $S = \{\langle s, t \rangle \mid \mathcal{S} \in MS\}$, the family of sets of equations and the set of pairs associated with the mated sets $\mathcal{S} = \{(s_1 \doteq t_1), \ldots, (s_m \doteq t_m), \neg(s \doteq t)\} \in MS$. It is obviously crucial to show that there is an algorithm for testing whether a family of mated sets forms a mating. From the above observation, this is equivalent to deciding whether a pair $\langle E, S\rangle$ is an equational premating. In the following sections, it will be shown that this problem is NP-complete. Actually, this result is an easy extension of a simpler problem, and we now focus on this problem.

Problem 3. Given a finite set $E = \{u_1 \doteq v_1, \ldots, u_n \doteq v_n\}$ of equations and a pair $\langle u, v\rangle$ of terms, is there a substitution θ such that, treating $\theta(E)$ as a set of ground equations, $\theta(u) \overset{*}{\cong}_{\theta(E)} \theta(v)$, that is, $\theta(u)$ and $\theta(v)$ are congruent modulo $\theta(E)$ (by congruence closure)? The substitution θ is called a *rigid E-unifier of u and v.*

It is interesting to observe, as pointed out by Jean Yves Girard, that the notion of rigid E-unification arises by *bounding* the resources, in this case, the number of available instances of equations in E. To be precise, only a single instance of each equation in E can be used, and in fact, these instances $\theta(u_1 \doteq v_1), \ldots, \theta(u_n \doteq v_n)$ must arise from the *same* substitution θ. Also, once these instances have been created, the remaining variables (if any) are considered rigid, that is, treated as constants, so that it is not possible to instantiate these instances. This is in the spirit of linear logic (Girard, 1987). The special case of rigid E-unification where E is a set of ground equations has been investigated by Kozen who has shown that this problem is NP-complete (Kozen, 1976, 1977). Thus, rigid E-unification is NP-hard. We will show that it is also in NP, and hence it is NP-complete.

Suppose we want to find a rigid E-unifier θ of u and v (actually, it can be assumed that $u = T$ and $v = F$ where T and F are two special constants). If we have n equations in E, then we have $n' \leq n$ instances in $\theta(E)$. Then there is a sequence $w_0, w_1, \ldots, w_p$ of term s.t. $w_0 = T$, $w_p = F$, and each w_{i+1} is obtained from w_i by equational replacement using an instance $\theta(l \doteq r)$ of an equation $l \doteq r \in E \cup E^{-1}$. Now, although we may only have n' instances in $\theta(E)$, this chain may be much longer since the same equation may be used many times.

The idea is to use a form of unfailing completion procedure (Knuth and Bendix (1970), Huet (1980), Bachmair (1987), Bachmair, Dershowitz, and Plaisted (1987)). This means essentially that there is some critical pair in the sequence (or that the sequence can be rearranged to make this true). Suppose that this critical pair is an overlap of $l_1 \doteq r_1$ and $l_2 \doteq r_2$ where l_2 is unifiable with a subterm of l_1. If l_2 is equal to the subterm of l_1, then we will get this critical pair essentially by simplification. Otherwise, some variable in l_1 or in l_2 gets bound to a term *not* containing this variable. Thus the total number of variables in E keeps decreasing. Therefore, after a polynomial number of steps (in fact, the number of variables in E) we must stop or fail. So we get membership in NP.

However, there are some significant departures from traditional Knuth-Bendix completion procedures, and this is for two reasons. The first reason is that we must ensure termination of the method. The second is that we want to show that the problem is in NP, and this forces us to be much more concerned about efficiency.

The proof that rigid E-unification is in NP requires quite a bit of machinery, and since this paper is already long, we will focus on the algorithmic aspect of the result, leaving out most proofs. Full details can be found in Gallier, Narendran, Plaisted, and Snyder (1987).

In order to show that our decision procedure is in NP, we will need the fact that if two terms u and v are unifiable, a mgu of u and v can be represented concisely in triangular form (the size of this system is linear in the number of symbols in u and v). This result can be obtained from the fast method using multiequations of Martelli and Montanari (1982) or the fast method using the graph unification closure of Paterson and Wegman (1978).

Definition 27. A *term pair* (or *pair*) is just a pair of two terms, denoted by $\langle s,t\rangle$, and a substitution θ is called a *unifier* of a pair $\langle s,t\rangle$ if $\theta(s)=\theta(t)$. A *term system* (or *system*) is a set of such pairs, and a substitution θ is a unifier of a system if it unifies each pair. A substitution σ is a *most general unifier*, or *mgu*, of a system S iff (i) $D(\sigma)\subseteq Var(S)$ and $D(\sigma)\cap I(\sigma)=\emptyset$ (σ is idempotent); (ii) σ is a unifier of S; (iii) For every unifier θ of S, $\sigma\leq\theta$ (where $\sigma\leq\theta$ iff $\theta=\sigma\circ\eta$ for some η).

Definition 28. Given an idempotent substitution σ (i.e., $D(\sigma)\cap I(\sigma)=\emptyset$) with domain $D(\sigma)=\{x_1,\ldots,x_k\}$, a *triangular form for* σ is a finite set T of pairs $\langle x,t\rangle$ where $x\in D(\sigma)$ and t is a term, such that this set T can be sorted (possibly in more than one way) into a sequence $\langle\langle x_1,t_1\rangle,\ldots,\langle x_k,t_k\rangle\rangle$ satisfying the following properties: for every i, $1\leq i\leq k$,

(1) $\{x_1,\ldots,x_i\}\cap Var(t_i)=\emptyset$, and

(2) $\sigma = [t_1/x_1] \circ \ldots \circ [t_k/x_k]$.

The set of variables $\{x_1, \ldots, x_k\}$ is called the *domain* of T. Note that in particular $x_i \notin Var(t_i)$ for every i, $1 \leq i \leq k$, but variables in the set $\{x_{i+1}, \ldots, x_k\}$ may occur in $t_1, \ldots, t_i$. It is easily seen that σ is an (idempotent) mgu of the term system T.

Example 5. Consider the substitution $\sigma = [f(f(x_3, x_3), f(x_3, x_3))/x_1, f(x_3, x_3)/x_2]$. The system $T = \{\langle x_1, f(x_2, x_2)\rangle, \langle x_2, f(x_3, x_3)\rangle\}$ is a triangular form of σ since it can be ordered as $\langle\langle x_1, f(x_2, x_2)\rangle, \langle x_2, f(x_3, x_3)\rangle\rangle$ and $\sigma = [f(x_2, x_2)/x_1] \circ [f(x_3, x_3)/x_2]$.

The triangular form $T = \{\langle x_1, t_1\rangle, \ldots, \langle x_k, t_k\rangle\}$ of a substitution σ also defines a substitution, namely $\sigma_T = [t_1/x_1, \ldots, t_k/x_k]$. This substitution is usually different from σ and not idempotent as can be seen from Example 5. However, this substitution plays a crucial role in our decision procedure because of the following property.

Lemma 6. *Given a triangular form $T = \{\langle x_1, t_1\rangle, \ldots, \langle x_k, t_k\rangle\}$ for a substitution σ and the associated substitution $\sigma_T = [t_1/x_1, \ldots, t_k/x_k]$, for every unifier θ of T, $\theta = \sigma_T \circ \theta$.*

Another important observation about σ_T is that even though it is usually not idempotent, at least one variable in $\{x_1, \ldots, x_k\}$ does not belong to $I(\sigma_T)$ (otherwise, condition (1) of the triangular form fails). We will assume that a procedure TU is available, which, given any unifiable term system S, returns a triangular form for an idempotent mgu of S, denoted by $TU(S)$. When S consists of a single pair $\langle u, v\rangle$, $TU(S)$ is also denoted by $TU(u, v)$.

Recall that we write $u \overset{*}{\cong}_E v$ to express that $u \overset{*}{\longleftrightarrow}_E v$, treating the equations in E as *ground equations.*

Definition 29. Let $E = \{(s_1 \doteq t_1), \ldots, (s_m \doteq t_m)\}$ be a finite set of equations, and let $Var(E) = \bigcup_{(s \doteq t) \in E} Var(s \doteq t)$ denote the set of variables occurring in E.[10] Given any two terms u and v,[11] a substitution θ is a *rigid unifier of u and v modulo E* (for short, a rigid *E-unifier of u and v*) iff

$\theta(u) \overset{*}{\cong}_{\theta(E)} \theta(v)$, that is, $\theta(u)$ and $\theta(v)$ are congruent modulo the set $\theta(E)$ considered as a set of *ground* equations.

[10]It is possible that equations have variables in common.

[11]It is possible that u and v have variables in common with the equations in E.

Definition 30. Let E be a (finite) set of equations, and W a (finite) set of variables. For any two substitutions σ and θ, $\sigma =_E \theta[W]$ iff $\sigma(x) \stackrel{*}{\cong}_E \theta(x)$ for every $x \in W$. The relation $\sqsubseteq_E$ is defined as follows. For any two substitutions σ and θ, $\sigma \sqsubseteq_E \theta[W]$ iff $\sigma =_{\theta(E)} \theta[W]$. The set W is omitted when $W = X$ (where X is the set of variables), and similarly E is omitted when $E = \emptyset$.

Intuitively speaking, $\sigma \sqsubseteq_E \theta$ iff σ can be generated from θ using the equations in $\theta(E)$. Clearly, $\sqsubseteq_E$ is reflexive. However, it is not symmetric as shown by the following example.

Example 6. Let $E = \{fx \doteq x\}$, $\sigma = [fa/x]$ and $\theta = [a/x]$. Then $\theta(E) = \{fa \doteq a\}$ and $\sigma(x) = fa \stackrel{*}{\cong}_{\theta(E)} a = \theta(x)$, and so $\sigma \sqsubseteq_E \theta$. On the other hand $\sigma(E) = \{ffa \doteq fa\}$, but a and fa are not congruent from $\{ffa \doteq fa\}$. Thus $\theta \sqsubseteq_E \sigma$ *does not* hold.

It is not difficult to show that $\sqsubseteq_E$ is also transitive. We also need an extension of $\sqsubseteq_E$ defined as follows.

Definition 31. Let E be a (finite) set of equations, and W a (finite) set of variables. The relation $\leq_E$ is defined as follows: for any two substitutions σ and θ, $\sigma \leq_E \theta[W]$ iff $\sigma \circ \eta \sqsubseteq_E \theta[W]$ for some substitution η (that is, $\sigma \circ \eta =_{\theta(E)} \theta[W]$ for some η).

Intuitively speaking, $\sigma \leq_E \theta$ iff σ is more general than some substitution that can be generated from θ using $\theta(E)$. Clearly, $\leq_E$ is reflexive. The transitivity of $\leq_E$ is also shown easily. When $\sigma \leq_E \theta[W]$, we say that *σ is (rigid) more general than θ over W*. It can be shown that if σ is a rigid E-unifier of u and v and $\sigma \leq_E \theta$, then θ is a rigid E-unifier of u and v. The converse is false.

Definition 32. Given a (finite) set E of equations, for any two terms u and v, letting $V = Var(u) \cup Var(v) \cup Var(E)$, a set U of substitutions is a *complete set of rigid E-unifiers for u and v* iff: For every $\sigma \in U$,

(i) $D(\sigma) \subseteq V$ and $D(\sigma) \cap I(\sigma) = \emptyset$ (idempotence),

(ii) σ is a rigid E-unifier of u and v,

(iii) For every rigid E-unifier θ of u and v, there is some $\sigma \in U$, such that, $\sigma \leq_E \theta[V]$.

It is very useful to observe that if a procedure P for finding sets of rigid E-unifiers satisfies the property stated in Definition 33 given next, then in order to show that this procedure yields complete sets, there is no loss of generality in showing completeness with respect to *ground* rigid E-unifiers whose domains contain V (that is, in clause (iii) of Definition 32, $\theta(x)$ is a ground term for every $x \in D(\theta)$, and $V \subseteq D(\theta)$).

Definition 33. A procedure P for finding sets of rigid E-unifiers is *pure* iff the following condition holds: For every ranked alphabet Σ, every finite set E of equations over $T_\Sigma(X)$ and every $u, v \in T_\Sigma(X)$, if $U = P(E, u, v)$ is the set of rigid E-unifiers for u and v given by procedure P, then for every $\sigma \in U$, for every $x \in D(\sigma)$, every constant or function symbol occurring in $\sigma(x)$ occurs either in some equation in E or in u or in v.

In other words, $P(E, u, v)$ does not contain constant or function symbols that do not already occur in the input (E, u, v). To prove what we claimed, we proceed as follows. We add countably infinitely many new (distinct) constants c_x to Σ, each constant c_x being associated with the variable x. The resulting alphabet is denoted by Σ_{SK}. If θ is not ground, we create the Skolemized version of θ, that is, the substitution $\widehat{\theta}$ obtained by replacing the variables in the terms $\theta(x)$ by new (distinct) constants.[12]

Lemma 7. *Given a rigid E-unification procedure P satisfying the property of Definition 33, assume that for every ranked alphabet Σ, every finite set E of equations over $T_\Sigma(X)$ and every $u, v \in T_\Sigma(X)$, the set $U = P(E, u, v)$ of rigid E-unifiers of u and v given by P satisfies conditions (i) and (ii) of Definition 32, and the new condition (iii'): for every rigid E-unifier θ of u and v such that $V \subseteq D(\theta)$ and $\theta(x) \in T_\Sigma$ for every $x \in D(\theta)$, there is some $\sigma \in U$ such that $\sigma \leq_E \theta[V]$ (where $V = Var(E) \cup Var(u, v)$). Then every set $U = P(E, u, v)$ is a complete set of rigid E-unifier for u and v.*

7. Minimal Rigid E-Unifiers

One of the reasons for the decidability of rigid E-unification is that if a pair $\langle u, v\rangle$ has some rigid E-unifier, then it has a rigid E-unifier that is minimal in a sense made precise in the sequel. Given a finite or countably infinite ranked alphabet Σ, it is always possible to define

[12] that is, $\widehat{\theta}$ is obtained from θ by replacing every variable y in each term $\theta(x)$ by the corresponding Skolem constant c_y, for each $x \in D(\theta)$.

a total simplification ordering $\preceq$ on T_Σ (the set of all ground terms) (Dershowitz, 1987). We use the total simplification ordering $\prec$ on T_Σ to define a well-founded partial order $\lll$ on ground substitutions. For this, it is assumed that the set of variables X is totally ordered as $X = \langle x_1, x_2, \ldots, x_n, \ldots \rangle$.

Definition 34. The partial order $\lll$ is defined on ground substitutions as follows. Given any two ground substitutions σ and θ such that $D(\sigma) = D(\theta)$, letting $\langle y_1 \ldots, y_n \rangle$ be the sequence obtained by ordering the variables in $D(\sigma)$ according to their order in X, then $\sigma \lll \theta$ iff

$$\langle \sigma(y_1), \ldots, \sigma(y_n) \rangle \preceq_{lex} \langle \theta(y_1), \ldots, \theta(y_n) \rangle, \tag{31}$$

where $\preceq_{lex}$ is the lexicographic ordering on tuples induced by $\preceq$.

Since $\preceq$ is well-founded and $\lll$ is induced by the lexicographic ordering $\preceq_{lex}$ which is well-founded, $\lll$ is also well-founded. In fact, given any finite set V of variables, note that $\lll$ is a total well-founded ordering for the set of ground substitutions with domain V.

Given a set E of equations and a total simplification ordering $\preceq$ on ground terms, for any ground substitution θ, we let $\theta(E)$ denote the set $\{\theta(l) \doteq \theta(r) \mid \theta(l) \succ \theta(r),\ l \doteq r \in E \cup E^{-1}\}$ of oriented instances of E. Thus, we can also view $\theta(E)$ as a set of rewrite rules.

The reason for considering the well-founded ordering $\lll$ on ground substitutions is that minimal rigid E-unifiers exist. This is one of the reasons for the decidability of rigid E-unification. The example below gives some motivation for the next definition and lemma.

Example 7. Let $E = \{fa \doteq a,\ x \doteq fa\}$, and $\langle u, v \rangle = \langle gx, x \rangle$. It is obvious that there is a simplification ordering total on ground terms such that $a \prec f \prec g$. The main point of this example is the fact that some rigid E-unifiers of gx and x are redundant, in the sense that they are subsumed by rigid E-unifiers that are smaller w.r.t. $\leq_E$. For instance, $\theta = [gf^{10}a/x]$ is a rigid E-unifier of gx and x, but so is $\sigma = [ga/x]$, and $\sigma \sqsubseteq_E \theta$.

An illustration of the redundancy of θ is the fact that $\theta(x) = gf^{10}a$ is reducible by the rule $fa \to a$. The fact that some term $\theta(x)$ may be reducible by some oriented instance $\theta(l) \to \theta(r)$ of an equation

$l \doteq r \in E \cup E^{-1}$ turns out to be a problem for the completeness of the method. In order to avoid such redundancies, for every rigid E-unifier θ of u and v, we consider the set $S_{E,u,v,\theta}$ of all ground rigid E-unifiers ρ of u and v that $\rho \sqsubseteq_E \theta$. The crucial fact is that the set $S_{E,u,v,\theta}$ has a smallest element σ under the ordering $\prec\!\!\!\prec$, and that this least substitution is nicely reduced w.r.t. $\sigma(E)$. Intuitively speaking, we find the least ground rigid E-unifier σ of u and v constructible from θ and $\theta(E)$ (least w.r.t. $\prec\!\!\!\prec$). Referring back to $\theta = [gf^{10}a/x]$, the substitution $\sigma = [ga/x]$ is the smallest element of $S_{E,u,v,\theta}$. It is not sufficient to simply consider all ground substitutions ρ such that $\rho \sqsubseteq_E \theta$, because some of them may not be rigid E-unifiers of u and v. For instance, we have $\rho \sqsubseteq_E \theta$ for $\rho = [a/x]$, but ρ is not a rigid E-unifier of ga and a since $\rho(E) = \{fa \doteq a\}$. Thus, we have to consider rigid E-unifiers of u and v such that $\rho \sqsubseteq_E \theta$.

The least element σ of the set $S_{E,u,v,\theta}$ enjoys some nice reduction properties w.r.t. $\sigma(E)$. These properties stated in the forthcoming lemma will be used in the proof that the method is complete.

Definition 35. Let E be a set of equations (over $T_\Sigma(X)$) and $u, v \in T_\Sigma(X)$ any two terms. For any ground rigid E-unifier θ of u and v, let

$$S_{E,u,v,\theta} = \{\rho \mid D(\rho) = D(\theta),\ \rho(u) \overset{*}{\cong}_{\rho(E)} \rho(v),\ \rho \sqsubseteq_E \theta,\ \text{and } \rho \text{ ground}\}. \tag{32}$$

Obviously, $\theta \in S_{E,u,v,\theta}$, so $S_{E,u,v,\theta}$ is not empty. Since $\prec\!\!\!\prec$ is total and well-founded on ground substitutions with domain $D(\theta)$, the set $S_{E,u,v,\theta}$ contains some least element σ (w.r.t. $\prec\!\!\!\prec$).

We shall now prove the following crucial result. For this, recall that a *degenerate equation* is of the form $x \doteq t$, where x is a variable and $x \notin Var(t)$, and that a *nondegenerate equation* is not a degenerate equation.

Lemma 8. *Let E be a set of equations (over $T_\Sigma(X)$) and $u, v \in T_\Sigma(X)$ any two terms. For any ground rigid E-unifier θ of u and v, if σ is the least element of the set $S_{E,u,v,\theta}$ of Definition 35, then the following properties hold:*

(1) $\sigma \sqsubseteq_E \theta$,

(2) *every term of the form* $\sigma(x)$ *is irreducible by every oriented instance* $\sigma(l) \rightarrow \sigma(r)$ *of a nondegenerate equation* $l \doteq r \in E \cup E^{-1}$, *and*

(3) *every proper subterm of a term of the form* $\sigma(x)$ *is irreducible by every oriented instance* $\sigma(l) \rightarrow \sigma(r)$ *of a degenerate equation* $l \doteq r \in E \cup E^{-1}$.

8. The Reduction Procedure

One of the major components of the decision procedure for rigid E-unification is a procedure for creating a reduced set of rewrite rules equivalent to a given (finite) set of ground equations. This procedure first presented in Gallier et al. (1987b) runs in polynomial time. However, due to the possibility that variables may occur in the equations, we have to make some changes to this procedure. Roughly speaking, given a "guess" $\mathcal{O}$ (which we call an *order assignment*) of the ordering among all subterms of the terms in a set of equations E, we can run the reduction procedure R on E and $\mathcal{O}$ to produce a reduced rewrite system $R(E, \mathcal{O})$ equivalent to E, and whose orientation is dictated by the ordering $\mathcal{O}$.

Definition 36. Given a set R of rewrite rules, we say that R *is rigid reduced* iff

(1) No lefthand side of any rewrite rule $l \rightarrow r \in R$ is reducible by any rewrite rule in $R - \{l \rightarrow r\}$ treated as a ground rule;

(2) No righthand side of any rewrite rule $l \rightarrow r \in R$ is reducible by any rewrite rule in R treated as a ground rule.

Definition 37. Given two sets E and E' of equations, we say that E and E' are *rigid equivalent* iff for every two terms u and v, $u \overset{*}{\cong}_E v$ iff $u \overset{*}{\cong}_{E'} v$ (treating E and E' as sets of ground equations).

For technical reasons, it will be convenient to view the problem of rigid E-unification as the problem of deciding whether two fixed constants are rigid E-unifiable. This can be achieved as follows (the idea is borrowed from Dershowitz). Let *eq* be a new binary function symbol not occurring in Σ, and T and F two new constants not occurring in Σ. The following simple but useful lemma holds.

Lemma 9. *Given a set E of equations and any two terms u and v, a substitution θ over $T_\Sigma(X)$ is a rigid E-unifier of u and v iff there is some substitution θ' over $T_\Sigma(X)$ such that $\theta = \theta'|_{D(\theta')-\{z\}}$ and $T \overset{*}{\cong}_{\theta'(E_{u,v})} F$, where $E_{u,v} = E \cup \{eq(u,v) \doteq F,\ eq(z,z) \doteq T\}$, and z is a new variable not in $Var(E) \cup Var(u,v)$.*

The total simplification ordering $\preceq$ can be extended to the set

$$T_\Sigma \cup \{T, F\} \cup \{eq(u,v) \mid u, v \in T_\Sigma\}. \tag{33}$$

For details, see Gallier *et al.* (1987c). We will need to show that in searching for rigid E-unifiers, it is always possible to deal with sets of equations that are rigid reduced. The proof of this fact uses the result shown elsewhere that every finite set E of ground equations is equivalent to a reduced set $R(E)$ of rewrite rules. We now review the procedure first presented in Gallier *et al.* (1987b) which, given a total simplification ordering $\prec$ on ground terms and a finite set E of ground equations returns a reduced rewrite system $R(E)$ equivalent to E.

Definition 38. (Basic reduction procedure) Let E be a finite set of ground equations, and $\prec$ a simplification ordering total on ground terms. The basic reduction procedure generates a finite sequence of triples $\langle \mathcal{E}_i, \Pi_i, \mathcal{R}_i \rangle$ where $\mathcal{E}_i$ is a finite set of ground equations, Π_i is a partition (associated with $\mathcal{E}_i$), and $\mathcal{R}_i$ is a set of ground rewrite rules. Given a triple $\langle \mathcal{E}_i, \Pi_i, \mathcal{R}_i \rangle$, we let $\mathcal{T}_i$ be the set of all subterms of terms occurring in equations in $\mathcal{E}_i$ or in rewrite rules in $\mathcal{R}_i$. The procedure makes use of the *congruence closure* of a finite set of ground equations (Kozen (1976, 1977), Nelson and Oppen (1980), Downey, Sethi, and Tarjan (1980)). Congruence closures are represented by their associated partition Π. Given an equivalence relation represented by its partition Π, the equivalence class of t is denoted by $[t]_\Pi$, or $[t]$. Recall that s, t are in the same equivalence class of Π iff s and t are subterms of the terms occurring in E and $s \overset{*}{\longleftrightarrow}_E t$ (for details, see Gallier, 1986). The congruence closure algorithm will only be run once on E to obtain Π_0, but the partition Π_i may change due to further steps (simplification steps).

begin algorithm

Initially, we set $\mathcal{E}_0 = E$, $\mathcal{R}_0 = \emptyset$, and run a congruence closure algorithm on the ground set E to obtain Π_0. $i := 0$;

while Π_i has some nontrivial equivalence class[13] **do** {Simplification steps}

Let ρ_{i+1} be the smallest element[14] of the set

$$\bigcup_{C \in \Pi_i, |C| \geq 2} C$$

of terms belonging to nontrivial classes in Π_i.[15] Let C_{i+1} be the nontrivial class that contains ρ_{i+1}, and write $C_{i+1} = \{\rho_{i+1}, \lambda_{i+1}^1, \ldots, \lambda_{i+1}^{k_{i+1}}\}$, where $k_{i+1} \geq 1$, since C_{i+1} is nontrivial. Let $\mathcal{S}_{i+1} = \{\lambda_{i+1}^1 \rightarrow \rho_{i+1}, \ldots, \lambda_{i+1}^{k_{i+1}} \rightarrow \rho_{i+1}\}$.

{Next, we use the rewrite rules in $\mathcal{S}_{i+1}$ to simplify the rewrite rules in $\mathcal{R}_i \cup \mathcal{S}_{i+1}$, the partition Π_i, and the equations in $\mathcal{E}_i$.}

To get $\mathcal{R}_{i+1}$, first, we get a canonical system equivalent to $\mathcal{S}_{i+1}$. For this, for every lefthand side λ of a rule in $\mathcal{S}_{i+1}$, replace every maximal redex of λ of the form λ^j by ρ, where $\lambda^j \rightarrow \rho \in \mathcal{S}_{i+1} - \{\lambda \rightarrow \rho\}$.[16] Let $\mathcal{S}'_{i+1}$ be the set of simplified rules. Also, let $\mathcal{R}'_{i+1}$ be the set obtained by simplifying the lefthand sides of rules in $\mathcal{R}_i$ using $\mathcal{S}_{i+1}$ (reducing maximal redexes only), and let

$$\mathcal{R}_{i+1} = \mathcal{R}'_{i+1} \cup \mathcal{S}'_{i+1}.$$

Finally, use $\mathcal{S}_{i+1}$ to simplify all terms in Π_i and $\mathcal{E}_i$, using the simplification process described earlier to obtain Π_{i+1} and $\mathcal{E}_{i+1}$.

$i := i + 1$

endwhile

{All classes of Π_i are trivial, and the set $\mathcal{R}_i$ is a canonical system equivalent to E.}

end algorithm

[13] that is, a class containing at least two elements, in which case $\mathcal{E}_i$ has at least one nontrivial equation.

[14] in the ordering $\prec$

[15] where $|C|$ denotes the cardinality of the set C

[16] By a maximal redex of λ, we mean a redex of λ that is not a proper subterm of any other redex of λ. The simplified term is irreducible w.r.t. $\mathcal{S}_{i+1}$, so these replacements are only done once, and they can be done in parallel because they apply to independent subterms of λ.

It is shown in Gallier *et al.* (1987b) that the above procedure always terminates with a system $\mathcal{R}_m$ equivalent to E that is reduced (and hence, canonical). Unfortunately, given a *nonground* set E of equations, the reduction procedure just presented may not be applicable since some of the equivalence classes may contain terms involving variables and the ordering $\prec$ may no longer be total on such a partition. We need to guess how terms containing variables compare to other terms in the partition in order to reduce the equations. However, it is useful to observe that the reduction algorithm applies, as long as at every stage of the algorithm, it is possible to determine the least element of each nontrivial equivalence class and to sort these least elements. This observation shows that in extending a simplification ordering $\prec$ total on ground terms to terms containing variables, it is sufficient to require this extension to be total on nontrivial equivalence classes and total on the set of least elements of these classes. Definition 41 will make use of this fact.

The key to extending ground orderings is that if some ground rigid E-unifier θ exists, since the ordering $\prec$ is total on ground terms, θ induces a preorder on the terms occurring in the congruence closure Π of E. For example, if $E = \{fa \doteq a, fa \doteq x\}$, $u = gx$, $v = x$, and $\theta = [ga/x]$, then Π has a single nontrivial class $\{fa, a, x\}$, and assuming that $a \prec f \prec g$, we have $a \prec fa \prec ga = \theta(x)$. Hence, we can extend $\prec$ so that $fa \prec x$. This way, the equations can be oriented as $fa \rightarrow a, x \rightarrow fa$.

We shall define the concept of an order assignment in order to formalize the above intuition. First, we define some relations induced by a ground substitution on a partition of a finite set of terms.

Definition 39. Given a finite set S of terms, let $ST(S)$ be the set of all subterms of terms in S (including the terms in S). Let $\preceq$ be a total simplification ordering on ground terms, Π a partition on $ST(S)$, and θ a ground substitution such that $Var(\Pi) \subseteq D(\theta)$. The relations $\equiv_{\theta,\Pi}$ and $\preceq_{\theta,\Pi}$ on $ST(S)$ are defined as follows: For every $u, v \in ST(S)$,

$$u \preceq_{\theta,\Pi} v \quad \text{iff} \quad \theta(u) \preceq \theta(v),$$

and

$$u \equiv_{\theta,\Pi} v \quad \text{iff} \quad \theta(u) = \theta(v). \tag{34}$$

As the next example shows, the equivalence relation $\equiv_{\theta,\Pi}$ may be nontrivial.

Example 8. Let $E = \{fx \doteq fgy,\ fgy \doteq gy,\ hgz \doteq gz\}$, $u = k(fx, gb)$, $v = k(ga, hgb)$, and $\theta = [ga/x, a/y, b/z]$. The nontrivial equivalence classes of the congruence closure Π of E are $\{fx, fgy, gy\}$, and $\{hgz, gz\}$. Then, since $\theta(x) = \theta(gy) = ga$, we have $x \equiv_{\theta,\Pi} gy$ and $fx \equiv_{\theta,\Pi} fgy$. Thus, $\equiv_{\theta,\Pi}$ has two nontrivial equivalence classes $\{x, gy\}$ and $\{fx, fgy\}$. Assuming that we have a total simplification ordering on ground terms such that $a \prec b \prec f \prec g \prec h$, we also have

$$y \preceq_{\theta,\Pi} z \preceq_{\theta,\Pi} x \preceq_{\theta,\Pi} gy \preceq_{\theta,\Pi} gz \preceq_{\theta,\Pi} fx \preceq_{\theta,\Pi} fgy \preceq_{\theta,\Pi} hgz. \quad (35)$$

The other pairs in $\preceq_{\theta,\Pi}$ are obtained by reflexivity and transitivity from $\equiv_{\theta,\Pi}$ and the above pairs.

This time, it is not obvious how to orient the equation $fx \doteq fgy$. This is because $\theta(fx) = \theta(fgy)$. One might think that this is a problem, but it can be overcome. Observe that since the ground equation $\theta(fx) \doteq \theta(fgy)$ is trivial, it does not help in any way in proving that $\theta(u)$ and $\theta(v)$ are congruent modulo $\theta(E)$. Also, observe that every equivalence class modulo $\equiv_{\theta,\Pi}$ has θ as a common unifier. The solution is to factor out the preorder $\preceq_{\theta,\Pi}$ by the equivalence relation $\equiv_{\theta,\Pi}$. This can be achieved by choosing representatives in the classes modulo $\equiv_{\theta,\Pi}$ and replacing every term in $E \cup \{u, v\}$, by the representative in its class modulo $\equiv_{\theta,\Pi}$. In order to keep track of this equivalence, we also form the triangular form of the common mgu of these classes. Referring to Example 8, the mgu of the nontrivial classes $\{x, gy\}$ and $\{fx, fgy\}$ of $\equiv_{\theta,\Pi}$ is $[gy/x]$, represented by the triangular form $\langle x, gy\rangle$. choosing fgy as the representative in $\{fx, fgy\}$, the set of equations becomes $E' = \{fgy \doteq gy,\ hgz \doteq gz\}$, and we have $u' = k(fgy, gb)$ and $v' = k(ga, hgb)$. The nontrivial classes of the congruence closure of E' are $\{fgy, gy\}$, and $\{hgz, gz\}$. Now, the order is forced by θ: $gy \prec fgy$, $gy \prec gz$, and $gz \prec hgz$. Note that $\theta = [ga/x, a/y, b/z]$ is a unifier of $\langle x, gy\rangle$ and a rigid E'-unifier of $u' = k(fgy, gb)$ and $v' = k(ga, hgb)$.

The partition Π induces an equivalence relation on the set of equivalence classes modulo $\equiv_{\theta,\Pi}$ defined as follows.

Definition 40. Given a set S and two equivalence relations Π and $\equiv$ on S, let $\Pi \sqcup \equiv\ =\ (\Pi \cup \equiv)^+$, the least equivalence relation containing Π and $\equiv$. The relation $\equiv / \Pi$ on the set of equivalence classes of $\equiv$ is defined as follows: for any two classes $[u]_\equiv$ and $[v]_\equiv$ of $\equiv$, $([u]_\equiv, [v]_\equiv) \in\ \equiv / \Pi$ iff $(u, v) \in \Pi \sqcup \equiv$.

Note that the sets of the form $\bigcup_{K \in \widehat{C}} K$ where $\widehat{C}$ is a class of $\equiv / \Pi$ are the equivalence classes of $\Pi \sqcup \equiv$. We will denote the set $\bigcup_{K \in \widehat{C}} K$ as $\bigcup \widehat{C}$. Actually, every class of $\Pi \sqcup \equiv$ is both the union of some classes of Π and the union of some classes of $\equiv$.

Example 9. Let $E = \{fx \doteq fgx,\ fy \doteq hy,\ kz \doteq fz\}$, and $\theta = [a/x, ga/y, b/z]$. The nontrivial classes of the the congruence closure Π of E are $\{fx, fgx\}$, $\{fy, hy\}$, and $\{kz, fz\}$, and the equivalence relation $\equiv_{\theta,\Pi}$ has two nontrivial equivalence classes $\{y, gx\}$ and $\{fy, fgx\}$ since $\theta(y) = \theta(gx) = ga$. The nontrivial equivalence classes of $\equiv_{\theta,\Pi} / \Pi$ are $\{\{fx\}, \{fgx, fy\}, \{hy\}\}$ and $\{\{kz\}, \{fz\}\}$. If $\widehat{C} = \{\{fx\}, \{fgx, fy\}, \{hy\}\}$, then $\bigcup \widehat{C} = \{fx, fgx, fy, hy\}$.

The above discussion leads to the following definition that makes use of the fact noted before Definition 38.

Definition 41. Let $\preceq$ be a total simplification ordering on ground terms. Given a finite set S of terms and a partition Π on $ST(S)$, an *order assignment for* Π is a preorder $\mathcal{O}$ on $ST(S)$ satisfying the properties listed below ($\mathcal{O}$ is also denoted as $\preceq_{\mathcal{O}}$). Let $\equiv_{\mathcal{O}}$ be the equivalence relation associated with $\preceq_{\mathcal{O}}$ defined such that

$$\equiv_{\mathcal{O}} = \{(u, v) \mid u, v \in ST(S),\ u \preceq_{\mathcal{O}} v \text{ and } v \preceq_{\mathcal{O}} u\}.$$

The following properties hold:

(1) $\preceq_{\mathcal{O}}$ has the subterm property and is monotonic on $ST(S)$, that is, for all $u_1, \ldots, u_n, v_1, \ldots, v_n \in ST(S)$, if $u_i \preceq_{\mathcal{O}} v_i$ for $i = 1, \ldots, n$ and $f(u_1, \ldots, u_n)$ and $f(v_1, \ldots, v_n) \in ST(S)$, then $f(u_1, \ldots, u_n) \preceq_{\mathcal{O}} f(v_1, \ldots, v_n)$;

(2) The restriction of $\preceq_{\mathcal{O}}$ to ground terms agrees with $\preceq$ (on $ST(S)$), $\preceq_{\mathcal{O}}$ is total on every set $\bigcup \widehat{C}$ where $\widehat{C}$ is a nontrivial equivalence class of $\equiv_{\mathcal{O}} / \Pi$, and $\preceq_{\mathcal{O}}$ is total on the union of the sets of minimal elements in the sets $\bigcup \widehat{C}$ (we say that u is minimal with respect to the preorder $\preceq_{\mathcal{O}}$ iff whenever $v \preceq_{\mathcal{O}} u$ then $u \preceq_{\mathcal{O}} v$, that is, $u \equiv_{\mathcal{O}} v$).

(3) There is some joint unifier of all equivalence classes modulo $\equiv_{\mathcal{O}}$. By this, we mean that there is some θ such that for every class K of $\equiv_{\mathcal{O}}$, for every pair of terms $u, v \in K$, $\theta(u) = \theta(v)$.

Note that condition (3) implies that each class of the equivalence relation $\equiv_{\mathcal{O}}$ contains at most one ground term. The set of all equivalence classes of $\equiv_{\mathcal{O}}$ on $ST(S)$ is denoted by $ST(S)/\equiv_{\mathcal{O}}$. Conditions (1)–(2) imply that the ordering on the set $ST(S)/\equiv_{\mathcal{O}}$ induced by $\preceq_{\mathcal{O}}$ (defined such that $[u] \preceq_{\mathcal{O}} [v]$ iff $u \preceq_{\mathcal{O}} v$ for every $[u], [v] \in ST(S)/\equiv_{\mathcal{O}}$) is monotonic on $ST(S)/\equiv_{\mathcal{O}}$. Furthermore, by (2), this ordering is total on each nontrivial equivalence class $\widehat{C}$ of $\equiv_{\mathcal{O}} / \Pi$, and the set of least elements these classes $\widehat{C}$ is totally ordered.

Given a finite set E of equations, if Π is the partition associated with the congruence closure of E, by an *order assignment for E* we mean an order assignment for Π.

The following lemma gives a useful method for obtaining order assignments.

Lemma 10. *Given a finite set S of terms and a partition Π on $ST(S)$, given any ground substitution θ such that $Var(\Pi) \subseteq D(\theta)$, the preorder $\preceq_{\theta,\Pi}$ is an order assignment for Π.*

In view of Lemma 10, the following definition is justified.

Definition 42. Given a finite set of terms S, an order assignment $\preceq_{\mathcal{O}}$ for a partition Π on $ST(S)$ is *realized* by a ground substitution θ such that $Var(\Pi) \subseteq D(\theta)$ iff $\preceq_{\mathcal{O}} = \preceq_{\theta,\Pi}$.

By condition (2) of Definition 40, the equivalence classes of $\equiv_{\mathcal{O}}$ have some common unifier. We now show how a triangular form of a joint mgu of these classes can be obtained.

Definition 43. Given an order assignment $\preceq_{\mathcal{O}}$ for a partition Π on $ST(S)$, for every nontrivial equivalence class C of $\equiv_{\mathcal{O}}$, let $S_C = \{\langle t_2, t_1\rangle, \langle t_3, t_1\rangle, \ldots, \langle t_n, t_1\rangle\}$, where t_1 is any chosen representative in C and $C = \{t_1, \ldots, t_n\}$, and let $S_{\mathcal{O}} = \bigcup S_C$ be the union of these systems. From the way the term system $S_{\mathcal{O}}$ is constructed, a substitution unifies $S_{\mathcal{O}}$ iff it unifies every class of $\equiv_{\mathcal{O}}$. Thus, we let $TU_{\mathcal{O}}$ denote the triangular form of the mgu of $S_{\mathcal{O}}$. We also denote by $\sigma_{\mathcal{O}}$ the substitution $[s_1/x_1, \ldots, s_k/x_k]$ defined by the triangular form $TU_{\mathcal{O}} = \{\langle x_1, s_1\rangle, \ldots, \langle x_k, s_k\rangle\}$, as explained after Definition 28.

Given two order assignments $\mathcal{O}$ on a partition Π for $ST(S)$ and $\mathcal{O}'$ on a partition Π' for $ST(S')$, we say that $\mathcal{O}$ and $\mathcal{O}'$ are *compatible* iff they coincide on $ST(S) \cap ST(S')$.

Example 10. Let $E = \{fx \doteq fgy,\ fgy \doteq gy,\ hgz \doteq gz\}$, as in Example 8. The nontrivial equivalence classes of the congruence closure Π of E are $\{fx, fgy, gy\}$, and $\{hgz, gz\}$. The preorder $\mathcal{O}_1$ on $\{x,\ y,\ z,\ fx,\ gy,\ gz,\ fgy,\ hgz\}$ of Example 8 whose only nontrivial equivalence classes are $\{x, gy\}$ and $\{fx, fgy\}$, and such that

$$y \preceq_{\theta,\Pi} z \preceq_{\theta,\Pi} x \preceq_{\theta,\Pi} gy \preceq_{\theta,\Pi} gz \preceq_{\theta,\Pi} fx \preceq_{\theta,\Pi} fgy \preceq_{\theta,\Pi} hgz \tag{36}$$

is an order assignment realized by $\theta = [ga/x, a/y, b/z]$.

Let $\mathcal{O}_2$ be the preorder on $\{x,\ y,\ z,\ fx,\ gy,\ gz,\ fgy,\ hgz\}$ whose equivalence relation is the identity relation, and such that $gy \preceq_{\mathcal{O}_2} gz$, $gy \preceq_{\mathcal{O}_2} fgy$, $fgy \preceq_{\mathcal{O}_2} fx$, and $gz \preceq_{\mathcal{O}_2} hgz$ (other pairs in $\preceq_{\mathcal{O}_2}$ are obtained by transitivity and reflexivity). It is immediately verified that $\mathcal{O}_2$ is an order assignment, but it is not realized by $\theta = [ga/x, a/y, b/z]$. What is true though is that $\preceq_{\mathcal{O}_2} \subseteq \preceq_{\theta,\Pi}$.

The next example arises from the problem of proving that every monoid such that $x \cdot x = 1$ (for all x) is commutative.

Example 11. Let $\mathcal{E}$ be the set of equations

$$\begin{aligned}
\mathcal{E} = \{u_1 \cdot 1 &\doteq u_1 \\
w_1 \cdot w_1 &\doteq 1 \\
x_1 \cdot (y_1 \cdot z_1) &\doteq (x_1 \cdot y_1) \cdot z_1 \\
x_2 \cdot (y_2 \cdot z_2) &\doteq (x_2 \cdot y_2) \cdot z_2 \\
w_2 \cdot w_2 &\doteq 1 \\
1 \cdot v_1 &\doteq v_1 \\
x_3 \cdot (y_3 \cdot z_3) &\doteq (x_3 \cdot y_3) \cdot z_3 \\
x_4 \cdot (y_4 \cdot z_4) &\doteq (x_4 \cdot y_4) \cdot z_4 \\
w_3 \cdot w_3 &\doteq 1 \\
eq(a \cdot b, b \cdot a) &\doteq F \\
eq(z, z) &\doteq T\}.
\end{aligned} \tag{37}$$

The nontrivial equivalence classes of the congruence closure Π of $\mathcal{E}$ are:

$$\begin{aligned}
&\{T,\ eq(z,z)\},\\
&\{F,\ eq(a \cdot b, b \cdot a)\},\\
&\{1,\ w_2 \cdot w_2,\ w_3 \cdot w_3,\ w_1 \cdot w_1\},\\
&\{u_1,\ u_1 \cdot 1\},\\
&\{v_1,\ 1 \cdot v_1\},\\
&\{x_2 \cdot (y_2 \cdot z_2),\ (x_2 \cdot y_2) \cdot z_2\},\\
&\{x_4 \cdot (y_4 \cdot z_4),\ (x_4 \cdot y_4) \cdot z_4\},\\
&\{x_3 \cdot (y_3 \cdot z_3),\ (x_3 \cdot y_3) \cdot z_3\},\\
&\{x_1 \cdot (y_1 \cdot z_1),\ (x_1 \cdot y_1) \cdot z_1\}.
\end{aligned}$$

We define the order assignment $\mathcal{O}$ on Π whose equivalence relation is the identity and such that the ordering $\preceq_{\mathcal{O}}$ is defined by the order in which the elements in each class of Π are listed, and for the least elements in these classes, the order in which the classes are listed. All other pairs in $\preceq_{\mathcal{O}}$ are determined by reflexivity and transitivity. It is easily seen that there is a total simplification ordering on ground terms such that $1 \prec a \prec b \prec \cdot$, and one can verify that $\preceq_{\mathcal{O}}$ is an order assignment.

Another order assignment $\preceq_{\mathcal{O}'}$ is defined as the preorder extending $\preceq_{\mathcal{O}}$ and whose nontrivial equivalence classes of $\equiv_{\mathcal{O}'}$ are

$$\begin{aligned}
&\{a,\ u_1,\ x_1,\ x_2,\ y_2,\ w_2,\ x_4\},\\
&\{b,\ z_2,\ v_1,\ x_3,\ z_3,\ y_4,\ z_4,\ w_3\},\\
&\{w_2 \cdot w_2,\ x_2 \cdot y_2\},\\
&\{a \cdot b,\ w_1,\ y_1,\ z_1,\ y_3,\ z,\ y_2 \cdot z_2,\ x_4 \cdot y_4\},\\
&\{w_3 \cdot w_3,\ y_4 \cdot z_4\},\\
&\{w_1 \cdot w_1,\ y_1 \cdot z_1\},
\end{aligned}$$

these classes being ordered as listed (since $a \prec b$). It is easy to verify that $\preceq_{\mathcal{O}'}$ is realized by the substitution

$$\begin{aligned}
\theta = [&a/u_1,\ a/x_1,\ a/x_2,\ a/y_2,\ a/w_2,\ a/x_4,\\
&b/z_2,\ b/v_1,\ b/x_3,\ b/z_3,\ b/y_4,\ b/z_4,\ b/w_3,\\
&a \cdot b/w_1,\ a \cdot b/y_1,\ a \cdot b/z_1,\ a \cdot b/y_3,\ a \cdot b/z].
\end{aligned} \tag{38}$$

Note that $\equiv_{\mathcal{O}'}$ causes the merging of some equivalence classes of Π, even some trivial ones.

One more issue that we would like to address before presenting a revised version of the procedure of Definition 38 is the simplification of equations using the equivalence relation $\equiv_{\mathcal{O}}$. This is primarily for efficiency reasons. The problem is illustrated by the order assignment $\preceq_{\mathcal{O}'}$ of Example 11.

Example 12. Recall that the nontrivial equivalence classes of $\equiv_{\mathcal{O}'}$ (from Example 11) are

$$
\begin{aligned}
&(1)\ \{a,\ u_1,\ x_1,\ x_2,\ y_2,\ w_2,\ x_4\},\\
&(2)\ \{b,\ z_2,\ v_1,\ x_3,\ z_3,\ y_4,\ z_4,\ w_3\},\\
&(3)\ \{w_2 \cdot w_2,\ x_2 \cdot y_2\},\\
&(4)\ \{a \cdot b,\ w_1,\ y_1,\ z_1,\ y_3,\ z,\ y_2 \cdot z_2,\ x_4 \cdot y_4\},\\
&(5)\ \{w_3 \cdot w_3,\ y_4 \cdot z_4\},\\
&(6)\ \{w_1 \cdot w_1,\ y_1 \cdot z_1\}.
\end{aligned}
$$

The problem is to simplify the equations by replacing subterms by equivalent terms modulo $\equiv_{\mathcal{O}'}$, in such a way that $\preceq_{\mathcal{O}'}$ is a partial order on the new partition associated with the set of simplified equations. Clearly, this is a problem of choice of representatives. For example, how do we simplify $x_1 \cdot (y_1 \cdot z_1) \doteq (x_1 \cdot y_1) \cdot z_1$? If we choose the first element of each class as a representative, then the above equation simplifies to $a \cdot (w_1 \cdot w_1) \doteq (a \cdot (a \cdot b)) \cdot (a \cdot b)$, if we replace maximal subterms (in the subterm ordering) by their representatives. But it is preferable to replace each variable by its representative in the class, since we obtain the ground equation $a \cdot ((a \cdot b) \cdot (a \cdot b)) \doteq (a \cdot (a \cdot b)) \cdot (a \cdot b)$. So, how do we proceed? A key observation is that the subterm ordering induces a strict order on the classes modulo $\equiv_{\mathcal{O}'}$. A class C precedes a class C', denoted as $C \ll C'$, iff C contains some term that is a proper subterm of some term in C'. Thus, (1) $\ll$ (3), (1) $\ll$ (4), (2) $\ll$ (4), (2) $\ll$ (5), (4) $\ll$ (6), and the other relations are obtained by transitivity and reflexivity. We propose to assign representatives from the bottom up, starting with the minimal classes (w.r.t. $\ll$) and proceeding up using the ordering $\ll$ on the classes. Furthermore, whenever possible, we pick ground representatives. For example, we would pick a in (1), b in (2), and then $a \cdot a$ in (3), $a \cdot b$ in (4), $b \cdot b$ in (5), and $(a \cdot b) \cdot (a \cdot b)$ in (6).

Before we proceed with rigorous definitions, let us observe that if $\mathcal{O}$ is an order assignment on a partition Π, since the classes modulo $\equiv_{\mathcal{O}}$ have some joint unifier, every nontrivial class contains at most one ground term, and all compound terms in a nontrivial class have the same root symbol. With a slight abuse of notation, we let $\equiv_{\mathcal{O}}$ denote the set of equivalence classes of the equivalence relation $\equiv_{\mathcal{O}}$.

Definition 44. Let $\mathcal{O}$ be an order assignment on a partition Π. The relation $\ll$ is defined on the set of classes modulo $\equiv_{\mathcal{O}}$ as follows: given any two classes $K, K' \in \equiv_{\mathcal{O}}$, $K \ll K'$ iff there are terms $t \in K$ and $t' \in K'$ such that t is a proper subterm of t'.

Lemma 11. *The relation $\ll$ given in Definition 44 is a strict order on the set of classes modulo $\equiv_{\mathcal{O}}$, and if $K \ll K'$ then $K \prec_{\mathcal{O}} K'$.*

We now use the strict order $\ll$ on the classes modulo $\equiv_{\mathcal{O}}$ to assign representatives inductively.

Definition 45. Let $\mathcal{O}$ be an order assignment on a partition Π on $ST(S)$. A function $\rho : \equiv_{\mathcal{O}} \to T_{\Sigma}(X)$ assigning a term $\rho(K)$ to every equivalence class K modulo $\equiv_{\mathcal{O}}$ is a *representative selector* iff for every minimal class K w.r.t. $\ll$, $\rho(K)$ is the unique ground term in K if it exists, or else any chosen element of K, and for every nonminimal class K, if

$$\widehat{K} = \{f(t_1^1, \dots, t_m^1), \dots, f(t_1^n, \dots, t_m^n)\} \tag{39}$$

is the subset of compound elements in K, then $\rho(K)$ is the unique ground term in

$$\begin{aligned}\rho(\widehat{K}) = \{&f(\rho([t_1^1]_{\equiv_{\mathcal{O}}}), \dots, \rho([t_m^1]_{\equiv_{\mathcal{O}}})), \dots,\\ &f(\rho([t_1^n]_{\equiv_{\mathcal{O}}}), \dots, \rho([t_m^n]_{\equiv_{\mathcal{O}}}))\}\end{aligned} \tag{40}$$

if it exists, or else any chosen element in $\rho(\widehat{K})$, where $[u]_{\equiv_{\mathcal{O}}}$ denotes the equivalence class of u modulo $\equiv_{\mathcal{O}}$.

The *reduced partition* $\rho(\Pi)$ is the partition whose classes are the sets of the form $\{\rho(K) \mid K \in \widehat{C},\ K \text{ is a class modulo } \equiv_{\mathcal{O}}, \widehat{C} \in \equiv_{\mathcal{O}} /\Pi\}$, and if Π is the congruence closure associated with a set E of equations, the *reduced set of equations* $\rho(E)$ is the set of equations $\{\rho([l]_{\equiv_{\mathcal{O}}}) \doteq \rho([r]_{\equiv_{\mathcal{O}}}) \mid l \doteq r \in E\}$. We also define the preorder

$\rho(\preceq_{\mathcal{O}})$ on $\rho(\Pi)$ such that $\rho(K)\ \rho(\preceq_{\mathcal{O}})\ \rho(K')$ iff $u \preceq_{\mathcal{O}} v$ for some $u \in K$, $v \in K'$. It is obvious that $\rho(\preceq_{\mathcal{O}})$ is a partial order on $\rho(\Pi)$ since K, K' are classes modulo $\equiv_{\mathcal{O}}$.

Note that $\rho(K) \in K$ if K is a minimal class (w.r.t. $\ll$), but it is possible that $\rho(K) \notin K$ if K is not minimal. However, as shown in the next lemma, ρ is injective and even though $\rho(K)$ may not be in K, this does not matter for our purposes as shown below.

Lemma 12. *Let $\mathcal{O}$ be an order assignment on a partition Π on $ST(S)$, and θ any joint unifier of the classes modulo $\equiv_{\mathcal{O}}$. (i) For every class K modulo $\equiv_{\mathcal{O}}$, $\theta(K) = \theta(\rho(K))$ (with the slight abuse of notation where $\theta(K)$ denotes the term s such that $\theta(u) = s$ for all $u \in K$). (ii) Every representative selector is injective.*

The following lemma shows that representative selectors always exist.

Lemma 13. *(Construction of representative selectors) Let $\mathcal{O}$ be an order assignment on a partition Π on $ST(S)$. There is an algorithm to construct representative selectors $\rho : \equiv_{\mathcal{O}} \to T_\Sigma(X)$.*

Lemma 14. *Let $\mathcal{O}$ be an order assignment on a partition Π on $ST(S)$. The strict order $\rho(\preceq_{\mathcal{O}})$ on $\rho(\Pi)$ is a simplification ordering that is total on the nontrivial classes of $\rho(\Pi)$ and total on the set of least elements of these nontrivial classes. If Π is the congruence closure associated with a set E of equations, then $\rho(\Pi)$ is the congruence closure associated with $\rho(E)$.*

We can now modify the procedure of Definition 38 in order to accommodate variables.

Definition 46. (Reduction procedure R) Let $\prec$ be a total simplification ordering on ground terms. Let $\mathcal{E} = \mathcal{E}_\Sigma \cup \{eq(u,v) \doteq F,\ eq(z,z) \doteq T\}$ be a finite set of equations, where $\mathcal{E}_\Sigma$ is a set of equations over $T_\Sigma(X)$, and $u, v \in T_\Sigma(X)$. Given any order assignment $\mathcal{O}$ on $\mathcal{E}$, the procedure R returns a rigid reduced rewrite system $R(\mathcal{E},\mathcal{O})$. To form the system $R(\mathcal{E},\mathcal{O})$, first, we use the algorithm of Lemma 13 to get a representative selector ρ for $\equiv_{\mathcal{O}}$ (if it is not the identity), and we let $\mathcal{E}'$ be the reduced set $\rho(\mathcal{E})$. Trivial equations are discarded. Let Π' be the congruence closure associated with $\mathcal{E}'$.

By Lemma 14, $\rho(\preceq_{\mathcal{O}})$ is a simplification ordering that is total on every nontrivial equivalence class of Π' and total on the set of least elements of these classes. From this point on, we apply to $\mathcal{E}'$ and Π' the procedure described in Definition 38, except that at the end of every round, it may be necessary to extend $\mathcal{O}$ since new terms may arise due to simplification. If at every round an extension of $\mathcal{O}$ can be found so that the next step can be performed, R succeeds and returns a rigid reduced rewrite system denoted as $R(\mathcal{E}, \mathcal{O})$. Otherwise, R returns failure.

It is useful to remark that since the reduction procedure deals with sets of equations of the form $\mathcal{E} = \mathcal{E}_\Sigma \cup \{eq(u, v) \doteq F,\ eq(z, z) \doteq T\}$, in the congruence closure Π of $\mathcal{E}$, the classes of T and F are always $\{eq(u, v), F\}$ and $\{eq(z, z), T\}$. From the way we have extended $\preceq$ to take care of T, F, and terms involving eq, it will be shown as a corollary of Theorem 4 that there is no loss of generality in choosing order assignments such that $T \preceq_{\mathcal{O}} F \preceq_{\mathcal{O}} s \preceq_{\mathcal{O}} eq(u, v)$ for all $s, u, v \in T_\Sigma(X)$. Using Lemma 10, we can show the following crucial result.

Lemma 15. *Let $\mathcal{E} = \mathcal{E}_\Sigma \cup \{eq(u, v) \doteq F,\ eq(z, z) \doteq T\}$ be a finite set of equations, where $\mathcal{E}_\Sigma$ is a set of equations over $T_\Sigma(X)$, $u, v \in T_\Sigma(X)$, and $\prec$ a total simplification ordering on ground terms.*

(i) *Given an order assignment $\mathcal{O}$ on $\mathcal{E}$, if a substitution θ (not necessarily ground) unifies $TU_{\mathcal{O}}$ and R does not fail, then $\theta(R(\mathcal{E}, \mathcal{O}))$ is rigid equivalent to $\theta(\mathcal{E})$.*

(ii) *Given an order assignment $\mathcal{O}$ on $\mathcal{E}$, if some ground substitution θ realizes $\mathcal{O}$ and R does not fail, then $\theta(R(\mathcal{E}, \mathcal{O}))$ is rigid equivalent to $\theta(\mathcal{E})$.*

It is important to note that part (i) of Lemma 15 holds even if θ is not ground. This fact will be used in the proof that the method is sound. We are now ready to define a procedure for finding rigid E-unifiers.

9. A Method for Finding Complete Sets of Rigid E-Unifiers

This method uses the reduction procedure of Section 8 and a single transformation on certain systems defined next. First, the following definition is needed.

Definition 47. Given a set E of equations and some equation $l \doteq r$, the set of equations obtained from E by deleting $l \doteq r$ and $r \doteq l$ from E is denoted by $(E - \{l \doteq r\})^\dagger$. Formally, we let $(E - \{l \doteq r\})^\dagger = \{u \doteq v \mid u \doteq v \in E,\ u \doteq v \neq l \doteq r, \text{ and } u \doteq v \neq r \doteq l\}$.

Definition 48. Let $\prec$ be a total simplification ordering on ground terms. We shall be considering finite sets of equations of the form $\mathcal{E} = \mathcal{E}_\Sigma \cup \{eq(u,v) \doteq F,\ eq(z,z) \doteq T\}$, where $\mathcal{E}_\Sigma$ is a set of equations over $T_\Sigma(X)$, and $u, v \in T_\Sigma(X)$. We define a transformation on systems of the form $\langle \mathcal{S}, \mathcal{E}, \mathcal{O}\rangle$, where $\mathcal{S}$ is a term system, $\mathcal{E}$ a set of equations as above, and $\mathcal{O}$ an order assignment:

$$\langle \mathcal{S}_0,\ \mathcal{E}_0,\ \mathcal{O}_0\rangle \Rightarrow \langle \mathcal{S}_1,\ \mathcal{E}_1,\ \mathcal{O}_1\rangle, \tag{41}$$

where $l_1 \doteq r_1,\ l_2 \doteq r_2 \in \mathcal{E}_0 \cup \mathcal{E}_0^{-1}$, either l_1/β is *not* a variable or $l_2 \doteq r_2$ is degenerate, $l_1/\beta \neq l_2$, $TU(l_1/\beta, l_2)$ represents an mgu of l_1/β and l_2 in triangular form,[17] $\sigma = [t_1/x_1, \ldots, t_p/x_p]$ where $TU(l_1/\beta, l_2) = \{\langle x_1, t_1\rangle, \ldots, \langle x_p, t_p\rangle\}$,

$$\mathcal{E}'_1 = \sigma((\mathcal{E}_0 - \{l_1 \doteq r_1\})^\dagger \cup \{l_1[\beta \leftarrow r_2] \doteq r_1\}), \tag{42}$$

$\mathcal{O}_1$ is an order assignment on $\mathcal{E}'_1$ compatible with $\mathcal{O}_0$, $\mathcal{S}_1 = \mathcal{S}_0 \cup TU(l_1/\beta, l_2) \cup TU_{\mathcal{O}_1}$, and $\mathcal{E}_1 = R(\mathcal{E}'_1, \mathcal{O}_1)$.

Observe that $\sigma(l_1[\beta \leftarrow r_2] \doteq r_1)$ looks like a critical pair of equations in $\mathcal{E}_0 \cup \mathcal{E}_0^{-1}$, but it is not. This is because a critical pair is formed by applying the mgu of l_1/β and l_2 to $l_1[\beta \leftarrow r_2] \doteq r_1$, but $[t_1/x_1, \ldots, t_p/x_p]$ is usually not a mgu of l_1/β and l_2. It is the composition $[t_1/x_1] \circ \ldots \circ [t_p/x_p]$ that is a mgu of l_1/β and l_2. The reason for not applying the mgu is that by repeated applications of this step, exponential size terms could be formed, and it would not be clear that the decision procedure is in NP. We have chosen an approach of "lazy" (or delayed) unification. Also note that we use the rigid reduced system $R(\mathcal{E}'_1, \mathcal{O}_1)$ rather than $\mathcal{E}'_1$, and so, a transformation step is defined only if R does not fail. The method then is the following.

Definition 49. (Method) Let $E_{u,v} = E \cup \{eq(u,v) \doteq F,\ eq(z,z) \doteq T\}$, $\mathcal{O}_0$ an order assignment on $E_{u,v}$, $\mathcal{S}_0 = TU_{\mathcal{O}_0}$, $\mathcal{E}_0 = R(E_{u,v}, \mathcal{O}_0)$,

[17] Note that we are requiring that l_1/β and l_2 have a *nontrivial* unifier. The triangular form of mgus is important for the NP-completeness of this method.

m the total number of variables in $\mathcal{E}_0$, and $V = Var(E) \cup Var(u, v)$. For any sequence

$$\langle \mathcal{S}_0, \mathcal{E}_0, \mathcal{O}_0 \rangle \Rightarrow^+ \langle \mathcal{S}_k, \mathcal{E}_k, \mathcal{O}_k \rangle \tag{43}$$

consisting of at most m transformation steps, if $\mathcal{S}_k$ is unifiable and $k \leq m$ is the first integer in the sequence such that $F \doteq T \in \mathcal{E}_k$, return the substitution $\theta_{\mathcal{S}_k}|_V$, where $\theta_{\mathcal{S}_k}$ is the mgu of $\mathcal{S}_k$ (over $T_\Sigma(X)$).

Example 13. Let E be the set of equations $E = \{fa \doteq a,\ fa \doteq x\}$, and $\langle u, v \rangle = \langle gx, x \rangle$. We have

$$E_{u,v} = \{fa \doteq a,\ x \doteq fa,\ eq(gx, x) \doteq F,\ eq(z, z) \doteq T\}. \tag{44}$$

The congruence closure Π of $E_{u,v}$ has three nontrivial classes $\{a, fa, x\}$, $\{eq(gx, x), F\}$, and $\{eq(z, z), T\}$. Let $\mathcal{O}_0$ be the order assignment on $E_{u,v}$ where every equivalence class is trivial and such that

$$\begin{aligned} &T \prec_{\mathcal{O}_0} eq(gx, x), \\ &F \prec_{\mathcal{O}_0} eq(z, z), \\ &a \prec_{\mathcal{O}_0} fa \prec_{\mathcal{O}_0} x, \end{aligned}$$

the least elements of classes being ordered in the order of listing of the classes. We have $\mathcal{S}_0 = \emptyset$, and the reduced system $\mathcal{E}_0 = R(E_{u,v}, \mathcal{O}_0)$ is

$$\mathcal{E}_0 = \{fa \doteq a,\ x \doteq a,\ eq(ga, a) \doteq F,\ eq(z, z) \doteq T\}. \tag{45}$$

Note that there is an overlap between $eq(ga, a) \doteq F$ and $x \doteq a$ at address 1 in $eq(ga, a)$, and we obtain the triangular system $\{\langle x, ga \rangle\}$ and the new equation $eq(a, a) \doteq F$. Thus, we have

$$\langle \mathcal{S}_0, \mathcal{E}_0, \mathcal{O}_0 \rangle \Rightarrow \langle \mathcal{S}_1, \mathcal{E}_1, \mathcal{O}_1 \rangle, \tag{46}$$

where $\mathcal{S}_1 = \{\langle x, ga \rangle\}$,

$$\mathcal{E}_1' = \{fa \doteq a,\ ga \doteq a,\ eq(a, a) \doteq F,\ eq(z, z) \doteq T\}, \tag{47}$$

$\mathcal{O}_1$ is the restriction of $\mathcal{O}_0$ to the subterms in $\mathcal{E}_1'$, and $\mathcal{E}_1 = \mathcal{E}_1'$ since $\mathcal{E}_1'$ is reduced. Next, we have an overlap between $eq(z, z) \doteq T$ and

$eq(a, a) \doteq F$, obtaining the new equation $F \doteq T$, and the new system $\{\langle z, a\rangle\}$. Thus,

$$\langle \mathcal{S}_1, \mathcal{E}_1, \mathcal{O}_1\rangle \Rightarrow \langle \mathcal{S}_2, \mathcal{E}_2, \mathcal{O}_2\rangle, \tag{48}$$

where $\mathcal{S}_2 = \{\langle x, ga\rangle,\ \langle z, a\rangle\}$,

$$\mathcal{E}_2' = \{fa \doteq a,\ ga \doteq a,\ eq(a, a) \doteq F,\ F \doteq T\}, \tag{49}$$

and $\mathcal{O}_2$ is the restriction of $\prec$ to the subterms in $\mathcal{E}_2'$. After reducing $\mathcal{E}_2'$, we have

$$\mathcal{E}_2 = \{fa \doteq a,\ ga \doteq a,\ eq(a, a) \doteq T,\ F \doteq T\}. \tag{50}$$

Since $F \doteq T \in \mathcal{E}_2$ and $\mathcal{S}_2$ is unifiable, the restriction $[ga/x]$ of the mgu $[ga/x, a/z]$ of $\mathcal{S}_2$ to $Var(E) \cup Var(u, v) = \{x\}$ is a rigid E-unifier of gx and x.

10. Soundness, Completeness, and Decidability of the Method

The main properties of the method are given in this section.

Theorem 3 (Soundness). *Let E be a set of equations over $T_\Sigma(X)$, u, v two terms in $T_\Sigma(X)$, $E_{u,v} = E \cup \{eq(z, z) \doteq T,\ eq(u, v) \doteq F\}$, $\mathcal{O}_0$ an order assignment on $E_{u,v}$, $\mathcal{S}_0 = TU_{\mathcal{O}_0}$, $\mathcal{E}_0 = R(E_{u,v}, \mathcal{O}_0)$, m the total number of variables in $\mathcal{E}_0$, and $V = Var(E) \cup Var(u, v)$. If*

$$\langle \mathcal{S}_0, \mathcal{E}_0, \mathcal{O}_0\rangle \Rightarrow^+ \langle \mathcal{S}_k, \mathcal{E}_k, \mathcal{O}_k\rangle \tag{51}$$

where $\mathcal{S}_k$ is unifiable, $F \doteq T \in \mathcal{E}_k$ and $F \doteq T \notin \mathcal{E}_i$ for all i, $0 \le i < k \le m$, then $\theta_{\mathcal{S}_k}|_V$ is a rigid E-unifier of u and v, where $\theta_{\mathcal{S}_k}$ is the mgu of $\mathcal{S}_k$ (over $T_\Sigma(X)$).

The proof of Theorem 3 does not use the fact that the systems $R(\mathcal{E}_i', \mathcal{O}_i)$ are rigid reduced, but only the fact that $\theta'(\mathcal{E}_i)$ and $\theta'(R(\mathcal{E}_i', \mathcal{O}_i))$ are rigid equivalent provided that θ' unifies $TU_{\mathcal{O}_i}$. However, the fact that the systems $R(\mathcal{E}_i', \mathcal{O}_i)$ are rigid reduced plays a crucial role in the proof of the completeness theorem. The $\mathcal{O}_i$'s are only needed for the completeness of the method, and to make sure that the reduction procedure terminates. We now turn to the completeness part. The main technique is roughly the removal of peaks by the use of critical pairs (Bachmair, 1987; Bachmair, Dershowitz, and Plaisted, 1987).

Theorem 4 (Completeness). *Let E be a set of equations over $T_\Sigma(X)$ and u, v two terms in $T_\Sigma(X)$. If θ is any rigid E-unifier of u and v, then there is an order assignment $\mathcal{O}_0$ on $E_{u,v}$, and letting $\mathcal{S}_0 = TU_{\mathcal{O}_0}$, $\mathcal{E}_0 = R(E_{u,v}, \mathcal{O}_0)$, m the number of variables in $R(E_{u,v}, \mathcal{O}_0)$, and $V = Var(E) \cup Var(u,v)$, there is a sequence of transformations*

$$\langle \mathcal{S}_0, \mathcal{E}_0, \mathcal{O}_0 \rangle \Rightarrow^+ \langle \mathcal{S}_k, \mathcal{E}_k, \mathcal{O}_k \rangle, \tag{52}$$

where $k \leq m$, $\mathcal{S}_k$ is unifiable, $F \doteq T \in \mathcal{E}_k$, $F \doteq T \notin \mathcal{E}_i$ for all i, $0 \leq i < k$, and $\theta_{\mathcal{S}_k}|_V \leq_E \theta[V]$, where $\theta_{\mathcal{S}_k}$ is the mgu of $\mathcal{S}_k$ over $T_\Sigma(X)$. Furthermore, $\theta_{\mathcal{S}_k}|_V$ is a rigid E-unifier of u and v.

Corollary 2. *If θ' is the mgu produced by a sequence of steps as in the soundness theorem, there is a ground substitution θ_1 such that $V \subseteq D(\theta_1)$ and a sequence of steps*

$$\langle \mathcal{S}_0, \mathcal{E}_0, \mathcal{O}_0 \rangle \Rightarrow^+ \langle \mathcal{S}_k, \mathcal{E}_k, \mathcal{O}_k, \rangle \tag{53}$$

such that $\theta_1 \sqsubseteq_E \widehat{\theta'}$, θ_1 is a unifier of $\mathcal{S}_k$, and θ_1 realizes all the $\mathcal{O}_i$'s in the above sequence. In particular, the method is still complete if we restrict ourselves to order assignments $\mathcal{O}$ such that $T \preceq_{\mathcal{O}} F \preceq_{\mathcal{O}} s \preceq_{\mathcal{O}} eq(u,v)$ for all $s, u, v \in T_\Sigma(X)$.

Theorem 4 also shows that rigid E-unification is decidable.

Corollary 3. *Rigid E-unification is decidable.*

Combining the results of Theorem 3 and 4, we also obtain the fact that for any E, u, v, there is always a finite complete set of rigid E-unifiers.

Theorem 5. *Let E be a set of equations over $T_\Sigma(X)$, u, v two terms in $T_\Sigma(X)$, m the number of variables in $E \cup \{u, v\}$, and $V = Var(E) \cup Var(u,v)$. There is a finite complete set of rigid E-unifiers for u and v given by the set*

$$\{\theta_{\mathcal{S}_k}|_V \mid \langle \mathcal{S}_0, \mathcal{E}_0, \mathcal{O}_0 \rangle \Rightarrow^+ \langle \mathcal{S}_k, \mathcal{E}_k, \mathcal{O}_k \rangle,\ k \leq m\}, \tag{54}$$

for any order assignment $\mathcal{O}_0$ on $E_{u,v}$, with $\mathcal{S}_0 = TU_{\mathcal{O}_0}$, $\mathcal{E}_0 = R(E_{u,v}, \mathcal{O}_0)$, and where $\mathcal{S}_k$ is unifiable, $F \doteq T \in \mathcal{E}_k$, $F \doteq T \notin \mathcal{E}_i$ for all i, $0 \leq i < k$, and $\theta_{\mathcal{S}_k}$ is the mgu of $\mathcal{S}_k$ over $T_\Sigma(X)$.

11. NP-Completeness of Rigid E-Unification

First, recall that rigid E-unification is NP-hard. This holds even for ground sets of equations, as shown by Kozen (1976, 1977). Using an idea of Kozen (1976), we show that rigid E-unification is NP-hard even when all equations in E are regular, all ground except one, and u and v are ground.

Definition 50. An equation $(l \doteq r)$ is *regular* iff $Var(l) = Var(r)$.

Theorem 6. *Rigid E-unification is NP-hard when all equations in E are regular, all ground except one, and u and v are ground.*

Proof. The satisfiability problem is reduced to rigid E-unification as follows. Let the set of function symbols consist of $\wedge$, $\vee$, $\neg$, and the constants $\top$ and $\bot$. Write down the set E_{bool} of 10 ground equations corresponding to the truth tables for $\wedge$, $\vee$, $\neg$. Given any clause A, if $Var(A) = \{x_1, \ldots, x_n\}$, let

$$B_A = (x_1 \wedge x_2 \wedge \ldots \wedge x_n \wedge \bot). \tag{55}$$

Finally, let $E_A = E_{bool} \cup \{A \doteq B_A\}$, $u = \top$ and $v = \bot$. Clearly, $A \doteq B_A$ is regular, and it is easy to see that a substitution σ such that $\top$ and $\bot$ are congruent modulo $\sigma(E_A)$ exists iff A is satisfiable, since B_A is false for every truth assignment. Hence, satisfiability is reduced to rigid E-unification. □

We now show that rigid E-unification is in NP.

Theorem 7. *Rigid E-unification is NP-complete.*

Proof. We already know that rigid E-unification is NP-hard. By Corollary 3, the problem is decidable. It remains to show that it is in NP. From Corollary 3, u and v have some rigid E-unifier iff there is some sequence of transformations

$$\langle \mathcal{S}_0, \mathcal{E}_0, \mathcal{O}_0 \rangle \Rightarrow^+ \langle \mathcal{S}_k, \mathcal{E}_k, \mathcal{O}_k \rangle \tag{56}$$

of at most $k \leq m$ steps where m is the number of variables in $\mathcal{E}_0$, and such that $\mathcal{S}_k$ is unifiable (over $T_\Sigma(X)$), $F \doteq T \in \mathcal{E}_k$ and $F \doteq T \notin \mathcal{E}_i$ for all i, $0 \leq i < k$. We need to verify that it is possible to ckeck these conditions in polynomial time. This is because the number of steps is at most the number of variables in $\mathcal{E}_0$, a triangular form can be computed in polynomial time, applying the substitutions associated with triangular forms can also be done in polynomial time, and checking that an order assignment is feasible can be done in polynomial time. □

If E is a set of ground equations, the $\mathcal{O}_i$'s are useless and the reduction procedure R needs only be applied once at the beginning to E. Thus, Theorem 7 provides another proof of a result first established by Kozen (1976, 1977).

12. Applications of Rigid E-Unification to Equational Matings

The method developed for one set of equations and one pair can be easily generalized to tackle Problem 1. In fact, an algorithm to decide whether a family of mated sets is an equational (pre)mating is obtained. The method of Definition 49 can be generalized to pairs $\langle E, S\rangle$ (as defined in Problem 1 in the introduction) by considering triples $\langle \mathcal{S}, \mathcal{E}, \mathcal{O}\rangle$, where $\mathcal{S}$ is a term system, and $\mathcal{E}$ is an n-tuple of sets of equations. The definition of a rigid E-unifier of a pair is generalized as follows.

Definition 51. Let $E = \{E_i \mid 1 \leq i \leq n\}$ be a family of n sets of equations (over $T_\Sigma(X)$) and $S = \{\langle u_i, v_i\rangle \mid 1 \leq i \leq n\}$ a set of n pairs of terms (over $T_\Sigma(X)$). A substitution θ (over $T_\Sigma(X)$) is a *rigid E-unifier of S* iff

$$\theta(u_i) \overset{*}{\cong}_{\theta(E_i)} \theta(v_i) \tag{57}$$

for every i, $1 \leq i \leq n$. A pair $\langle E, S\rangle$ such that S has some rigid E-unifier is called an *equational premating.*

The suitable generalization of the preorder $\leq_E$ to a family $E = \{E_i \mid 1 \leq i \leq n\}$ of n sets of equations turns out to be the following.

Definition 52. Given a family $E = \{E_i \mid 1 \leq i \leq n\}$ of n sets of equations, for any (finite) set of variables W, for any two substitutions σ and θ, $\sigma \leq_E \theta$ iff there is some η such that $\sigma \circ \eta \sqsubseteq_{E_i} \theta[V]$ for every i, $1 \leq i \leq n$.

Note that this condition is stronger than the condition $\sigma \leq_{E_i} \theta[V]$ for every i, $1 \leq i \leq n$, because with this second condition we only know that there are substitutions $\eta_1, \ldots, \eta_n$ such that $\sigma \circ \eta_i \sqsubseteq_{E_i} \theta[V]$ for every i, $1 \leq i \leq n$. In Definition 52, it is required that $\eta_1 = \ldots = \eta_n$. The generalization of Theorem 4 goes through with the stronger Definition 52, which is obviously preferable.

Complete sets of rigid E-unifiers for S are defined as follows.

Definition 53. Let $E = \{E_i \mid 1 \leq i \leq n\}$ and $S = \{\langle u_i, v_i \rangle \mid 1 \leq i \leq n\}$ as in Definition 51, and let $V = Var(E) \cup Var(S)$. A set U of substitutions is a *complete set of rigid E-unifiers for S* iff: For every $\sigma \in U$,

(i) $D(\sigma) \subseteq V$ and $D(\sigma) \cap I(\sigma) = \emptyset$ (idempotence),
(ii) σ is a rigid E-unifier of S,
(iii) For every rigid E-unifier θ of S, there is some $\sigma \in U$ such that $\sigma \leq_E \theta[V]$.

Minimal rigid E-unifiers also exist and are defined as follows.

Definition 54. Let E be a set of sets of equations and S a term system as in Definition 51. For any ground rigid E-unifier θ of S, let

$$\begin{aligned} S_{E,S,\theta} = \{\rho \mid D(\rho) = D(\theta),\ & \rho(u_i) \overset{*}{\cong}_{\rho(E_i)} \rho(v_i), \\ & \rho \sqsubseteq_{E_i} \theta,\ 1 \leq i \leq n, \quad \text{and } \rho \text{ ground}\}. \end{aligned} \tag{58}$$

Since $\prec\!\!\prec$ is total and well-founded on ground substitutions with domain $D(\theta)$, the set $S_{E,S,\theta}$ contains some least element σ (w.r.t. $\prec\!\!\prec$).

It is easy to see that Lemma 8 can be generalized as follows.

Lemma 16. *Let E be a set of sets equations and S a term system as in Definition 51. For any ground rigid E-unifier θ of S, if σ is the least element of the set $S_{E,S,\theta}$ of Definition 54, then the following properties hold:*

(1) *$\sigma \sqsubseteq_{E_i} \theta$ for every i, $1 \leq i \leq n$,*
(2) *every term of the form $\sigma(x)$ is irreducible by every oriented instance $\sigma(l) \to \sigma(r)$ of a nondegenerate equation $l \doteq r \in E \cup E^{-1}$, and*
(3) *every proper subterm of a term of the form $\sigma(x)$ is irreducible by every oriented instance $\sigma(l) \to \sigma(r)$ of a degenerate equation $l \doteq r \in E \cup E^{-1}$.*

Lemma 9 is easily generalized as follows. We let $eq_1, \ldots, eq_n$ be n new distinct binary function symbols not in Σ (and distinct from T and F).

Lemma 17. *Let E be a set of sets equations and S a term system as in Definition 51. A substitution θ over $T_\Sigma(X)$ is a rigid E-unifier of S iff there is some substitution θ' over $T_\Sigma(X)$ such that $\theta = \theta'|_{D(\theta')-\{z_1,\ldots,z_n\}}$ and $T \overset{*}{\cong}_{\theta'(E^i)} F$ for every i, $1 \leq i \leq n$, where $E^i = E_i \cup \{eq_i(u_i, v_i) \doteq F,\ eq_i(z_i, z_i) \doteq T\}$, and $\{z_1, \ldots, z_n\}$ is a set of new variables not in $Var(E) \cup Var(S)$.*

The total simplification ordering $\prec$ is extended to the set

$$T_\Sigma \cup \{T, F\} \cup \bigcup_{i=1}^{i=n} \{eq_i(u, v) \mid u, v \in T_\Sigma\} \tag{59}$$

as follows:

For any terms $s, t, u, v \in T_\Sigma$,

(a) $T \prec F \prec u \prec eq_i(s, t)$;
(b) $eq_i(s, t) \prec eq_i(u, v)$ iff $\{s, t\} \prec_{lex} \{u, v\}$, where $\prec_{lex}$ is the lexicographic extension of $\prec$;
(c) $eq_i(s, t) \prec eq_j(u, v)$ iff $1 \leq i < j \leq n$.

Clearly, this extension of $\prec$ is a total simplification ordering. We define a transformation on systems as follows. We shall be considering n-tuples $\mathcal{E} = \langle \mathcal{E}^1, \ldots, \mathcal{E}^n \rangle$ of finite sets of equations of the form $\mathcal{E}^i = \mathcal{E}^i_\Sigma \cup \{eq_i(u, v) \doteq F,\ eq_i(z_i, z_i) \doteq T\}$, where $\mathcal{E}_\Sigma$ is a set of equations over $T_\Sigma(X)$ and $u, v \in T_\Sigma(X)$. We define a transformation on systems of the form $\langle \mathcal{S}, \mathcal{E}, \mathcal{O} \rangle$, where $\mathcal{S}$ is a term system, $\mathcal{E}$ an n-tuple of sets of equations as above, and $\mathcal{O}$ an order assignment:

$$\langle \mathcal{S}_0,\ \mathcal{E}_0,\ \mathcal{O}_0 \rangle \Rightarrow \langle \mathcal{S}_1,\ \mathcal{E}_1,\ \mathcal{O}_1 \rangle, \tag{60}$$

where $l_1 \doteq r_1$, $l_2 \doteq r_2 \in \mathcal{E}^i_0 \cup (\mathcal{E}^i_0)^{-1}$ for some i, $1 \leq i \leq n$, either l_1/β is *not* a variable or $l_2 \doteq r_2$ is degenerate, $l_1/\beta \neq l_2$, $TU(l_1/\beta, l_2)$ represents a mgu of l_1/β and l_2 in triangular form, $\sigma = [t_1/x_1, \ldots, t_p/x_p]$ where $TU(l_1/\beta, l_2) = \{\langle x_1, t_1 \rangle, \ldots, \langle x_p, t_p \rangle\}$,

$$\begin{aligned} &\mathcal{E}'^i_1 = \sigma((\mathcal{E}^i_0 - \{l_1 \doteq r_1\})^\dagger \cup \{l_1[\beta \leftarrow r_2] \doteq r_1\}) \qquad \text{and} \\ &\mathcal{E}'^j_1 = \sigma(\mathcal{E}^j_0) \text{ for every } j \neq i\ , \end{aligned} \tag{61}$$

$\mathcal{O}_1$ is an order assignment on $\mathcal{E}'_1$ compatible with $\mathcal{O}_0$, $\mathcal{S}_1 = \mathcal{S}_0 \cup TU(l_1/\beta, l_2) \cup TU_{\mathcal{O}_1}$, and $\mathcal{E}_1 = \langle \mathcal{E}^1_1, \ldots, \mathcal{E}^n_1 \rangle$, where $\mathcal{E}^j_1 = R(\mathcal{E}'^j_1, \mathcal{O}_1)$ for all j, $1 \leq j \leq n$.

The method for finding rigid E-unifiers of S is the following.

Definition 55. (Method) Let $E = \{E_i \mid 1 \leq i \leq n\}$ and $S = \{\langle u_i, v_i\rangle \mid 1 \leq i \leq n\}$ as in Definition 51, let $E^i = E_i \cup \{eq_i(u_i, v_i) \doteq F,\ eq_i(z_i, z_i) \doteq T\}$ for every i, $1 \leq i \leq n$, $\mathcal{O}_0$ an order assignment on $\langle E^1, \dots, E^n\rangle$, $\mathcal{S}_0 = TU_{\mathcal{O}_0}$, $\mathcal{E}_0^i = R(E^i, \mathcal{O}_0)$ for every i, $1 \leq i \leq n$, $\mathcal{E}_0 = \langle \mathcal{E}_0^1, \dots, \mathcal{E}_0^n\rangle$, m the total number of variables in $\mathcal{E}_0$, and $V = Var(E) \cup Var(S)$. For any sequence

$$\langle \mathcal{S}_0, \mathcal{E}_0, \mathcal{O}_0\rangle \Rightarrow^+ \langle \mathcal{S}_k, \mathcal{E}_k, \mathcal{O}_k\rangle$$

consisting of at most m transformation steps, if $\mathcal{S}_k$ is unifiable and $k \leq m$ is the first integer in the sequence such that $F \doteq T \in \mathcal{E}_k^i$ for every i, $1 \leq i \leq n$, return the substitution $\theta_{\mathcal{S}_k}|_V$, where $\theta_{\mathcal{S}_k}$ is the mgu of $\mathcal{S}_k$ (over $T_\Sigma(X)$).

The proofs of Theorem 3 and Theorem 4 can be easily adapted to prove that the finite set of all substitutions returned by the method of Definition 55 forms a complete set of rigid E-unifiers for S. In particular, the method provides a decision procedure for deciding whether a family of mated sets is an equational premating that is in NP.

Theorem 8 (Soundness). *Let $E = \{E_i \mid 1 \leq i \leq n\}$ and $S = \{\langle u_i, v_i\rangle \mid 1 \leq i \leq n\}$ as in Definition 51, let $E^i = E_i \cup \{eq_i(u_i, v_i) \doteq F,\ eq_i(z_i, z_i) \doteq T\}$ for every i, $1 \leq i \leq n$, $\mathcal{O}_0$ an order assignment on $\langle E^1, \dots, E^n\rangle$, $\mathcal{S}_0 = TU_{\mathcal{O}_0}$, $\mathcal{E}_0^i = R(E^i, \mathcal{O}_0)$ for every i, $1 \leq i \leq n$, $\mathcal{E}_0 = \langle \mathcal{E}_0^1, \dots, \mathcal{E}_0^n\rangle$, m the total number of variables in $\mathcal{E}_0$, and $V = Var(E) \cup Var(S)$. If*

$$\langle \mathcal{S}_0, \mathcal{E}_0, \mathcal{O}_0\rangle \Rightarrow^+ \langle \mathcal{S}_k, \mathcal{E}_k, \mathcal{O}_k\rangle, \tag{62}$$

where $\mathcal{S}_k$ is unifiable, $F \doteq T \in \mathcal{E}_k^j$ and $F \doteq T \notin \mathcal{E}_i^j$ for all i and j, $0 \leq i < k \leq m$, $1 \leq j \leq n$, then $\theta_{\mathcal{S}_k}|_V$ is a rigid E-unifier of S, where $\theta_{\mathcal{S}_k}$ is the mgu of $\mathcal{S}_k$ (over $T_\Sigma(X)$).

Theorem 9 (Completeness). *Let $E = \{E_i \mid 1 \leq i \leq n\}$ and $S = \{\langle u_i, v_i\rangle \mid 1 \leq i \leq n\}$ as in Definition 51, and let $E^i = E_i \cup \{eq_i(u_i, v_i) \doteq F,\ eq_i(z_i, z_i) \doteq T\}$ for every i, $1 \leq i \leq n$. If θ is any rigid E-unifier of S, then there is an order assignment $\mathcal{O}_0$ on $\langle E^1, \dots, E^n\rangle$, and letting $\mathcal{S}_0 = TU_{\mathcal{O}_0}$, $\mathcal{E}_0^i = R(E^i, \mathcal{O}_0)$ for every i,*

$1 \leq i \leq n$, $\mathcal{E}_0 = \langle \mathcal{E}_0^1, \ldots, \mathcal{E}_0^n \rangle$, m the total number of variables in $\mathcal{E}_0$, and $V = Var(E) \cup Var(S)$, there is a sequence of transformations

$$\langle \mathcal{S}_0, \mathcal{E}_0, \mathcal{O}_0 \rangle \Rightarrow^+ \langle \mathcal{S}_k, \mathcal{E}_k, \mathcal{O}_k \rangle, \tag{63}$$

where $k \leq m$, $\mathcal{S}_k$ is unifiable, $F \doteq T \in \mathcal{E}_k^j$, $F \doteq T \notin \mathcal{E}_i^j$ for all i and j, $0 \leq i < k$, $1 \leq j \leq n$, and $\theta_{\mathcal{S}_k}|_V \leq_E \theta[V]$, where $\theta_{\mathcal{S}_k}$ is the mgu of $\mathcal{S}_k$ over $T_\Sigma(X)$. Furthermore, $\theta_{\mathcal{S}_k}|_V$ is a rigid E-unifier of S.

Actually, Theorem 9 can be sharpened. Examination of the induction proof reveals that for any rigid E-unifier θ of S, a rigid E-unifier more general than θ can be found, even if the transformations are applied in a certain order.

Definition 56. We say that a derivation

$$\langle \mathcal{S}_0, \mathcal{E}_0, \mathcal{O}_0 \rangle \Rightarrow^+ \langle \mathcal{S}_m, \mathcal{E}_m, \mathcal{O}_m \rangle \tag{64}$$

is an *lr-derivation* iff for every subderivation

$$\langle \mathcal{S}_0, \mathcal{E}_0, \mathcal{O}_0 \rangle \Rightarrow^* \langle \mathcal{S}_i, \mathcal{E}_i, \mathcal{O}_i \rangle \Rightarrow \langle \mathcal{S}_{i+1}, \mathcal{E}_{i+1}, \mathcal{O}_{i+1} \rangle, \tag{65}$$

in the step from i to $i+1$ $(0 \leq i < m)$, the equations $l_1 \doteq r_1$ and $l_2 \doteq r_2$ are chosen in the set $\mathcal{E}_i^j$ such that $j \geq 1$ is the least index such that $F \doteq T \in \mathcal{E}_i^l$ for every $l < j$ and $F \doteq T \notin \mathcal{E}_i^j$.

In some sense, such derivations compute rigid E-unifiers incrementally from left to right.

Theorem 10 (Incremental Completeness). *Theorem 9 holds with lr-derivations instead of arbitrary derivations.*

This sharpening of Theorem 9 is very useful in practice, because it yields an incremental way of finding rigid E-unifiers. From Theorem 9, it is obvious that Theorem 5 also holds for a set of sets of equations E and a term system S.

Theorem 11. *Let $E = \{E_i \mid 1 \leq i \leq n\}$ and $S = \{\langle u_i, v_i \rangle \mid 1 \leq i \leq n\}$ as in Definition 51, $E^i = E_i \cup \{eq_i(u_i, v_i) \doteq F,\ eq_i(z_i, z_i) \doteq T\}$ for every i, $1 \leq i \leq n$, m the number of variables in $E \cup S$, and*

$V = Var(E) \cup Var(S)$. *There is a finite complete set of rigid E-unifiers for S given by the set*

$$\{\theta_{\mathcal{S}_k}|_V \mid \langle \mathcal{S}_0, \mathcal{E}_0, \mathcal{O}_0 \rangle \Rightarrow^+ \langle \mathcal{S}_k, \mathcal{E}_k, \mathcal{O}_k \rangle,\ k \leq m\}, \qquad (66)$$

for any order assignment $\mathcal{O}_0$ *on* $\langle E^1, \ldots, E^n \rangle$, *with* $\mathcal{S}_0 = TU_{\mathcal{O}_0}$, $\mathcal{E}_0^i = R(E^i, \mathcal{O}_0)$ *for every* i, $1 \leq i \leq n$, $\mathcal{E}_0 = \langle \mathcal{E}_0^1, \ldots, \mathcal{E}_0^n \rangle$, *and where* $\mathcal{S}_k$ *is unifiable,* $F \doteq T \in \mathcal{E}_k^j$, $F \doteq T \notin \mathcal{E}_i^j$ *for all* i *and* j, $0 \leq i < k$, $1 \leq j \leq n$, *and* $\theta_{\mathcal{S}_k}$ *is the mgu of* $\mathcal{S}_k$ *over* $T_\Sigma(X)$.

Finally, it is obvious that Theorem 9 yields a generalization of Theorem 7 to equational prematings.

Theorem 12. *Finding whether a pair* $\langle E, S \rangle$ *(as in Definition 51) is an equational premating is NP-complete.*

As a consequence, since the problem of deciding whether a family of mated sets forms an equational mating is precisely the problem of finding whether a pair $\langle E, S \rangle$ is an equational premating, the former problem is also NP-complete.

In the next section, we present a procedure based on the method of equational matings. The basic idea of such a procedure is straightforward, as suggested by Theorem 2: compute incrementally amplifications of a formula in nnf, and at each stage, test whether such an amplification has a p-acceptable mating. The efficiency related issues here are the same as in Andrews's non-equational case, except that they are harder: in addition to efficient data structures which save information between stages, we must identify mated sets instead of mated pairs, and use rigid E-unification instead of standard unification.

While implementation issues are of importance for a practical procedure, we don't feel they are as new as the ideas of equational matings and rigid E-unification, and thus, we only give a high level description of the procedure.

13. A Refutation Procedure

We now consider defining a refutation procedure based on equational matings and rigid E-unification. As mentioned in Section 4, such a procedure is suggested by Theorem 2: compute incrementally amplifications of a formula in nnf, and at each stage, test whether such an amplification has a p-acceptable mating. This idea can be formalized in the following non-deterministic definition, which uses the incremental rigid E-unification algorithm *E_UNIF* given by Theorem 10.

Definition 57. Let A be a rectified universal sentence in nnf and D an amplification of A. An *EQ-derivation* R is a sequence of tuples $\langle (D_0, \Pi_0, MS_0, \theta_0), \ldots, (D_p, \Pi_p, MS_p, \theta_p) \rangle$, such that for $0 \leq i \leq p$, D_i is an amplification of A, Π_i is a set of vertical paths in D_i, MS_i is a set of mated sets, each such set of the form $\{(s_1 \doteq t_1), \ldots, (s_n \doteq t_n), \neg(s \doteq t)\}$, and θ_i is a substitution, such that

(1) D_0 is the quantifier-free form of A, $\Pi_0 = vp(D_0)$, $MS_0 = \emptyset$, and $\theta_0 = Id$, and

(2) For every i, $0 \leq i < p$, if $MS_i = \{S_1, \ldots, S_m\}$, then either

  (i) there is some vertical path π_{i+1} in Π_i, some subset S_{i+1} of π_{i+1} such that $S_{i+1} = \{(s_1 \doteq t_1), \ldots, (s_n \doteq t_n), \neg(s \doteq t)\}$, and some rigid E-unifier σ_{i+1} for S_{i+1} given by the procedure E_UNIF. Then $D_{i+1} = D_i$, $\Pi_{i+1} = \sigma_{i+1}(\Pi_i - \{\pi_{i+1}\})$, $MS_{i+1} = MS_i \cup \{S_{i+1}\}$, and $\theta_{i+1} = \theta_i \circ \sigma_{i+1}$;

  or

  (ii) if D_i is obtained from the rectified form of a sentence C_i by deleting quantifiers, where C_i is a sentence in a sequence $\langle C_1, \ldots, C_i \rangle$ $(i \geq 1)$ of formulae resulting from quantifier duplications, then D_{i+1} is obtained from the rectified form of a sentence C_{i+1}, obtained by quantifier duplication from C_i, by deleting quantifiers. Then $\Pi_{i+1} = vp(D_{i+1})$, $MS_{i+1} = \emptyset$, and $\theta_{i+1} = Id$.

If in addition, $\langle MS_p, \theta_p \rangle$ is a p-acceptable mating, we say that R is an *EQ-refutation*.

Since the method outlined in Definition 57 non-deterministically enumerates all equational matings for potentially all amplifications of A, it is immediate from Theorem 2, that since A is unsatisfiable iff some amplification of A has a p-acceptable mating, A is unsatisfiable iff there is an EQ-refutation for A.

There are a number of implementation problems with Definition 57 as the basis for a practical method for showing the unsatisfiability of a formula in nnf:

(1) The p-acceptable mating found is maximal, since no attempt is made to identify overlapping vertical paths that are spanned by common mated sets.
(2) Every time a subformula of A is amplified, the entire computation up until that point is discarded.
(3) No effort is made to use any failure in any step as a source of information for the next step.

These points are closely related to the acceptability criteria given by Andrews in section 2.3 of Andrews (1981), where he defines a procedure for finding (non-equational) p-acceptable matings. We tried to adapt the notion of a connection graph used by Andrews in section 3 of Andrews (1981), but unfortunately with no success. The difficulty in the presence of equality is that a vertical path π is closed iff it contains some mated *set* $\{(s_1 \doteq t_1), \ldots, (s_n \doteq t_n), \neg(s \doteq t)\}$ such that s and t have some rigid E-unifier (where $E = \{(s_1 \doteq t_1), \ldots, (s_n \doteq t_n)\}$), but there is no guarantee that $n \leq 1$. For languages without equality, a mated set is of the form $\{L, \neg L'\}$ where L and L' are unifiable. In order to determine which pairs of literals are unifiable it is necessary to examine $O(n^2)$ pairs, where n is the total number of literals in D. Hence, for languages without equality, it is advantageous to precompute a connection graph recording the pairs of literals $\{L, \neg L'\}$ where L and L' are unifiable, since every closed path must contain such a pair. However, for languages with equality, if D contains $n = q + r$ literals where q literals are positive and r literals are negative ($r \geq 1$), to form a mated set there are r choices for the negative literal and for each such choice, any subset of the positive literals can be chosen. Thus, there are potentially $r2^q$ mated sets, that is, an exponential number of mated sets. In addition, rigid E-unification is NP-complete.

Thus, the cost of determining which sets of literals are mated sets is exponential and there does not seem to be any advantage in computing such sets. Since our investigations on this subject are still very preliminary, we will not elaborate any further. However, this is a very interesting topic that clearly requires more work.

We conclude this section with a naive procedure written in pseudocode implementing the method of equational matings. We have made no efforts towards improving efficiency of the basic method. This aspect should be addressed in further work.

Let us now turn our attention to identifying mated sets in the set $vp(D)$ of vertical paths in D. Since mated sets are of the form $\{(s_1 \doteq t_1), \ldots, (s_n \doteq t_n), \neg(s \doteq t)\}$, the search is organized around the negative literals. Observe that if some path $\pi \in vp(D)$ does not contain a negated literal, it cannot contain an unsatisfiable mated set and D is satisfiable. In this case, the procedure stops with failure. Suppose now that for some vertical path π, there is a mated set S of the form $\{(s_1 \doteq t_1), \ldots, (s_n \doteq t_n), \neg(s \doteq t)\}$ but it has no rigid E-unifier. Then we must perform an amplification step. We would like this step to supply information that was missing in the attempt to find an unsatisfiable mated set. Unfortunately, any arbitrary duplication may fail to do this, and may even introduce *new* vertical paths. As Andrews says in discussing this problem in the context of looking for non-equational p-acceptable matings, "One would like to develop a set of heuristics for duplicating quantifiers." However, it is beyond the scope of this paper to consider this issue in detail (as it was beyond the scope of Andrews's paper), and we will use a straightforward breadth-first outermost duplication strategy: in the lexical order of occurrences of subformulae in A, perform an outermost duplication of the first non-ground subformula, then the second, and so on until the last non-ground subformula in the lexical order has been duplicated, and then start again from the top. A breadth-first strategy clearly generates a complete search space of outermost duplications, however, it can clearly result in superfluous paths.[18]

Example 14. Consider the following formula A, formed from the union of Examples 3 and 4, where $\hat{x}, \hat{y}, \hat{z}, u, v, w, x, y, z$ denote variables:

---

[18]Andrews has also shown that outermost duplication itself can generate superfluous literals.

$$(a \doteq b) \vee \neg(*(a,b) \doteq *(b,a)) \wedge \tag{1}$$
$$\forall \hat{x}((f^3\hat{x} \doteq \hat{x}) \vee \neg(f\hat{x} \doteq fb)) \wedge \tag{2}$$
$$(Qa \vee \neg(f^3a \doteq a)) \wedge \tag{3}$$
$$\forall \hat{y}((f^5\hat{y} \doteq \hat{y}) \vee \neg Q\hat{y}) \wedge \tag{4}$$
$$(Ra \vee \neg(fa \doteq a) \vee \neg Pfa) \wedge \tag{5}$$
$$\forall \hat{z} \neg Rf\hat{z} \wedge \tag{6}$$
$$Pa \wedge \tag{7}$$
$$\forall x \forall y \forall z(*(x,*(y,z)) \doteq *(*(x,y),z))) \wedge \tag{8}$$
$$\forall u(*(u,1) \doteq u) \wedge \tag{9}$$
$$\forall v(*(1,v) \doteq v) \wedge \tag{10}$$
$$\forall w(*(w,w) \doteq 1) \tag{11}$$

The problem is to find the right sequence of duplications for A. We know from Example 4 that 3 duplications of subformula (8) and 2 duplications of subformula (11) are necessary for the existence of a p-acceptable mating. But an obvious breadth-first outermost sequence of duplications results in 21 duplications before these duplications are generated. Subformulae (1),(3),(5), and (7) are ground and thus not subject to duplication.

Let A be a rectified universal formula in nnf. A breadth-first outermost amplification sequence $\langle D_1, \ldots, D_i \rangle$ $(i \geq 1)$ for A is a sequence such that D_k is obtained from the rectified form of a sentence C_k by deleting quantifiers, where C_k is a sentence in a sequence $\langle C_1, \ldots, C_i \rangle$ $(i \geq 1)$ of formulae resulting from quantifier duplications $(1 \leq k \leq i)$. The C_k's are defined such that $C_1 = A$, and C_{k+1} is obtained by quantifier duplication of an outermost universal subformula of C_k using a breadth-first strategy $(1 \leq k \leq i-1)$. Given a counter $k \geq 1$, a call $amplify(k+1, A, D)$ to the procedure $amplify$ returns the amplification $D = D_{k+1}$ obtained from C_k as explained above. For $k = 0$, it is assumed that $amplify(1, A, D)$ returns the quantifier free formula D_1 obtained by deleting quantifiers from A. We are assuming that the E-unification algorithm *E_UNIF* (given by Theorem 10) takes as input a mated set $S = \{(s_1 \doteq t_1), \ldots, (s_n \doteq t_n), \neg(s \doteq t)\}$ and returns a finite complete set of rigid E-unifiers of s and t where $E = \{(s_1 \doteq t_1), \ldots, (s_n \doteq t_n)\}$, in the form of a set of triangular forms, where each triangular form T represents a substitution σ_T. For presenting the refutation procedure, we also assume that a mating $\mathcal{M}$ is represented by a pair $\langle MS, U\rangle$, where MS is a collection of mated sets, and U is the triangular representation of a substitution. Also, given a set of paths Π and a substitution σ, let

$$\begin{aligned} apply(\sigma, \Pi) = \{\pi' \mid \pi = \{P_1, \ldots, P_k\} \in \Pi \quad \text{and} \\ \pi' = \{\sigma(P_1), \ldots, \sigma(P_k)\}\}. \end{aligned} \tag{67}$$

We assume that the application of σ to Π is done intelligently, i.e., since σ is the identity substitution on almost all literals in π, some table lookup mechanism is available to identify the literals which have variables in the domain of σ.

We collect the information discussed in this section into the following pseudo-code version of a refutation procedure for formulae in nnf which uses the following variables and procedures: A is a rectified universal sentence in nnf; i is a counter for vertical paths in an amplification; j is a counter for negative literals in a vertical path; k is an amplification counter; $\mathcal{M}$ is an equational mating; *p_acceptable* is a boolean value which is true iff $\mathcal{M}$ is *p*_acceptable; *found* is a boolean value which is true if an unsatisfiable mated set is identified in some path; $select\_path(i, vp(D))$ returns the ith path associated with amplification D; $select\_negative\_literal(j, \pi)$ returns the jth negative literal in π; $choose\_positive\_subset(\pi)$ returns the set of positive literals for some path π; This procedure must be understood as a nondeterministic procedure. A deterministic version can be written by implementing explicitly the backtracking needed to handle the choice of literal, path, etc. However, we feel that it would not be as clear as the present version.

A Refutation Procedure

procedure $equational\_refutation(A)$;

begin
 $k \leftarrow 1$;
 $amplify(k, A, D)$; (D is thus the quantifier-free form of A)
 $\mathcal{M} \leftarrow \langle \emptyset, \emptyset \rangle$;
 $i\_limit \leftarrow \#paths(vp(D))$; $i \leftarrow 1$;
 $p\_acceptable \leftarrow \bot$;
 while $i \leq i\_limit \land \neg p\_acceptable$ **do**
  $\pi \leftarrow select\_path(i, vp(D))$;

```
        j_limit ← #negative_literals(π); found ← ⊥;
        if j_limit ≠ 0 then
            [j ← 1;
             while j ≤ j_limit ∧ ¬found do
                 N ← select_negative_literal(j, π);
                 S ← choose_positive_subset(π) ∪ {N};
                 if ∃T ∈ E_UNIF(S) then
                     [found ← ⊤;
                      if i = i_limit then p_acceptable ← ⊤ else i ← i + 1;
                      let ⟨MS, U⟩ = 𝓜;  𝓜 ← ⟨MS ∪ S, U ∪ T⟩;
                      vp(D) ← apply(σ_T, vp(D))]
                 else j ← j + 1
             endwhile ]
        else return; {A is satisfiable}
        if ¬found then
            [k ← k + 1;
             amplify(k, A, D);
             i_limit ← #paths(vp(D));  i ← 1]
    endwhile;
    return(𝓜)
end;
```

14. Conclusion and Further Work

We have generalized Andrews's method of matings to first-order languages with equality. This new method is sound and complete, and uses a decidable form of E-unification, rigid E-unification. We have shown that both rigid E-unification and finding whether a pair $\langle E, S\rangle$ is an equational premating are NP-complete problems. We also have shown that finite complete sets of rigid E-unifiers always exist. Theorem 12 has important implications regarding the computational complexity of theorem proving for first-order languages with equality using the method of matings. It shows that there is an algorithm for finding equational matings, but not only is the problem of deciding whether an equational mating is p-acceptable co-NP-complete, the problem of deciding that a family of mated sets is an equational mating is NP-complete. For languages without equality, the first problem is still co-NP-complete, but the second can be solved in polynomial time using standard unification, and in fact in linear time.

It is essential to find ways of trimming the search space of feasible order assignments if we want the method to be practical. When a reduction ordering $\prec$ is available and all subterms in $\mathcal{E}'_i$ are ordered by $\prec$, $\mathcal{O}_i$ is completely determined. It would be interesting to investigate subcases where feasible order assignments can be found quickly. An actual implementation of the refutation procedure would also interesting, as well as a comparison with other methods, those based on Knuth-Bendix completions in particular. The above questions are left for further research.

Acknowledgement

We wish to thank Peter Andrews, Jean Yves Girard, Tomas Isakowitz, Leo Bachmair, Dale Miller, Frank Pfenning, David Plaisted, and Richard Statman, for some helpful suggestions and for their encouragements.

References

Andrews, P. (1981). "Theorem Proving via General Matings." *J. ACM* **28**(2), 193–214.

Andrews, P. (1986). *An Introduction to Mathematical Logic and Type Theory: To Truth Through Proof.* Academic Press, New York.

Bachmair, L. (1987). *Proof Methods for Equational Theories*, Ph.D Thesis, University of Illinois, Urbana Champaign, Illinois.

Bachmair, L., Dershowitz, N., and Plaisted, D. (1987). "Completion without Failure," Proceedings of CREAS, Lakeway, Texas. Also submitted for publication to *Journal of Symbolic Computation.*

Bibel, W. (1979). "Tautology Testing With a Generalized Matrix Reduction Method," *TCS* **8**, 31–44.

Bibel, W. (1981). "On Matrices With Connections," *J. ACM* **28**, 633–645.

Bibel, W. (1982). *Automated Theorem Proving.* Friedr. Vieweg & Sohn, Braunschweig.

Dershowitz, N,. (1987). "Termination of Rewriting," *Journal of Symbolic Computation* **3**, 69–116.

Downey, Peter J., Sethi, Ravi, and Tarjan, Endre R. (1980). "Variations on the Common Subexpressions Problem," *J. ACM* **27**(4), 758–771.

Fitting, M. A. (1986). Tableaux Based Automated Theorem Prover for Classical Logic. Technical Report, CUNY.

Gallier, J. H. (1986). *Logic for Computer Science: Foundations of Automatic Theorem Proving.* New York, Harper and Row.

Gallier, J. H., Raatz, S., and Snyder, W., (1987a). "Theorem Proving using Rigid *E*-Unification: Equational Matings," *LICS'87*, Ithaca, New York, 338–346.

Gallier, J. H., Narendran, P., Plaisted, D., Raatz, S., and Snyder, W., (1987b). "Finding canonical rewriting systems equivalent to a finite set of ground equations in polynomial time," submitted to *J. ACM.*

Gallier, J. H., Narendran, P., Plaisted D., and Snyder, W., (1987c). "Rigid E-Unification: NP-completeness and Applications to Theorem Proving," submitted to *J. ACM.*

Gallier, J. H., Raatz, S., and Snyder, W., (1988). "Theorem Proving Using Equational Matings and Rigid *E*-Unification," submitted for publication to *J. ACM.*

Girard, J. Y., (1987). "Linear Logic," *Theoretical Computer Science* **50**:1, 1–102.

Huet, G., (1980). "Confluent Reductions: Abstract Properties and Applications to Term Rewriting Systems," *JACM* **27**:4, 797–821.

Huet, G. and Oppen, D. C., (1982). "Equations and Rewrite Rules: A Survey," In *Formal Languages: Perspectives and Open Problems* (R.V.Book, ed.), Academic Press, New York.

Knuth, D. E. and Bendix, P. B.,(1970). "Simple Word Problems in Universal Algebras," In *Computational Problems in Abstract Algebra* (J. Leech, ed.), Pergamon Press.

Kozen, D. (1976). Complexity of Finitely Presented Algebras. Technical Report TR 76–294, Department of Computer Science, Cornell University, Ithaca, NY.

Kozen, D. (1977). Complexity of Finitely Presented Algebras. *9th STOC Symposium*, Boulder Colorado, 164–177.

Kozen, D. (1981). "Positive First-Order Logic is NP-Complete," *IBM Journal of Research and Development*, **25**(4), 327–332.

Levy, A., (1979). *Basic Set Theory.* Springer-Verlag, New York.

Martelli, A. and Montanari, U. (1982). An Efficient Unification Algorithm. *TOPLAS* **4**(2), 258–282.

Miller, D. A. (1984). Expansion Trees and Their Conversion to Natural Deduction Proofs. In *7th International Conference on Automated Deduction, Napa, CA* (edited by R. E. Shostak), L.N.C.S, No. 170, New York, Springer Verlag.

Nelson G. and Oppen, D. C. (1980). "Fast Decision Procedures Based on Congruence Closure," *J. ACM* **27**(2), 356–364.

Paterson, M. S., Wegman, M. N. (1978). "Linear Unification," *Journal of Computer and System Sciences*, **16**, 158–167.

Pfenning, F. (1987). Proof Transformations in Higher-Order Logic. Ph.D. Thesis, Department of Mathematics, Carnegie Mellon University, Pittsburgh, PA.

Plotkin, G. (1972) "Building in Equational Theories," *Machine Intelligence* **7**, 73–90.

Smullyan, R. M. (1968). *First-order Logic.* Berlin, Heidelberg, New York, Springer-Verlag.

Szabo, M. E. (1970). *The Collected Papers of Gerhard Gentzen.* Studies in Logic. New York, Elsevier North-Holland.

6 What Is Unification?

A Categorical View of Substitution, Equation and Solution

JOSEPH A. GOGUEN[1]
Oxford University Computing Laboratory
Programming Research Group
Oxford, England

and

SRI International, Menlo Park California, and
Center for the Study of Language and Information
Stanford University
Stanford, California

1. Introduction

From a general perspective, a *substitution* is a transformation from one space to another, an *equation* is a pair of such substitutions, and a *solution* to an equation is a substitution that yields the same value when composed with (i.e., when sustituted into) the substitutions that constitute the given equation. In some special cases, solutions are called *unifiers*. Unification has become important in computer science because of its connections to mechanical theorem proving and so-called "logic" programming, and in linguistics because of its connections to so-called "unification grammar." Other examples include Scott domain equations, linear programming, type inference, and differential equations.

[1]Supported by Office of Naval Research Contracts N00014-85-C-0417 and N00014-86-C-0450, NSF Grant CCR-8707155, and a gift from the System Development Foundation.

ISBN 0-12-046370-9

The intuition that the composition of substitutions should be associative when defined, and should have identities, motivates a general concept of *substitution system* based on category theory. Although certain simple ideas about substitutions, equations and solutions have been implicit in the categorical literature at least since Lawvere's thesis (Lawvere, 1963), it has been difficult for computer scientists, linguists, and even mathematicians, to relate this work to their own concerns, and appreciate how simple it really is. This paper provides an introduction to these ideas, emphasizing the applications to unification, by developing a very general notion of "substitution system" that not only includes algebraic theories in the sense of Lawvere, but also seems able to encompass virtually every kind of equation found in mathematics, computer science, and linguistics. Notions of morphism, congruence, and quotient are given for substitution systems, each with the expected properties, and some general cardinality bounds are proved for most general solution sets (which are minimal sets of solutions with the property that any other solution is a sustitution instance of one in the set). The notions of equation and solution also generalize to systems of equations, i.e., to constraint solving, and are used to clarify the notions of "compositionality" and "unification" in linguistics. This material can perhaps be considered the beginnings of a General Unification Theory, extending the classical framework of term unification, as developed for example by Siekmann (1988). The paper is self-contained as regards category theory, and indeed, could be used as an introductory tutorial on that subject. The only prerequisites are some elementary set theory and an interest in the subject matter.

Some readers may wish to see some motivation for pursuing a subject that is so very abstract and general. Here are four reasons:

1. Relationships among different applications may be revealed; sometimes these can be surprising and/or suggestive (for example, solving polynomials, inferring types, and understanding natural language can all be seen as solving equations in substitution systems).
2. The need to prove essentially the same results in many different contexts may be avoided (for example, our very general results in Section 6.2 about the uniqueness of most general solution sets apply to all the examples mentioned above).

3. New algorithms and applications may be suggested as extensions of existing special cases (for example, Section 7.5 suggests using order sorted unification to extend Milner's polymorphic type inference to handle subtypes).
4. Old concepts may be simplified, clarified, and/or generalized. For example, Section 5 shows that many inadequacies of traditional syntactic approaches to term substitution are overcome by systematically exploiting the freeness of term algebras, and by making the source and target variable sets explicit. The precise sense in which unification grammars involve unification is clarified in Section 7.8. Other examples include the very general results on floorings in Section 6.2, and on quotients and colimits of arbitrary substitution systems in Section 6.5.

Elegant category theoretic expositions often require sophisticated technical apparatus; since this paper assumes no previous familiarity with category theory, it can reach only a modest level of sophistication. Our discussions often start with informal motivation, so if something seems too vague, please read on until you see it in **boldface**, which marks a formal definition; *italics* are used for informal emphasis. Also, there are many unproved assertions; unless explicitly stated otherwise, they have short proofs, and the reader is urged to use them as exercises to test and improve his/her understanding. Some may find this a relatively painless way to learn some basic category theory. The standard advanced introduction to category theory is Mac Lane (1971), while Goldblatt (1979) provides a gentler approach that includes topoi; neither book discusses Lawvere theories, but (for example) Schubert (1972) does.

I thank Timothy Fernando and Peter Rathmann for their comments on this paper, and Dr. José Meseguer both for his valuable comments and for his collaboration on many of its ideas.

1.1. An Example

Let us consider the simple case of polynomials with integer coefficients, such as $p = x^2 + 2y + 3$. We will use the notation $\{p_1 \Rightarrow x_1, p_2 \Rightarrow x_2, \ldots, p_n \Rightarrow x_n\}$ for the *substitution* that substitutes the polynomial p_i for the variable x_i for $i = 1, ..., n$; for example,

$f = \{2z \Rightarrow x,\ 0 \Rightarrow y\}$ substitutes $2z$ for x and 0 for y. Now let[2] "$f;p$" denote the result of applying the substitution f to p, and then simplifying according to the usual laws of polynomial algebra; for example, with f and p as above, we have $f;p = 4z^2 + 3$. We may consider the application $f;p$ to be well defined provided that only the variables of f occur in p (this intuition is further refined in Section 2 below).

Next, given also $g = \{2x \Rightarrow u,\ y+1 \Rightarrow z\}$, we can form $g;(f;p) = 4y^2 + 8y + 7$ as well as $g;f = \{2y+2 \Rightarrow x, 0 \Rightarrow y\}$ and, of course, $(g;f);p = 4y^2 + 8y + 7$. It is no coincidence that $g;(f;p) = (g;f);p$. For uniformity of notation, we can also regard polynomials as substitutions, e.g., $\{x^2 + 2y + 3 \Rightarrow p\}$ for the above p.

If we are also given $\{x^2 + y + z \Rightarrow q\}$, what does it mean to solve the equation $p = q$? It means to find a substitution h such that $h;p = h;q$. There are many, including $h_1 = \{3 \Rightarrow x, -1 \Rightarrow y, 2 \Rightarrow z\}$, $h_2 = \{-5 \Rightarrow x,\ u-1 \Rightarrow y,\ u+2 \Rightarrow z\}$, $h_3 = \{u \Rightarrow x,\ v \Rightarrow y,\ v+3 \Rightarrow z\}$, and $h_4 = \{1 \Rightarrow x,\ y \Rightarrow y,\ y+3 \Rightarrow z,\ x+y \Rightarrow w\}$. Notice that h_1 and h_2 are both "substitution instances" of h_3, and that h_4 is invalid, because it substitutes for a variable w that doesn't occur in either p or q. Also notice what it means that h_2 is a substitution instance of h_3: there is a substitution j such that $h_2 = j;h_3$, namely $j = \{-5 \Rightarrow u,\ u-1 \Rightarrow v\}$. In fact, h_3 is a "most general solution," in the sense that *any* other solution h (that does not substitute for extraneous variables) is a substitution instance of h_3.

This polynomial example is used informally as motivation throughout the paper, and finally developed more formally in Section 6.3.3. But first, we greatly generalize this example, and also clarify our conventions regarding variables in the composition of substitutions.

2. Substitutions and Categories

The most basic intuitions about substitutions concern their *composition*. In particular, composition should satisfy associative and

[2]Mathematics traditionally writes composition in the opposite order from that used in this paper, often using the notation "$\circ$". Here, we follow the computer science notation ";" for sequential composition of operations. While it would be an exaggeration to say that mathematics "got it wrong," it does seem fair to say that many things go smoother in the "backwards" notation, and indeed, many category theorists ordinarily use it.

identity laws. Since we wish to capture the idea that there may be *constraints* on what can be substituted where, the composition operation should be *partial* rather than total, reflecting the idea that $f;g$ is defined only when the target type of f, denoted $\partial_1 f$, equals the source type of g, denoted $\partial_0 g$.

Thus, a substitution system should have a set $\mathcal{S}$ of **substitutions**, a set $|\mathcal{S}|$ of **types**, a partial **composition** operation on $\mathcal{S}$ denoted ";" plus **source** and **target** operations denoted $\partial_0, \partial_1\colon \mathcal{S} \Rightarrow |\mathcal{S}|$ such that:

(1) $f;g$ is defined iff $\partial_1 f = \partial_0 g$.
(2) $\partial_0(f;g) = \partial_0 f$ and $\partial_1(f;g) = \partial_1 g$.
(3) $(f;g);h = f;(g;h)$ whenever all compositions are defined.
(4) For each $T \in |\mathcal{S}|$, there is a substitution id_T such that $\partial_0 id_T = \partial_1 id_T = T$ and such that $f;id_T = f$ and $id_T;g = g$ whenever these compositions are defined.

Axiom (4) implies that there is exactly one identity on each object S: any two are equal since $id_S;id'_S = id_S$ and $id_S;id'_S = id'_S$.

Many readers will not be surprised to learn that these axioms define neither more nor less than **categories**, and from here on we will introduce and then use standard bits of terminology and notation from category theory. In particular, we will say "morphism," "map" or "arrow" instead of "substitution," and to say "object" instead of "type." Also, we will say "category" instead of "substitution system," particularly because these axioms are only a first step toward a more complete definition to be given later. Also, we write $f\colon R \to S$ when $\partial_0 f = R$ and $\partial_1 f = S$, and we let $\mathcal{S}[R,S]$ denote the collection of all maps from R to S in $\mathcal{S}$.

The following discussion will show that even this very simple machinery is sufficient to define some interesting concepts, including isomorphism, substitution instance, and ground substitution.

A morphism $f\colon S \to T$ in a category $\mathcal{C}$ is an **isomorphism** iff there is another morphism $g\colon T \to S$ such that $f;g = id_S$ and $g;f = id_T$. We may write f^{-1} for g and g^{-1} for f, and we may also write $S \cong T$, and call f and g **invertible morphisms**. We will see some substitution systems where the isomorphisms are just one-to-one renamings of variables.

Isomorphisms have a number of simple, useful and easily verified properties, including the following, which assume that $f\colon R \to S$ and $g\colon S \to T$ are isomorphisms in a category $\mathcal{C}$:

(a) $(f^{-1})^{-1} = f$.
(b) $(f;g)^{-1} = g^{-1}; f^{-1}$.
(c) $id_S^{-1} = id_S$.
(d) $\cong$ is an equivalence relation on $|\mathcal{C}|$.

Let us again consider our polynomial substitution system, which from here on will be denoted $\mathcal{P}$. Its types are finite sets of variables, and if f: $R \to S$, then S contains the variables that f substitutes *into*, while R contains the variables that may be used in the polynomials that are substituted for the variables in S. The constraint on when a composition $f;g$ is defined guarantees that f must tell what happens to all and only the variables that can occur in polynomials in g. The identity on T is the substitution $\{x \Rightarrow x \mid x \in T\}$. It is an interesting exercise to determine the isomorphisms of $\mathcal{P}$, but perhaps it is better to delay this until the details have been developed formally in Section 6.3.3.

Returning now to the general case (but remembering the polynomials for motivation), we say that a substitution h': $R' \to S$ is an **instance** of a substitution h: $R \to S$ iff there is a substitution j: $R' \to R$ such that $h' = j;h$. This defines a relation on substitutions called **subsumption**, and we write $h' \preceq h$ to indicate that h **subsumes** h'. It is easy to check that subsumption is transitive and reflexive, i.e., a pre-ordering.

Fig. 1 shows a so-called *commutative diagram* that renders the definition of substitution instance graphically. In such a diagram, the edges are morphisms in some category $\mathcal{C}$ and the nodes are objects in $\mathcal{C}$. The diagram is said to *commute* iff, given any two nodes R and S, the composition of the morphisms along any two paths from R to S in the diagram are equal as morphisms in $\mathcal{C}$. Such diagrams can often help with visualizing a set of equations among morphisms, or as in this case, substitutions.

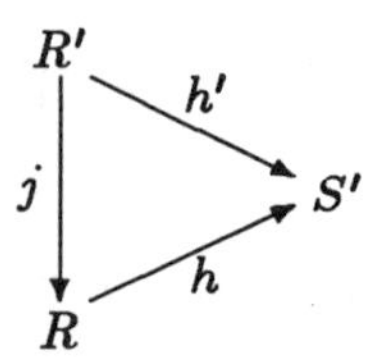

Fig. 1. Substitution Instance.

Returning again to the polynomial substitution system $\mathcal{P}$, some substitutions are *constants*, in the sense that what is substituted for variables does not involve any variables, i.e., is an integer. We can capture this concept categorically by postulating a "constant type" 1 that is the source for all constants; for $\mathcal{P}$, it is the *empty* set of variables. Then a constant of sort S is just a substitution c: $1 \to S$, e.g., for $\mathcal{P}$, a substitution that involves no variables. But how can we characterize 1? In fact, 1 has a simple categorical property that (we will soon see) characterizes it uniquely up to isomorphism: there is exactly one substitution with target 1 for each possible source object, i.e.,

(5) There is a type $1 \in |\mathcal{S}|$ such that for each $S \in |\mathcal{S}|$, there is one and only one substitution $S \to 1$ in $\mathcal{S}$, which we may denote $!_S$ or even just $!$.

In category theory, an object 1 satisfying axiom (5) is called a **final** or **terminal** object. In $\mathcal{P}$, the arrows $!_S$ are empty functions. The basic fact about final objects is the following:

Proposition 1. *Let F and F' be final objects in a category $\mathcal{C}$. Then the unique morphisms i: $F \to F'$ and i': $F' \to F$ are inverse isomorphisms.*

Proof. First notice that we have $i; i'$: $F \to F$ and $i'; i$: $F' \to F'$ in $\mathcal{C}$. Since F and F' are both final, and since there can be only one morphism each $F \to F$ and $F' \to F'$, and since id_F and id'_F are such morphisms, the given compositions must equal these identities. $\square$

Let us call a substitution f: $S \to T$ a **constant** if $S = 1$, and let us call it a **ground substitution** if it factors through 1, i.e., if $f = !; c$, i.e., if $f \preceq c$, for some constant c: $1 \to S$. Then

Fact 1. $f; g; h$ is ground whenever g is ground. The only invertible ground morphisms are the boring ones $!$: $1 \to F$ where F is another final object. If $h \preceq g$ and g is ground, then h is also ground. $\square$

To summarize, this section develops categories with final objects as a model for substitution systems. Although this model has very few axioms and concepts, it already allows us to define renaming and ground substitutions, as well as subsumption, and to develop some of their basic properties. However, it is only a first step toward the definition of substitution system finally given in Section 4.

3. Equations and Solutions

What is an equation? Our view is that it is just a pair of substitutions with the same source and the same target, i.e., an **equation** is a pair $\langle f, g\rangle$ with $\partial_0 f = \partial_0 g$ and $\partial_1 f = \partial_1 g$; we will often use the notation $f, g\colon S \to T$. Sections 5 and 6 will show that this definition includes the expected examples, and in particular, that terms can be seen as substitutions.

Given this view of equations, a **solution** to an equation $f, g\colon S \to T$ is just a substitution $h\colon R \to S$, for some type R, such that $h; f = h; g$.

Usually, we prefer a **most general** solution if there is one, i.e., a solution such that any other solution is a substitution instance of it, i.e., a solution $h\colon R \to S$ such that for any solution $h'\colon R' \to S$ there is a "factoring" substitution $j\colon R' \to R$, such that $j; h = h'$.

In many examples, including our polynomial example, a most general solution h can be chosen so that whenever $h' \preceq h$ there is a *unique* j such that $h' = j; h$. Let us say that a substitution h such that $f; h = g; h$ implies $f = g$ is **monic** (another term is **monomorphic**). If h is monic and most general, then the factoring substitution j is necessarily unique. We will soon see that this uniqueness of the factoring substitution guarantees the uniqueness of h (up to isomorphism), which is a deepening of our intuition about most general substitutions. This discussion motivates the following

Definition 1. A **final solution**[3] to an equation $f, g\colon S \to T$ is a substitution $h\colon R \to S$ such that any other solution h' to $\langle f, g\rangle$ is of the form $j; h$ for a *unique* substitution j. □

Recall that a **maximum** element m for a partial order $\preceq$ has the property that $a \preceq m$ for all a.

Fact 2. A solution is most general iff it is maximum with respect to $\preceq$. Moreover, any monic most general solution is final. □

We now prove the uniqueness of final solutions by using abstract

[3] I regret the connotations of this phrase, but "terminal solution" seemed almost equally unfortunate, yet less suggestive. I also considered reversing all arrows, so that the phrase "initial solution" could be used, and I might well have done so, had it not required rewriting virtually the entire paper.

nonsense[4]. Although the construction is certainly abstract, it is really rather simple, and is also useful in many other situations. Given f, g: $S \to T$, we construct a category $\mathcal{SOL}(f,g)$ of solutions, in which final solutions are final objects:

- Let $\mathcal{SOL}(f,g)$ have the solutions of $\langle f,g\rangle$ as its objects, i.e., the substitutions h: $R \to S$ (for some R) such that $h;f = h;g$.
- Given solutions h: $R \to S$ and h': $R' \to S$ to $\langle f,g\rangle$, we define a **morphism of solutions** $h' \to h$ to be a substitution j: $R' \to R$ such that $j;h = h'$. See Fig. 2.

Notice that $h' \preceq h$ iff $h' \Rightarrow h$. We immediately have

Proposition 2. *$\mathcal{SOL}(f,g)$ as defined above is a category, and h is a final solution for $\langle f,g\rangle$ iff it is final in $\mathcal{SOL}(f,g)$.* $\square$

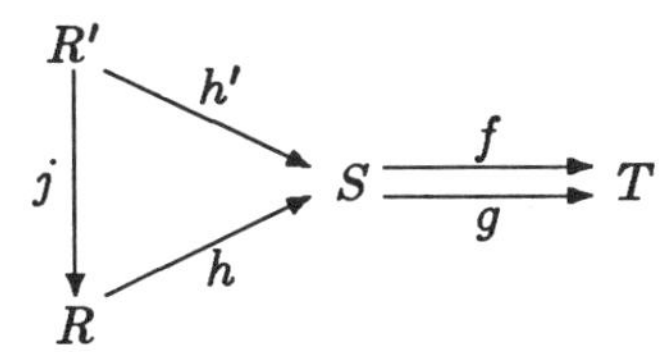

Fig. 2. A Morphism of Solutions.

Now Proposition 1 implies that final solutions are unique up to isomorphism in $\mathcal{SOL}$, which means in particular that their source types are isomorphic. We cannot, of course, expect that every equation has a solution, let alone a final solution, but it is comforting to know that final solutions are unique when they exist. Section 6.1 will show that the final solution concept includes "most general unifier." In the standard language of category theory, a final solution for $\langle f,g\rangle$ is an **equalizer** of f and g.

A property of the form "there exists a unique morphism such that ..." is called a **universal property**, and we have now seen two of these: the characterizing properties for equalizers and for final objects. We will see other examples soon.

[4]This is a sort of technical term in category theory, referring to cases where something is proved without actually looking at how it has been constructed.

We close this section with an easy generalization of the notion of *equation*: instead of a pair $\langle f, g \rangle$, we may consider any set Γ of morphisms all having the same source and the same target. A solution is then a substitution that makes them all equal, and we get a category $\mathcal{SOL}(\Gamma)$ of solutions of Γ, in which final solutions are again final objects, and therefore unique (up to isomorphism). All our results generalize to this situation, but this will not be explicitly developed.

4. Types, Variables and Products

Let us return to the example of Section 1, polynomials with integer coefficients, and examine more closely how it fits into the framework of Sections 2 and 3. Our goal is to define a substitution system $\mathcal{P}$ whose morphisms are the polynomial substitutions, and our first step is to describe its types, the elements of $|\mathcal{P}|$. To this end, let $X_\omega = \{x_0, x_1, x_2, \ldots\}$ and let $|\mathcal{P}|$ be the set of all *finite* subsets of X_ω including the empty subset $\emptyset$. Also, we let u, v, w, x, y, z denote elements of X_ω and call them **variables**.

We (informally) let the morphisms $f\colon S \to T$ in $\mathcal{P}$ be the substitutions of the form $\{f_x \Rightarrow x \mid x \in T\}$ where f_x is a polynomial with integer coefficients involving only variables from S. Composition $f;g$ is of course substitution of the polynomials in f for the variables in the polynomials of g, and the identity $T \to T$ is $\{x \Rightarrow x \mid x \in T\}$. The final object is the empty type, $\emptyset$, because the only substitution $T \to \emptyset$ is the empty one. The morphisms $\emptyset \to T$ are (tuples of) constants, $\{i_x \Rightarrow x \mid x \in T\}$ with each i_x an integer. (All this will be developed more formally in Section 6.3.3.)

Among the morphisms $S \to T$ in $\mathcal{P}$ are some very simple ones, called **type morphisms**, that are of the form $\{f_x \Rightarrow x \mid x \in T\}$ where $f_x \in S$, i.e., where each f_x is just a variable in S. It is easy to be convinced that the composition of two type morphisms is again a type morphism, and also that identity morphisms are again type morphisms. Then types with type morphisms form a category, which we will denote by $\mathcal{X}$ from here on.

Now $\mathcal{X}$ is a subcategory of $\mathcal{P}$, in the following sense: A category $\mathcal{C}'$ is a **subcategory** of a category $\mathcal{C}$ if the objects of $\mathcal{C}'$ are also objects of $\mathcal{C}$, and if the morphisms of $\mathcal{C}'$ from R to S are also morphisms of $\mathcal{C}$ from R to S, such that compositions and identities in $\mathcal{C}'$ agree with those in $\mathcal{C}$; let us write $\mathcal{C}' \subseteq \mathcal{C}$. A subcategory $\mathcal{C}'$ of $\mathcal{C}$ is **broad** iff $|\mathcal{C}'| = |\mathcal{C}|$, and is **full** iff $\mathcal{C}'[R, S] = \mathcal{C}[R, S]$ for all $R, S \in |\mathcal{C}'|$.

Fact 3. If $\mathcal{C}' \subseteq \mathcal{C}$ is broad and full, then $\mathcal{C}' = \mathcal{C}$. □

Notice that a type morphism $f\colon R \to S$ in $\mathcal{P}$ is just an assignment of an element of R to each element of S, i.e., a *function* $S \to R$ that we will denote f^o. Moreover, given type morphisms $f\colon R \to S$ and $g\colon S \to T$, their composition $f;g\colon R \to T$ corresponds to the functional composition $g^o; f^o$, i.e.,

$$(f;g)^o = g^o; f^o.$$

Now let us construct $\mathcal{X}$ quite precisely: its objects (called "types") are the finite subsets of X_ω; its morphisms $f\colon R \to S$ are functions $f^o\colon S \to R$; its composition is defined by the formula above; and $id_S = 1_S^o$ where $1_S\colon S \to S$ is the identity function on the set S.

We could entirely dispense with variables if we wanted, by making them implicit in types. This illustrates an important theme in category theory, that isomorphic objects are "essentially the same." Under this view, we could use *any* finite set as a type, i.e., we could use anything at all as a variable symbol. In particular, instead of the finite subsets of $X_\omega = \{x_0, x_1, x_2, \dots\}$ we could use the finite sets of integers[5], i.e., the finite subsets of $\omega = \{0, 1, 2, \dots\}$. In fact, *variables play only an auxiliary role*, as a notational convenience. The real issue, as we will see below, is the product structure of types.

Sometimes it is convenient to use the following "canonical" subsets of X_ω

$$X_n = \{x_0, x_1, x_2, \dots, x_{n-1}\},$$

with $X_0 = \emptyset$ by convention. Let $\mathcal{N}$ denote the category with objects X_n and with the functions $f^o\colon X_n \to X_m$ as the arrows $X_m \to X_n$. $\mathcal{N}$ is a full subcategory of $\mathcal{X}$.

The type morphisms in $\mathcal{P}$ can be used to "untuple" substitutions. Given a type T and $x \in T$, let $p_x\colon T \to \{x\}$ denote the type morphism corresponding to the inclusion function $\{x\} \to T$. Now given $f\colon S \to T$, the composition $f;p_x$ "pulls out" the x component of f. For example, if $f = \{2x \Rightarrow w,\ y + 1 \Rightarrow x,\ x + 2y \Rightarrow y\}$, then

[5]However, to avoid ambiguous notation, it would be necessary to somehow distinguish integers used as variable symbols from integers used as coefficients, for example by enclosing them in brackets.

$f;p_x = f_x = \{y + 1 \Rightarrow x\}$. It is also very useful to "tuple up" substitutions, e.g., to form $\{2x \Rightarrow w,\ y + 1 \Rightarrow x,\ x + 2y \Rightarrow y\}$ from its three components. Suppose we are given a substitution $f_x\colon S \to \{x\}$ for each $x \in T$; then what we want is $f\colon S \to T$ such that $f;p_x = f_x$ for each $x \in T$. As it happens, there is always a *unique* such f for polynomial substitutions, and our intuition suggests that satisfying the equations $f;p_x = f_x$ should completely characterize f in any substitution system.

It is convenient to use language and notation for tupling that is inspired by Cartesian products of sets, which will turn out to be a special case of the general construction given below. Thus, we will call $p_x\colon T \to \{x\}$ for $x \in T$ the x-**projection** of T; also, we will write $\langle f_x \mid x \in T\rangle$ for the construction of f from its components f_x and call it the **tupling** of the f_x (all this is formally defined below).

These considerations motivate a general construction, in the setting of an arbitrary category $\mathcal{C}$, for tupling an arbitrary family of $f_i\colon S \to T_i$ of morphisms, for $i \in I$. To characterize the tupled morphism $f = \langle f_i \mid i \in I\rangle$, we must also characterize its target T, which is the **product object** of the T_i. We do this through its family of **projections**, which are the basic feature of products:

- Given a family of objects T_i for $i \in I$, let us call a family $p_i\colon T \to T_i$ of morphisms a **cone** over the T_i.
- Given also another cone over the T_i, say $p'_i\colon T' \to T_i$, a **cone morphism** $f\colon T' \to T$ over the T_i is a morphism $f\colon T' \to T$ such that $f;p_i = p'_i$ for each $i \in I$. See Fig. 3.

We thus obtain a category $\mathcal{C}(T_i)$ of cones over the family T_i, and we define a **product** of the T_i to be a *final object* in this category. The "apex" of this cone is the product object, and its morphisms are the projections. Once again, Proposition 1 gives uniqueness up to isomorphism. Moreover, given a cone $f_i\colon S \to T_i$ over the T_i, the unique morphism $S \to T$, where T is the apex of the product cone, is the tupled substitution $\langle f_i \mid i \in I\rangle$ of the f_i.

Of course, a given category may fail to have certain products. When $I = \emptyset$, any final object is a product for the empty family. When $I = \{1, 2\}$, we get a **binary** product, and write $T_1 \times T_2$ for its object. For the general case, we use the notation $\prod_{i \in I} T_i$ for the product object.

We will see in the subsection below that, for our polynomial example, the product of types is given by their disjoint union.

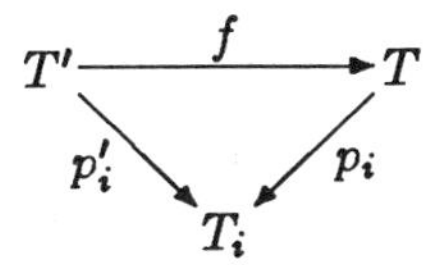

Fig. 3. A Morphism of Cones.

Proposition 3. *A category has all finite products iff it has all binary products and a final object. In such a category,*

- $1 \times T \cong T \times 1 \cong T$ *when* 1 *is final.*
- $T \times T' \cong T' \times T$.
- $(T \times T') \times T'' \cong T \times (T' \times T'')$.
- *Given a finite set* I *and* T_i *for* $i \in I$, *let* $T = \prod_i T_i$. *Then each* $f\colon S \to T$ *is of the form* $\langle f_i \mid i \in I\rangle$ *for uniquely determined* f_i: $S \to T_i$. *In particular, if* $p_i\colon T \to T_i$ *are the projections, then*
  - $f_i = f; p_i$
  - $id_T = \langle p_i \mid i \in I\rangle$.

(Only the first two results need a final object.) □

We can summarize our discussion so far with the following: A **type system** $\mathcal{T}$ is a category with finite products, and a **substitution system** is a category $\mathcal{S}$ with finite products having a type system $\mathcal{T}$ as a broad subcategory such that products in $\mathcal{S}$ are the same[6] as products in $\mathcal{T}$. By abuse of notation, we will denote the pair $\langle \mathcal{S}, \mathcal{T}\rangle$ by just $\mathcal{S}$; also, we may call $\mathcal{T}$ the **base** of $\mathcal{S}$. (These definitions are still a bit vague, and will be refined further in Section 6.5.)

The following sections will show that many interesting notions of type and substitution fall within this general framework. (However, the reader is warned that some models of these axioms have morphisms that look nothing like substitutions.)

4.1. Finite Sets and Duality

Let us look more closely at types and their products in our polynomial example. Recall that this type system $\mathcal{X}$ has as its objects

[6]This means that their final objects are the same and their product cones are the same; in particular, their projections are identical.

the finite subsets of X_ω and that its morphisms $f\colon S \to T$ are given by functions $f^o\colon T \to S$, with composition $f;g$ defined by $(f;g)^o = g^o; f^o$.

We can understand $\mathcal{X}$ better by studying the closely related category $\mathcal{FSET}$ of finite sets; this will also shed light on the strange reversal of composition order in the above equation. $\mathcal{FSET}$ has finite sets as its objects and functions as their morphisms. This category (or more precisely, a subcategory of it) is *dual* to $\mathcal{X}$, in the following sense:

Given a category $\mathcal{C}$, its **dual category** $\mathcal{C}^o$ has the same objects as $\mathcal{C}$ (i.e., $|\mathcal{C}| = |\mathcal{C}^o|$), with morphisms $f^o\colon S \to T$ in $\mathcal{C}^o$ being morphisms $f\colon T \to S$ in $\mathcal{C}$, with $id^o_T = id_T$ and with $f^o; g^o = (g;f)^o$. In our example, $\mathcal{X}^o$ is the full subcategory of $\mathcal{FSET}$ having as its objects the finite subsets of X_ω. In general, we have

Fact 4. *If $\mathcal{C}$ is a category, so is its opposite $\mathcal{C}^o$. Moreover, $\mathcal{C}^{oo} = \mathcal{C}$.* □

A very nice feature of category theory is that every categorical concept has a *dual* concept. For example, the dual of final object is **initial object**, an object such that there is a unique morphism *from* it to any other. Similarly, the dual of product is **coproduct**: Given two objects T_1 and T_2, their coproduct in $\mathcal{C}$ consists of an object C and two **injections** $j_i\colon T_i \to C$ with a co-universal property. We can make this more precise and general by using the category $\mathcal{C}(T_i)$ of cones in $\mathcal{C}$ over a family T_i of objects in $\mathcal{C}$ for $i \in I$: A coproduct of the T_i is a final object in $\mathcal{C}^o(T_i)$, i.e., a product of the T_i in $\mathcal{C}^o$. It now follows that coproducts are determined uniquely up to isomorphism, as usual by Proposition 1.

Epics are dual to monics; specifically, $h\colon X \to Y$ is an **epimorphism** iff whenever $f, g\colon Y \to Z$ satisfy $h; f = h; g$ then $f = g$. Returning to $\mathcal{FSET}$, if we are given disjoint sets T_1 and T_2, then their union $T_1 \cup T_2$ is a coproduct, with the inclusions as injections. More generally, given any finite sets T_1 and T_2, their coproduct object in $\mathcal{FSET}$ is their **disjoint union**, which could be any finite set T whose cardinality is the sum of those of T_1 and T_2. One such set is $T = T_1 \times \{1\} \cup T_2 \times \{2\}$, with injections $j_i\colon T_i \to T$ sending t in T_i to the pair $\langle t, i\rangle$ in T.

An advantage of not being committed to any particular construction for disjoint union is that, since any isomorphic set will do, we are not tied down to any particular choice of variables. In particular, the awkward phenomenon of "renaming away," which appears for example in the usual approaches to narrowing, is entirely avoided.

Returning now to the polynomial substitution system $\mathcal{P}$ and its type system $\mathcal{X}$, we see for example that $\{x_0, x_1, x_2\} = \{x_0\} \times \{x_1\} \times \{x_2\}$, so that we can write

$$\{2x_1 + x_2^2 + x_0x_1 \Rightarrow x_0\}\colon \{x_0\} \times \{x_1\} \times \{x_2\} \to \{x_0\},$$

which seems rather neat.

$\mathcal{FSET}$ also has finite products, given by the usual Cartesian product construction, i.e., $T_1 \times T_2 = \{\langle t_1, t_2 \rangle \mid t_1 \in T_1, t_2 \in T_2\}$ with $p_i\colon T \to T_i$ sending $\langle t_1, t_2 \rangle$ to t_i for $i = 1, 2$. Any one point set is a final object.

5. Classical Term Substitution

Both mathematics and computer science provide many interesting substitution systems besides the polynomial example that we have been using for motivation. This section is devoted to the case of classical terms.

5.1. Terms

We first define terms over a given fixed set of function symbols. Terms appear as a basic data structure in many functional and logic programming languages, for example, OBJ (Futatsugi *et al.*, 1985), Hope (Burstall *et al.*, 1980) and Prolog (Colmerauer, 1979). An unsorted (or one sorted) **signature** consists of a set Σ whose elements are called **function symbols** and a function $\alpha\colon \Sigma \to \omega$ assigning an **arity** to each symbol; $\sigma \in \Sigma$ is a **constant** symbol when $\alpha(\sigma) = 0$. Let Σ_n be $\{\sigma \in \Sigma \mid \alpha(\sigma) = n\}$. Now define the set T_Σ of all Σ-**terms** to be the smallest set of *strings* over the $\Sigma \cup \{\underline{(}, \underline{)}\}$ (where $\underline{(}$ and $\underline{)}$ are special symbols disjoint from Σ) such that

- $\Sigma_0 \subseteq T_\Sigma$ and
- given $t_1, \ldots, t_n \in T_\Sigma$ and $\sigma \in \Sigma$, then $\sigma\underline{(}t_1 \ldots t_n\underline{)} \in T_\Sigma$.

The discussion below will show that the Σ-terms form a Σ-algebra, and that a simple universal property characterizes this algebra.

A Σ-**algebra** is a set A with a function A_σ: $A^n \to A$ for each $\sigma \in \Sigma_n$; note that if $\sigma \in \Sigma_0$, then A_σ is essentially an element of A, since A^0 is a one point set. Given Σ-algebras A and B, a Σ-**homomorphism** h: $A \to B$ is a function h such that

$$h(A_\sigma(a_1, \ldots, a_n)) = B_\sigma(h(a_1), \ldots, h(a_n)).$$

Then Σ-algebras and Σ-homomorphisms form a category, denoted $\mathcal{ALG}_\Sigma$. We now view T_Σ as a Σ-algebra as follows:

- For $\sigma \in \Sigma_0$, let $(T_\Sigma)_\sigma$ be the string σ.
- For $\sigma \in \Sigma_n$, let $(T_\Sigma)_\sigma$ be the function sending $t_1, \ldots, t_n$ to the string $\sigma\underline{(}t_1 \ldots t_n\underline{)}$.

Thus, $\sigma(t_1, \ldots, t_n) = \sigma\underline{(}t_1 \ldots t_n\underline{)}$, and from here on we prefer to use the first notation. The key property of T_Σ is its *initiality* in $\mathcal{ALG}_\Sigma$:

Theorem 1. *For any Σ-algebra A, there is a unique Σ-homomorphism $T_\Sigma \to A$.* $\square$

Initiality is dual to finality, and the proof of Proposition 1 applies to show that any two initial objects in a category are not only isomorphic, but are isomorphic by a unique homomorphism. In particular, this shows that our use of strings was not essential in T_Σ: any construction giving an isomorphic algebra would be just as good, for example, lists, trees, and many other representations could have been used. The proof of Theorem 1 is by induction over the construction of T_Σ and is not so simple as previous proofs that have been omitted. It may be found, for example, in Goguen *et al.* (1976).

5.2. Freebies

We can extend Theorem 1 to a concept called "free algebras" using some tricks on signatures. Given a signature Σ and a set X of elements called "variable symbols," let $\Sigma(X)$ denote the new signature with constants $\Sigma(X)_0 = \Sigma_0 \cup X$ and with $\Sigma(X)_n = \Sigma_n$ for $n > 0$. We can now form the initial $\Sigma(X)$-algebra $T_{\Sigma(X)}$ and then view $T_{\Sigma(X)}$ as a Σ-algebra just by forgetting about the additional constants in X; let us denote this Σ-algebra by $T_\Sigma(X)$, and let us also denote the set inclusion function of X into $T_\Sigma(X)$ by i_X.

Corollary 1. *$T_\Sigma(X)$ is the* **free Σ-algebra on** *X, in the sense that any function $f\colon X \to A$ from X to a Σ-algebra A extends uniquely to a Σ-homomorphism, denoted $\overline{f}\colon T_\Sigma(X) \to A$. (See Fig. 4.)*

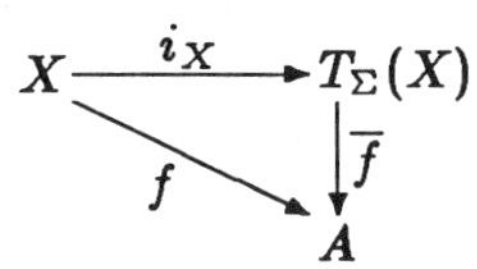

Fig. 4. The Free Σ-Algebra on X.

Proof. We construct a category $\mathcal{F}_X$ whose objects are functions $f\colon X \to A$ from X to some Σ-algebra A, and whose morphisms from $f\colon X \to A$ to $g\colon X \to B$ are functions $j\colon A \to B$ such that $f;j = g$. The composition in $\mathcal{F}_X$ of $j\colon A \to B$ from $f\colon X \to A$ to $g\colon X \to B$ with $j'\colon B \to C$ from $g\colon X \to B$ to $h\colon X \to C$ is then just $j;j'\colon A \to C$. (See Fig. 5.)

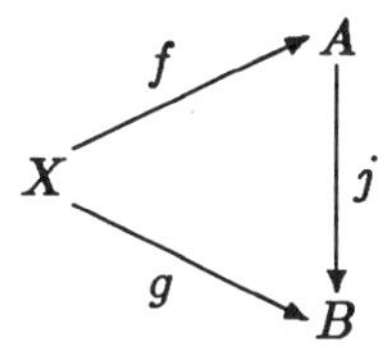

Fig. 5. Composition in $\mathcal{F}_X$.

It suffices to show that $i_X\colon X \to T_\Sigma(X)$ is initial in $\mathcal{F}_X$, which follows from these observations: (1) objects in $\mathcal{F}_X$ are in one-to-one correspondence with $\Sigma(X)$-algebras, such that (2) $\mathcal{F}_X$-homomorphisms correspond to $\Sigma(X)$-homomorphisms, and (3) $i_X\colon X \to T_\Sigma(X)$ corresponds to $T_{\Sigma(X)}$, and finally, (4) $T_{\Sigma(X)}$ is initial in $\mathcal{ALG}_{\Sigma(X)}$. $\square$

Given $t \in T_{\Sigma(X)}$, let $var(t)$, the set of variables that occur in t, be the least $Y \subseteq X$ such that $t \in T_{\Sigma(Y)}$.

5.3. Term Substitution Systems

We are now ready to define a substitution system $\mathcal{S}_\Sigma$ whose morphisms are the substitutions on the terms over a given signature Σ.

Using the same base type system $\mathcal{X}$ as in the polynomial example (it has the finite subsets of $X_\omega = \{x_0, x_1, x_2, \ldots\}$ as its objects), we define a **substitution** $X \to Y$ in $\mathcal{S}_\Sigma$ to be a function $Y \to T_\Sigma(X)$. Given substitutions f: $Y \to T_\Sigma(X)$ and g: $Z \to T_\Sigma(Y)$, we define their **composition** as substitutions, denoted $f; g$: $X \to Z$, to be the composition $g; \overline{f}$: $Z \to T_\Sigma(X)$ of functions where $\overline{f}$: $T_\Sigma(Y) \to T_\Sigma(X)$ is the free extension given by Corollary 1. Finally, we define a base type morphism f: $X \to Y$ to be a function $Y \to T_\Sigma(X)$ that factors through the inclusion i_X: $X \to T_\Sigma(X)$; thus a base type morphism $X \to Y$ corresponds to a function $Y \to X$, as expected.

It is now fun to prove the associative law for the composition of term substitutions, $f; (g; h) = (f; g); h$, i.e.,

$$h; \overline{g; \overline{f}} = (h; \overline{g}); \overline{f},$$

for which it suffices to show that

$$\overline{g; \overline{f}} = \overline{g}; \overline{f}$$

because the composition of functions is associative. But this last equation follows directly from the universal property of $T_\Sigma(X)$, as can be seen from contemplating the diagram in Fig. 6. Moreover, i_X is clearly the identity on X, and thus $\mathcal{S}_\Sigma$ is indeed a category.

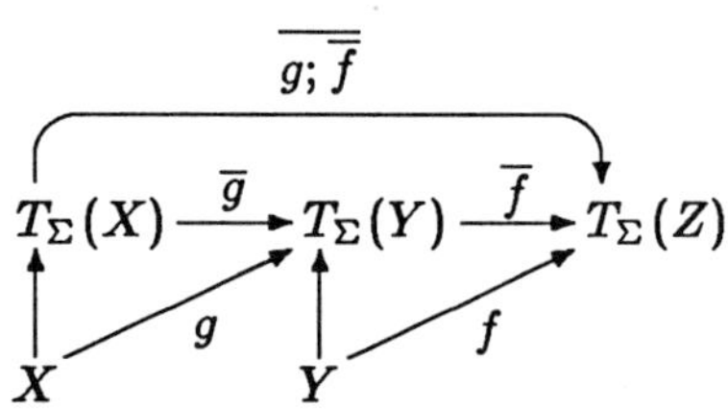

Fig. 6. The Associative Law for Substitution.

Notice that a term $t \in T_\Sigma(X)$ can appear as a morphism t: $Y \to \{x\}$ for any Y containing X and any $x \in X_\omega$. However, if we use the canonical variable sets $X_n = \{x_0, x_1 \ldots, x_{n-1}\}$ then $\sigma \in \Sigma_n$ appears only as σ: $X_n \to X_1$.

Let us check that $\mathcal{S}_\Sigma$ has products, and that these products agree with those in its base type system $\mathcal{X}$. First notice that $\emptyset$ is a final object (actually the only final object), since there is a unique function $\emptyset \to T_\Sigma(X)$ for any X. By Proposition 3, it now suffices to construct binary products. In fact, X_{m+n} is a product of X_m and X_n, i.e.,

$$X_m \times X_n \cong X_{m+n}$$

with projection p: $X_{m+n} \to X_m$ given by the inclusion function $X_m \to X_{m+n}$ and with projection p': $X_{m+n} \to X_n$ given by the function $X_n \to X_{m+n}$ sending i to $i + m$. We leave this as an exercise.

Now let's look at a simple example.

Example 1. Let $\Sigma_0 = \{0\}, \Sigma_1 = \{s\}, \Sigma_2 = \{+\}$, and $\Sigma_n = \emptyset$ for $n > 2$. Taking the liberty of writing + with infix syntax, two typical Σ-terms are $x_0 + s(x_1 + 0)$ and $s(x_1 + s(0)) + x_2$, and a typical substitution is

$$\{x_0 + s(x_1 + 0) \Rightarrow x_1,\\ s(x_1 + s(0)) + x_2 \Rightarrow x_3\}\colon \{x_0, x_1, x_2\} \to \{x_1, x_3\}.$$

Traditional syntactic approaches define a substitution to be something like a finite partial function from an infinite set of variables to the set of all terms on those variables; Lassez *et al.* (1988) gives a nice technical and historical survey of such approaches, also pointing out some of their difficulties. Our discussion above shows that algebra can greatly ease proofs about substitution; for example, it is simple to prove associativity by exploiting the freeness of term algebras (see Fig. 6). Moreover, many technical annoyances about "renaming away" variables, etc., disappear by making source and target variable sets explicit. Finally, many issues are clarified, such as most general unifiers and the sense in which they are unique (again, see Lassez *et al.* (1988) for some history and discussion of the problem).

6. What Is Unification?

This section contains the technical core of the paper. We first consider classical term unification, which is the inspiration for the whole subject. This is followed by a very general discussion of solution sets, giving useful sufficient conditions for several pleasant properties, including some cardinality bounds. Next, congruences and quotients are developed for substitution systems, motivated by substitution and solution modulo equations. Then we consider families of substitution systems, and give our "official" definition of substitution system. This material can be considered the beginnings of a General Unification Theory. Many sorted algebra and order sorted algebra are also considered.

6.1. Classical Term Unification

Although term unification goes back at least[7] to Herbrand's thesis (Herbrand, 1930), it rose to its present importance in computer science because of the close connections with mechanical theorem proving developed by Alan Robinson (1965) and systematically exploited in so-called "logic" programming (for example, see Lloyd (1984)).

Using the framework of Section 5, let us define a **most general unifier** of two given terms f, g that only involve variables in $X \subseteq X_\omega$ to be a most general solution of the equation $f, g\colon X \to \{x_0\}$. Although most general solutions are not always final, there is always a closely related final solution in $\mathcal{S}_\Sigma$. Following Meseguer *et al.* (1987), call a substitution $f\colon X \to Y$ in $\mathcal{S}_\Sigma$ **sober** iff $X = \bigcup_{y \in Y} vars(f_y)$ where $f = \{f_y \mid y \in Y\}$. Then,

Proposition 3. *A substitution $h\colon X \to Y$ in $\mathcal{S}_\Sigma$ is monic iff it is sober. Moreover, if $f, g\colon Y \to Z$ in $\mathcal{S}_\Sigma$ has a most general solution, then it has a final solution.*

Proof. The first assertion is clear from knowing that substitution works by "sticking in" terms, but it seems awkward to prove from the formal definition based on free algebras, and I would be grateful to see a simple proof along those lines.

For the second assertion, given a most general solution $h\colon X \to Y$ of f, g, let $h'\colon X' \to Y$ be the substitution with X' containing exactly the variables that actually occur in the components h_y of h, and with $h'_y = h_y$ for $y \in Y$. Then h' is most general because h was, and it is monic because it is sober. Fact 2 now implies that h' is final. □

The most important result about classical term unification, which may be called the **Herbrand-Robinson Theorem**, can be stated as follows:

Theorem 2. *If a pair $\langle f, g\rangle$ has a solution, then it has a final solution.* □

A **unification algorithm** finds a most general unifier when there is one, and the hard part of the traditional proofs of the Herbrand-

[7]According to Siekman (1988), suggestions can also be found in the notebooks of Emil Post.

Robinson theorem is to show that some unification algorithm necessarily terminates. A very nice categorical formulation of a unification algorithm is given by Rydeheard and Burstall (1988), who also sketch a largely categorical proof of its correctness, and explain the obstacles that they found to a purely categorical proof. Meseguer, Goguen and Smolka (1987) develop general categorical results about unification for application to the order sorted case. The results in this and the next subsection somewhat improve those, but unfortunately are still not quite sufficient to prove the classical Herbrand-Robinson Theorem.

I realized that most general unifiers can be seen as equalizers in 1972, but have only recently appreciated what could be done with the viewpoint implicit in this observation; this mostly arose from working on Meseguer *et al.* (1987), and from reading Rydeheard and Burstall (1988) and Siekmann (1988). The usual definitions of unifier are (in my view) much too syntactic, since they depend on a notion of substitution that is too syntactic, e.g., an endo-function on the set of all terms in all variables, as in Schmidt-Schauss (1986). Taking too syntactic a view of substitution can also give misleading results. For example, Walther (1986) showed that most general unifiers only exist in order sorted algebra when the subsort graph is a forest; however, this result only holds for his notion of unifier, and is perhaps best seen as demonstrating the inadequacy of that notion; see Meseguer *et al.* (1987) for further discussion of this point. Another advantage of the categorical view is that it generalizes very naturally, as demonstrated below.

6.2. Floorings and Solution Sets

We can summarize and generalize the above discussion by saying that a unifier is a solution in some substitution system, and a most general unifier is a most general solution. However, there are many examples beyond classical term unification that do *not* have most general unifiers in this simple sense, including our ongoing integer polynomial example[8]. Such examples require a more general notion of most general solution. In keeping with our whole viewpoint and development, it is convenient to do this by generalizing the notion

[8]For example, consider the equation $x^2 = 4$.

of final object. This material, which extends and clarifies aspects of Meseguer *et al.* (1987), is somewhat more difficult than what we have seen so far, and some readers may prefer to skip it the first time through.

Definition 2. A **flooring**[9] in a category $\mathcal{C}$ is a set $\mathcal{F}$ of objects of $\mathcal{C}$ such that for each $C \in |\mathcal{C}|$ there is some $F \in \mathcal{F}$ with a morphism $C \to F$ in $\mathcal{C}$. A **minimal flooring** is a flooring such that no proper subset is a flooring. A minimal flooring $\mathcal{F}$ is **uniquely minimal** if for each $C \in |\mathcal{C}|$ and each $F \in \mathcal{F}$, there is *at most one* morphism $C \to F$ in $\mathcal{C}$. Then a **solution set** for an equation $\langle f, g \rangle$ is a flooring in $\mathcal{SOL}(f, g)$, a **most general solution set** is a minimal flooring in $\mathcal{SOL}(f, g)$, and a **final solution set** is a uniquely minimal flooring in $\mathcal{SOL}(f, g)$. □

The terminology of Mac Lane (1971) would say *weak final set* instead of flooring.

Lemma 1. *A flooring $\mathcal{F}$ is minimal iff whenever there is a morphism $F \to F'$ in $\mathcal{C}$ for $F, F' \in \mathcal{F}$, then $F = F'$.*

Proof. We first show by contradiction that the first condition implies the second. If $F \to F'$ for $F, F' \in \mathcal{F}$ with $F \neq F'$, then $\mathcal{F} - \{F'\}$ is also a flooring, contradicting the minimality of $\mathcal{F}$. Conversely, let us assume of $\mathcal{F}$ that $F \to F'$ in $\mathcal{C}$ implies that $F = F'$, and also assume that $\mathcal{F} - \{G\}$ is a flooring for some $G \in \mathcal{F}$. Then there is some $H \in \mathcal{F} - \{G\}$ such that $G \to H$. Therefore $G = H$, another contradiction. □

Not every flooring contains a minimal subflooring; however, any finite flooring certainly has a minimal subflooring. Moreover,

Proposition 5. *If a category $\mathcal{C}$ has a minimal flooring $\mathcal{F}$, then any flooring $\mathcal{G}$ contains a subset $\mathcal{G}'$ that is a minimal flooring, and moreover, there is an isomorphism $m\colon \mathcal{F} \to \mathcal{G}'$ such that there are morphisms $F \to m(F)$ and $m(F) \to F$ in $\mathcal{C}$.*

Proof. Since $\mathcal{G}$ is a flooring, for each $F \in \mathcal{F}$ there is some $G \in \mathcal{G}$ such that $F \to G$ in $\mathcal{C}$; let $G = m(F)$ and let $\mathcal{G}' = \{m(F) \mid F \in \mathcal{F}\}$.

[9]This term is intended to suggest the dual of "covering," since "cocovering" and "overing" seem unsuitable.

Then $m: \mathcal{F} \to \mathcal{G}'$ is surjective and $\mathcal{G}'$ is also a flooring. Since $\mathcal{F}$ is a flooring, for each $G \in \mathcal{G}'$, there is some $F \in \mathcal{F}$ such that $G \to F$; let $F = m'(G)$. Then we have $F \to m(F) \to m'(m(F))$ and so minimality of $\mathcal{F}$ implies that $m'(m(F)) = F$. Therefore m is also injective, and hence an isomorphism.

To show the minimality of $\mathcal{G}'$, let us assume some morphism $m(F) \to m(F')$. Then injectivity of m implies that $F = F'$, and so we get minimality by Lemma 1. □

The following says that any two minimal floorings are essentially the same; in particular, they have the same cardinality.

Corollary 2. *If $\mathcal{F}$ and $\mathcal{G}$ are both minimal floorings of $\mathcal{C}$, then there is an isomorphism $m: \mathcal{F} \to \mathcal{G}$ such that $F \to m(F)$ and $m(F) \to F$ for each $F \in \mathcal{F}$.* □

We can use this result to classify equations $\langle f, g \rangle$ by the cardinality of a most general solution set $\mathcal{F}$ as follows: the equation is **unitary** if $\mathcal{F}$ has exactly one element (i.e., it has a most general solution); it is **finitary** if $\mathcal{F}$ is finite[10]; and it is **infinitary** if $\mathcal{F}$ is infinite. Then, a substitution system $\mathcal{S}$ is: **unitary** if every solvable equation is unitary; is **finitary** if every solvable equation is finitary; is **infinitary** if every solvable equation is either finitary or infinitary, and some equation is infinitary; and otherwise, is **nullary** (i.e., some equation does not have a most general solution set). Siekmann (1988) gives this classification for classical term unification modulo equations, and the present discussion lifts it to General Unification Theory; Adachi (1984) applies similar results in a similar way.

A nullary substitution system is given in Section 6.3.4. below. However, most substitution systems of practical interest are not nullary, by the following proposition. Given a category $\mathcal{C}$, let $C < C'$ mean that $C \to C'$ but not $C' \to C$. Then call $\mathcal{C}$ **Noetherian** if there is no infinite ascending chain $C_1 < C_2 < \ldots$ in $\mathcal{C}$. An object M is **maximal** with respect to $<$ iff there is no object M' with $M < M'$.

Proposition 6. *Any Noetherian category has a minimal flooring.*

Proof. Let $\mathcal{M}$ be the set of all maximal objects in $\mathcal{C}$, and let $C \in |\mathcal{C}|$. If C is not maximal, then there is some C_1 with $C < C_1$;

[10]Notice that unitary equations are also finitary.

if C_1 is not maximal, there is some C_2 with $C_1 < C_2$; etc. In this way, either $C < M$ for some $M \in \mathcal{M}$, or else we get an infinite ascending chain, contradicting the Noetherian assumption. Thus, every $C \in |\mathcal{C}|$ is $<$ some $M \in \mathcal{M}$. Therefore, $\mathcal{M}$ is a flooring of $\mathcal{C}$, and it is minimal by construction. □

Recall that $f \preceq g$ means that $f = j; g$ for some $j \in \mathcal{C}$, and let $f \prec g$ mean that $f \preceq g$ but not $g \preceq f$. Now define $\mathcal{C}$ to be **finite factoring** if there is no infinite ascending chain $f_1 \prec f_2 \prec \ldots$, and notice that a substitution system $\mathcal{S}$ is finite factoring iff each $\mathcal{SOL}(f, g)$ is Noetherian.

Corollary 3. *No finite factoring substitution system $\mathcal{S}$ is nullary.*

Uniquely minimal flooring might seem a very specialized concept, but actually it is not. We already know that final objects and most general term unifiers are special cases. Moreover, in the category with all fields as objects and field homomorphisms as morphisms, the fields $\mathcal{Z}_p$ for $p \geq 0$ (with $\mathcal{Z}_0$ the rational numbers) are an initial covering (the dual concept to uniquely minimal flooring). Let us call a substitution system $\mathcal{S}$ **mono solution factoring** iff any solution h of any pair f, g has a factorization $e; m$ with m monic in $\mathcal{S}$ and a solution of f, g. Then

Proposition 7. *In a mono solution factoring substitution system, if $\langle f, g\rangle$ has a most general solution set, then it has a final solution set.*

Proof. First, notice that if $\mathcal{F}$ is a most general solution set for f, g, i.e., a minimal flooring of $\mathcal{SOL}(f, g)$, and if each $F \in \mathcal{F}$ is monic, then $\mathcal{F}$ is uniquely minimal. For, if there were some solution C with $a, b \colon C \to F$ (for $F \in \mathcal{F}$) with $a; F = b; F = c$ then $a = b$ since F is monic.

Now let $\{F_i \mid i \in I\}$ be a most general solution set, and let $F_i = E_i; M_i$ be a factorization for each F_i. Then $\{M_i \mid i \in I\}$ is a solution set because $\{F_i \mid i \in I\}$ is. Since there is a most general solution set, Proposition 5 gives us a subflooring of $\{M_i \mid i \in I\}$ that is monic, minimal, and therefore uniquely minimal. □

For example, since $\mathcal{S}_\Sigma$ is mono solution factoring by the proof of Proposition 4, and is easily seen to be finite factoring, it follows from Propositions 7 and 6 that every solvable equation in $\mathcal{S}_\Sigma$ has a final solution set. In order to prove the Herbrand-Robinson Theorem, it would now suffice to show that each $\mathcal{SOL}(f,g)$ has upper bounds with respect to $\preceq$. Unfortunately, I do not know any simple abstract algebraic proof of this fact.

6.3. Substitution Modulo Equations

Many interesting examples arise by imposing a fixed set $\mathcal{E}$ of equations on the function symbols in Σ, and then considering other equations *modulo* $\mathcal{E}$. In particular, our ongoing polynomial example, and so-called AC (for associative-commutative) unification, both arise this way. We will let the given set $\mathcal{E}$ of Σ-equations define a congruence $\equiv_{\mathcal{E}}$ on $\mathcal{S}_\Sigma$, and then the substitution system that we want will appear as the quotient $\mathcal{S}_{\Sigma,\mathcal{E}}$ of $\mathcal{S}_\Sigma$ by $\equiv_{\mathcal{E}}$.

The definitions are really no more complex for the general case. Given a substitution system $\mathcal{S}$, a **substitution congruence** on $\mathcal{S}$ is a family $\{\equiv_{R,S} | \ R, S \in |\mathcal{S}|\}$ of equivalence relations on the morphism sets $\mathcal{S}[R,S]$ of $\mathcal{S}$, such that

(1) $f \equiv_{R,S} f'$ and $g \equiv_{S,T} g'$ imply $f;g \equiv_{R,T} f';g'$ (if the compositions are defined), and

(2) $f;p_i \equiv_{R,S_i} g;p_i$ for $f,g\colon R \to S$ with $S = \prod_i S_i$ where $p_i\colon S \to S_i$ are its projections, implies $f \equiv_{R,S} g$ (if the compositions are defined).

Then the **quotient substitution system** $\mathcal{S}/\!\equiv$ has the same objects $|\mathcal{S}|$ as $\mathcal{S}$ (and as its base type system $\mathcal{T}$), and has for its morphisms in $\mathcal{S}/\!\equiv[R,S]$ the equivalence classes $[f]$ of the morphisms f in $\mathcal{S}[R,S]$, with their composition defined by

(3) $[f];[g] = [f;g]$.

We must now show that this gives a substitution system, and in particular, that

Proposition 8. *$\mathcal{S}/\!\equiv$ is a category with finite products.*

Proof. Let $\mathcal{Q}$ denote $\mathcal{S}/\!\equiv$ in this proof. That the composition defined by (3) is really well-defined follows from (1), and is left to the reader, who should also check that $\mathcal{Q}$ has identities. Next, we check that $\mathcal{Q}$ has a final object. Since each $\mathcal{S}[S,1]$ has just one element, namely !, each $\mathcal{S}/\equiv[S, 1]$ also has just one element, namely $[!] = \{!\}$. Using Proposition 3, it suffices to show that $\mathcal{Q}$ has binary products. Let T_1, T_2 be objects, and let T be their product in $\mathcal{S}$, with projections p_1, p_2; then T is also their product in $\mathcal{S}/\equiv$, with projections $[p_1], [p_2]$. For, if $[q_i]$: $S \to T_i$ (for $i = 1, 2$) is another cone in $\mathcal{Q}$, then q_i: $S \to T_i$ is a cone in $\mathcal{S}$, and so we get $q = \langle q_i \mid i \in \{1, 2\}\rangle$: $S \to T$ such that $q; p_i = q_i$. Therefore $[q]$ satisfies $[q]; [p_i] = [p_i]$ by (3), and the uniqueness of $[q]$ satisfying these two equations follows directly from (2). □

Now let $\mathcal{T}'$ be the subcategory of $\mathcal{S}/\equiv$ with its objects the same as those of $\mathcal{S}$, and with its morphisms the $\equiv$-classes in $\mathcal{S}$ of morphisms from $\mathcal{T}$, i.e., with $\mathcal{T}'[R, S] = \{[f] \mid f \in \mathcal{T}[R, S]\}$. Then $\mathcal{T}'$ is a broad subcategory of $\mathcal{S}/\equiv$, and $\mathcal{S}/\equiv$ and $\mathcal{T}'$ together form a substitution system that we will denote $\mathcal{S}/\equiv$. Let us call $\equiv$ **non-degenerate** iff no two distinct morphisms of $\mathcal{T}$ are identified by $\equiv$. If $\equiv$ is non-degenerate, then $\mathcal{S}/\equiv$ can be seen as still having the base type system $\mathcal{T}$. All the standard examples are non-degenerate.

The lemma below shows that any family of relations generates a congruence. We will use this result to define the congruence $\equiv_{\mathcal{E}}$ generated by a set $\mathcal{E}$ of Σ-equations. Then we let $\mathcal{S}_{\Sigma,\mathcal{E}}$ be $\mathcal{S}_{\Sigma}/\equiv_{\mathcal{E}}$.

Lemma 2. *Given a family $\{R_{S,T} \mid S, T \in |\mathcal{S}|\}$ of relations on the morphism sets $\mathcal{S}[S, T]$ of a substitution system $\mathcal{S}$, there is a least substitution congruence $\equiv$ on $\mathcal{S}$ such that $R_{S,T} \subseteq \equiv_{S,T}$ called the congruence* **generated** *by $R_{S,T}$.*

Proof. It is well-known that the intersection of equivalence relations is an equivalence relation, and it is easy to check that the intersection of any family of equivalence relations satisfying (1) also satisfies (1), and of equivalence relations satisfying (2) also satisfies (2). Moreover, since the family of relations each of which identifies everything, is certainly a substitution congruence, there is at least one substitution congruence containing the given relation $R_{S,T}$. Therefore, it makes sense to take the intersection of *all* substitution congruences containing the given relation family $R_{S,T}$ and the result is guaranteed to be the least substitution congruence containing $R_{S,T}$. □

So now, we only need to define the relation family $R_{S,T}$ generated by $\mathcal{E}$ to get the substitution congruence $\equiv_{\mathcal{E}}$ generated by $\mathcal{E}$. In fact, we let each $R_{S,T}$ be the set of pairs $\langle f, g \rangle$ such that $f, g\colon S \to T$ is an equation in $\mathcal{E}$. Then $\mathcal{S}_{\Sigma,\mathcal{E}} = \mathcal{S}_\Sigma/{\equiv_\mathcal{E}}$. The next few subsections give examples of this construction.

6.3.1. *Associative Unification*

This example assumes a signature Σ containing a binary function symbol, say $*$, satisfying the associative equation,

$$x * (y * z) = (x * y) * z,$$

where we have used infix notation for clarity. Plotkin (1972) shows that this substitution system theory is infinitary, i.e., all solvable equations have a most general solution set, but some solvable equation has an infinite most general solution set. For example, if Σ also has a binary function symbol f, then the two terms

$$x * f(a, b)$$
$$f(y, b) * x$$

have the following infinite most general solution set,

$$\{f(a,b) \Rightarrow x, a \Rightarrow y\}$$
$$\{f(a,b) * f(a,b) \Rightarrow x, a \Rightarrow y\}$$
$$\{f(a,b) * (f(a,b) * f(a,b)) \Rightarrow x, a \Rightarrow y\}$$
$$\ldots$$

Plotkin (1972) also gives a unification algorithm for this case.

6.3.2. *AC Unification*

This case assumes that Σ contains a binary function symbol $*$ satisfying not only the associative equation, but also the commutative equation,

$$x * y = y * x.$$

This system is finitary, and Stickel (1981) and others have given unification algorithms for it.

6.3.3. Polynomials

Our ongoing polynomial example now can be developed rigorously, by giving a signature and some equations that define polynomials. Here both the signature and the equation set are infinite. Let $\Sigma_0 = \mathcal{Z}$ (the integers), $\Sigma_1 = \emptyset$, $\Sigma_2 = \{+, *\}$, and $\Sigma_n = \emptyset$ for $n > 2$, where $+$ and $*$ are both AC, with the following additional equations

$$x + 0 = x$$
$$x * 0 = 0$$
$$x * (y + z) = (x * y) + (x * z)$$

plus an infinite number of equations that are the complete addition and multiplication tables for $\mathcal{Z}$, including for example, $2+2 = 4$ and $329 * -13 = -4277$. (There are various ways to do the same job with a finite signature and equation set, but the above approach seems more amusing.)

6.3.4. A Nullary Example

The following example is due to Fages (unpublished): let $\Sigma_0 = \{1, a\}, \Sigma_1 = \{g\}, \Sigma_2 = \{*\}$, $\Sigma_n = \emptyset$ for $n > 2$, and let $\mathcal{E}$ contain the equations

$$1 * x = x$$
$$g(y * x) = g(x).$$

Then each of the following is a solution to the equation $g(x) = g(a)$,

$$\{a \Rightarrow x\}$$
$$\{y_1 * a \Rightarrow x\}$$
$$\{(y_2 * y_1) * a \Rightarrow x\}$$
$$\ldots$$

and since each is a substitution instance of the one below, there is no most general solution, i.e., no maximum element, and $\mathcal{SOL}(g(x), g(a))$ is not Noetherian. Moreover, in considering Corollary 3, notice that $\mathcal{S}_{\Sigma,\mathcal{E}}$ is not finite factoring, since we have

$$\{a \Rightarrow x\} \prec \{y_1 * a \Rightarrow x\} \prec \{(y_2 * y_1) * a \Rightarrow x\} \prec \ldots.$$

6.4. Lawvere Theories

All the above examples are *Lawvere theories.* Lawvere's original formulation is perhaps less intuitive but more elegant: substitution systems with base type system $\mathcal{N}$. But the choice of base system is inessential, since any substitution system with base $\mathcal{X}$ has an equivalent substitution system with base $\mathcal{N}$ (Mac Lane (1971), page 91 gives the precise definition of equivalence for categories); moreover, this system is unique up to isomorphism.

Although it took us a rather long journey to construct the systems $\mathcal{S}_{\Sigma,\mathcal{E}}$, what is wonderful is that, in a certain sense, this journey was unnecessary, by Theorem 3 below. But first, we must define functor and isomorphism of categories.

A functor is essentially a morphism of categories. Given categories $\mathcal{C}$ and $\mathcal{C}'$, a **functor** F: $\mathcal{C} \to \mathcal{C}'$ consists of a function $|F|$: $|\mathcal{C}| \to |\mathcal{C}'|$ plus for each $A, B \in |\mathcal{C}|$, a function $F_{A,B}$: $\mathcal{C}[A, B] \to \mathcal{C}'[|F|(A), |F|(B)]$ such that

1. $F(id_A) = id_{|F|(A)}$ and
2. $F(f; g) = F(f); F(g)$ whenever $f; g$ is defined.

The composition of F: $\mathcal{C} \to \mathcal{C}'$ with G: $\mathcal{C}' \to \mathcal{C}''$ is defined the obvious way, and turns out to be associative and to have identities $id_{\mathcal{C}}$ given by $|id_{\mathcal{C}}|(A) = A$ and $(id_{\mathcal{C}})_{A,B}(f) = f$, for any category $\mathcal{C}$. Thus (ignoring any possible difficulties with set theoretic foundations, as is now quite usual), we get a category $\mathcal{CAT}$ of all categories and functors. An **isomorphism of categories** is an isomorphism in $\mathcal{CAT}$, which turns out to be just a functor F: $\mathcal{C} \to \mathcal{C}'$ with $|F|$ and each $F_{A,B}$ an isomorphism.

Theorem 3. *Any substitution system with base $\mathcal{X}$ is isomorphic to some $\mathcal{S}_{\Sigma,\mathcal{E}}$.* □

This is one of the main results about Lawvere theories; it is not trivial (Lawvere, 1963; Schubert, 1972).

6.5. Categories and Algebras of Substitution Systems

Actually, our definition of substitution system is not quite right, since the category of substitution systems with a given base type system does not have all coequalizers; the correct definition generalizes the inclusion of the base type system to a finite product preserving functor, and we will see that everything goes at least as smoothly as before.

Let us be more precise. If $\mathcal{C}$ and $\mathcal{C}'$ are categories with finite products (i.e., type systems), then a functor F: $\mathcal{C} \to \mathcal{C}'$ is **finite product preserving** if 1 is final in $\mathcal{C}'$ implies that $|F|(1)$ is final in $\mathcal{C}$, and if $\{f_i \mid i \in I\}$ is a product cone in $\mathcal{C}$ implies that $\{F(f_i) \mid i \in I\}$ is a product cone in $\mathcal{C}'$. Moreover, F: $\mathcal{C} \to \mathcal{C}'$ is **broad** if $|F|$ is surjective.

Definition 3. Let $\mathcal{B}$ be a category with finite products. Then a **substitution system** with **base** $\mathcal{B}$ is a broad finite product preserving functor S: $\mathcal{B} \to \mathcal{S}$, and a **morphism** of substitution systems with base $\mathcal{B}$, from S: $\mathcal{B} \to \mathcal{S}$ to S': $\mathcal{B} \to \mathcal{S}'$, is a product preserving functor F: $\mathcal{S} \to \mathcal{S}'$ such that $F; S' = S$. □

This definition gives a category $\mathcal{SUBS}_{\mathcal{B}}$ of substitution systems with base $\mathcal{B}$ that contains the degenerate substitution systems, and from here on, Definition 3 is the "official" notion of substitution system. In particular, $\mathcal{SUBS}_{\mathcal{N}}$ is the category $\mathcal{LAW}$ of Lawvere theories. Moreover, these categories are *cocomplete*, i.e.,

Proposition 9. *The categories $\mathcal{SUBS}_{\mathcal{B}}$ of substitution systems have all coproducts and coequalizers.* □

This follows by general nonsense, using the fact that the categories in question are comma categories (see MacLane (1971) or Goguen and Burstall (1984) for this material; Goguen and Burstall (1984) also gives some computer science examples). The identity functor $\mathcal{B} \to \mathcal{B}$ is of course initial in $\mathcal{SUBS}_{\mathcal{B}}$.

It is worth noticing that substitution systems could equally well have been developed in a dual manner, as categories with finite coproducts and an initial object.

Given a substitution system S: $\mathcal{B} \to \mathcal{S}$, an S-**algebra** is a finite product preserving functor A: $\mathcal{S} \to \mathcal{SET}$, and an S-**homomorphism** is a natural transformation between two such functors. This gives rise to a category $\mathcal{ALG}_S$ of S-algebras and S-homomorphisms, which is a subcategory of a functor category.

6.6. Many Sorted Algebra

Many sorted algebra assumes a set S of **sorts**, which are used to restrict the arguments and values of function symbols. This subsection will use notation introduced in lectures that I gave at the University of Chicago in 1969. It was once thought that many sorted algebra could be developed in the framework of classical Lawvere theories, but this is not so. Some generalization is required, and the way of viewing many sorted substitution systems as functors was worked out at IBM Research in 1972. Essentially, we will follow the same path as for $\mathcal{S}_{\Sigma,\mathcal{E}}$ in Section 6.3, but using S-indexed families in the base category instead of just sets. (Similar work was done even earlier by Benabou (1968).)

Let $\mathcal{X}^s_\omega = \{x^s_0, x^s_1, x^s_2, \ldots\}$ be an infinite set of "variable symbols" for each $s \in S$, and let $\mathcal{X}_S$ be the type system whose objects are S-indexed families $\{X_s \mid s \in S\}$ of subsets X_s of X^s_ω and whose morphisms $\{X_s \mid s \in S\} \to \{X'_s \mid s \in S\}$ are S-indexed families $\{f_s\colon X_s \to X'_s \mid s \in S\}$ of functions. (It is an exercise to check that $\mathcal{X}_S$ is a category with finite products.)

An S-**sorted signature** is a set Σ with an **arity** function $\alpha\colon \Sigma \to S^* \times S$; now let $\Sigma_{w,s} = \{\sigma \in \Sigma \mid \alpha(\sigma) = \langle w, s\rangle\}$. The intuition is that if $w = s_1 \ldots s_n$ then $f \in \Sigma_{w,s}$ takes n arguments of sorts $s_1, \ldots, s_n$ and yields a value of sort s. We now extend the initial algebra construction of Section 5.1 to the many sorted context, defining the sets $T_{\Sigma,s}$ of all Σ-**terms** of sort s to be the smallest S-sorted family of sets of strings over the alphabet $\Sigma \cup \{\underline{(}, \underline{)}\}$ such that

- $\Sigma_{\lambda,s} \subseteq T_{\Sigma,s}$ and
- given $w = s_1 ... s_n$, $t_i \in T_{\Sigma,s_i}$ and $\sigma \in \Sigma_{w,s}$ then $\sigma\underline{(}t_1...t_n\underline{)} \in T_{\Sigma,s}$.

As in Section 5.1, these sets of terms form an algebra, and it is again characterized by initiality. To make this more precise, we need the following: Given a many sorted signature Σ, a Σ-**algebra** A consists of an S-indexed family $\{A_s \mid s \in S\}$ of sets and a function $A_\sigma\colon A_w \to A_s$ for each $\sigma \in \Sigma_{w,s}$ where $A_{s_1 \ldots s_n} = A_{s_1} \times \cdots \times A_{s_n}$ and where in particular, A_λ is some one point set $\{*\}$, so that $A_\sigma\colon \{*\} \to A_s$ for $\sigma \in \Sigma_{\lambda,s}$ gives a constant of sort s. Also, a Σ-**homomorphism** $h\colon A \to B$ is an S-indexed family $\{h_s\colon A_s \to B_s \mid s \in S\}$ such that

$$h_s(A_\sigma(a_1, ..., a_n)) = B_\sigma(h_{s_1}(a_1), ..., h_{s_n}(a_n))$$

for each $s \in S$, where $\sigma \in \Sigma_{w,s}$ with $w = s_1 ... s_n$ and $a_i \in A_{s_i}$ for $i = 1, ..., n$. Then we get, as before, a category $\mathcal{ALG}_\Sigma$ of Σ-algebras and Σ-homomorphisms. Also just as in Section 5.1, we can make T_Σ into a Σ-algebra, and it is initial in $\mathcal{ALG}_\Sigma$.

The next step is to define the **free** Σ-algebra $T_\Sigma(X)$, for X an S-indexed set, to be $T_{\Sigma(X)}$ viewed as a Σ-algebra, where $\Sigma(X)_{\lambda,s} = \Sigma_{\lambda,s} \cup X_s$ for $s \in S$, and $\Sigma(X)_{w,s} = \Sigma_{w,s}$ for $w \neq \lambda$. The proof of Corollary 1 works to show the same unique extension universal property.

Analogous to Section 5.3, given X and Y in $\mathcal{X}_S$ let us define a substitution $X \to Y$ to be an S-indexed function f: $T \to T_\Sigma(X)$, and also define composition and identity just as before, getting a substitution system that we will again denote $\mathcal{S}_\Sigma$. Using the results of Section 6.3, we can generate a congruence from a given set $\mathcal{E}$ of equations, and then form the quotient substitution system.

We can also argue that since $\mathcal{S}_\Sigma$ with Σ many sorted, is mono solution factoring and finite factoring, solvable equations have final solution sets. (In fact, this substitution system, like the unsorted case, is unitary, but our very general tools are not quite strong enough to prove it.)

6.7. Order Sorted Algebra

Order sorted algebra and order sorted unification can also be developed using substitution systems. Order sorted algebra was created in 1978 (Goguen, 1978) to solve an outstanding problem in algebraic specification: it is difficult to deal with exceptions and partially defined functions using many sorted algebra; the difficulties are spelled out in some detail in Goguen *et al.* (1976). The 1978 approach to order sorted algebra accomplished this goal, but was more complex than necessary. Goguen and Meseguer (1988) gives a thorough development of basic theory for our current approach, while Meseguer *et al.* (1987) discuss unification; Smolka *et al.* (1988) gives a slightly different approach. Goguen *et al.* (1985) gives details about the operational semantics that is the basis for the OBJ2 programming language. Futatsugi *et al.* (1985, 1987), and Kirchner (1988) does the same for OBJ3. Order sorted unification has interesting applications to knowledge representation with *is-a* hierarchies, and to polymorphic type inference where there are subtypes.

Three essential ideas of order sorted algebra are:

1. Impose a partial ordering $\leq$ on the set S of sorts.
2. Define an S-indexed family $\{F_s \mid s \in S\}$ where S is a poset, to be a family of sets such that $s \leq s'$ implies $F_s \subseteq F_{s'}$.
3. Define an S-sorted signature[11] to be a family $\{\Sigma_{w,s} \mid w \in S^*, s \in S\}$ satisfying the condition that $s \leq s'$ implies $\Sigma_{w,s} \subseteq \Sigma_{w,s'}$.

Believe it or not, the exposition given in Section 6.6 goes through for this generalization without any change at all, except that signatures must satisfy a technical condition called **regularity** (see Goguen and Meseguer 1988) to insure that terms always have a well-defined *least parse*:

> given $w_0 \leq w$ and $\sigma \in \Sigma_{w,s}$ there must be some $\overline{w} \in S^*$ and $\overline{s} \in S$ such that $\sigma \in \Sigma_{\overline{w},\overline{s}}$ and $w_0 \leq \overline{w}$ such that $\sigma \in \Sigma_{w',s'}$ and $w_0 \leq w'$ implies $\overline{w} \leq w'$ and $s \leq s'$.

Additional technical conditions are needed for some results, for example, that every connected component of S has a maximum element (Goguen and Meseguer, 1988; Meseguer *et al.* 1987) In particular, we can have unification for order sorted algebra modulo equations. Meseguer *et al.* (1987) shows that under some simple syntactic conditions, $\mathcal{S}_\Sigma$ is unitary and has a linear unification algorithm.

7. Further Explorations

An introductory paper is not a good place to discuss difficult examples in detail, but it may be nice to notice that they are examples. Consequently, what this section provides is little more than some sketches and relevant citations; there are probably also bugs. Readers are urged to work out further details.

[11]Viewing a signature as a family rather than as an arity function $\alpha\colon \Sigma \to S^* \times S$ has the advantage of allowing *overloading* of function symbols, since the *same* symbol can occur in several different sets $\Sigma_{w,s}$.

7.1. Infinite Terms

Infinite terms are important in many areas, including logic programming (Colmerauer, 1982), concurrency, and natural language processing (Shieber, 1986); Section 7.5 gives an example involving type inference. Our approach is to generalize the construction of $\mathcal{S}_\Sigma$ given in Section 5.3 above in such a way that both finite and infinitary unsorted terms will be special cases.

We begin by formalizing the notion of a free construction, since free algebras are the main device used in building $\mathcal{S}_\Sigma$. The crucial relationship is that between the free algebras and their generating sets. We can give a more precise description of this relationship using the "forgetful" functor U: $\mathcal{ALG}_\Sigma \to \mathcal{SET}$ which takes each Σ-algebra A to its "underlying" set, also denoted A, and each Σ-homomorphism h: $A \to B$ to its underlying function h: $A \to B$. Then $T_\Sigma(X)$ is characterized (up to isomorphism) by the following universal property:

> given any Σ-algebra A and any (set) function f: $X \to U(A)$, there is a unique Σ-homomorphism $\overline{f}$: $T_\Sigma(X) \to A$ such that $i_X; U(\overline{f}) = f$, where i_X: $X \to U(T_\Sigma(X))$ is the inclusion function.

Now let us generalize: given a functor U: $\mathcal{A} \to \mathcal{SET}$ and a set X, an object $T(X) \in |\mathcal{A}|$ is **freely generated** by X in $\mathcal{A}$ (with respect to U) iff

> there is a function i_X: $X \to U(A)$ such that given any object A of $\mathcal{A}$ and any function f: $X \to U(A)$, there is a unique $\mathcal{A}$-morphism $\overline{f}$: $T(X) \to A$ such that $i_X; U(\overline{f}) = f$.

Given U: $\mathcal{A} \to \mathcal{SET}$ such that each set X generates a free object $T(X)$ in $\mathcal{A}$, the next step is to generalize the construction of $\mathcal{S}_\Sigma$ to that of a category $\mathcal{S}_U$ whose objects are sets, and whose morphisms from X to Y are functions $Y \to U(T(X))$. Everything proceeds just as in Section 5.3: The composition $f; g$ in $\mathcal{S}_\Sigma$ is defined to be $g; \overline{f}$; the type morphisms from X to Y are functions $Y \to X$, and they induce morphisms $X \to Y$ that factor through i_X: $X \to T(X)$, thus giving rise to a functor $\mathcal{SET}^o \to \mathcal{S}_U$ which can be shown finite product preserving.

In fact, $\mathcal{S}_A$ is $\mathcal{S}_\Sigma$ if we take $\mathcal{A}$ to be the category of Σ-algebras. But if we take $\mathcal{A}$ to be the category of continuous algebras (Goguen *et al.*, 1977) (or rational algebras, or some other variant), we will get various kinds of infinite terms as morphisms.

If we want to avoid uncountable sets of variables, we can restrict to the full subcategory whose objects are just the subsets of X_ω.

Also, notice that we can generalize the target category of the forgetful functor, from $\mathcal{SET}$ to the category $\mathcal{SET}_S$ of all S-indexed sets to get substitution systems with many sorted, or order sorted, infinite terms as morphisms.

Constructions of this kind have been rather thoroughly explored in category theory; for example, see Manes (1976).

7.2. Fixpoint Equations

Fixpoint equations are used in computer science to define many different structures, and least fixpoints are particularly used, because their existence can often be guaranteed by the well-known Tarski fixpoint theorem. A general context for these considerations is a function $f\colon X \to X$ on a poset X. Then a *fixpoint* of f is a solution $a\colon \{*\} \to X$ (in our sense) with one-point source, to the equation $f = id_X$.

When f is monotone, a most general solution for this equation is the inclusion function for the subset of all fixpoints of f. So the least fixpoint, if there is one, is the least element of the most general solution. It seems to me that the set of all fixpoints is a better starting place for discussion than the least fixpoint, because least fixpoints do not always exist, and because other fixpoints can sometimes be of equal or greater interest, e.g., the greatest fixpoint may be the one we want. Also, in this setting, there is no reason to restrict to equations of the form $f = id_X$ because an equation of the form $f = g$ also has an equalizer when f, g are monotone, and this again turns out to be the poset of all one-point solutions, including the least one-point solution if there is one.

Aït-Kaci (1986) gives a fixpoint-based approach to solving what he calls "type equations" for the semantics of a language called KBL. However, this approach seems a bit awkward since infinitary structures are not really needed, and it has been subsumed by some elegant later work of Smolka and Aït-Kaci (1987), relating what they call "feature unification" to a kind of order sorted unification.

7.3. Scott Domain Equations

The so-called "domain equations" introduced by Scott for the Scott-Strachey "denotational" model-oriented semantics are an important technique in the semantics of programming languages (Scott, 1972). Smyth and Plotkin (1982) have developed an attractive categorical generalization of the fixpoint approach and shown that it encompasses Scott's approach, perhaps even with some advantages.

As a first step, let us generalize the setup of Section 7.2 by noticing that a pre-ordered set is essentially the same thing as a category $\mathcal{C}$ with the following property:

(PO) Given $A, B \in |\mathcal{C}|$, there is at most one morphism between A and B in $\mathcal{C}$.

This follows by defining $A \to B$ iff $A \leq B$. Moreover, a monotone map between pre-ordered sets corresponds to a functor between the corresponding categories. (More precisely, there is an equivalence between the category of pre-ordered sets with monotone maps and the category of small categories satisfying condition (PO).) Therefore, we may as well use categories instead of just pre-orders. Generalizing Section 7.2, least fixpoints correspond to initial objects in most general solution categories; this works nicely, because $\mathcal{CAT}$ has equalizers.

The Smyth-Plotkin approach (1982) uses an attractive generalization of the Tarski theorem, following Wand (1975). Let us say a category $\mathcal{C}$ is **ω-continuous** iff $\mathcal{C}$ has an initial object and every diagram of the form

$$C_0 \to C_1 \to \ldots C_n \to \ldots$$

has an **ω-colimit** in $\mathcal{C}$ (that is, an initial cone). Also, let us call an endofunctor $F\colon \mathcal{C} \to \mathcal{C}$ on $\mathcal{C}$ **ω-continuous** iff it preserves whatever ω-colimits exist in $\mathcal{C}$. Now, given an endofunctor $F\colon \mathcal{C} \to \mathcal{C}$, we define an **$F$-algebra** to be an object $C \in |\mathcal{C}|$ and a morphism $A\colon F(C) \to C$ in $\mathcal{C}$, and we define a **homomorphism** of F-algebras, from A to A' to be a morphism $h\colon C \to C'$ in $\mathcal{C}$ such that $A; h = F(h); A'$. Then F-algebras form a category, and

Theorem 4. *Any ω-continuous endofunctor $F\colon \mathcal{C} \to \mathcal{C}$ on an ω-continuous category $\mathcal{C}$ has an initial F-algebra. Moreover, if A:*

$F(C) \to C$ is an initial F-algebra, then A is an isomorphism in $\mathcal{C}$. □

It is interesting to examine what happens here when $\mathcal{C}$ is a poset. ω-continuity of $\mathcal{C}$ means that countable chains have least upper bounds, and ω-continuity of F means that F preserves these least upper bounds. Then an F-algebra is some C in $\mathcal{C}$ such that $F(C) \leq C$, and an initial F-algebra is a fixpoint of F, since isomorphisms in a poset are necessarily equalities. Of course, there may also be other F-algebras that are isomorphisms, i.e., that are fixpoints, but the initial one is the least fixpoint.

There are many different kinds of "domain," but all of them are at least partially ordered sets. Readers who have gotten this far will realize that various "bootstrapping" tricks are commonplace in applying category theory, and in particular, that categorical ideas are often applied to themselves. In particular, by taking $\mathcal{C}$ to be various suitable categories of domains, Smyth and Plotkin show that everything works, including the famous "arrow" domain constructors. However, the details are more complex than you might like, and I personally feel that an approach using a substitution system whose types are Cartesian closed categories would be more satisfactory; one would then take interpretations (algebras) into categories of domains.

7.4. Algebraic Domain Equations

Ehrich and Lipeck (1983) have developed an algebraic analog of Scott's domain equations. We can give a rather elegant treatment of this example by using the substitution system whose types are the S-sorted substitution systems, for all S, with morphisms that allow changing and identifying sorts (actually, this example works even better using order sorted substitution systems). Another curious reversal of arrows occurs, and most general solutions appear as coequalizers rather than as equalizers. This approach allows treating any number of simultaneous equations in any number of variables. Of course, we do not get all the expressive power of Scott domain equations, because there are no infinitary structures. However, it is possible to carry out the same constructions with S-sorted infinitary substitution systems.

7.5. Polymorphic Type Inference

Milner's lovely paper on type polymorphism (1978) shows how to infer the types of expressions in certain higher order programming languages by using classical term unification to solve systems of type equations; the key insight is to identify polymorphic types with ordinary first order terms that contain type variables. The exclusion of subtypes from this approach can be overcome by using type expressions in order sorted algebra. A first analysis leads to systems of *inequations* among type expressions; but any such system is equivalent to a system of equations among more general type expressions that allow type disjunction, because of the fact that

$$t \leq t' \text{ iff } t \vee t' = t'.$$

The resulting equations can then be solved using order sorted unification modulo the semilattice equations for $\vee$, which exactly capture the fact that $\leq$ is a partial ordering. This technique can also be applied to solving systems of inequalities over other structures.

The approach of Milner (1978) also excludes types like $Stream(\alpha)$, as defined by

$$\text{Stream}(\alpha) = \text{cons}(\alpha, \text{Stream}(\alpha)),$$

that (in some sense) are infinite. However, the methods of Section 7.1 seem adequate for this purpose, and I would venture to suggest that substitution systems might provide an appropriate framework for exploring even more general notions of type inference.

7.6. Database Query and Logical Programming Languages

There is a rather large body of work showing how unification, as embodied in Prolog and related languages, can be made the basic inference mechanism for sophisticated database query systems; for example, the papers in Gallaire and Minker (1978) represent an early stage in this evolution. I believe that it is very important to have a precise semantic foundation for any such endeavor. A more recent approach combines logic and functional programming to extend the Prolog framework with features such as abstract data types, multiple inheritance, generic modules, and both forward and backward chaining; some recent work in this area is collected in DeGroot and Lindstrom (1986). For example, Goguen and Meseguer (1986) describe the Eqlog language, which has a rigorous semantics based upon (order sorted) Horn clause logic with equality. Some still more recent work extends this approach with object-oriented features; for example, see the FOOPS and FOOPlog languages in Goguen and Meseguer (1987). FOOPlog is a very powerful integrated database/programming system with a rigorous logical semantics, in which the solving of equations plays a basic role in answering database queries.

7.7. Generalizing the Notion of Equation

This subsection generalizes the notions of equation and solution in a way that can be very convenient in practice, even though, in theory, it does not add any expressive power. The end of Section 3 suggested generalizing the notion of equation to a set Γ of morphisms having the same source and the same target. Adachi (1984) suggests a different generalization: an equation is a pair of morphisms with the same target. Actually, we may as well go the whole way and consider an equation to be an arbitrary **diagram** D in a category $\mathcal{C}$, i.e., a graph D with each node $n \in |D|$ labelled by an object $D_n \in |\mathcal{C}|$, and each edge $e\colon n \to n'$ of D labelled by a morphism $D_e\colon D_n \to D_{n'}$ in $\mathcal{C}$. Now we can generalize the definitions and results of Section 6.2. A **solution** F of D is a family $\{f_n \mid n \in |D|\}$ of morphisms $F_n\colon C \to D_n$ in $\mathcal{C}$ such that $f_n; D_e = f_{n'}$ for each $e\colon n \to n'$ in D; we call C the **apex** of F. Similarly, a **morphism of solutions** $m\colon \{f_n \mid n \in D\} \to \{f'_n \mid n \in |D|\}$ is a morphism $m\colon C \to C'$ of the apices such that $m; f'_n = f_n$ for each $n \in |D|$. This gives a category $\mathcal{SOL}(D)$ of solutions of D, and a **most general solution** is a final object in it. These notions of solution and most general solution are actually *identical* with the usual notions of **cone** over a diagram D and **limit** cone of D; see Mac Lane (1971).

Continuing along this line, a **solution set** for a diagram D is a flooring in $\mathcal{SOL}(D)$, i.e., a set $\mathcal{F}$ of solutions of D such that for any solution G of D, there is some $F \in \mathcal{F}$ with a morphism $G \to F$. Then of course, a **minimal solution set** is a solution set such that no proper subset is a solution set, and any two of these are isomorphic by

Lemma 1, so that cardinality is a well defined invariant for minimal solution sets.

Notice that this notion of equation as diagram is general enough to include an arbitrary conjunction of (ordinary) equations, with variables shared among equations in any desired way; i.e., we can think of it as a system of equations, or a system of constraints. This framework also seems general enough to consider the Herbrand style of unification algorithm, which involves manipulating just such sets of equations. Adachi (1984) also defines a notion of "kite" that generalizes Horn clauses, and gives a resoltuion algorithm that specializes to a Prolog-like interpreter. The above could be used to simplify this a bit.

It is well known in category theory that an arbitrary limit problem can be reduced to an equalizer problem, by first taking suitable products; for example, see Mac Lane (1971), page 109. This implies that the generalization of equation and solution developed above adds no expressive power, so long as the appropriate products can be formed in $\mathcal{C}$.

7.8. Unification Grammars

So-called "unification grammar" formalisms have recently become important in linguistics. In these formalisms, a "meaning" is some kind of parameterized record structure, probably with sharing (i.e., it is a graph rather than a tree), and possibly with cycles; see Shieber (1986) for a general overview. Of course, these structures are nothing like "meanings" in the sense that human beings experience meanings, or even in the sense that sentences in formal logic have meanings in models; they are purely syntactic, without any blood or any given denotation. In linguistics, these structures have been called *feature structures*, or also "functional structures," or "dags," while in Artificial Intelligence, they have been called "semantic networks" and "frame systems;" there are also many other names. Such structures have nodes (or slots) that may contain "logical variables" (in the sense of Prolog, or more exactly, of Eqlog (Goguen and Meseguer, 1986)) that can unify with fragments of other structures to represent complex relationships; moreover, cyclic graphs represent what amount to infinite structures. Order sorted algebra is useful in this context, for providing so-called *partiality*, the possibility that further record fields can be added later. These topics will be further explored in forthcoming work with Dr. José Meseguer; see also Smolka and Aït-Kaci (1987).

The ideas developed in this paper suffice to explain the meaning of "unification" in unification grammar formalisms. To avoid commitment to any particular formalism, let us just assume that we are given a free **meaning algebra** $M(X)$ for each set X of parameters; more general types than sets can also be used. Plausible choices for M are given in Smolka and Aït-Kaci (1987) and Mukai (1985), and others have been hinted at in this paper: We can now construct a substitution system whose objects are the parameter types, and whose morphisms are the (parameterized) meanings, by using the recipe of Section 7.1; let us denote this category $\mathcal{M}$. Then given a sentence (or more generally, a discourse), a unification grammar provides a **diagram** in $\mathcal{M}$, and the **limit** (possibly weak) of this diagram is the meaning of the sentence (or discourse) as a whole; this process is called **unification** in linguistics, and we can now see that it *really is unification* in the precise sense of this paper. We are solving the equation (generalized, in the sense of Section 7.7) given by the diagram, i.e., we are solving a system of local "constraints" to determine the meaning of the whole.

This process is *compositional*, since the meaning of the whole is composed from the meaning of the parts, but in a more general sense than (for example) in Montague grammar (Montague, 1974), where meaning is compositional in the sense that it is given by the unique homomorphism from an initial algebra, i.e., it is given by "initial algebra semantics" in the sense of Goguen (1974) and Goguen *et al.* (1988) (see van Emde Boas and Janssen (1979) for a general discussion of the relation of initial algebra semantics to compositionality in linguistics). It is also interesting to note that this approach is consistent with the "General Systems Theory" doctrine that the limit of a diagram solves an arbitrary system of constraints (Goguen, 1971). Also note that ambiguity can arise naturally in this setting, because unification is not necessarily unitary.

7.9. Differential Equations

Standard techniques of functional analysis, such as differential operators on function spaces (e.g., see Yosida (1968)), seem ideally suited for developing an exposition of differential equations within the framework of this paper. Standard cases may require restricting the morphisms allowed in the equations and/or in the solutions.

7.10. Cartesian Closed Categories and Topoi

Cartesian closed categories and topoi represent substantial extensions of the point of view developed here in somewhat different directions. Each has an underlying category of base types with finite products, as well as much more. Cartesian closed categories capture typed lambda calculi, and solving equations in a Cartesian closed category is higher order unification. Topoi, which were introduced by Lawvere and Tierney, are Cartesian closed categories with additional structure that captures set-like structures that arise in geometry, algebra and logic, thus providing a surprising unification of many areas of mathematics; Goldblatt (1979) provides an introduction to these topics, and of course, we now know what it would mean to solve equations in topoi.

8. Summary

This paper develops a very general approach to the solution of equations, starting from the simple example of polynomials with integer coefficients, and gradually working toward a General Unification Theory that encompasses much more sophisticated examples such as order sorted unification, Scott domain equations, type inference, natural language semantics, and differential equations. A fair amount of basic category theory is introduced along the way, including initial and final objects, products and coproducts, equalizers and coequalizers, duality, limits, free constructions, and Lawvere theories. Substitution systems are a unifying theme, and some of their basic theory is developed, including morphisms, quotients, algebras, and cardinality bounds for most general solution sets.

References

Adachi, Takanori, (1984). Unification in categories. In *Several Aspects of Unification*, (Toshiaki Kurokawa, ed.) ICOT, Technical Report TM-0029, 35–43.

Aït-Kaci, Hassan, (1986). "An algebraic semantics approach to the effective resolution of type equations," *Theoretical Computer Science*, **45**, 293–351.

Benabou, Jean, (1968). "Structures algébriques dans les catégories," *Cahiers de Topologie et Géometrie Différentielle*, **10**, 1–126.

Burstall, Rod, MacQueen, David, and Sannella, Donald, (1980). "Hope: an experimental applicative language," In *Proceedings, First LISP Conference*, 136–143, Stanford University, Stanford, California.

Colmerauer, Alain, (1982). "Prolog and infinite trees," In *Logic Programming* (Keith Clark and Sten-Åke Tärnlund, eds.), Orlando, Academic Press, 231–251.

Colmerauer, Alain, Kanoui, H., and vanCaneghem, M. (1979). *Etude et Réalisation d'un Système Prolog.* Technical Report, Groupe d'Intelligence Artificielle, U.E.R. de Luminy, Université d'Aix-Marseille II.

DeGroot, Douglas, and Lindstrom, Gary, (1986). *Logic Programming: Functions, Relations, and Equations*, Englewood Cliffs, New Jersey, Prentice-Hall.

Ehrich, Hans-Dieter, and Lipeck, Udo, (1983). "Algebraic domain equations," *Theoretical Computer Science*, **27**, 167–196.

Futatsugi, Kokichi, Goguen, Joseph, Jouannaud, Jean-Pierre, and Meseguer, José, (1985). Principles of OBJ2. In *Proceedings of 12th ACM Symposium on Principles of Programming Languages* (Brian Reid, ed.), 52–66, Association for Computing Machinery.

Futatsugi, Kokichi, Goguen, Joseph, Meseguer, José, and Okada, Koji, (1987). "Parameterized programming in OBJ2," In *Proceedings, Ninth International Conference on Software Engineering* (Robert Balzer, ed.), 51–60, IEEE Computer Society Press.

Gallaire, Hervé, and Minker, Jack, (1978). *Logic and Data Bases*, New York, Plenum Press.

Goguen, Joseph, (1971). "Mathematical representation of hierarchically organized systems." In *Global Systems Dynamics* (E. Attinger, ed.), S. Karger, 112–128.

Goguen, Joseph, (1978). Order Sorted Algebra. Technical Report 14, UCLA Computer Science Department, Semantics and Theory of Computation Series.

Goguen, Joseph, (1974). "Semantics of Computation." In *Proceedings, First International Symposium on Category Theory Applied to Computation and Control* (Ernest G. Manes, ed.), University of Massachusetts at Amherst, 234–249. Also published in Lecture Notes in Computer Science, Volume 25, New York, Springer-Verlag, 1975, 151–163.

Goguen, Joseph, and Burstall, Rod, (1984). "Some fundamental algebraic tools for the semantics of computation, Part 1: Comma categories, colimits, signatures and theories," *Theoretical Computer Science*, **31**(2), 175–209.

Goguen, Joseph, Jouannaud, Jean-Pierre, and Meseguer, José, (1985). Operational semantics of order-sorted algebra. In *Proceedings, 1985 International Conference on Automata, Languages and Programming* (W. Brauer, ed.), New York Springer-Verlag, Lecture Notes in Computer Science, Volume 194.

Goguen, Joseph, and Meseguer, José, (1986). "Eqlog: equality, types, and generic modules for logic programming." In *Logic Programming: Functions, Relations, and Equations* (Douglas DeGroot and Gary Lindstrom, eds.), Englewood Cliffs, New Jersey Prentice-Hall, 295–363. An earlier version appears in *Journal of Logic Programming*, Volume 1, Number 2, pages 179-210, September 1984.

Goguen, Joseph, and Meseguer, José, (1987). "Extensions and foundations for object-oriented programming." In *Research Directions in Object-Oriented Programming* (Bruce Shriver and Peter Wegner, eds.), MIT Press, 417–477. Preliminary version in *SIGPLAN Notices*, Volume 21, Number 10, pages 153-162, October 1986; also, Technical Report CSLI-87-93, Center for the Study of Language and Information, Stanford University, March 1987.

Goguen, Joseph, and Meseguer, José, (1988). Order-Sorted Algebra I: Partial and Overloaded Operations, Errors and Inheritance. Technical Report to appear, SRI International, Computer Science Lab. Given as lecture at Seminar on Types, Carnegie-Mellon University, June 1983.

Goguen, Joseph, Thatcher, James, and Wagner, Eric, (1976). An Initial Algebra Approach to the Specification, Correctness and Implementation of Abstract Data Types. Technical Report RC 6487, IBM T. J. Watson Research Center. Appears in *Current Trends in Programming Methodology, IV* (Raymond Yeh, ed.), Englewood Cliffs, New Jersey, Prentice-Hall, 80–149.

Goguen, Joseph, Thatcher, James, Wagner, Eric, and Wright, Jesse, (1977). "Initial algebra semantics and continuous algebras," *Journal of the Association for Computing Machinery,* **24**(1), 68–95.

Goldblatt, Robert, (1979). *Topoi, The Categorial Analysis of Logic.* New York, North-Holland, New York.

Herbrand, Jaques, (1930). Recherches sur la théorie de la démonstration. *Travaux de la Société des Sciences et des Lettres de Varsovie, Classe III,* **33**(128).

Kirchner, Claude, Kirchner, Hélène, and Meseguer, José, (1988). Operational semantics of OBJ3. In preparation, SRI International.

Lassez, Jean-Louis, Maher, Michael, and Marriott, Kimball, (1988). "Unification revisited." In *Foundations of Deductive Databases and Logic Programming* (Jack Minker, ed.), Morgan Kaufmann, 587– 625.

Lawvere, F. William, (1963). "Functorial semantics of algebraic theories," *Proceedings, National Academy of Sciences,* **50**. Summary of Ph.D. Thesis, Columbia University, New York, New York.

Lloyd, John, (1984). *Foundations of Logic Programming*, New York, Springer-Verlag.

Mac Lane, Saunders, (1971). *Categories for the Working Mathematician*, New York, Springer-Verlag.

Manes, Ernest, (1976). *Algebraic Theories*, New York, Springer-Verlag. Graduate Texts in Mathematics, Volume 26.

Meseguer, José, Goguen, Joseph, and Smolka, Gert, (1987). Order-Sorted Unification. Technical Report CSLI-87-86, Center for the Study of Language and Information, Stanford University, Standford, California. To be submitted for publication.

Milner, Robin, (1978). "A theory of type polymorphism in programming," *Journal of Computer and System Sciences,* **17**(3), 348–375.

Montague, Richard, (1974). *Formal Philosoph: Selected Papers of Richard Montague.* New Haven, Yale University Press. Edited and with an introduction by Richard Thomason.

Mukai, Kuniaki, (1985). Unification over Complex Indeterminates in Prolog. Technical Report TR-113, ICOT.

Plotkin, Gordon, (1972). "Building-in equational theories," *Machine Intelligence,* **7**, 73–90.

Robinson, J. Alan, (1965). "A machine-oriented logic based on the resolution principle," *Journal of the Association for Computing Machinery,* **12**, 23–41.

Rydeheard, David, and Burstall, Rod, (1988). *Computational Category Theory*, Englewood Cliffs, New Jersey, Prentice-Hall. To appear.

Schmidt-Schauss, Manfred, (1986). "Unification in many-sorted equational theories," *Proceedings, 8th International Conference on Automated Deduction*, 538–552, New York, Springer-Verlag. Lecture Notes in Computer Science, Volume 230.

Schubert, Horst, (1972). *Categories*, New York, Springer-Verlag.

Scott, Dana, (1972). "Lattice theory, data types and semantics." In *Formal Semantics of Algorithmic Languages* (Randall Rustin, ed.), Englewood Cliffs, New Jersey, Prentice Hall, 65–106.

Shieber, Stuart, (1986). *An Introduction to Unification-Based Approaches to Grammar*. Center for the Study of Language and Information.

Siekmann, Jörg, (1988). "Unification theory," *Journal of Symbolic Computation*, to appear. Preliminary Version in *Proceedings*, European Conference on Artificial Intelligence, Brighton, 1986.

Smolka, Gert, and Aït-Kaci, Hassan, (1987). Inheritance Hierarchies: Semantics and Unification. Technical Report Report AI-057-87, MCC. To appear in *Journal of Symbolic Computation*, special issue on unification (Claude Kirchner, editor), 1988.

Smolka, Gert, Nutt, Werner, Goguen, Joseph, and Meseguer, José, (1988). "Order-sorted equational computation." In *Resolution of Equations in Algebraic Structures* (Hassan Aït-Kaci and Maurice Nivat, eds.), Academic Press. Preliminary version in *Proceedings*, Colloquium on the Resolution of Equations in Algebraic Structures, held in Lakeway TX, May 1987; also appears as SEKI Report SR-87-14, Universitat Kaiserslautern, December 1987.

Smyth, Michael, and Plotkin, Gordon, (1982). "The category-theoretic solution of recursive domain equations," *SIAM Journal of Computation*, **11**, 761–783. Also Technical Report D.A.I. 60, University of Edinburgh, Department of Artificial Intelligence, December 1978.

Stickel, Mark, (1981). "A unification algorithm for associative-commutative functions," *Journal of the Association for Computing Machinery*, **28**, 423–434.

van Emde Boas, Peter, and Janssen, Theo, (1979). The Impact of Frege's Principle of Compositionality for the Semantics of Programming and Natural Languages. Technical Report 79-07, University of Amsterdam, Department of Mathematics.

Walther, Christoph, (1986). "A classification of many-sorted unification theories." In *Proceedings, 8th International Conference on Automated Deduction*, 525–537, New York, Springer-Verlag. Lecture Notes in Computer Science, Volume 230.

Wand, Mitchell, (1975). On the recursive specification of data types. In *Proceedings, Symposium on Category Theory Applied to Computation and Control* (Ernest Manes, ed.), New York, Springer-Verlag, 214–217. Lecture Notes in Computer Science, Volume 25.

Yosida, Kosaku, (1968). *Functional Analysis*, Second Edition, New York, Springer-Verlag.

7 Some Fixpoint Techniques in Algebraic Structures and Applications to Computer Science

IRÈNE GUESSARIAN
C.N.R.S - L.I.T.P, UER de Mathématiques
Université Paris VII, Paris, France

1. Introduction

The trend in Computer Science recently switched towards integrating recursion in programming. This movement was initiated with LISP and now is growing stronger and stronger, even in practical applications. Functional programming languages, logic programming languages, deductive data bases, all include recursion as an essential feature. This comes from the fact that recursion allows one to express in a straightforward way the essential ideas of an algorithm, or a definition, without useless implementation details. However, if this straightforward recursive definition must then be converted into a complex imperative program, with lots of stack manipulations and book keeping, most of the gain is lost. So, once a recursive definition corresponding to a program, or describing a relation in a deductive data base, is given, usually in the form of a set of mutually recursive equations, we are interested in

1) finding the final state (result of a computation of the program, elements belonging to the relation in the data base), which

ISBN 0-12-046370-9

corresponds to a "stable state", namely a fixpoint of the set of recursive equations, and

2) finding this stable state in the most effective and efficient way, i.e. computing the fixpoint as fast as possible and/or using as little space as possible.

The above mentioned two points form the core of the fixpoint theory, and in the past few years, researchers in domains going from semantics to data bases, and from automated proof theory to logic programming have been using various fixpoint theorems and implementations thereof; we try to show in the present paper the connections and common points between the various fixpoint methods which are used. As we will see, nearly all the fixpoint techniques which are usable rely on an induction principle: they all compute the fixpoint by induction, using successive substitutions. Surprisingly enough, this very simple method leads to pretty efficient results. We show that in some cases, fixpoint techniques can even be used to transform "cheaply" a system whose solution, or stable state, is hard to find, into an equivalent (i.e. having the same stable state) system which can be easily solved (see Example 8). In other cases, e.g. a wide class of recursive queries in deductive data bases, fixpoint techniques allow one to solve the queries efficiently, without computing irrelevant or duplicate facts.

We first recall in Section 2 the classical fixpoint theorem, then some variants of this theorem which are more handy for use in computer science. Then, in Section 3, we show, via examples, some applications of the various fixpoint theorems to some areas in computer science, namely semantics and proof theory, logic programming, efficient strategies for answering recursive queries in deductive data bases.

2. Fixpoint Theorems

Fixpoint theorems are used in quite a few areas of computer science such as giving semantics to programs and/or programming languages, describing models for Prolog programs, answering recursive queries in deductive data bases, etc. .... According to the problem to be solved, variants of fixpoint theorems (Birkhoff, 1979; Lassez et al., 1982) can be used: we recall herein some of these variants. The first variant is Tarski's classical lattice theoretical fixpoint theorem (Tarski, 1955). A similar theorem, in a more restricted setting, had been previously published (Knaster, 1928).

Fixpoint Theorem 1 (Classical Fixpoint Theorem). *Let L be a complete lattice and $f : L \to L$ a monotone function; then f has a least fixpoint* $\mathrm{fix}\, f$ *such that:* $f(\mathrm{fix}\, f) = f$ *and for all e in L,* $f(e) = e \Rightarrow \mathrm{fix}\, f \leq e$ *(Birkhoff, 1979; Tarski, 1955).*

Proof. Recall that a complete lattice is a partially ordered set in which every subset has a least upper bound and a greatest lower bound. Let S be the set of all x in L such that $x \geq f(x)$ and let $a = \inf S$. By monotonicity, $x \geq f(x) \geq f(a)$ for all x in S. Hence $a = \inf S \geq f(a)$, and thus $a \in S$. By monotonicity, $a \geq f(a) \Rightarrow f(a) \geq f(f(a))$; hence $f(a) \in S$, and $f(a) \geq \inf S = a$. Together with $a \geq f(a)$, this implies $a = f(a)$. Hence a is a fixpoint of f; moreover any fixpoint e of f satisfies $e = f(e)$, hence belongs to S, and thus $a \leq e$. Letting fix $f = a$ thus proves the theorem. □

This proof, even though it is very elegant, is purely existential and perfectly non constructive: we show the least fixpoint exists, but give no hints on how to obtain it. Slightly more constructive, but still non effective, is a proof by transfinite induction. Namely define $a_\circ = \perp$ (the least element of L) and, for any ordinal $\alpha : a_{\alpha+1} = f(a_\alpha)$, and if α is a limit ordinal $a_\alpha = \sup\{a_\beta | \beta < \alpha\}$. Since L is a set, the sequence a_α must stop at some point, and so there exists α such that $a_\alpha = a_{\alpha+1} = f(a_\alpha)$. Another potential liability of this classical fixpoint theorem lies in its hypotheses, namely completeness of the lattice and monotonicity of the function, which might in some cases be too exacting. Henceforth, computer scientists looked for variants of the fixpoint theorem which could:

- on one hand provide a simple inductive construction of the least fixpoint,
- on the other hand relax either the monotonicity or the completeness hypotheses (obviously not both hypotheses can be relaxed together since a lattice in which every monotone function has a fixpoint must be complete (Davis, 1955)).

Recall first that:

(1) a c.p.o., or complete partial order, is a partially ordered set endowed with a least element $\perp$ and such that every chain has a least upper bound, and

(2) a function f is said to be continuous iff it preserves least upper bounds of chains, namely if $f(\sup\{e_n|n \in N\}) = \sup \{f(e_n)|n \in N\}$, for all chains $\{e_n|n \in N\}$.

(3) a c.p.o. is said to be a complete semilattice if every non empty subset has a greatest lower bound.

Fixpoint Theorem 1′. *Let L be a complete semilattice and $f : L \rightarrow L$ a monotone function; then f has a least fixpoint fix f (Fitting, 1985).* □

The least fixpoint is constructed by transfinite induction. For the proof we refer the reader to (Fitting, 1985) where it is also shown that f admits a family of so-called intrinsic fixpoints, i.e. fixpoints which are in some sense minimal, and which are also of interest.

Fixpoint Theorem 2. *Let E be a complete partial order, in short c.p.o., and $f{:}E \rightarrow E$ a continuous function; then f has a least fixpoint fix f such that $f(\mathit{fix}\ f) = \mathit{fix}\ f$, and for all e in E, $f(e) = e \longrightarrow e \geq \mathit{fix}\ f$. Moreover, fix f can be computed by $\mathit{fix}\ f = \sup\{f^n(\perp),\ n \in N\}$.*

Proof of Theorem 2. Define $a = \sup\{f^n(\perp)|n \in N\}$. By continuity $f(a) = \sup\{f^{n+1}(\perp)|n \in N\} = a$; hence, a is a fixpoint of f. It is the least fixpoint of f because if e is any fixpoint of f, then $\perp \leq$ e, hence, for all n: (1) $f^n(\perp) \leq e$ by induction; assuming (1) holds for n, we obtain by the monotonicity of f (every continuous function is also monotone), $f^{n+1}(\perp) \leq f(e) = e$; whence $a = \sup\{f^n(\perp)|n \in N\} \leq e$. □

We gave here the simplest variant of the Fixpoint Theorems 1′ and 2; this variant refers to *chain*-complete partially ordered sets, together with *chain*-continuous functions. Other forms of continuity and completeness can also be accommodated, mainly continuity and completeness with respect to directed sets (Guessarian, 1981); the results and proof techniques are exactly similar to the case where we restrict our attention to chains; moreover the latter case captures the flavor of the fixpoint methods and gives, to our belief, a more simplified and better understanding of the relationship between fixpoints and successive approximations.

This second fixpoint theorem achieves two of our goals which were: (1) to obtain an effectively computable fixpoint, and (2) to relax the requirement that the underlying domain be a complete lattice. This fixpoint theorem provided the basis for most of the work done in semantics, denotational semantics as well as algebraic semantics; it is used to solve recursive equations; the method is to compute by induction successive approximations to the solution of a recursive equation, let it represent a program or a value domain, or even a recursive query. However, in the case of equations corresponding to recursive queries, or of equations corresponding to Prolog, or more generally, logic programs, another constraint proves cumbersome, namely the monotonicity requirement on the functions: for Prolog programs for instance, one is often tempted to use negation, which then makes everything non-monotonic. Similarly, in Database theory one would like to ask for data bases where one is able to add informations *and also suppress* information which has become obsolete, or false: in the case of deductive data bases, this requires for some tools to deal with negation and hence non monotonicity. A first step in that direction is achieved by the following theorem.

Fixpoint Theorem 3. *Let E be a c.p.o. and f_i,$i = 1, \ldots, n$ be functions $f_i : E^i \to E$, such that each f_i is continuous in its last argument only, namely: $f_i(x_1, \ldots, x_{i-1}, \sup\{x_p | p \in N\}) = \sup\{f_i(x_1, \ldots, x_{i-1}, x_p) | p \in N\}$. Then $f_1, \ldots, f_n$ has a minimal fixpoint $\mathrm{fix}_1, \ldots, \mathrm{fix}_n$ such that:*

(i) *for $i = 1, \ldots, n$ $\quad f_i(\mathrm{fix}_1, \ldots, \mathrm{fix}_i) = \mathrm{fix}_i$*

(ii) *for all $e_1, \ldots, e_n$ such that: $f_i(e_1, \ldots, e_i) = e_i$ and $e_i \leq \mathrm{fix}_i$ for $i = 1, \ldots, n$, then $\mathrm{fix}_i = e_i$ for $i = 1, \ldots, n$.*

(iii) *fix^i is defined inductively, using $\mathrm{fix}_1, \ldots, \mathrm{fix}_{i-1}$, as follows: letting $f_i''(x) = f_i(\mathrm{fix}_1, \ldots, \mathrm{fix}_{i-1}, x), \mathrm{fix}_i = \sup\{f_{in}''(\perp) | n \in N\}$.*

Equivalently, Theorem 3 provides a minimal, and in some sense canonical, solution to the system of recursive equations:

$$\begin{aligned} f_1(x_1) &= x_1 \\ f_2(x_1, x_2) &= x_2 \\ &\dots \\ f_n(x_1, \dots, x_n) &= x_n \end{aligned} \tag{1}$$

where each f_i is monotonic and continuous in its last argument x_i only.

Proof 3. by induction on n. For $n = 1$ this is Fixpoint Theorem 2. Assume the result is true for $n - 1$ functions $f_1, \dots, f_{n-1}$ and add and nth function f_n. Let $\mathrm{fix}_1, \dots, \mathrm{fix}_{n-1}$ be the minimal fixpoint of $f_1, \dots, f_{n-1}$. Define $f_n''(x_n) = f_n(\mathrm{fix}_1, \dots, \mathrm{fix}_{n-1}, x_n)$; $f_n'' : E \to E$ is continuous, hence, by Theorem 2 f_n'' has a least fixpoint fix_n defined by $\mathrm{fix}_n = \sup\{f_n''^p(\bot) | p \in N\}$. fix_n thus satisfies $f_n''(\mathrm{fix}_n) = f_n(\mathrm{fix}_1, \dots, \mathrm{fix}_n) = \mathrm{fix}_n$, whence conditions (i) and (iii) of the theorem, since they are true by the inductive hypothesis for $i \leq n - 1$, and by construction for $i = n$. Now condition (ii) is also true by the inductive hypothesis for $i \leq n - 1$, and has to be checked only for $i = n$: since for $i = 1, \dots, n-1$ $e_i = \mathrm{fix}_i$, e_n satisfies $f_n''(e_n) = f_n(\mathrm{fix}_1, \dots, \mathrm{fix}_{n-1}, e_n) = e_n$;$e_n$ is thus also a fixpoint of f_n'' and fix_n being the least fixpoint of f_n'',$\mathrm{fix}_n \leq e_n$, whence $\mathrm{fix}_n = e_n$ (since we assumed $e_n \leq \mathrm{fix}_n$). □

This Fixpoint Theorem 3 is optimal in the sense that $(f_1, \dots, f_n)$ does not have a least fixpoint and can only have minimal fixpoints as shown by the next example.

Example 1. Let D be a set and $E = \mathbf{P}(D)$ be the powerset of D. E ordered by inclusion is a c.p.o. and even a complete lattice, the least element being the empty set, $\sup = \cup$, $\inf = \cap$.

Define

$$\begin{aligned} f_1(x_1) &= x_1 \\ f_2(x_1, x_2) &= x_1^{\sim} \cup x_2 \end{aligned} \tag{2}$$

where $f_1 : E \to E$, $f_2 : E^2 \to E$ and $x^{\sim}$ denotes the complement of x. f_1 is continuous, but f_2 is continuous only in its second argument because the presence of complementation destroys monotonicity, and a fortiori continuity for the first argument. The minimal fixpoint defined in Theorem 3 is $\mathrm{fix}_1 = \emptyset$, $\mathrm{fix}_2 = D$. But $(\emptyset, D)$ is *not* a least fixpoint for (f_1,f_2). $x_1 = D$, $x_2 = \emptyset$ provides us with another minimal fixpoint of f_1, f_2, since (f_1, f_2) $(D,\emptyset)=$ $(D,\emptyset)$; in addition these two fixpoints are incomparable since their least upper bound is $(\emptyset, D) \cap (D,\emptyset)= (\emptyset, \emptyset)$ which is *not* a fixpoint of (f_1,f_2). □

This example may seem extreme, but it is in fact quintessential of what is going on in e.g. logic programs with negation: there are no more least fixpoints around, the best we can do is find minimal fixpoints, and then, we are faced with the problem of choosing one among the minimal fixpoints. It is shown in (Apt et al., 1986) that the most natural choice, from the standpoint of computer science, is the minimal fixpoint computed in Theorem 3: they show that this fixpoint is in some sense naturally generated by successive approximations from the program itself.

3. Applications

Whenever one has a recursive or iterative equation to solve, or a function to compute by successive approximations, then a fixpoint theorem is usually the soundest and most convenient tool to be used. This explains why fixpoints are so often seen in computer science.

The present section is devoted to a (necessarily subjective) guided tour of some of the most attractive sights in the fixpoint world. We begin with the first place where fixpoints were encountered on a large scale in computer science, and where they opened the way to the whole area of denotational semantics. Historically, least fixpoints were used already before (Ginsburg-Rice, 1961; Schützenberger, 1963) to characterize context-free languages; the ideas first laid out in (Schützenberger, 1963) were applied later on to algebraic semantics. We then show other applications of fixpoint theorems to semantics: algebraic semantics, logics of programs, semantics of Prolog programs. Finally, we show how to (efficiently) evaluate recursive queries in a deductive data base by using fixpoints.

3.1. Domain Equations and Denotational Semantics

For a mathematical description of the denotational semantics of programming languages (Scott, 1971; Scott and Strachey, 1971; Stoy, 1977], we need to construct domains which may contain themselves;

such domains are defined recursively, or circularly, and are said to be *reflexive domains.*

Consider for instance the domain V of binary trees whose leaves are labeled by integers, and whose internal nodes are not labeled. Such a domain V must contain: the set N of integers, $V \times V$, and nothing else. Hence V is defined by:

$$V = N + V \times V. \tag{3}$$

V is constructed inductively by using the Fixpoint Theorem 2; letting $V_0 = \emptyset$, $V_{i+1} = N + V_i \times V_i$ for all integers i, we have: $V = \sup\{V_n | n \in N\}$. Of course this presupposes that Theorem 2 can be applied, namely that there is around a c.p.o. available in which Equation (3) is defined. This universal c.p.o. is the domain D_∞ constructed by Scott and to which we come back below. Note before that the domain V we just constructed contains all binary trees, finite or infinite. For instance, $x =< 1, x >$ and $y =< y,y >$ define the infinite trees pictured in Fig. 1.

The basis of the entire theory of denotational semantics is the most celebrated domain D_∞, which provides a domain for the λ-calculus, together with the fundamentals of nearly all subsequent work in denotational semantics. Because of the η-conversion rule of the λ-calculus, namely:

$$\text{for } x \text{ not free in } M: \quad \lambda x.Mx \equiv M,$$

every term M of D_∞ is required to represent at the same time a value and a function. Whence the equation representing D_∞:

$$E = [E \to E] \tag{4}$$

where $[E \to E]$ denotes the continuous functions from E into E. Now, the previous inductive method applied to $E = [E \to E]$ yields $\emptyset$ as least solution, i.e. the trivial solution consisting of the empty domain. The solution method is then slightly modified to obtain a nontrivial solution to Equation (4), which provides us with the D_∞ model of the λ-calculus.

We thus need more elaborate methods to construct a fixpoint of (4) which will be minimal among nontrivial fixpoints. The first construction of a domain D_∞ satisfying (4), i.e. isomorphic to its function

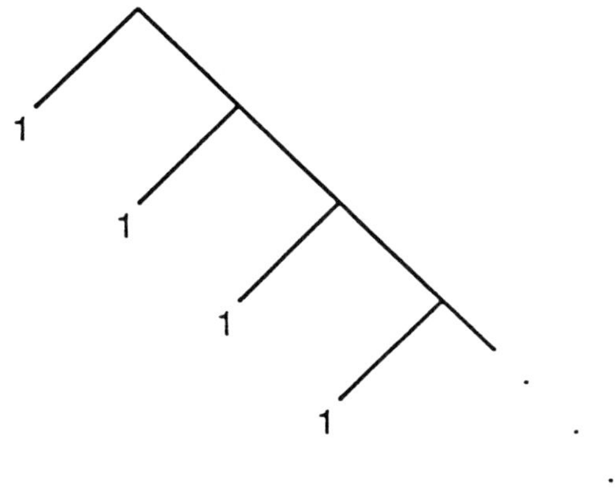

x = <1,x>

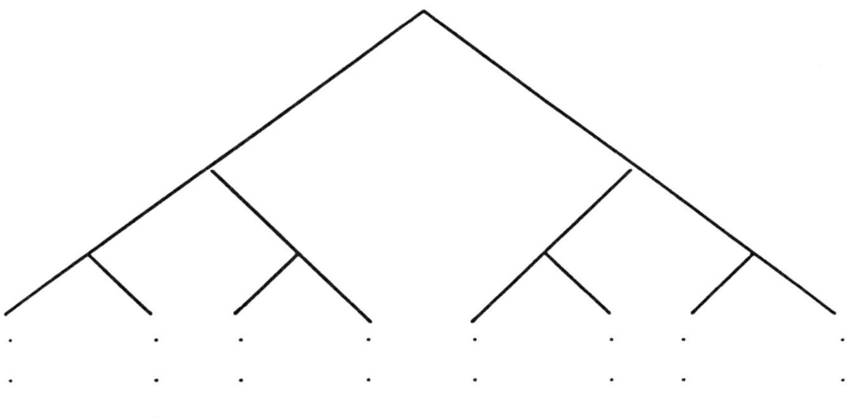

y = <y,y>

Fig. 1.

space, is due to Scott. The domain D_∞ is obtained as an inverse limit of function spaces (Scott, 1973; Scott, 1971).

Recall first that the inverse limit of the spaces $(E_n, j_n)_{n=o}^{\infty}$, where the $j_n : E_{n+1} \to E_n$ are continuous maps, is the subspace E_∞ of the product space $\prod_{n=0}^{\infty} E_n$, consisting exactly of those infinite sequences $e = (e_n)_{n=0}^{\infty}$ such that for each n, $e_n = j_n(e_{n+1})$. E_∞ is equipped with continuous maps $j_{\infty n} : E_\infty \to E_n$, such that for each n, $e_n = j_{\infty n}(e)$, and $j_{\infty n} = j_n \circ j_{\infty(n+1)}$.

The domain D_∞ is now constructed as follows:

D_0 is the two-element complete lattice $D_0 =, \{\bot, \top\}$. We have to start this construction with a lattice having at least two elements, because, as noted above, the standard construction initiating the sequence with the one element lattice $\{\bot\}$, gives a trivial solution for Equation (4). We thus let: $D_1 = [D_0 \to D_0]$, and define the embedding $i_0 = D_1 \to D_0$ as follows: for each x in D_0, $i_0(x)$ is the constant function with value x. The mapping $j_0 = D_0 \to D_1$ defined by: $j_0(x') = x'(\bot)$ for each x' in D_1 turns out to be the partial inverse of i_0, i.e. $j_0 \circ i_0 = id_{D_0}$, and $i_0 \circ j_0 \sqsubseteq id_{D_1}$.

We then define inductively:

$$D_{n+1} = [D_n \to D_n], \tag{5}$$

and if $i_{n-1} : D_{n-1} \to D_n$ and $j_{n-1} : D_n \to D_{n-1}$, we define the continuous functions $i_n : D_n \to D_{n+1}$ and $j_n : D_{n+1} \to D_n$, by: $i_n(f) = i_{n-1} \circ f \circ j_{n-1}$, and $j_n(f') = j_{n-1} \circ f' \circ i_{n-1}$ for $f \in D_n = [D_{n-1} \to D_{n-1}]$ and $f' \in D_{n+1}$. Each i_n is an embedding and each j_n is a projection.

The model of the λ-calculus we are looking for is then D_∞ the inverse limit of the sequence $(D_n, j_n)_{n=0}^{\infty}$. D_∞ is also the direct limit of the sequence $(D_n, i_n)_{n=0}^{\infty}$. Any $x = (x_n)_{n=0}^{\infty} \in D_\infty$ satisfies: $i_n(x_n) \sqsubseteq x_{n+1}$, $\forall n \in N$; hence the x_n's can be considered to be better and better finite approximations to x. Finally, it can be shown that D_∞ is homeomorphic to its own function space, hence solves the Equation (4) (Scott, 1971) we were concerned with.

3.2. Algebraic Semantics and Proof Theory

3.2.1. Algebraic Semantics of Recursive Program Schemes

The fixpoint theory is most suitable to describe the algebraic semantics of program schemes; algebraic semantics is halfway in between denotational and operational semantics: the fixpoint description captures the ideas of denotational semantics, while on the other hand, the fixpoint computation corresponding to Theorem 2 models closely the operational semantics. We now show on an example how the operational semantics proceeds to compute the solution of a program, by successive unwindings of the program, and how this process is modeled by fixpoint theory.

Example 2. Having defined a binary operation *mult* on a data type, we want to extend that operation to an operation *lmult* on lists of elements of that data type by means of the recursive LISP like program P:

lmult (L,L′) = if L = NIL then NIL else
if L′ = NIL then NIL else
cons(mult(car L, car L′), lmult (cdr L, cdr L′))

where car L (resp. cdr L) is the first element (resp. the rest) of the list L.

To this recursive program, corresponds a program scheme, naturally obtained by abstracting the meanings of the functions if-then-else, car, etc.,..., and replacing as much as possible of the right hand side of the above equation by a single function name. We obtain here the scheme S

S: lmult(L,L′) = g(L,L′,lmult(cdr L, cdr L′))

The solution of scheme S (and the corresponding program) is then computed by successive approximations. The nth approximation σ_n to that solution is defined by:

1) unwinding first the scheme S n times, i.e. replacing n times all occurrences of lmult by the righthand side of S, starting with the term lmult (L, L'). This gives here, after n unwindings $s_n(L, L') = g(L, L', g(cdrL, cdrL', g(\ldots, g(cdr^{n-1}L, cdr^{n-1}L',$ $\text{lmult}(cdr^n L, cdr^n L'))\ldots)))$.
2) replacing then all occurrences of lmult in s_n by $\perp$ (the totally undefined function) yielding here: $\sigma_n(L, L') = g(L, L', g(cdrL,$ $cdrL', g(\ldots, g(cdr^{n-1}L, cdr^{n-1}L', \perp)\ldots)))$.

In the present case, evaluating σ_n yields :

$$\sigma_n(a_1a_2\ldots a_p, a'_1a'_2\ldots a'_{p'}) = \text{mult}(a_1,a'_1)\text{mult}(a_2,a'_2)\ldots\text{mult}(a_q,a'_q) \quad \text{if } q = \inf(p,p') < n$$
$$= \perp \text{ (the undefined element) otherwise.} \tag{6}$$

We choose here the simpler but ambiguous notation of christening of the same name $\perp$ the undefined elements, or equivalently least elements, of the various domains that we consider.

Thus the σ_n make their domain of definition grow larger and larger. That results in a chain $\sigma_0 < \sigma_1 < \cdots < \sigma_n < \sigma_{n+1} < \ldots$ where the ordering $<$ represents the relation : "to be less defined than."

A tree like representation of S, the s_n's and σ_n's which might help the intuition is given in Fig. 2.

What we did here was roughly the following: we solved the program P operationally and we obtained its solution as the least upper

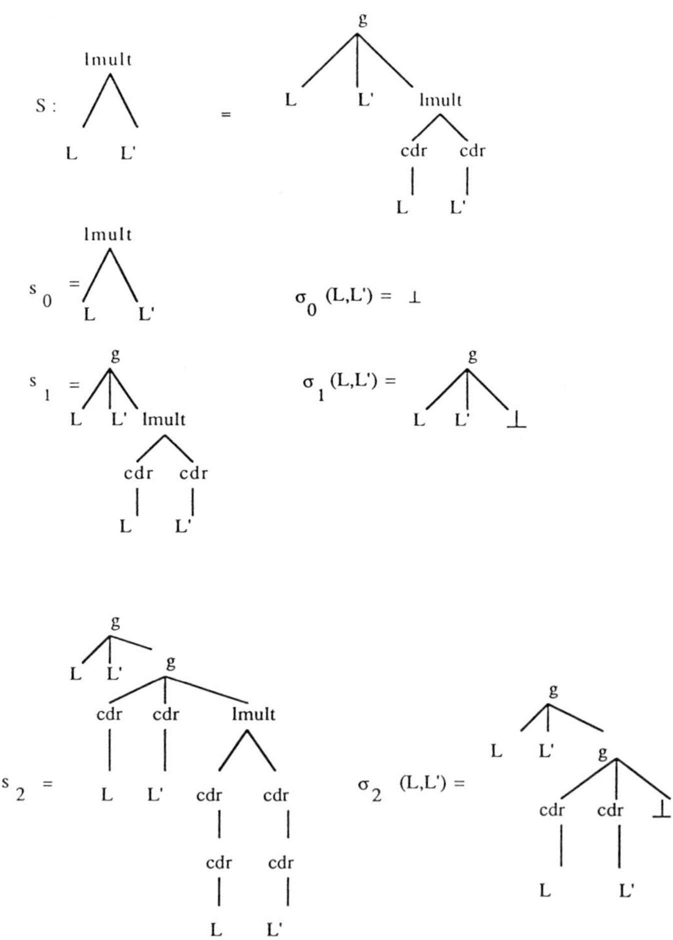

Fig. 2.

bound of the chain σ_n. Algebraic semantics goes one step further in that it solves the scheme S at the purely syntactic level, by being interested only in the symbolic computations σ_n and not in their evaluation. This is done using fixpoint methods at the purely syntactic level; then evaluation occurs only as very last resort, if needed.

Formally, let Σ be the base alphabet, here $\Sigma = \{\perp, \text{cdr}, g\}$. Define T_Σ (resp. CT_Σ) the algebra of finite (resp. finite and infinite) terms on Σ with variables {L,L′ }. T_Σ (resp. CT_Σ) are also denoted by M(F, {L,L′ }) (resp. M^∞ (F, {L,L′})), and CT_Σ is called the Herbrand model or Herbrand interpretation for Σ. S can be viewed as a fixpoint equation on the power set $\mathbf{P}(\text{T}_\Sigma)$. Define $\varphi_S(t(t_1,t_2)) = \{\perp, g(t_1,t_2,t(\text{cdr}(t_1),\text{cdr}(t_2)))\}$; roughly speaking $\varphi_S(t(t_1,t_2))$ substitutes t for lmult in the righthand side of S. So $\varphi_S : T_\Sigma \to \mathbf{P}(T_\Sigma)$. Extend φ_S to $\mathbf{P}(T_\Sigma)$ elementwise (see Guessarian, 1981, for details). Then $\mathbf{P}(\text{T}_\Sigma)$ ordered by inclusion is a complete lattice ; φ_S is continuous, hence has a least fixpoint fix $\varphi_S = \sup\{\varphi^n(\emptyset) | n \in N\}$. On the other hand, CT_Σ, with $\perp$ as least element, and the least ordering which preserves the operations as ordering, is a c.p.o. Recall that the operations are defined in the standard way, e.g., $\text{cdr}_{\text{CT}_\Sigma}(t_1)$ is simply the term $\text{cdr}(t_1)$, $g_{\text{CT}_\Sigma}(t_1,t_2,t_3)$ is the term $g(t_1,t_2,t_3)$. Moreover, the $\varphi_S^n(\emptyset)$ naturally correspond to a chain of CT_Σ, actually the chain formed of the σ_n. Hence, fix φ_S can be (and is) represented by an infinite tree T_S in CT_Σ. Thus solving S on CT_Σ is equivalent to finding the least fixpoint of φ_S in $\mathbf{P}(\text{T}_\Sigma)$ and results in an infinite tree T_S in CT_Σ. T_S is the least upper bound of the sequence σ_n, which is obtained by cutting T_S after the nth iteration, and glueing $\perp$ to the cutpoint. Fig. 3 might help.

Once S is solved in CT_Σ, then the algebraic tools enable us to derive for free the solution of S in any other algebra, together with its properties. Let D be any domain and $\mu_D(\text{S})$ the program corresponding to S on the domain D. $\mu_D(\text{S})$ defines a fixpoint equation which we want to solve on D. D is a complete Σ algebra, and CT_Σ is the free complete Σ algebra on generators {L,L′}; hence, there is a unique morphism μ'_D making Fig. 4 commute.

Then, fix $\mu_D(\text{S}) = \mu'_D(\text{T}_S) = \mu'_D$ (fix S); this shows that we can either first solve syntactically S and obtain T_S, and then interpret via μ, or first interpret S into $\mu_D(\text{S})$, and then solve the resulting equation; both choices will lead to the same result; the former corresponds to the technique of algebraic semantics, the latter to the

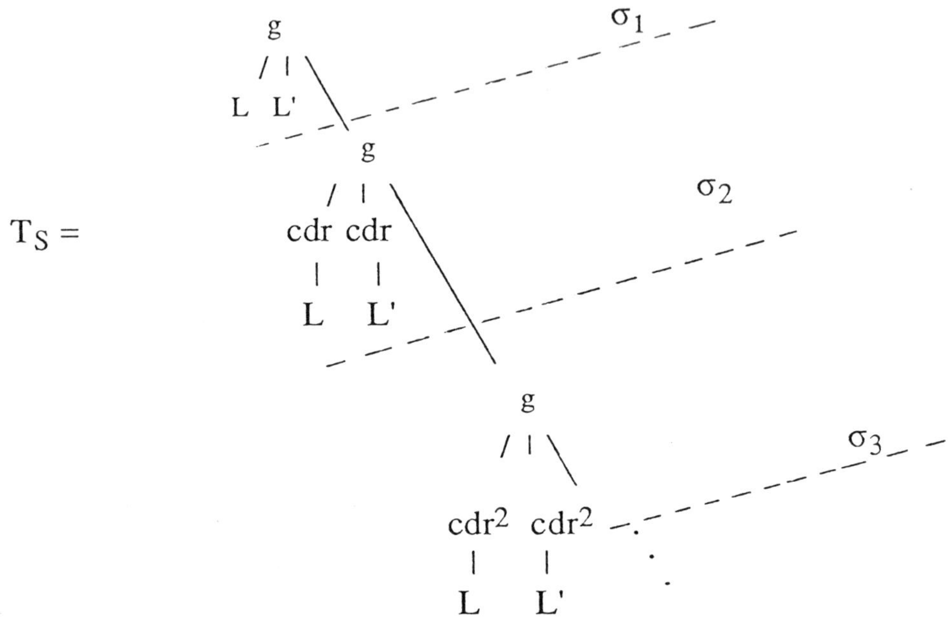

Fig. 3.

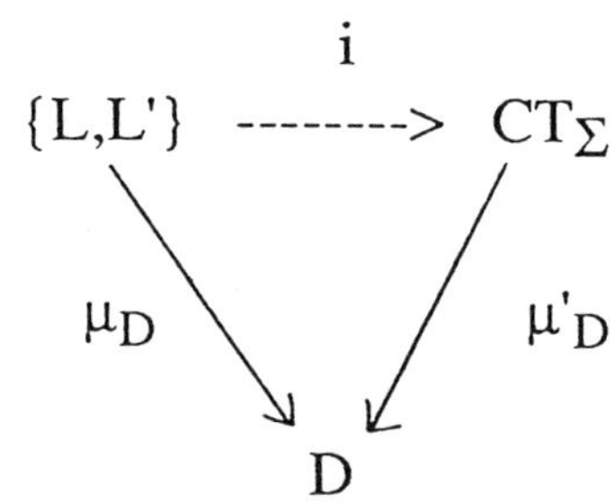

Fig. 4.

technique of denotational semantics. The advantage of algebraic semantics is that we prove properties at the syntactic level for T_S once and for all, and then the same properties follow trivially in all domains D via the morphism μ'_D. Moreover, it is usually simpler to prove properties at the purely syntactic level, because then we do not have to bother with irrelevant facts. Examples thereof will be found in the next subsection.

The general ideas and methods outlined in the present subsection apply in quite a few domains. A small sampling is listed below:

- in semantics (Andréka and Németi, 1981; Goguen et al., 1977; Gallier, 1985; Nivat, 1975) and abstract data type specifications, (Broy, 1986; Goguen et al., 1985; Wirsing and Broy, 1980), by providing initial, or sometimes terminal, models in which fixpoint theorems are subsequently used,
- in constructing algebraic models for parallelism (Baeten et al., 1985; Darondeau and Gamatié, 1987; Degano et al., 1985; Hennessy, 1985; Main and Benson, 1984), or non determinism (Arnold et al., 1985),
- in describing the actual semantics of a knowledge based language for artificial intelligence and data bases (Aït-Kaci, 1986).

The metric approach (Arnold et al., 1985) is also inspired by the approach presented in the present subsection, and is quite often, but not always, amenable to it. The metric approach consists in endowing the free algebra CT_Σ with a metric structure, and finding unique fixpoints instead of least fixpoints (Courcelle, 1985; Tiuryn, 1980). Bloom also investigates the cases when there are no unique fixpoints (Bloom, 1983). In general however, neither greatest fixpoints (Arnold et al., 1985) nor the various metric fixpoints have effective characterizations, and usually only the least, or in some cases unique, fixpoints do have an effectively computable characterization.

3.2.2. Proving Program Properties with Fixpoint Techniques

We will show examples illustrating two, among many, of the proof techniques for verifying program properties which are derived from fixpoint theory. Both proof techniques can be implemented as strategies for automated deduction systems. The first proof technique is know as *computation induction*, the second as *fixpoint induction.*

Example 2 (continued). Let mult and lmult be defined as previously. Assuming mult is associative and commutative, we would like to prove that e.g. lmult is also commutative; this can be done very easily in an equational framework, using the equations characterizing the commutativity of mult and the properties of the if-then-else (Guessarian and Meseguer, 1987), together with computation induction. Using the commutativity of mult and the axioms characterizing if-then-else, we can show easily that for any n, $\sigma_n(\mathrm{L},\mathrm{L}') = \sigma_n(\mathrm{L}',\mathrm{L})$; whence by continuity $\mathrm{T}_S(\mathrm{L},\mathrm{L}') = \mathrm{T}_S(\mathrm{L}',\mathrm{L})$. Now, let D be an actual domain, l, l' be lists of elements of the domain: $l = \mu_D(\mathrm{L})$ and $l' = \mu_D(\mathrm{L}')$ for some valuation μ_D. Then, by morphism: lmult $(l, l') = \mu_D(\mathrm{S})(l, l') = \mu'_D(\mathrm{T}_S(\mathrm{L},\mathrm{L}')) = \mu'_D(\mathrm{T}_S(\mathrm{L}',\mathrm{L})) = \mu_D(\mathrm{S})(l', l) = \mathrm{lmult}(l', l)$.

The proof technique used in the previous example proceeds from a general method, called *computation induction*, and whose most general from is as follows:

Computation induction rule. *Let* $T = \sup\{t_n | n \in N\}$, $T' = \sup\{t'_n | n \in N\}$, *assume* $\forall n \exists m \; t_n \leq t'_m$, *then* $T \leq T'$.

That is the most general, and most guaranteed, form of computation induction. Its sole drawback is the existential quantifier, which makes it less easily automatizable. Its asset is its wide range of applicability; it will succeed even in tricky cases where other, simplified but more restricted, rules of computation induction failed. We next give a much simpler rule, also derived from the fixpoint theory, the *fixpoint induction* rule. This rule succeeds less often, but when it does succeed, its application is quite straightforward, and it *is* in fact implemented in most automated deduction proof systems (Lescanne, 1984).

Fixpoint induction rule. *Let* Q *be a predicate, assume (1)* $Q(\perp)$ *holds and (2)* $Q(x) \rightarrow Q(f(x))$, *then* $Q(fixf)$ *holds.*

Note. Not every Q and f are admissible : See (Stoy, 1977) for details on the restrictions to be assumed for Q and f.

Example 3. Let f,g,h be functions such that $f(\perp) = \perp$ and $f \circ h = g \circ f$, then: fix $g = f(\text{fix } h)$.

Proof. 4. Define

$$S = \begin{cases} x = h(x) \\ y = g(y) \end{cases} \tag{7}$$

and $Q(x, y) = \{f(x) = y\}$.

We apply fixpoint induction : $Q(\perp, \perp)$ holds becauses of the strictness assumtion $f(\perp) = \perp$. We assume $Q(x, y)$, wherefrom we deduce

$Q(h(x), g(y))$.

$$\begin{aligned} f(h(x)) &= g(f(x)) && \text{(since } f \circ h = g \circ f) \\ &= g(y) && \text{(induction hypothesis).} \quad \square \end{aligned} \tag{8}$$

3.3. Prolog and Logic Programs Semantics

Let **L** be a first-order language, A, B_1, ...,B_n atoms, i.e., atomic formulae. A clause C is a formula $A_1 \leftarrow B_1\& \ldots \&B_n$. The Herbrand base U of **L** is the set of variable free atoms of **L**. An interpretation I of **L** is a subset of U. An interpretation I is said to be a model of a clause C if C is true in I. See van Emden-Kowalski (1976) and Lloyd (1984) for more details. A program P is a finite set of clauses. A Herbrand model I of P is an interpretation I of P such that each clause of P is true in I. The intersection $\cap\{I_j | \ j \in J\}$ of any non-empty set of Herbrand models I_j of P is again a Herbrand model of P; P thus admits a least Herbrand model which describes the semantics of P. Now the way to effectively compute this least model, is again via the use of a fixpoint theorem, and was described first by van Emden and Kowalski. They define a function T_P mapping interpretations of P into interpretations of P. For an interpretation I, $T_P(I) = \{A|$ for some clause $A_1 \leftarrow B_1\& \ldots \&B_n$ in P, and substitution Θ, $B_1\Theta\& \ldots \&B_n\Theta$ is true in I and $A = A_1\Theta\}$. Intuitively, $T_P(I)$ is the set of immediate consequences of I using P, i.e., the set of conclusions obtainable by applying an instance of a rule of P only once. $\mathbf{P}(U)$, the powerset of U which is the set of all interpretations, ordered by inclusion is a complete lattice, and T_P is continuous; hence, both Fixpoint Theorems 1 and 2 can be applied, and T_P has a least fixpoint defined by : fix $T_P = \sup\{T_P^n(\emptyset) | n \in N\}$. Now, by the definition of the truth of a formula in an interpretation, I is a model of P iff $I \supseteq T_P(I)$. Referring now to the proof of the Fixpoint Theorem 1, we deduce that fix $T_P = \inf\{I | I \supseteq T_P(I)\}$ is thus the least model of P. We henceforth have an inductively computable characterization of that least model.

Example 4. Consider the archetype program

$$P : \begin{cases} \text{Fact}(1, 0) \\ \text{Fact}(n \times s(y), s(y)) \leftarrow \ \text{Fact}(n, y) \end{cases}$$

Then the previous method gives as least model for P: $M_p = \{\text{Fact}(1, 0), \text{Fact}(1 \times s(0), s(0)), \text{Fact}(1 \times s(0) \times s^2(0), s^2(0)), \ldots\}$.
□

Example 5. Letting now

$$P : \begin{cases} q(y, b) \leftarrow \\ r(x) \leftarrow q(a, x) \end{cases} \tag{9}$$

We obtain

$$\begin{aligned} &T_P(\emptyset) = \{q(a, b), q(b, b)\} = I_1 \\ &T_P^2(\emptyset) = T_P(I_1) = \{q(a, b), q(b, b), r(b)\} = I_2 \\ &T_P^3(\emptyset) = T_P(I_2) = I_2 \text{ which is thus the least model of } P. \quad \Box \end{aligned}$$

If we now want to consider slightly more general programs allowing for negations in the clauses, then most of the previous results collapse, ... but fixpoint theory still can provide an answer to our problems. We now allow for clauses of the form $A_1 \leftarrow L_1 \;\&\; \ldots \;\&\; L_n$ where the L_i must not be only atoms, but may be also negations of atoms. Then, most unpleasant phenomena occur, namely that T_P is not even monotonic and that the intersection of two Herbrand models of P needs not any more be a model of P. Henceforth T_P has no least fixpoint and P has no least model. This unfortunate situation is illustrated by the next example.

Example 6. Let P be the program with one only clause $p \leftarrow\sim q$ where $\sim q$ denotes the negation of q, i.e. $\sim q$ is *not* q. Then $\{p\}$ and $\{q\}$ are two minimal and incomparable Herbrand models whose intersection $\{p\} \cap \{q\} = \emptyset$ is not a model of P. □

So in the case of programs with negative literals in their clauses, we are faced with the problem of choosing one suitable fixpoint, or one suitable minimal model for the program. Apt et al. (1986) addressed that problem and showed that a quite natural minimal model can be defined in the case of stratified programs. Intuitively, a program is stratified iff recursion and negation are not too badly mixed; more precisely, this means that there is no recursion which passes "through" negation, or, alternatively, that the program can be divided into successive strata $P_1, \ldots, P_n$, such that in stratum P_i only the symbols which were previously defined in the strata $P_1, \ldots, P_{i-1}$ can be negated. For instance,

$$P : \begin{cases} q \leftarrow p \\ p \leftarrow \sim q \end{cases} \tag{10}$$

is not stratified, whereas

$$\begin{cases} r \leftarrow \\ q \leftarrow q \& \sim r \\ p \leftarrow \sim q \end{cases} \tag{11}$$

is stratified by $P_1 = \{r\}$, $P_2 = \{q \leftarrow q\& \sim r\}, P_3 = \{p \leftarrow\sim q\}$.

By rather tricky computations, it is shown in Apt et al. (1986) that the models of stratified programs correspond to fixpoints of systems of functions of the type treated in the Fixpoint Theorem 3. It can also be shown that the minimal fixpoint computed in Theorem 3 corresponds to a minimal Herbrand model of the program P which is naturally generated by the program P itself. See Apt et al (1986) for further details. This result gives a good first step towards using fixpoint theory to study programs with negations.

An alternate approach to the semantics of logic programs, which avoids the pitfalls of nonmonotonicity, is Fitting's (Fitting, 1985). It also relies on fixpoints, the main difference with the classical method being the domains on which the fixpoints operate. In van Emden and Kowalski (1976) and their followers, I-models are considered, which are suitable subsets of the Herbrand base, and the underlying logic is a two-valued logic. In Fitting (1982), S-models are considered, which are suitable subsets of the Herbrand base *together with* a valuation in a three-valued logic, with truth-values {*true, false, undefined*}, which is made explicit. I-models are contained in the set of interpretations, and, similarly, S-models are contained in the set of basic sets (we follow Fitting's terminology). Now, the explicit use of the valuations and the undefined value allows one to treat negation while staying monotone.

Basic sets form a complete semilattice, and we associate to each program P a monotone mapping Φ_P transforming basic sets into basic sets. By the Fixpoint Theorem 1$'$, Φ_P has a least fixpoint which corresponds to the least S-model of P, and is computable by transfinite induction.

In the case of classical Prolog programs without negation, the two methods lead to very close results : the least I-model is the set of atomic formulae whose valuation is *true* in the least S-model. The minimal fixpoint computed for stratified programs can be similarly deduced from one of the S-models.

Fitting's method is thus more general in the sense that it gives a least fixpoint semantics to arbitrary logic programs, possibly involving negation and disjunction; it is however less effective, in the sense that it definitely needs transfinite induction. The problem then is to find classes of logic programs for which this transfinite induction stops after ω, or $\omega + \cdots + \omega$ steps, for instance, programs with a continuous Φ_P mapping, or stratified programs, will meet such requirements.

3.4. Recursive Queries in Deductive Data Bases

Deductive data bases (see Gallaire et al., 1984) extend relational data bases in the following sense: they consist not of only the extensional definitions of the facts in the data base, but also of an intensional part, namely rules which allow to deduce new facts to be added to the data base. Henceforth, answering queries is not any more a mere retrieval of information, but may require computations. Very often, the rules of the deductive data base are simple transcripts of the properties of the data items one is trying to define, and hence are recursive rules. Then, the problem of efficiently evaluating the recursive queries associated to such rules becomes crucial. Many strategies try to optimize the evaluation of recursive queries (see Bancilhon-Ramakrishnan (1986) for a complete and readable survey), by avoiding as much as possible irrelevant computations. One such strategy, efficient for most usual cases of recursive queries, is the computation of a well-chosen least fixpoint (Gardarin-de Maindreville, 1986; Guessarian, 1986). The basic idea of the method is to "push the selection through recursion", thus transforming binary relational queries into unary fixpoint equations. We again illustrate this method on examples.

Example 7. Consider the PARENT data base, which is described by a two column relation, the columns *"parent"* and *"child"*; an instance of this relation is defined below.

PARENT	*parent*	*child*
	John	John Jr
	Mary	John Jr
	Rosalie	Mary
	Rosalie	Martin
	Peter	Martin

Define now the *ancestor* relation by the rules:

$$(i) \quad \begin{cases} \text{ancestor } (x,y) \Leftarrow \text{parent } (x,y) \\ \text{ancestor } (x,y) \Leftarrow \text{parent } (x,z) \quad \& \quad \text{ancestor } (z,y) \end{cases}$$

i.e., by reducing to a single rule:
(ii) ancestor $(x,y) \Leftarrow$ parent (x,y) + parent (x,z) & ancestor (z,y) where + denotes the set-theoretic union, i.e. the "or" relation, and & denotes the "and" relation, i.e. the "join" of the relations parent and ancestor here.

In PARENT assume the query: *select ancestor where child = John Jr*; we wish to answer that query by evaluating only relevant facts, without computing the whole ancestor relation. To this end, we will transform the definition (i) above into a fixpoint equation ; then the definition of the least fixpoint according to Theorem 2 naturally leads to a computation which evaluates only the relevant ancestors, namely those of John Jr. Note that the above query can be shortened into *ancestor (?, John Jr)*.

Define $p(Y) = \{x \mid \text{parent}(x,y) \text{ for some } y \text{ in } Y\}$= parent $(?,Y)$, and $a(Y) = \{x|$ ancestor (x,y) for some y in Y $\}$ = ancestor (?,Y). Then (ii) is equivalent to:

$$a(y) = p(y) + p(a(y)), \tag{12}$$

henceforth, letting $\mathbf{a}$ = p(Y), $\mathbf{a}$ is the least fixpoint of the function φ defined on $\mathbf{P}$(D) by $\varphi(\mathbf{a}) = p(\mathrm{Y}) + p(\mathbf{a})$, where + denotes the set-theoretic union, D is the set of all the elements in the columns *parent* or *child* , and $\mathbf{P}$(D) is ordered by inclusion. Now, if we are interested in the evaluation of a particular query, e.g., a(John Jr.), then again a least fixpoint answers all our problems: letting $\alpha = a$(John Jr.), α is the least fixpoint of $\varphi' : \varphi'(\alpha) = p$ (John Jr.) + $p(\alpha)$; the computation method of Theorem 2 gives the answer to that query in two inductive steps, computing only relevant facts; i.e.,

$$\begin{aligned} \varphi'(\emptyset) &= p(\text{John Jr.}) = \{\text{John, Mary}\} \\ \varphi'^2(\emptyset) &= p(\text{John Jr.}) + p(\{\text{John, Mary}\}) \\ &= \{\text{John, Mary, Rosalie}\} = \varphi'^3(\emptyset). \end{aligned} \tag{13}$$

In the case of the present example, this computation is optimal. Note that again, both Fixpoint Theorems 1 and 2 can be applied.

On the same data base PARENT, consider now the query: *select child from ancestor where parent = Rosalie*, i.e., in shorthand notation *ancestor(Rosalie, ?)*. Obviously, Equation (12) cannot help in solving the present query. However, let

$$\begin{aligned} p'(X) &= \text{parent}(X, ?) = \{y | \text{for some } x \text{ in } X,\ \text{parent}(x, y)\}, \text{ and} \\ a'(X) &= \text{ancestor}(X, ?) = \{y | \text{for some } x \text{ in } X,\ \text{ancestor}(x, y)\}. \end{aligned}$$

Then (ii) is equivalent to:

$$a'(x) = p'(x) + a'(p(x)), \tag{14}$$

and the query ancestor(Rosalie, ?) can be solved by computing a'(Rosalie) by fixpoint methods. However, there is a change in the domains to which we apply the fixpoint methods. We can no more apply them directly to the domain D of elements, where D is the set of all the elements in the columns *parent* or *child*. We must now apply the fixpoint method in the domain $E = \mathbf{P}((\mathrm{D}) \to \mathbf{P}(\mathrm{D}))$ of monotone functions from $\mathbf{P}(\mathrm{D})$ into $\mathbf{P}(\mathrm{D})$; E, ordered by inclusion, i.e. $e < e'$ iff for all X: $e(X) \subset e'(X)$, is also a complete lattice, and the function φ defined by: $\varphi(e) = p' + e \circ p'$, for any e in E, is a continuous function from E into E; henceforth, $a' = \text{fix } \varphi = \sup\{\varphi^n(\emptyset) | n \in N\} = p' + p'^2 + p'^3 + \ldots$; this immediately yields $a'(\text{Rosalie}) = p'(\text{Rosalie}) + p'^2(\text{Rosalie}) + p'^3(\text{Rosalie})\ldots.$

This example shows that it is fundamental that the projection a' of the ancestor relation operate on a power set, i.e., may take sets as arguments rather than mere elements: if we would restrict our attention to functions which take only elements as arguments, then we could not deal with the query ancestor(Rosalie, ?) by using fixpoint methods. □

Fixpoint techniques can even do more for you than you would expect! In the previous example, we were interested in optimizing the computation of an answer to a query: we can also use the fixpoint techniques to transform a query whose solution is too complicated, or for which we have no algorithm for finding the solution, into a simpler equivalent query, whose solution we know how to compute.

Example 8. Consider again the data base PARENT of Example 7, and define the n-generations ancestor relation by:

$$\begin{cases} n\text{-ancestor}(x,y) & \Leftarrow & \text{parent}(x,x_1)\&\text{parent}(x_1,x_2)\&\ldots\& \\ & & \text{parent}(x_{n-1},y) \\ n\text{-ancestor}(x,y) & \Leftarrow & n\text{-ancestor}(x,x_1)\&n\text{-ancestor}(x_1,x_2)\&\ldots\& \\ & & n\text{-ancestor }(x_{n-1},y). \end{cases} \tag{15}$$

Letting as above $n - a(x) = n\text{-ancestor}(?,\ x)$, Equation (15) is equivalent to: $n - a(x) = p^n(x) + n - a^n(x)$. This last fixpoint equation, being exponential, is much harder to solve. However, using computation induction, we can prove that the fixpoint equations :

$$a = q + a^n, \text{ and} \tag{16}$$

$$b = q + c \circ b, \tag{17}$$

where c is suitably defined, have the same solution. If $\alpha = \sup \alpha_n$ (resp. $\beta = \sup \beta_n$) is the solution of Equation (16) (resp. Equation (17)), then we can check easily that: $\forall n\ \beta_n \leq \alpha_n$, and $\forall n \exists m\ \alpha_n \leq \beta_m$, whence $\alpha = \beta$ by computation induction. Now Equation (17) is a regular equation, hence, its least fixpoint is very easily computed with a WHILE loop; henceforth the evaluations of the queries corresponding to Equation (16) are solved also by the same WHILE loop (Guessarian, 1988). □

The above method is quite general; it can be used to answer via fixpoint techniques most queries of the form $A(x,?)$ or $A(?,y)$, where A is a recursively defined binary relation. The idea is to propagate the selections through recursion, as in (Bancilhon et al., 1887; Gardarin and de Maindreville, 1986), in order to avoid useless computations. However, we allow for more flexibility in that our functions can take sets as arguments and perform operations on sets if so needed; this enables us to answer via unary recursion and fixpoints a class of queries wider than the usually considered chain queries. When the functions can take only basic elements as arguments, instead of sets thereof, it is shown in (Bancilhon et al., 1987) that this propagation of selection through recursion can be performed if and only if the query is a "regular" chain query.

The difference between the methods of (Bancilhon et al., 1987) and ours is twofold:

1) The unary recursion we advocate concerns functions which can take sets as arguments, hence, is more powerful than the monadic programs considered in (Bancilhon et al., 1987); technically, the computations corresponding to our unary recursions can no more be described by the finite derivation trees of (Bancilhon et al., 1987): this explains why we can translate into unary recursions queries more general than the "regular" ones.
2) We do not restrict our attention to chain queries; the fact that the queries are chain queries is also essential in proving part of the fundamental results of (Bancilhon et al., 1987).

The method we propose proceeds as follows:

1) Associate to each deductive rule its connection graph (Gardarin and de Maindreville, 1986) defined as follows: vertices are labeled by the variables occurring in the rule, and for each relation $R(x, y)$ occurring in the rule, there is an edge going from vertex x to vertex y.
2) Use the connection graphs associated with the various deductive rules to write down, for each recursively defined relation A, a pair of fixpoint equations describing the queries $A(x, ?)$ and $A(?, y)$. For most usual queries, one only of these two equations will suffice; both equations are needed solely for the cases of crossed recursions.

A precise description of the above method will be found in (Gardarin et al., in preparation); its flavor is illustrated by the next example, which also gives an idea of its generality.

Example 9. Consider again the data base PARENT of Example 7, abbreviate parent(x, y) into $\mathrm{P}(x, y)$ and define the query $\mathrm{Q}(x, y)$ by:

$$\begin{cases} \mathrm{Q}(x,y) & \leftarrow & \mathrm{P}(x,z)\ \&\ \mathrm{Q}(z,y)\ \&\ \mathrm{Q}(y,x)\ \&\ \mathrm{P}(z,u) \\ \mathrm{Q}(x,y) & \leftarrow & \mathrm{P}(z,x)\ \&\ \mathrm{Q}(\text{Rosalie},z) \\ \mathrm{Q}(x,y) & \leftarrow & \mathrm{P}(x,z)\ \&\ \mathrm{P}(z,y). \end{cases} \tag{18}$$

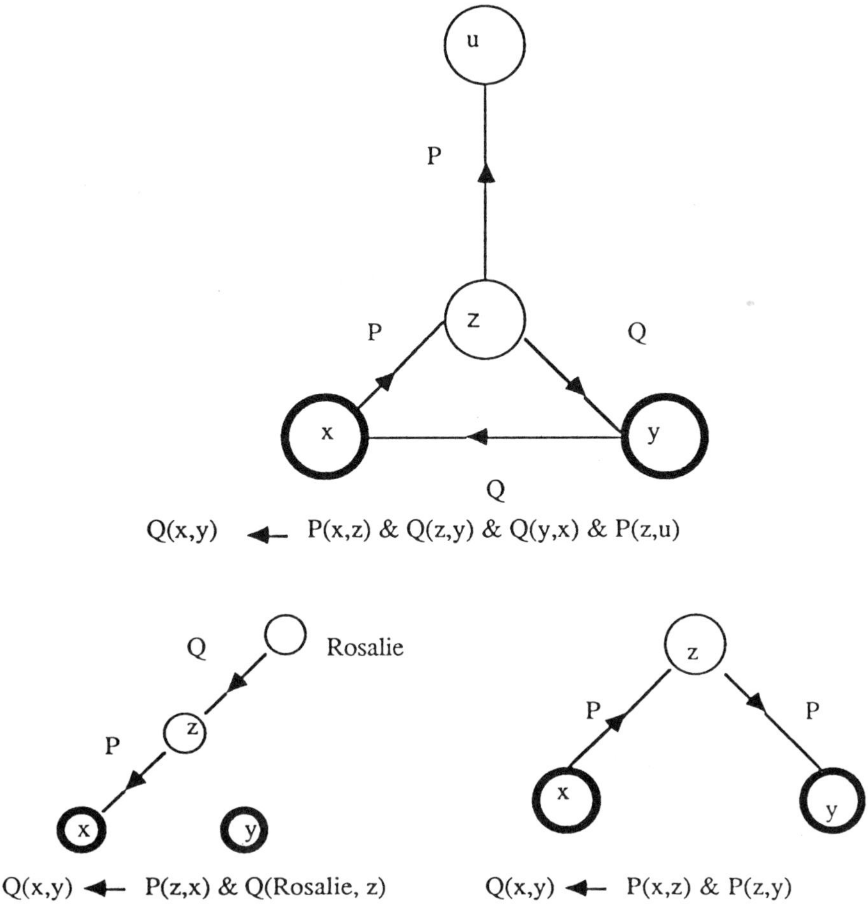

Fig. 5.

The connection graphs corresponding to the rules of Q are pictured in Fig. 5.

Let D = {John, Mary, Rosalie, Peter, John Jr., Martin} be the set of all the elements in the data base, and for X, Y included in D:

$$\begin{aligned} p(Y) &= P(?, Y) \qquad & p'(X) &= P(X, ?) \\ q(Y) &= Q(?, Y) \qquad & q'(X) &= Q(X, ?) \end{aligned} \tag{19}$$

Then, using the connection graphs pictured in Fig. 5, the deductive rules (Equation 18) can immediately be translated into the following fixpoint equations:

$$\left\{\begin{array}{rcl} q(Y) & = & q'(Y) \cap p(q(Y) \cap p(D)) + p'(q'(\text{Rosalie})) + p^2(Y) \\ q'(X) & = & q(X) \cap q'(p'(X) \cap p(D)) \\ & & +\{\text{if } X \cap p'(q'(\text{Rosalie})) \neq \emptyset \\ & & \quad \text{then } D \text{ else } \emptyset\} + p'^2(X) \end{array}\right. \tag{20}$$

These equations are then solved in the domain $E = (\mathbf{P}(\mathrm{D}) \rightarrow \mathbf{P}(\mathrm{D}))$ of monotone functions from D into D by the usual fixpoint techniques: we note that the functionals corresponding to our fixpoint equations are in fact continuous from E into E. We can thus define a sequence of approximations by:

$$q_0(Y) = q'_0(X) = \emptyset \tag{21}$$

and for n in N

$$\begin{array}{rl} q_{n+1}(Y) & = q'_n(Y) \cap p(q_n(Y) \cap p(D)) + p'(q'_n(\text{Rosalie })) + p^2(Y) \\ q'_{n+1}(X) & = q_n(X) \cap q'_n(p'(X) \cap p(D)) \\ & \quad + \{\text{if } X \cap p'(q'_n(\text{Rosalie })) \neq \emptyset \\ & \quad \text{then } D \text{ else } \emptyset\} + p'^2(X) \end{array} \tag{22}$$

and the least fixpoint of Equation(20) is $(q_\omega,\, q'_\omega)$ defined by:

$$q_\omega = \sup\{q_n | n \in N\}, \quad \text{and } q'_\omega = \sup\{q'_n | \in N\}. \tag{23}$$

Assume now we are interested in the query $q(\text{John Jr}) = Q(?,$ John Jr). We need to evaluate the following subqueries: $q'(\text{Rosalie})$, $q'(\text{John Jr})$ and $q(\text{Rosalie})$, which are all relevant for the computation of $q(\text{John Jr})$. Using the above method, we obtain:

$$\begin{array}{l} q_\omega(\text{Rosalie}) = q_2(\text{Rosalie}) = \emptyset \\ q'_\omega(\text{Rosalie}) = q'_2(\text{Rosalie}) = \text{John Jr.} \end{array} \tag{24}$$

Whence we deduce immediately (in fact just one computation step is enough) that:

$$\begin{array}{l} q'_\omega(\text{John Jr}) = q'_2(\text{John Jr}) = \emptyset \\ q_\omega(\text{John Jr}) = q_2(\text{John Jr}) = \text{Rosalie.} \end{array} \tag{25}$$

Slightly modifying our set of deductive rules (18) by substituting the rule $Q^0(x,y) \leftarrow P(x,z)$ & Q^0(Rosalie,z) for $Q(x,y) \leftarrow P(z,x)$ & Q(Rosalie,z) will result in substituting $p(q'$(Rosalie)) for $p'(q'$(Rosalie)) in the fixpoint Equations (20). Evaluating q^0(Rosalie), we would then generate $q^{0'}$(Rosalie) and $q^{0'}$(Mary), and obtain: q^0_ω(Rosalie) $=$ q^0_5(Rosalie) $=$ {John, Mary, Rosalie, Peter}. □

4. Conclusion and Discussion

In the present paper, we showed two examples of least fixpoint theorems: we tried to highlight the fact that these fixpoint theorems, or variants thereof, apply to quite a wide variety of domains in computer science. Fixpoint methods also yield usually simple solutions to the problems one has to treat, or even can yield solutions when other methods fail, e.g. in the n-generations ancestor problem. Moreover, fixpoint techniques allow the use of recursive tools and recursive equations very easily: such tools are usually both more powerful and simpler, in the following sense: using recursion, one can express things in a way which is simpler and closer to "reality", by forgetting inessential details. Thus, especially in the data base field, fixpoint techniques yield nearly optimal strategies for evaluating queries, because they naturally do not introduce irrelevant facts and avoid duplicate computations.

We of course were nonexhaustive, because we left out of our study the whole dual theory of greatest fixpoints, and the study of all other fixpoints: usually a whole lattice of them! The main reason is that from a computational standpoint, all other fixpoints are much "harder" to compute than the least fixpoint, and usually there are no effective characterizations of these fixpoints, with the exception of one attempt in Cousot and Cousot (1979). Consequently, greatest fixpoints and other fixpoints are less handy from computer science and computation standpoints than least fixpoints. However, effective characterizations of fixpoints other than the least one would be of great interest, mainly in the fields of non monotonic reasoning and logic programming for the cases when least fixpoints cannot apply: we saw one such case with Example 6, cf. also Apt et al. (1986). A study of fixpoints for some non-monotonic functions can be found in Erné (1986); it should be investigated whether these results apply to the functions T_P corresponding to logic programs with negations; moreover, only existence theorems are given in Erné (1986); effective construction methods should be described in order to make these results usable.

Another example is the case of Prolog programs where it is shown that the negation by (finite) failure evaluation rule naturally corresponds to greatest fixpoints (Apt and Van Emden, 1982; Lloyd, 1984), and other evaluations might lead to fixpoints in between the least and greatest ones (Delahaye, 1986), cf. also the intrinsic fixpoints approach of Fitting (1985), which generalizes the optimal fixpoint of Manna and Shamir (1977); henceforth, good effective algorithms to compute such fixpoints would be very helpful. Such fixpoints could also be used to study deductive data bases where one can delete, as well as add, information, e.g. by allowing for negations in the recursive queries.

Fixpoints other than least ones are also used in proofs of parallel program properties, see e.g. Cousot and Cousot (1980); Darondeau and Gamatié (submitted): a problem quite often encountered is to characterize greatest fixpoints in factor algebras, which is unsolved to our knowledge. This shows that many problems remain to be investigated in the fixed points world.

Acknowledgments

Many thanks to Krzysztof Apt and Jean-Paul Delahaye for enlightening discussions on logic programming, and to Michel de Rougemont for very insightful comments, especially bringing to my attention the second part of Example 7, and some features of Example 9. Thanks also to Arlette Dupont who typed the paper with unfailing patience.

References

Aït-Kaci, H., (1986). "An algebraic semantics approach to the effective resolution of type equations," To appear in *Theor. Comput. Sc.*

Andréka, H., and Németi, I., (1981). "Applications of Universal Algebra, Model theory and Categories in Computer Science," Proc. FCT'81, *Lect. Notes in Comput. Sc.* **117**, Springer-Verlag, Berlin, 16–23.

Apt, K. R., Blair, H., Walker, A., (1986). "Towards a theory of declarative knowledge," in *Foundations of Deductive Data Bases and Logic Programming* (J. Minker, ed.), Morgan-Kaufmann Publishers, Palo Alto, 89–148.

Apt, K. R., Van Emden, M. H., (1982). "Contributions to the theory of logic programming," *Jour. Assoc. Comput. Mach.* **29**, 841–862.

Arnold, A., Naudin, P., Nivat, M., (1985). "On semantics of nondeterministic recursive program schemes," In **Algebraic Methods in Semantics** (Nivat & Reynolds, eds.) Cambridge Univ. Press, London, 1–33.

Baeten, J. C. M., Bergstra, J. A., Klop, J. W., (1985). "An operational semantics for process algebra," Rep. CS-R8522, CWI, Amsterdam.

Bancilhon, F., Beeri, C., Kanellakis, P., Ramakrishnan, R.,(1987). "Bounds on the propagation of selection into logic programs," 6th ACM Symp. on principles of data base systems, San Diego.

Bancilhon, F., and Ramakrishnan, R., (1986). "An amateur's introduction to recursive query processing strategies," Proc. Sigmod 86, Washington D.C.

Birkhoff, G., (1979). *Lattice theory, 3rd Edition*, Americ. Math. Soc., New York.

Bloom, S. L., (1983). "All solutions on a system of recursion equations in infinite trees and other contraction theories," *Jour. Comput. Sys. Sci.* **27**, 225–255.

Broy, M., (1986). "Partial interpretations of higher order algebraic types," Proc. MFCS'86, *Lect. Notes in Computer Sc.* **233**, Springer Verlag, Berlin, 29–43.

Courcelle, B., (1983). "Fundamental properties of infinite trees," *Theor. Comput. Sci.* **25**, 95–169.

Cousot, P., Cousot, R., (1979). "Constructive versions of Tarski's fixed point theorems," *Pacific Jour. of Math.* **82** 43–57.

Cousot, P., Cousot, R., (1980). "Semantic analysis of communicating sequential processes," Proc. ICALP 80, *Lect. Notes in Comput. Sci.* **85**, Springer Verlag, Berlin, 119–133.

Darondeau, P., Gamatié, B. "A fully observational model for infinite behaviors of communicating systems." Submitted to *Fundamenta Informaticae* (1987).

Davis, A. C., (1955). "A characterization of complete lattices," *Pacific Jour. of Math.* **5**, 311–319.

Degano, P., de Nicola, R., Montanari, U., (1985). "Partial ordering derivations for CCS," FCT'85, *Lect. Notes in Comput. Sc.* **199**, Springer-Verlag, Berlin, 520–533.

Delahaye, J. P., (1986). Sémantiques logiques et dénotationnelles des interpréteurs Prolog, LIFL Report, Lille Univerity (1986), to appear.

Van Emden, M. H., Kowalski, R. A., (1976). "The Semantics of Predicate Logic as a Programming Language," *Jour. Assoc. Comput. Mach.* **23**, 733–742.

Erné, M., (1986). "W-completeness and fixed point properties," Tech. Rep. Hannover Univers. Presented at the 33rd Workshop on Universal Algebra, Darmstadt.

Fitting, M., (1985). "A Kripke-Kleene semantics for logic programs," *J. Logic Programming*, **4**, 295–312.

Gallaire, H., Minker, J., Nicolas, J.M., (1984). "Logic and data bases: a deductive approach," *Assoc. Comput. Mach. Comput. Surveys*, **16**, 153–185.

Gallier, J. H., (1985). "The semantics of recursive programs with function parameters of finite types: n-rational algebras and logic of inequalities," *In Algebraic methods in semantics* (Nivat & Reynolds eds.), Cambridge Univ. Press, London, 313–362.

Gardarin, G., Guessarian, I., de Maindreville, C. Translation of "Datalog rules into functional fixpoint equations," to appear in *Theor. Comput. Sci.*

Gardarin, G., and de Maindreville, C., (1986). "Evaluation of data base recursive logic programs as recurrent function series," Proc. Sigmod 86, Washington D.C.

Ginsburg, S., Rice, H. G., (1961). Two families of languages related to Algol, *Jour. Assoc. Comput. Mach.* **9**, 350–370.

Goguen, J., Jouannaud, J. P.. Meseguer, J., (1985). "Operational semantics of order sorted algebras," Proc. ICALP'85, *Lect. Notes in Comput. Sc.* **194**, Springer Verlag, Berlin.

Goguen, J., Thatcher, J., Wagner, E., Wright, J., (1977). "Initial algebra semantics and continuous algebras," *J. Assoc. Comput. Mach.* **24**, 68–95.

Guessarian, I., (1981). "Algebraic Semantics," *Lect. Notes in Comput. Sci.* **99**, Springer-Verlag, Berlin.

Guessarian, I., (1988). "A note on fixpoint techniques in data base recursive logic programs," *RAIRO Infor. Théor.* **22**, 49–56.

Guessarian, I., and Meseguer, J., (1987). "On the axiomatisation of if-then-else," *SIAM J. Comput.* **16**, 332–357.

Hennessy, M., (1985). "Acceptance trees," *Jour. Assoc. Comput. Mac.* **32**, 896–928.

Knaster, B., (1928). "Un théorème sur les fonctions d'ensembles," *Ann. Soc. Polon. Math.* **6**, 133–134.

Lassez, J. L., Nguyen, V. L., Sonenberg, E. A., (1982). "Fixed point theorems and semantics: a folk tale," *Infor. Proces. Letters* **14**, 112–116.

Lescanne, P., (1984). "Term rewriting systems and algebra," Proc. 7th Inter. Conf. on Autom. Deduction, Napa Valley.

Lloyd, J. W., (1981). *Foundations of Logic Programming*, Springer Verlag, Berlin.

Main, M., Benson, D. B., (1984). "Functional behavior of nondeterministic and concurrent programs," *Inform. & Control* **62**, 144–189.

Manna, Z., Shamir, A., (1977). "The optimal approach to recursive programs," *Comm. Assoc. Comput. Mach.* **20** 824–831.

Nivat, M., (1975). "On the interpretation of recursive polyadic program schemes," *Symp. Math.* **15**, Rome, 255–281.

Schützenberger, M. P., (1963). "On context-free languages and pushdown automata," *Inform. & Control* **6**, 246–264.

Scott, D., (1971). "The lattice of flowdiagrams," *Symp. on Semantics of Algorithmic Languages, Lect. Notes Math.* **188**, Springer Verlag, Berlin, 311–366.

Scott, D., (1973). *Models for various type-free calculi*, In *Logic, Methodology and Philosophy of Science IV*, P. Suppes, L. Henkin, A. Joja, and G. C. Moisil, eds.) North-Holland, Amsterdam, 157–187.

Scott, D., (1971). Continuous Lattices, Technical Monograph PRG-7, University of Oxford.

Scott, D., Strachey, C., (1971). *Toward a mathematical semantics for computer languages*, Techn. Monograph PRG-6, Oxford Univ. Computing Lab.

Stoy, J., (1977). *Denotational Semantics: The Scott-Strachey Approach to Programming Language Theory*, MIT Press, Cambridge Mass - London.

Tarski, A., (1955). "A lattice theoretical fixpoint theorem and its applications," *Pacific Jour. of Math.* **5**, 285–309.

Tiuryn, J., (1980). "Unique fixed points vs. least fixed points," *Theor. Comput. Sci.* **12**, 229–254.

Wirsing, M., Broy, M., (1980). "Abstract data types as lattices of finitely generated modes," Proc. MFCS'80, *Lect. Notes in Comput. Sc.* **88**, Springer-Verlag, Berlin, 673-685.

8 Canonical Representatives for Observational Equivalence Classes

UGO MONTANARI

Dipartimento di Informatica
Università di Pisa
Pisa, Italy [1]

MARCELLO SGAMMA

Olivetti DOR
Pisa, Italy

1. Introduction

The goal of an observation method is to determine the observable behaviours of a system under various conditions. The conditions may involve different observers, different stimuli to the system, or simply range over all possible executions of a nondeterministic system. Observable behaviours may be useful in expressing the requirement that a system satisfies a specification, which may require or exclude certain observable behaviours. Furthermore, an observation method may be used for defining an equivalence notion between systems: two systems are equivalent whenever no observation can distinguish them.

Many different observation methods and observation equivalences have been proposed for nondeterministic systems, and especially for nondeterministic concurrent systems. This is mainly due to the large

[1]Research performed in part when on leave at SRI International, Menlo Park, USA, while supported by Office of Naval Research Contract N00014-86-C-0450.

ISBN 0-12-046370-9

number of properties which may be relevant to their analysis. Most of these methods are defined for suitable classes of *labeled transition systems.* For a rather complete description and comparison of proposed equivalences for transition systems, see DeNicola (1987).

Here we consider in particular the notion of *observational equivalence* (Park, 1981; Milner, 1985). Conceptually, the behaviour of a transition system is represented as a *synchronization tree* with arcs labeled by actions, defined as the unfolding of the transition system itself. Two trees are equivalent if, for every path from the root in the first tree, a path from the root can be found in the second tree such that the two paths: i) are labeled by the same sequence of *visible* actions, and ii) end up in equivalent subtrees. The same property must be true, when the roles of the first and the second tree are exchanged. This is an equation on tree equivalences, whose solutions are called *bisimulations.* The maximal fixpoint does exist (since the functional in the left member of the equation is monotone over the inclusion ordering on relations) and is called observational equivalence.

In many interesting cases, the set of paths labeled by a sequence of visible actions may be infinite, making the above functional noncontinuous. Therefore, proving properties of observational equivalence may require transfinite induction. However, it is often easy to prove that two trees are observationally equivalent (Sanderson, 1982): it is enough to exhibit a bisimulation relating them.

When nondeterminism arises in the context of concurrent systems, a number of new issues become relevant. In particular, if the behaviour of the parallel composition of two systems is defined as the interleaving of their behaviours, no distinction is drawn between causal and temporal dependencies.

Causal dependencies play a crucial role in defining important properties (e.g., fairness, starvation freedom) of distributed systems and it is usually impossible to recover causal dependencies from temporal dependencies when needed. Thus causal dependencies are often assumed as directly observable. A natural representation is obtained using *partial orderings of events*, where ordered events are considered as causally related and unordered events as concurrent/independent.

It is tempting to refine observational equivalence to deal with observations consisting of partial orderings of events instead of sequences of actions. However a severe difficulty arises, which makes synchronization trees inadequate: it is not clear at all how to recover from a path in the given tree the information which is necessary for constructing the partial ordering observation. Whenever two consecutive events are considered, there is no way to tell whether they are causally related or concurrent.

A new class of trees called *Nondeterministic Measurement Systems* (*NMS*) was introduced in previous papers by the first author in collaboration with Pierpaolo Degano (1987, 1987a) and Rocco De Nicola. Given a transition system, its NMS consists of its finite computations ordered by prefix and labeled with *observations* taken from a set D. Different choices of the set D (e.g. sequences, partial orderings, etc.) and of the labeling function capture different notions of observation. NMS are a useful tool when dealing with partial orderings since they extract entire observations directly from computations instead of composing incremental observations.

In the same papers (Degano et al., 1987, 1987a), a notion of *NMS observational equivalence* was defined on NMS and Milner's *Calculus of Communicating Systems (CCS)* was used as a test case. A new *partial ordering observational equivalence* was given on CCS agents, and it was proved finer than classical observational equivalence since it distinguishes the interleaving of sequential nondeterministic processes from their concurrent execution. Moreover, the two equivalences coincide when sequential processes only are considered.

A drawback of equivalences as semantic definitions (for both synchronization trees and NMS) is that the elements of the semantic domains are sets of trees: they are difficult to visualize and to prove properties about. A satisfactory solution is to select a *canonical representative* for every equivalence class and to define a domain consisting of canonical representatives.

Canonical representatives for CCS observational equivalence (in fact observational congruence) classes have been proposed in Castellani (1987). Given any element of the class, the canonical representative can be obtained as the result of a simplification operation called *abstraction homomorphism*. The notion of abstraction homomorphism was introduced for the first time in Castellani et al. (1983) for a model called *Labeled Event Structures*.

However, two limitations reduce the applicability of the results in Castellani (1987). First, the observations must be sequences of actions. Thus no partial ordering observations are possible.

Second, the functional in the bisimulation equation must be continuous. Therefore the result is restricted to a special case: in particular, it does not apply to languages, like CCS, with both recursion and invisible actions.

Benson and Ben-Shachar (1986, 1987) define a category of automata and simulations between them. They show that the epimorphisms between pairs of automata partition the category into bisimulation equivalence classes and that there is a unique canonical representative for each bisimulation equivalence class. However, their notion of bisimulation is defined, as usual, by steps consisting of single events, and thus is inadequate to express partial ordering bisimulation. The connection between bisimulation and finality has also been noticed (for SCCS) by Aczel (1987).

In this paper we prove under very general assumptions the existence of canonical representatives for observational equivalence classes. The NMS model is considered, and it is only required that the set D of observations be a partial ordering and that the observation function be monotone. The result applies to both interleaving and partial ordering CCS semantics.

The paper is organized as follows. In Section 2 and 3 we present the necessary background and outline some relevant published results. In particular, in Section 2 we define NMS and our notion of NMS observational equivalence; Subsection 3.1 contains a brief survey of CCS syntax and interleaving semantics. In Subsection 3.2 we recall some simple notions about transition systems, and define a classical and a distributed transition system for CCS; then we define two observation functions on them, corresponding to interleaving and partial ordering observations. Finally, we define two NMS accordingly. NMS observational equivalence on the interleaving NMS coincides with Milner observational equivalence, while NMS observational equivalence on the partial ordering NMS defines a new equivalence. In Section 4, node observational equivalence is defined and is shown to be a convex equivalence relation. In Section 5 we define abstraction homomorphisms. An abstraction homomorphism between an *original* NMS and a *simplified* NMS is just a surjective map preserving labels and successor sets. Calling *terminal* an omomorphism such that its simplified NMS is *reduced* (i.e., only an isomorphism is possible from it), we show that, quite unexpectedly, several terminal homomorphisms are possible from the same NMS.

In Section 6 we have the main theorem of the paper. It proves that all the reduced NMS in the terminal homomorphisms are isomorphic. A simple corollary then states that two NMS are NMS observational equivalent iff their reduced NMS are isomorphic.

2. Nondeterministic Measurement Systems

In this section we introduce Nondeterministic Measurement Systems and the corresponding notion of NMS observational equivalence, as defined in Degano and Montanari (1987) and Degano et al. (1987a).

Definition 1 (NMS). Given a partial ordering D of *observations*, a *Nondeterministic Measurement System* (NMS) is a triple $t = \langle N, \leq, o\rangle$ where

- $\langle N, \leq \rangle$ is a tree-like, countable partial ordering, with all nodes at finite depth;
- $o : N \to D$ is a monotone *observation function.*

NMS are defined up to isomorphism (a isomorphism is a bijective node map preserving orderings and observations).

Given an NMS t, the set of all successors of a node n (which includes n) will be denoted by $succ(n)$. A *subtree* $t' = \langle N', \leq', o'\rangle$ of $t = \langle N, \leq, o\rangle$ is an NMS consisting of a node n of t and all its successors, namely $N' = \{m \mid n \leq m\}$ and $\leq'$ and o' are $\leq$ and o restricted to N'. Note that there is an obvious surjection σ between nodes and subtrees of t, such that $n \leq n'$ implies that $\sigma(n')$ is a subtree of $\sigma(n)$. □

Definition 2 (NMS Bisimulation). If R is a relation over NMS, then Φ, a function from relations to relations, is defined as follows.

$\langle t, u\rangle \in \Phi(R)$ iff

- the labels of the roots of t and u are the same
- for every subtree t' of t there exists a subtree u' of u such that $\langle t', u'\rangle \in R$
- for every subtree u' of u there exists a subtree t' of t such that $\langle t', u'\rangle \in R$

A relation R is an *NMS bisimulation* iff $R \subseteq \Phi(R)$.

Relation $\cong$, defined as $\cong = \bigcup\{R \mid R \subseteq \Phi(R)\}$ is called *NMS observational equivalence.* □

Proposition 1 (NMS observational equivalence is the maximal NMS bisimulation).

- *Function Φ is monotone on the lattice of relations under inclusion.*
- *Relation $\cong$ is an NMS bisimulation.*
- *Relation $\cong$ is an equivalence relation.* □

3. Observing Concurrency Models

In this section we give some background on the syntax and on the classical, interleaving semantics of CCS. Following Degano and Montanari (1987), and Degano et al. (1987a), (1987b), (1988), we then define a partial ordering semantics for CCS in terms of NMS constructed via partial ordering observations.

3.1. The Calculus for Communicating Systems

This subsection contains a brief survey of CCS syntax and interleaving semantics. First, we shall recall the main operators of the calculus, then we will present the traditional semantics and the observational equivalence of Milner (1985). The concrete syntax of pure CCS (i.e. CCS without value passing) is defined as follows.

Definition 3 (Agents). Let

- $\Delta = \{\alpha, \beta, \gamma, \ldots\}$ be a fixed set and $\bar{\Delta} = \{\bar{\alpha} \mid \alpha \in \Delta\}$;
- $\Lambda = \Delta \cup \bar{\Delta}$ (ranged over by λ) be the set of *visible actions*;
- $\tau \notin \Lambda$ be a distinguished *invisible action*, and $\Lambda \cup \{\tau\}$ be ranged over by μ.

The CCS agents are closed terms (i.e., terms without free variables) which can be generated by the following BNF-like grammar

$$E ::= x \mid NIL \mid \mu E \mid E \backslash \alpha \mid E[\phi] \mid E + E \mid E|E \mid rec\, x \,.\, E$$

where x is a variable and ϕ is a permutation of $\Lambda \cup \{\tau\}$ which preserves τ and the operation of complementation. We will let $PROC$ denote the set of all CCS agents, ranged over by E. □

CCS has a two level semantics: the first level describes the behaviour of agents through an abstract machine and the second level forgets their internal structure by identifying those machines which all exhibit the same external behaviour.

The first level, i.e., the interleaving operational semantics, is based on a labeled transition system, the transition relation of which is defined by a set of transition rules over agents. A relation $\xrightarrow{\mu}$, called derivation relation, is defined, with the intuition that agent E_1 may evolve to become agent E_2 either by reacting to a stimulus from its environment ($E_1 \xrightarrow{\lambda} E_2$) or by performing an internal action which is independent of the environment ($E_1 \xrightarrow{\tau} E_2$).

Definition 4 (Transitions). Milner's derivation relation $E_1 \xrightarrow{\mu} E_2$ is defined as the least relation satisfying the following axiom and inference rules.

Act) $\mu E \xrightarrow{\mu} E$

Res) $$\frac{E_1 \xrightarrow{\mu} E_2 \text{ and } \mu \notin \{\alpha, \bar{\alpha}\}}{E_1 \backslash \alpha \xrightarrow{\mu} E_2 \backslash \alpha}$$

Rel) $$\frac{E_1 \xrightarrow{\mu} E_2}{E_1[\phi] \xrightarrow{\phi(\mu)} E_2[\phi]}$$

Sum) $$\frac{E_1 \xrightarrow{\mu} E_2}{E_1 + E \xrightarrow{\mu} E_2 \text{ and } E + E_1 \xrightarrow{\mu} E_2}$$

Com) $$\frac{E_1 \xrightarrow{\mu} E_2}{E_1 \mid E \xrightarrow{\mu} E_2 \text{ and } E \mid E_1 \xrightarrow{\mu} E \mid E_2}$$

$$\frac{E_1 \xrightarrow{\lambda} E_2 \text{ and } E_1' \xrightarrow{\bar{\lambda}} E_2'}{E_1 \mid E_1' \xrightarrow{\tau} E_2 \mid E_2'}$$

Rec) $$\frac{E_1[\text{rec } x \,.\, E_1/x] \xrightarrow{\mu} E_2}{\text{rec } x.\, E_1 \xrightarrow{\mu} E_2}.$$

□

Hereto, we will use the following conventions to talk about sequences of actions and sequences of visible actions:

- $E \stackrel{\varepsilon}{\Longrightarrow} E'$, ε being the null string of Λ^*, stands for $E \stackrel{\tau^n}{\longrightarrow} E'$, $n \geq 0$;
- $E \stackrel{\lambda}{\Longrightarrow} E'$, stands for $\exists E_1, E_2$ such that $E \stackrel{\varepsilon}{\Longrightarrow} E_1 \stackrel{\lambda}{\longrightarrow} E_2 \stackrel{\varepsilon}{\Longrightarrow} E'$;
- $E \stackrel{s}{\Longrightarrow} E'$, $s = \lambda_1 \ldots \lambda_n \in \Lambda^+$, means $\exists E_i, 0 < i < n$, such that $E = E_0 \stackrel{\lambda_1}{\Longrightarrow} E_1 \stackrel{\lambda_2}{\Longrightarrow} \ldots \stackrel{\lambda_n}{\Longrightarrow} E_n = E'$;
- the relation $\stackrel{s}{\Longrightarrow}$, $s \in \Lambda^*$ will be referred to as many step derivation.

The derivation relation of Definition 6 completely specifies the operational semantics of CCS; the second level of CCS semantics is obtained by abstracting from unwanted details. To this purpose, a notion of observational equivalence is introduced which is based on the notion of bisimulation (Park, 1981). Below we introduce the notion of bisimulation following Milner (1985). This definition is formally different from that of NMS bisimulation given in Definition 2: the equivalence between the two will be stated later.

Definition 5 (Milner bisimulation). If R is a relation over CCS agents, then Ψ, a function from relations to relations, is defined as follows.

$\langle E_1, E_2 \rangle \in \Psi(R)$ iff $\forall s \in \Lambda^*$:

- whenever $E_1 \stackrel{s}{\Longrightarrow} E_1'$, there exists E_2' such that $E_2 \stackrel{s}{\Longrightarrow} E_2'$ and $\langle E_1', E_2' \rangle \in R$
- whenever $E_2 \stackrel{s}{\Longrightarrow} E_2'$, there exists E_1' such that $E_1 \stackrel{s}{\Longrightarrow} E_1'$ and $\langle E_1', E_2' \rangle \in R$

A relation R is a *Milner bisimulation* iff $R \subseteq \Psi(R)$.

Relation $\approx$, defined as $\approx = \bigcup\{R \mid R \subseteq \Psi(R)\}$ is called *Milner observational equivalence.* □

Proposition 2 (Milner observational equivalence is the maximal Milner bisimulation).

- *Function Ψ is monotone on the lattice of relations under inclusion.*

- *Relation $\approx$ is a Milner bisimulation.*
- *Relation $\approx$ is an equivalence relation.* □

Loosely speaking, two CCS agents E_1 and E_2 are considered as equivalent, written $E_1 \approx E_2$, iff there exists a relation R, called Milner bisimulation, which contains the pair $\langle E_1, E_2 \rangle$ and guarantees that E_1 and E_2 must be able to perform equal sequences of visible actions evolving to equal (up to R) agents. Below we propose two pairs of equivalent processes. The first shows that the bisimulation-based equivalence succeeds in ignoring the internal structure of agents; the second shows that concurrent and nondeterministic processes are sometimes identified.

Example 1 (Two pairs of observationally equivalent agents).

1. $\alpha(\beta NIL + \tau\gamma NIL) + \alpha\gamma NIL \approx \alpha(\beta NIL + \tau\gamma NIL)$;
2. $\alpha NIL \mid \beta NIL \approx \alpha\beta NIL + \beta\alpha NIL$.

Here, the relevant bisimulations are

1. $\{\langle \alpha(\beta NIL+\tau\gamma NIL)+\alpha\gamma NIL,\ \alpha(\beta NIL+\tau\gamma NIL)\rangle, \langle \beta NIL+\tau\gamma NIL,\ \beta NIL + \tau\gamma NIL\rangle, \langle \gamma NIL, \gamma NIL\rangle, \langle NIL, NIL\rangle\}$;
2. $\{\langle \alpha NIL \mid \beta NIL,\ \alpha\beta NIL + \beta\alpha NIL\rangle, \langle NIL \mid \beta NIL,\ \beta NIL\rangle, \langle \alpha NIL \mid NIL,\ \alpha NIL\rangle, \langle NIL \mid NIL,\ NIL\rangle\}$. □

3.2. Transition Systems and NMS for CCS

We first recall some simple notions about transition systems, and define a classical and a distributed transition system for CCS; then we define two observation functions on them, corresponding to interleaving and partial ordering observations. Finally, we define two NMS accordingly. NMS observational equivalence on the interleaving NMS coincides with Milner observational equivalence, while NMS observational equivalence on the partial ordering NMS defines a new *partial ordering* observational equivalence. The latter is finer than classical observational equivalence since it distinguishes the interleaving of sequential nondeterministic processes from their concurrent execution. Moreover, the two equivalences coincide when sequential processes only are considered.

Definition 6 (Transition systems). A *transition system* is a quadruple $\langle Q, T, c, q_0 \rangle$ where

- Q is a countable set of *states*;
- T is a countable set of *transitions*;
- $c : T \longrightarrow Q \times Q$ is a function giving for every transition its initial and its final state;
- $q_0 \in Q$ is the initial state.

A *computation* is a finite path (understood as a sequence of occurrences of both states and transitions) starting from q_0. □

Transition system are purely operational, intensional models. To make transition systems more extensional, we define equivalence relations on them.

Definition 7 (Observation function). Given a transition system $\mathcal{T}$, an observation function o is a monotone function from the computations of $\mathcal{T}$, ordered by prefix, to a partial ordering D of observations. □

Here, the purpose of the observation function is to extract the relevant informations from the computation.

Definition 8 (From transition systems to NMS). Given a transition system $\mathcal{T}$ and an observation function o from its computations to a partial ordering D, the *derived* NMS is the node-labeled tree consisting of the computations of $\mathcal{T}$ ordered by prefix, labeled via o. □

Having defined a notion of NMS observational equivalence in Section 2, our construction yields an equivalence relation on transition systems. But a transition system can be naturally associated to any CCS term as follows.

Definition 9 (The classical transition system for CCS). Given a CCS agent E, the transition system $\mathcal{T}_{CCS}(E)$ for CCS is $\langle Q, T, c, q_0 \rangle$, where:

- $Q = PROC$;

- $T = \{E_1 \xrightarrow{\mu} E_2\}$;
- $c(E_1 \xrightarrow{\mu} E_2) = \langle E_1, E_2 \rangle$;
- $q_0 = E$.

□

Definition 10 (Observation function for CCS). Given a computation of $\mathcal{T}_{CCS}(E_1)$ corresponding to $E_1 \overset{s}{\Longrightarrow} E_2$, its observation, through the observation function *obs*, is the string s. □

Obviously, function *obs* is monotone when strings of actions are ordered by prefix.

Definition 11 (NMS observational equivalence via obs for CCS). Given two CCS agents E_1 and E_2, E_1 is NMS observationally equivalent to E_2 via *obs*, written $E_1 \cong_{obs} E_2$, whenever the NMS derived from $\mathcal{T}_{CCS}(E_1)$ and $\mathcal{T}_{CCS}(E_2)$ via *obs* are in $\cong$. □

Theorem 1 (NMS observational equivalence via *obs* is Milner's observational equivalence). *Given two CCS agents E_1 and E_2, $E_1 \approx E_2$ if and only if $E_1 \cong_{obs} E_2$.* □

The operational semantics for CCS presented above does not consider the operator | for parallel composition of processes as primitive: given any finite agent containing |, there always exists another agent without | which exhibits the same behaviour. In the following, we outline a new operational semantics for CCS which allows us to treat the parallel operator as a first class one. The observation function is the same of Degano et al. (1987a). The transition system we consider here is fully described in Degano et al. (1987b, 1988), where it is proved consistent in a suitable sense with the classical transition system of the language and capable of expressing full parallelism. The proof has been given only for *guarded* CCS agents, i.e. terms where every variable occurrence is prefixed, in its recursive context, by at least an action. A simpler, less satisfactory, version of the transition system is described in Degano and Montanari (1987) and Degano et al. (1987a); a related development is contained in Olderog (1987).

Introducing the new partial ordering transitions requires the following steps. First, agents are decomposed into sets of sequential processes, called grapes. Then the new relation between sets of grapes is defined: it specifies both the actions an agent E may perform and which sequential processes of E actually perform them.

Intuitively speaking, a grape represents a sub-agent of a CCS agent, together with its access path. The latter is used to take into account the context in which sequential processes operate. We have an operator on grapes for each CCS operator and we keep the same name for all the operators apart from that for parallel composition. This is replaced by two unary operators, $|id$ and $id|$, which are tags recording that their arguments do not operate in isolation, but are the left and right components of a concurrent agent.

The transitions we use here have the form $H_1 \xrightarrow{\mu} H_2$, and their intuitive meaning is that set of grapes H_1 may become set H_2 by performing action μ.

Definition 12 (Grapes). A *grape* is a term defined by the following BNF-like grammar:

$$G ::= NIL \mid \mu E \mid G\backslash\alpha \mid G[\phi] \mid G + G \mid id|G \mid G|id$$

□

We will use I, J, H, K (possibly indexed) to denote sets of grapes, and g to denote single grapes.

Grapes are obtained from CCS agents through the decomposition function *Dec* defined below. Function *Dec* goes inside the structure of agents and only stops when a process prefixed by an action or the NIL process are encountered. Hereto, we understand constructors as extended to operate elementwise on sets, e.g., $J\backslash\alpha = \{g\backslash\alpha \mid g \in J\}$.

Definition 13 (Decomposition function). The *decomposition function Dec* is defined by structural induction as follows:

$$\begin{aligned}
\mathrm{Dec}(NIL) &= \{\mathrm{NIL}\}\\
\mathrm{Dec}(\mu E) &= \{\mu E\}\\
\mathrm{Dec}(E\backslash\alpha) &= \mathrm{Dec}(E)\backslash\alpha\\
\mathrm{Dec}(E[\phi]) &= \mathrm{Dec}(E)[\phi]\\
\mathrm{Dec}(E' + E'') &= \mathrm{Dec}(E') + \mathrm{Dec}(E'')\\
\mathrm{Dec}(E' \mid E'') &= \mathrm{Dec}(E')|\,\mathrm{id} \cup \mathrm{id}\,|\,\mathrm{Dec}(E'')\\
\mathrm{Dec}(\mathrm{rec}\ x.\,E) &= \mathrm{Dec}(E[(\mathrm{rec}\ x.\,E)/x]).
\end{aligned}$$

A set I such that $I = Dec(E)$ for some E is called *complete.* □

Notice that if E is guarded, the recursive computation of $dec(E)$ always terminates, since only recursive subterms not prefixed by actions are unfolded, and only once.

Example 2 (Decomposing an agent into grapes). Let $E_1 = rec\ x.\alpha x + \beta x$, $E_2 = rec\ x.\alpha x + \gamma x$ and $E_3 = rec\ x.\bar{\alpha} x$. Then $Dec(((E_1 \mid E_2) \mid E_3)\backslash\alpha) = \{g_0, g_1, g_2\}$, where

$$\begin{aligned} g_0 &= (((\alpha E_1 + \beta E_1)|\,\mathrm{id})|\,\mathrm{id})\backslash\alpha \\ g_1 &= ((\mathrm{id}\,|(\alpha E_2 + \gamma E_2))|\,\mathrm{id})\backslash\alpha; \\ g_2 &= (\mathrm{id}\,|\bar{\alpha} E_3)\backslash\alpha. \end{aligned}$$

□

The following definition introduces transitions $H_1 \xrightarrow{\mu} H_2$. Variables H_1 and H_2 denote sets of grapes; μ is an action.

Definition 14 (Partial ordering derivation relation). The partial ordering derivation relation is defined as the least relation satisfying the following inference rules:

act) $\{\mu E\} \xrightarrow{\mu} \mathrm{Dec}(E)$

res) $$\frac{H_1 \xrightarrow{\mu} H_2 \text{ and } \mu \notin \{\alpha, \bar{\alpha}\}}{H_1\backslash\alpha \xrightarrow{\mu} H_2\backslash\alpha}$$

rel) $$\frac{H_1 \xrightarrow{\mu} H_2}{H_1[\phi] \xrightarrow{\phi(\mu)} H_2[\phi]}$$

sum) $$\frac{H_1 \cup K \xrightarrow{\mu} H_2 \text{ and } \mathrm{Dec}(E) = I}{(H_1 + I) \cup K \xrightarrow{\mu} H_2 \text{ and } (I + H_1) \cup K \xrightarrow{\mu} H_2}$$

com) $$\frac{H_1 \xrightarrow{\mu} H_2}{H_1|\,\mathrm{id} \xrightarrow{\mu} H_2|\,\mathrm{id} \text{ and } \mathrm{id}\,|H_1 \xrightarrow{\mu} id|H_2}$$

$$\frac{H_1 \xrightarrow{\lambda} H_2 \text{ and } H_1' \xrightarrow{\bar{\lambda}} H_2'}{H_1|\,\mathrm{id} \cup \mathrm{id}\,|H_1' \xrightarrow{\tau} H_2|\,\mathrm{id} \cup \mathrm{id}\,|H_2'}$$

□

The basic meaning of $H_1 \xrightarrow{\mu} H_2$ is that set of grapes H_1 evolves to set H_2 by performing action μ. In rule *act*), a single grape is rewritten as a set of grapes, since the occurrence of the action makes the (possible) parallelism of E explicit. The rules *res*) and *rel*) and the first rule for *com*) simply state that if a set of grapes H_1 can be rewritten as H_2 via μ, then we can combine the access paths of the grapes in both sets with path constructors $.\backslash\alpha$, $.[\phi]$, $.|id$ or $id|.$, and still obtain a derivation, labeled, say, by μ'. Clearly, when dealing with restriction we have that μ' is μ, but the inference is possible only when $\mu \notin \{\alpha, \bar{\alpha}\}$; in *rel*) μ' is $\phi(\mu)$ and in the first rule of *com*) μ' is again μ. The second rule for *com*) is the synchronization rule. To comment *sum*), let us first consider the case $K = \emptyset$. If so, before enabling H_1 to perform its action, we must make sure that alternative H_1 has not been discarded by the occurrence of actions of I. Certainly, we must avoid situations where, given two pairs of non intersecting sets of grapes $J_1', J_2' \subseteq J'$ and $J_1'', J_2'' \subseteq J''$, the two sets $J_1' + J_1''$ and $J_2' + J_2''$ (both contained in the sum $J' + J''$) perform successive actions via both the right and the left operands of $+$. For this reason, a set of grapes is enabled to perform an action only if it is alternative to a complete set of grapes I obtained by decomposing an agent E. When $K \neq \emptyset$, a similar reasoning applies; we only have to consider that the transition is the result of a communication between H_1 and K and that the presence of alternative I does not influence the overall outcome.

Indeed, we will never have to define the tra.sitions of a process whose top-level operator is *rec*, since we consider only guarded terms and we perform unwinding of recursive processes whenever decomposing them into grapes, after an action.

Definition 15 (From CCS agents to distributed transition systems). Given a CCS agent E, let $\mathcal{D}_{CCS}(E) = \langle Q, T, c, q_0 \rangle$ be its associated *distributed* transition system, where

- the states in Q are the finite sets of grapes;
- $T = \{H_1 \cup K \xrightarrow{\mu,K} H_2 \cup K \mid H_1 \xrightarrow{\mu} H_2,\ H_1 \cap K = \emptyset\}$;
- $c(I_1 \xrightarrow{\mu,K} I_2) = \langle I_1, I_2 \rangle$;
- $q_0 = Dec(E)$. □

In $I_1 \xrightarrow{\mu,K} I_2$, the grapes in K represent the sequential processes which stay idle in the move. The definition of T above expresses the asynchrony of CCS.

Example 3 (A distributed transition system for Example 2). Let us consider the CCS agent of Example 2: $E = (((E_1 \mid E_2) \mid E_3)\backslash\alpha)$ where $E_1 = rec\ x.\, \alpha x + \beta x$, $E_2 = rec\ x.\, \alpha x + \gamma x$ and $E_3 = rec\ x.\, \bar{\alpha} x$. The transition system $\mathcal{D}_{CCS}(E) = \langle Q, T, c, Dec(E)\rangle$ has as reachable state only the initial state $q_0 = \{g_0, g_1, g_2\}$, where

$$g_0 = (((\alpha E_1 + \beta E_1)|id)|id)\backslash\alpha;$$
$$g_1 = ((id|(\alpha E_2 + \gamma E_2))|id)\backslash\alpha;$$
$$g_2 = (id|\bar{\alpha} E_3)\backslash\alpha,$$

and the following four transitions:

$$t_0 : q_0 \stackrel{\tau,\{g_1\}}{\longrightarrow} q_0;$$
$$t_1 : q_0 \stackrel{\beta,\{g_1,g_2\}}{\longrightarrow} q_0;$$
$$t_2 : q_0 \stackrel{\gamma,\{g_0,g_2\}}{\longrightarrow} q_0;$$
$$t_3 : q_0 \stackrel{\tau,\{g_0\}}{\longrightarrow} q_0.$$

□

We now define our partial ordering observations on the computations of $\mathcal{D}_{CCS}$. We generate the events together with their causal dependencies. An observation is obtained in three steps: first an event is associated to every transition in the computation; then the causal dependencies among the events are determined; and finally all unobservable events (i.e. the events labeled by τ) are removed.

Definition 16 (Partial orderings of events). A partial ordering of events h is a triple $\langle S, l, \leq\rangle$, where

- S is a finite set of *events*;
- $l : S \longrightarrow \Lambda \cup \{\tau\}$ is a labeling function;
- $\leq$ is a partial ordering relation on S, called *causal relation.*

Two events e_1 and e_2 are *concurrent* if neither $e_1 \leq e_2$ nor $e_2 \leq e_1$. Two partial orderings of events will be identified if isomorphic, i.e. if there is a label- and order-preserving bijection between their events.

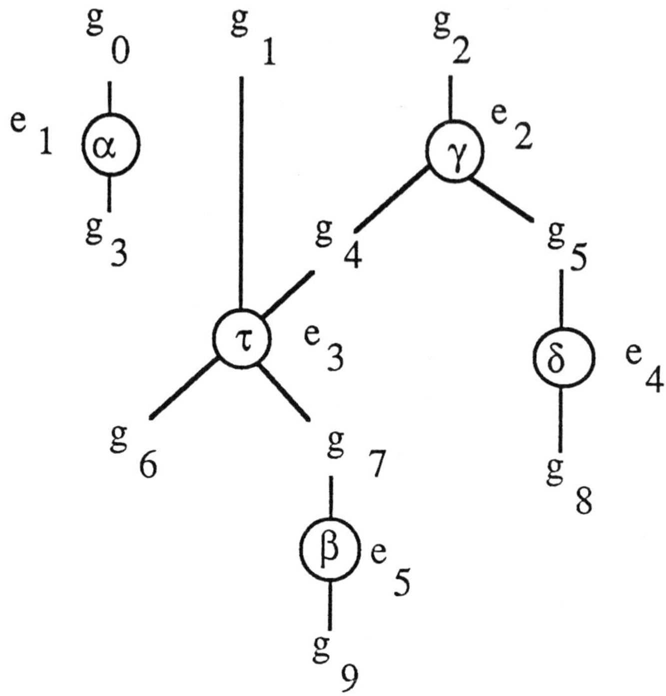

Fig. 1. An intermediate step in getting the partial ordering observation of a computation.

Partial orderings of events form in turn a partial ordering under the following relation $\sqsubseteq$:
$\langle S', l', \leq' \rangle \sqsubseteq \langle S, l, \leq \rangle$ iff: i) $S' \subseteq S$; ii) $e' \in S'$ and $e \leq e'$ implies $e \in S'$; and iii) l' and $\leq'$ are the restrictions to S' of l and $\leq$. □

Fig. 2 shows a partial ordering of events, with the conventions that events are represented by circles with their labels inside, and that the partial ordering $\leq$ is represented by its Hasse diagram growing downwards. For example, we have that the event labeled by α has no relation with the other three events, thus is concurrent with them all. Event labeled by γ dominates, i.e. causes, both the remaining events, which in turn are concurrent.

Definition 17 (Partial ordering observation function). Given a finite computation ξ of the transition system $\mathcal{D}_{CCS}(E)$

associated to CCS agent E,

$$\xi = \{I_0 \, (I_0 \overset{\mu_1, K_1}{\longrightarrow} I_1) \, I_1 \, (I_1 \overset{\mu_2, K_2}{\longrightarrow} I_2) I_2 \ldots$$
$$\ldots \, I_{n-1} \, (I_{n-1} \overset{\mu_n, K_n}{\longrightarrow} I_n) \, I_n\},$$

where $Dec(E) = I_0$, its partial ordering observation is the partial ordering of events $h = \langle S, l, \leq \rangle$ defined by the following procedure, which implements the observation function po.

1. Let $S' = \{e_1, \ldots, e_n\}$, where $e_1, \ldots, e_n$ are new elements, and let $B = \{\langle g, i \rangle \mid g \in I_i\}$;
2. let us define a relation F on $S' \cup B$. We have the following rules:
   - $\dfrac{g \in K_i}{\langle g, i-1 \rangle F \langle g, i \rangle}$
   - $\dfrac{g \in I_{i-1} - K_i}{\langle g, i-1 \rangle F e_i}$
   - $\dfrac{g \in I_i - K_i}{e_i F \langle g, i \rangle}$;
3. let $S = \{e_i \mid \mu_i \neq \tau\}$, let $l(e_i) = \mu_i$, and let $\leq$ be the transitive and reflexive closure of F restricted to S. □

Notice that function po defined above is obviously monotone with respect to the ordering $\sqsubseteq$ specified in Definition 16.

Example 4 (An example of partial ordering observation). Let $E = \alpha NIL \mid (\bar{\beta} NIL \mid \gamma(\beta\beta NIL \mid \delta NIL))$, and let

$g_0 = \alpha NIL|id$; $\quad g_1 = id|(\bar{\beta} NIL|id)$;
$g_2 = id|(id|\gamma(\beta\beta NIL|\delta NIL))$; $\quad g_3 = NIL|id$;
$g_4 = id|(id|(\beta\beta NIL|id))$; $\quad g_5 = id|(id|(id|\delta NIL))$;
$g_6 = id|(NIL|id)$; $\quad g_7 = id|(id|(\beta NIL|id))$;
$g_8 = id|(id|(id|NIL))$; $\quad g_9 = id|(id|(NIL|id))$.

Finally, let

$I_0 = \{g_0, g_1, g_2\}$; $\quad I_1 = \{g_3, g_1, g_2\}$
$I_2 = \{g_3, g_1, g_4, g_5\}$; $\quad I_3 = \{g_3, g_6, g_7, g_5\}$
$I_4 = \{g_3, g_6, g_7, g_8\}$; $\quad I_5 = \{g_3, g_6, g_9, g_8\}$

The following is a computation of $\mathcal{D}_{CCS}(E)$:

$$\{I_0(I_0 \stackrel{\alpha,\{g_1,g_2\}}{\longrightarrow} I_1)I_1(I_1 \stackrel{\gamma,\{g_3,g_1\}}{\longrightarrow} I_2)I_2(I_2 \stackrel{\tau,\{g_3,g_5\}}{\longrightarrow} I_3)$$
$$I_3\ (I_3 \stackrel{\delta,\{g_3,g_6,g_7\}}{\longrightarrow} I_4)I_4(I_4 \stackrel{\beta,\{g_3,g_6,g_8\}}{\longrightarrow} I_5)I_5\}.$$

Its partial ordering observation is $h = (S, l, \leq)$, where

- $S = \{e_1, e_2, e_4, e_5\}$;
- $l(e_1) = \alpha$, $l(e_2) = \gamma$, $l(e_4) = \delta$, $l(e_5) = \beta$;
- $e_2 \leq e_4$, $e_2 \leq e_5$.

Fig. 2 shows h and Fig. 1 shows an intermediate snapshot in getting the observation, in which also grape instances contained in B are depicted. Notice that the two instances of a grape which stays idle in a step are identified. Notice also that only the Hasse diagram is shown, instead of the full partial ordering $\leq$, thus e.g., the arc between events e_2 and e_5 is not drawn in Fig. 1. □

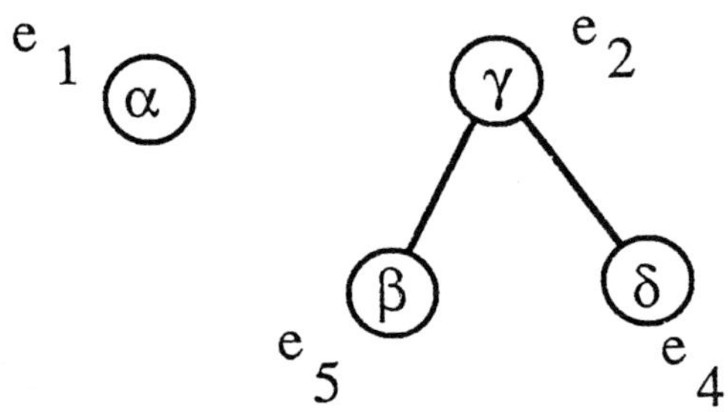

Fig. 2. A partial ordering observation.

Definition 18 (NMS partial ordering equivalence for CCS). Given two CCS agents E_1 and E_2, E_1 is NMS observationally equivalent to E_2 via *po*, written $E_1 \cong_{po} E_2$, whenever the NMS derived from $\mathcal{D}_{CCS}(E_1)$ and $\mathcal{D}_{CCS}(E_2)$ via *po* are in $\cong$. □

Theorem 2 (NMS *po* equivalence is finer than Milner observational equivalence). *Given two CCS agents E_1 and E_2, $E_1 \cong_{po} E_2$ implies $E_1 \approx E_2$. The inclusion is strict, as shown by Example 5. Finally, $\cong_{po}$ coincides with $\approx$ on terms constructed without the $|$ operator.* □

Example 5 (Two Milner equivalent agents which are not po equivalent). We have

$\alpha NIL \mid \beta NIL \approx \alpha\beta NIL + \beta\alpha NIL$
but
$\alpha NIL \mid \beta NIL \not\cong_{po} \alpha\beta NIL + \beta\alpha NIL.$ □

4. Some Properties of Observational Equivalence

In this section, we apply the notion of bisimulation to the nodes of an NMS, and prove that node observational equivalence, the maximal node bisimulation, is convex, namely that its equivalence classes are partially ordered.

We need the following proposition and definition.

Definition 19 (Bisimulation on nodes). Given two nodes n and n' of an NMS t, a relation nRn' is a node bisimulation iff there exists an NMS bisimulation P with nRn' iff $\sigma(n)P\sigma(n')$. We define $n \simeq n'$ iff $\sigma(n) \cong \sigma(n')$, where $\sigma(n)$ is the subtree rooted in n. □

Definition 20 (Convex equivalence relation). Given a partial ordering $\langle N, \leq \rangle$, an equivalence relation E on N is *convex* iff

$$\equiv_{(E \cup \leq)^*} = E$$

namely iff the equivalence produced on N by the preorder obtained by taking the transitive closure of the union of E and of $\leq$ is still E.

Notice that $\sqsubseteq_{\leq,E} = (E\cup \leq)^*/E$ is a partial ordering on the equivalence classes of E. □

We can prove the following theorem.

Theorem 3 (Node observational equivalence is convex). *Given an NMS $t = \langle N, \leq, o \rangle$, the equivalence relation $\simeq$ defined on its nodes is convex.* □

Proof. If $n \equiv_{(\simeq \cup \leq)^*} m$, then there are nodes n_i, $1 \leq i \leq r$, and m_j, $1 \leq j \leq s$, such that $n = n_1 \simeq n_2 \leq n_3 \simeq n_4 \leq \ldots \leq n_{r-1} \simeq n_r = m = m_1 \simeq m_2 \leq m_3 \simeq m_4 \leq \ldots \leq m_{s-1} \simeq m_s = n$. We show that $\sigma(n) \cong \sigma(m)$ and thus $n \simeq m$. In fact, $o(n) = o(n_1) = o(n_2) \sqsubseteq o(n_3) = o(n_4) \sqsubseteq \ldots \sqsubseteq o(n_{r-1}) = o(n_r) = o(m) = o(m_1) = o(m_2) \sqsubseteq o(m_3) = o(m_4) \sqsubseteq \ldots \sqsubseteq o(m_{s-1}) = o(m_s) = o(n)$. Since D

is a partial ordering, we have $o(n) = o(m)$. Furthermore, given any subtree t' of $\sigma(n)$ it is possible to find subtrees $u'(j)$ of $\sigma(m(j))$ such that $t' = u'(s) \cong u'(s-1) = u'(s-2) \cong u'(s-3) = \ldots = u'(1) = u'$, i.e. u' is a subtree of $\sigma(m)$ and $t' \cong u'$. Indeed, given any $u'(s)$, by definition we can find $u'(s-1)$, since $\sigma(m(s)) \cong \sigma(m(s-1))$; moreover, we can choose $u'(s-1) = u'(s-2)$ since $m(s-1) \geq m(s-2)$ and thus any subtree of $\sigma(m(s-1))$ is also a subtree of $\sigma(m(s-2))$; and so on. Similarly, given any subtree u' of $\sigma(m)$ it is possible to find a subtree t' of $\sigma(n)$ such that $t' \cong u'$. □

Definition 21 (Regular equivalence relation). Given an NMS $t = \langle N, \leq, o \rangle$, a convex relation E on it is *regular* iff nEn' implies $o(n) = o(n')$ and partial ordering $\sqsubseteq_{\leq,E}$ is a tree. We will define the quotient NMS t/E as $t/E = \langle N/E, \sqsubseteq_{\leq,E}, o/E \rangle$. □

Example 6 (Node observational equivalence $\simeq$ may be non-regular). Let the observation partial ordering D be the set of finite strings on an alphabet, ordered by prefix. In Fig. 3 we see a finite NMS t (in a) and the partial ordering on the bisimulation equivalence classes of its nodes (in d), which is not a tree. In Fig. 3 b and c we see the quotient NMS t/E' and t/E'' for two regular equivalences E' and E''. Equivalences E' and E'' are not node bisimulations. In fact we have $n_2 E' n_5$, but not $n_2 E' n_7$, even if $\sigma(n_5) = \sigma(n_7)$. Similarly for E''. □

5. Abstraction Homomorphisms

In this section we introduce a notion of NMS simplification called abstraction homomorphism. We also show the relation between abstraction homomorphisms and node bisimulation equivalences. Our notion of abstraction homomorphism is similar to the notion described in Castellani (1987). However, the labels of our NMS can be taken from any partial ordering of observations, while the labels in the trees described in Castellani (1987) are strings of actions. Furthermore, differently from Castellani (1987), our results apply also to the infinite branching case.

Definition 22 (Abstraction homomorphism). Given two NMS $t_1 = \langle N_1, \leq_1, o_1 \rangle$ and $t_2 = \langle N_2, \leq_2, o_2 \rangle$, a surjective function

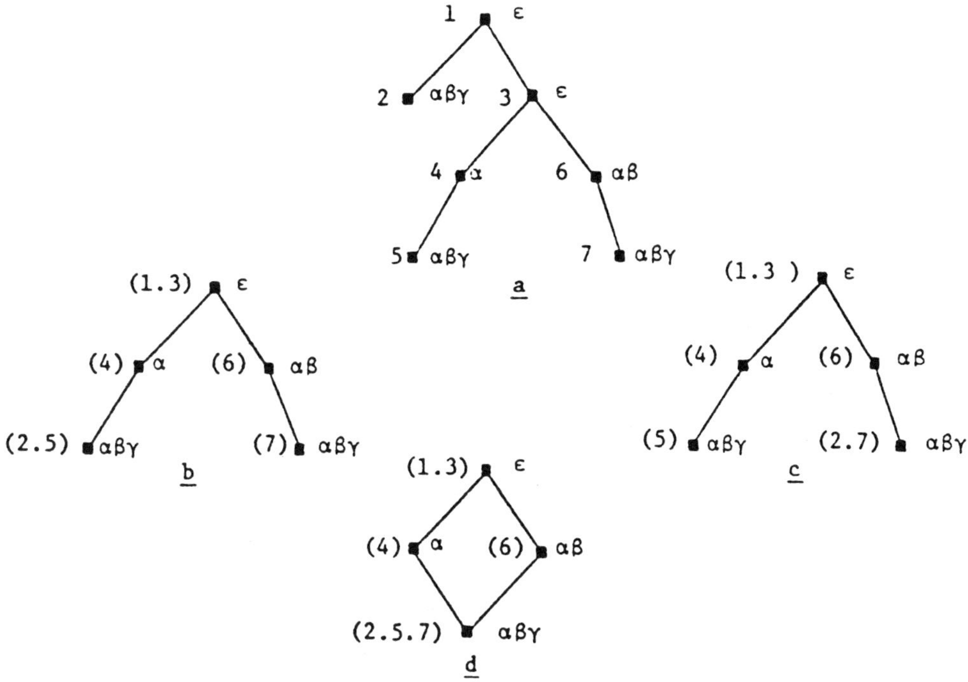

Fig. 3. An NMS (a), two regular equivalences (b and c) and the node bisimulation equivalence (d).

$h : N_1 \longrightarrow N_2$ is an *abstraction homomorphism* iff, for every node n_1 of t_1, we have

- $o_2(h(n_1)) = o_1(n_1)$
- $succ_2(h(n_1)) = h(succ_1(n_1))$

NMS t_1 and t_2 are called the *original* and the *simplified* NMS respectively.

An NMS from which only injective homomorphisms exist, is called *reduced.* A homomorphism, whose simplified NMS is reduced, is called *terminal* □

Proposition 3 (Injective abstraction homomorphism). *An injective abstraction homomorphism is an isomorphism.* □

Definition 23 (From regular node bisimulations to homomorphisms). Given a regular bisimulation E on the nodes of an NMS t, the corresponding mapping is $h : t \longrightarrow t/E$, $h(n) = [n]/E$. □

Theorem 4 (Regular node bisimulations → homomorphisms). *The above defined mapping is an abstraction homomorphism.* □

Proof. We have $o([n]/E) = o(n)$ by Definition 21. Furthermore, for every n' such that $n \leq n'$, we have that $[n]/E \sqsubseteq_{\leq,E} [n']/E$ by Definition 20. Finally, for every m such that $[n] \leq m$ there exists an n' such that $[n'] = m$ and $n \leq n'$. In fact, by Definition 20 we have that $[n] \leq m$ implies the existence of n'' and n''' with $n''En$, $n'' \leq n'''$ and $[n'''] = m$. But since E is a node bisimulation, there exists n' such that $n \leq n'$ and $n'En'''$, i.e. $[n'] = m$. Thus $succ(h(n)) = h(succ(n))$. □

Note that, given an original NMS, there may be more than one terminal homomorphism from it. In Fig. 3 *b* and *c* we see two terminal homomorphisms from the same NMS, shown in *a*. However, the two reduced trees corresponding to these homomorphisms are isomorphic. This is not a special case, as shown in the next section. Note also that the nonregular node observational equivalence in Fig. 3 *d* has no corresponding abstraction homomorphism (since $\sqsubseteq_{\leq,\simeq}$ is not a tree ordering).

6. Canonical Representatives

In this section we show that it is possible to define a canonical representative for each NMS observational equivalence class.

Definition 24 (Canonical representatives). Given a NMS $t = \langle N, \leq, o\rangle$, its *canonical representative* $\hat{t}$ is an NMS defined as follows. Let us consider the triple $p = \langle N/\simeq, \sqsubseteq_{\leq,\simeq}, o/\simeq\rangle = \langle N_p, \leq_p, o_p\rangle$. Notice that $\leq_p$ is a partial ordering on N_p by Theorem 3.

Let $\hat{t} = \langle N_{\hat{t}}, \leq_{\hat{t}}, o_{\hat{t}}\rangle$ be the unfolding of p, i.e. an NMS where every element $n_{\hat{t}} \in N_{\hat{t}}$ is a path in the Hasse diagram of p from the root to a node n_p, then $\leq_{\hat{t}}$ is path ordering by prefix and $o_{\hat{t}}(n_{\hat{t}}) = o_p(n_p)$. □

Lemma 1 (Mapping an NMS to its canonical representative). *Given a NMS* $t = \langle N, \leq, o \rangle$*:*

1. *it is possible to construct an abstraction homomorphism from* t *to its canonical representative* $\hat{t}$*;*
2. *there is no noninjective abstraction homomorphism from* $\hat{t}$ *to* $\hat{t}$*.*

□

Proof. To prove (1), let h_{tp} from t to p be the homomorphism corresponding to node observational equivalence. We construct an abstraction homomorphism $h_{t\hat{t}}$ from t to $\hat{t}$ inductively as follows. We let $h_{t\hat{t}}(r_t) = r_{\hat{t}}$, namely the root of t is mapped in the root of $\hat{t}$. Furthermore, if $h_{tp}(n_t) = n_p$ and $h_{t\hat{t}}(n_t) = \pi = n_{\hat{t}}$, where π is some path in p from the root to n_p, then, given an immediate descendent m_t of n_t with $h_{tp}(m_t) = m_p$, we will have $h_{t\hat{t}}(m_t) = \pi\pi' = m_{\hat{t}}$, where π' is *any* path in p from n_p to m_p. The possibility of choosing different values for π' accounts for the possible existence of many abstraction homomorphisms from an NMS to its canonical representative.

To prove (2), we remark that if a noninjective abstraction homomorphism $h_{\hat{t}\hat{t}}$ from $\hat{t}$ to $\hat{t}$ existed, it would be possible to derive from it an abstraction homomorphism $h'_{t\hat{t}} = h_{t\hat{t}} \circ h_{\hat{t}\hat{t}}$ from t to $\hat{t}$ and a node bisimulation R defined as nRn' iff $h'_{t\hat{t}}(n) = h'_{t\hat{t}}(n')$ or $h_{tp}(n) = h_{tp}(n')$ or $\sigma(n) = \sigma(n')$, that would be larger than the node observational equivalence corresponding to $h_{t\hat{t}}$, which is absurd. □

Theorem 5 (The canonical representative is the unique reduced NMS). *Given an NMS* t *and its canonical representative* $\hat{t}$*, we have*

1. $\hat{t}$ *is reduced;*
2. *if* h_{tu} *is a terminal abstraction homomorphism from* t *to a reduced NMS* u*, then* u *is isomorphic to* $\hat{t}$*.*

□

Proof. We prove (1) by *reductio ad absurdum*: let $h_{\hat{t}u}$ be a noninjective abstraction homomorphism from $\hat{t}$ to some NMS u. Since there would be an abstraction homomorphism $h_{u\hat{t}}$ from u to $\hat{t}$ for Lemma 1 (1), then the existence of the noninjective abstraction homomorphism obtained by composing $h_{u\hat{t}}$ and $h_{\hat{t}u}$ would contradict Lemma 1 (2).

Part (2) immediately follows from Lemma 1 (1), since the abstraction homomorphism $h_{u\hat{t}}$ must be an isomorphism. □

We now prove that canonical representatives characterize NMS observational equivalence classes.

Corollary 1 (Same representative iff $\cong$). *Given two NMS t and u, we have $t \cong u$ iff $\hat{t} = \hat{u}$* □

Proof. Let us construct an NMS v as the disjoint union of t and u plus a new root r. It is easy to see that $t \cong u$ iff the root of t and the root of v are mapped together in $\hat{v}$. In turn, the root of t and the root of v are mapped together in $\hat{v}$ iff $\hat{t} = \hat{u}$. □

We give some examples below.

Example 7 (Abstraction homomorphisms may create infinite branching). In Fig. 4 a we see a finite branching NMS whose canonical representative (in b) is infinite branching. Here the observation domain is the flat domain of natural numbers, and $\{n_i\}$ is any enumeration of the set of natural numbers where every number appears infinitely many times.

In Fig. 4 c we see an NMS (on strings ordered by prefix) having only infinite node bisimulations. In fact it has only one node bisimulation, corresponding to the canonical representative shown in d. □

Example 8 (Observ. equiv. distinguishes infinite NMS with the same finite approximations). In Fig. 5 a and b we see two different, reduced NMS. They are the canonical representatives of the observational equivalence classes of the following two CCS agents p and q respectively.

$$\begin{aligned} &p = (\mathrm{rec}\, x.\ C(x))\backslash\beta\backslash\gamma \\ &q = (\mathrm{rec}\, x.\ C(x) + \tau s)\backslash\beta\backslash\gamma \\ &\text{where } C(x) = \tau((\beta\alpha + \gamma) \mid x) + \bar{\gamma}x + \tau\, \mathrm{rec}\, y.\ \bar{\beta}y \\ &\ \text{and } s = \mathrm{rec}\, z.\ \alpha z \end{aligned}$$

To show this, let

$$q_i = (\underbrace{(\beta\alpha + \gamma) \mid \ldots \mid (\beta\alpha + \gamma)}_{i} \mid \mathrm{rec}\, x.\ C(x) + \tau s)\backslash\beta\backslash\gamma$$

$$r_i = (\underbrace{(\beta\alpha + \gamma) \mid \ldots \mid (\beta\alpha + \gamma)}_{i} \mid \mathrm{rec}\, y.\ \bar{\beta}y)\backslash\beta\backslash\gamma$$

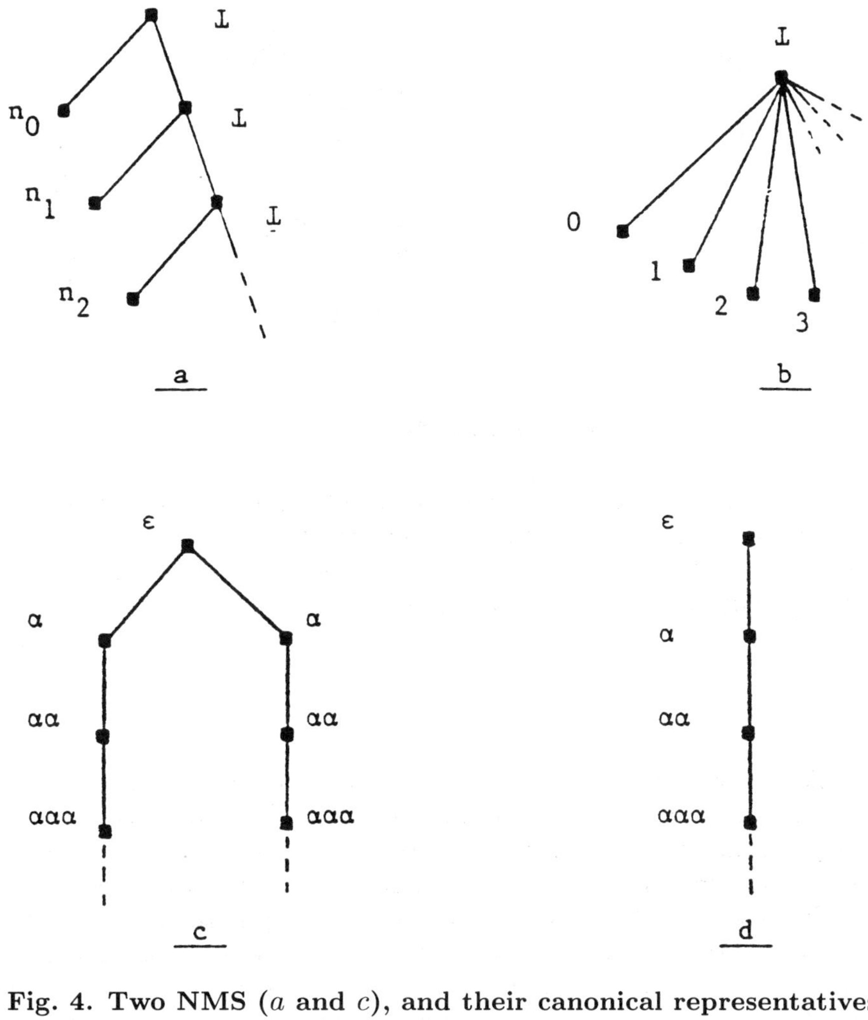

Fig. 4. Two NMS (a and c), and their canonical representatives (b and d, respectively).

It is easy to see that $q = q_0$; $[q_i] = [q]$; and $[r_0] = [Nil]$, where $[x]$ denotes the NMS observational equivalence class of x. Furthermore, the possible moves are

$$[q] \stackrel{\varepsilon}{\Longrightarrow} [r_i];\ [r_{i+1}] \stackrel{\alpha}{\Longrightarrow} [r_i]$$
$$[q] \stackrel{\varepsilon}{\Longrightarrow} [s]; [s] \stackrel{\alpha}{\Longrightarrow} [s].$$

Similarly for p.

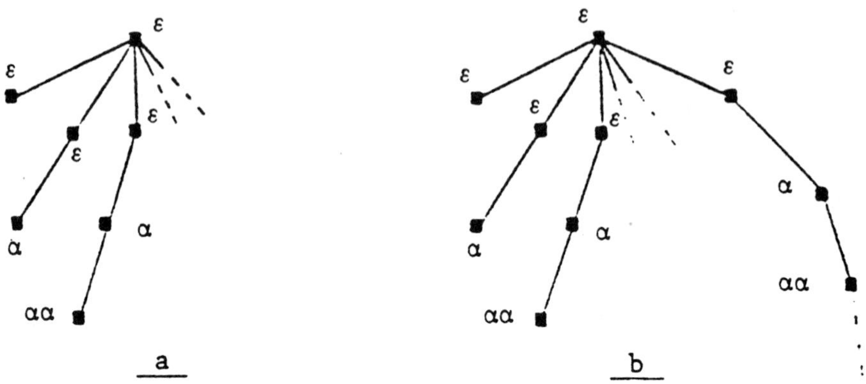

Fig. 5. Two reduced NMS.

The infinite NMS in Fig. 5 *a* and *b* are identified in most common completion constructions. The fact that they are distinguished here shows the potential of NMS observational equivalence for representing noncontinuous properties like fairness. □

7. Conclusion

In this paper we have shown that the notion of observational equivalence, originally defined using bisimulation when observations are strings of actions, can be generalized to observations which form a partial ordering. Furthermore, an independent, algebraic notion of simplification has been given, such that a unique, reduced canonical representative exists. Finally, it has been proved that the two notions coincide, i.e., that two systems are observationally equivalent iff they have the same canonical representative. No finiteness requirement on the set of successors with the same label is necessary. The results apply to both standard and partial ordering semantics of CCS.

Acknowledgements

The authors are glad to acknowledge interesting discussions on the subject of this paper with Ilaria Castellani, Pierpaolo Degano and

Rocco De Nicola. Sections 2 and 3 are survey results taken from Degano et al. (1987a) and Degano et al. (1988). Common work by the first author with Maria Victoria Cengarle led to a better formulation of Definition 24.

References

Aczel, Peter, (1987). *Lecture Notes on Non-well-founded Sets*, Stanford University Press, Stanford.

Benson, David B., and Ben-Shachar, Ofer, (1986). "Strong Bisimulation of State Automata," *Proc. Symp. on Logic in Computer Science*, IEEE, 77–81.

Benson, David B., and Ben-Shachar, Ofer, (1987). *Bisimulation of Automata*, Technical Report CS-87-167, Computer Science Department, Washington State University, Pullman, Washington.

Castellani, Ilaria, (1987). "Bisimulation and Abstraction Homomorphisms," *Journal of Computer and System Sciences*, **34**.

Castellani, Ilaria, Franceschi, Paola, and Montanari, Ugo, (1983). "Labelled Event Structures: A Model for Observable Concurrency," In *IFIP TC2 Workshop on Formal Description of Programming Concepts II* (D. Bjørner, ed.), North-Holland, New York, 383–400.

Degano, Pierpaolo, and Montanari, Ugo, (1987). "Concurrent Histories: A Basis for Observing Distributed Systems," *Journal of Computer and System Sciences*, **34**, 442–461.

Degano, Pierpaolo, De Nicola, Rocco, and Montanari, Ugo, (1987a). "Observational Equivalences for Concurrency Models," In *Formal Description of Programming Concepts III* (M. Wirsing, ed.), North-Holland, New York, 105–132.

Degano, Pierpaolo, De Nicola, Rocco, and Montanari, Ugo, (1987b). *A Distributed Operational Semantics for CCS Based On Condition/Event Systems*, Nota Interna B4-21, IEI-CNR, Pisa. To appear in *Acta Informatica.*

Degano, Pierpaolo, De Nicola, Rocco, and Montanari, Ugo, (1988). "On the Consistency of 'Truly Concurrent' Operational and Denotational Semantics," *Proc. Symp. on Logics in Computer Science 1988*, Edinburgh, IEEE.

Milner, Robin, (1985). "Lectures on a Calculus for Communicating Systems," In *Control Flow and Data Flow: Concepts of Distributed Programming* (M. Broy, ed.), Springer NATO ASI Series F, Springer-Verlag, New York, **14**, 205–228.

DeNicola, Rocco, (1987). "Extensional Equivalences for Transition Systems," *Acta Informatica*, **24**, 211–237.

Olderog, Ernst-Rüdiger, (1987). "Operational Petri Net Semantics for CCSP," In *Advances in Petri Nets 1987* (G. Rozenberg, ed.), Lecture Notes in Computer Science 266, Springer-Verlag, New York, 196–223.

Park, David, (1981). "Concurrency and Automata on Infinite Sequences," In *Proc. Theoretical Computer Science* (P. Deussen, ed.), Lecture Notes in Computer Science 104, Springer-Verlag, New York, 167–183.

Sanderson, M. T., (1982). *Proof Techniques for CCS*, PhD Thesis, Computer Science Department, University of Edinburgh, Technical Report CST-19-82.

9 Minimizing Expansions of Recursions

JEFFREY F. NAUGHTON[1]

Department of Computer Science
Princeton University
Princeton, New Jersey

YEHOSHUA SAGIV[2]

Department of Computer Science
Hebrew University of Jerusalem
Jerusalem, Israel

1. Introduction

In recent years there has been a growing acknowledgement among database researchers that relational query languages are not sufficiently powerful. One proposed alternative to relational query languages is query languages built upon horn-clause logic. While such logic-based query languages are more expressive than relational languages, this extra expressive power does not come for free, as logic-based languages are more difficult to implement efficiently. Particularly problematic for efficient implementation are recursive definitions.

There is a large body of literature proposing efficient evaluation algorithms for recursive definitions (for a survey, see Bancilhon and

[1] This work was done at Stanford University, and was supported by NSF grant IST-84-12791 and a grant from the IBM Corporation.

[2]Work supported in part by grant 85-00082/1 from the United States-Israel Binational Science Foundation (BSF), Jerusalem, Israel.

ISBN 0-12-046370-9

Ramakrishnan (1986).) There has been less work on rewriting recursive definitions to remove redundancy. This paper summarizes and unifies some earlier work in that direction, and proves some new results.

In this paper we will consider the language of function-free horn clauses, which is popularly known as "Datalog" because it is the subset of prolog that obeys the first-normal form assumptions of relational databases. Datalog has none of the extra-logical features present in Prolog, such as the "cut" operator or a built-in procedural interpretation. However, we will use Prolog notation and write the horn clauses as rules, with the positive literal in the head or consequent, and the conjunction of the negative literals as the body or antecedent.

We will use the terms "predicate" and "relation" interchangeably, and assume that the relations in the system can be separated into two types. The EDB or extensional database predicates appear in the head of no rule and are defined completely by their extent, that is, by the tuples stored in the relations corresponding to the predicate. The IDB or intensional database predicates are defined in terms of other predicates by the rules.

If an IDB predicate is defined either directly or transitively in terms of itself, we say that the definition of that predicate is recursive. The relation for a recursively defined IDB predicate is the least fixpoint that includes the given values for the EDB predicates in the definition. The *expansion* of a recursively defined predicate will be defined precisely in Section 2; for now it suffices to state that by an infinite sequence of rule applications to a recursive predicate one can produce an infinite set of conjunctions of EDB predicates that also defines the recursive predicate. This infinite set of conjunctions is the expansion of the recursive predicate.

We say that two programs P and P' are equivalent if they compute the same relations for the IDB predicates. In this paper except where noted otherwise we consider programs consisting of a single, linear recursive rule. A rule is linear recursive if the predicate in the head appears exactly once in the body. We further assume that the head of the rule contain no repeated variables or constants, and that all nonrecursive predicates in the rule be EDB predicates.

Such a definition must have some initialization; we assume it is provided by a nonrecursive rule

$$t \coloneq t_0.$$

where t_0 does not appear in the recursive rule, has the same number of arguments as t, and contains the same variables in the same position as the predicate instance in the rule head.

We say that the expansion of a recursively defined predicate contains *inter-element redundancy* if there exist elements of the expansion s_1 and s_2 such that for all possible values of the EDB relations, evaluating s_1 will return a superset of the tuples returned by s_2. We say that the expansion contains *intra-element redundancy* if there is some element s such that there is a conjunction of some proper subset of the predicate instances in s, say s', such that s' and s define the same relation for all possible values of the EDB relations.

Evaluation algorithms for recursive definitions must at least implicitly evaluate enough elements of the expansion to construct the fixpoint or some portion thereof. If the expansion contains inter-element redundancy, then in the course of evaluating the expansion the evaluation algorithm will evaluate redundant elements of the expansion; if the expansion contains intra-element redundancy, then in the course of evaluating some element of the expansion the evaluation algorithm will evaluate redundant predicate instances. The efficiency of the evaluation algorithm will be improved if these redundancies are first removed — so a natural goal is to try to develop algorithms to convert recursive definitions to equivalent definitions that produce redundancy free expansions.

Example 1. Suppose that we have the following EDB relations: $l(X, Y)$ of consumers X and products they like Y; $c(Y)$ of inexpensive products Y; and $k(X, W)$ of pairs of people X, W such that X knows W. Also, suppose that a person will buy a product if it's cheap and they like it; or if it's cheap and they know someone who bought the product. Then we can define the relation $b(X, Y)$ of consumers X and products Y that they buy with the following rules:

$$\begin{aligned} r_1: \quad & b(X, Y) \coloneq l(X, Y), c(Y). \\ r_2: \quad & b(X, Y) \coloneq k(X, W), b(W, Y), c(Y). \end{aligned} \tag{1}$$

The first three elements of the expansion of *buys* are

$$\begin{gathered} l(X,Y)c(Y) \\ k(X,W_0)l(W_0,Y)c(Y)c(Y) \\ k(X,W_0)k(W_0,W_1)l(W_1,Y)c(Y)c(Y)c(Y) \end{gathered} \tag{2}$$

One can show that there is no inter-element redundancy in this expansion. However, there is intra-element redundancy: every element after the first contains redundant instances of $c(Y)$. The recursion can be re-written equivalently as

$$\begin{aligned} r_1: \quad & b(X,Y) :- l(X,Y), c(Y). \\ r_2': \quad & b(X,Y) :- k(X,W), b(W,Y). \end{aligned} \tag{3}$$

Here the first three elements of the expansion are

$$\begin{gathered} l(X,Y)c(Y) \\ k(X,W_0)l(W_0,Y)c(Y) \\ k(X,W_0)k(W_0,W_1)l(W_1,Y)c(Y) \end{gathered} \tag{4}$$

and there is no intra-element redundancy.

To see that this is not completely trivial, consider the rules

$$\begin{aligned} r_3: \quad & b(X,Y) :- r(X), l(X,Y). \\ r_4: \quad & b(X,Y) :- r(X), k(X,W), b(W,Y). \end{aligned} \tag{5}$$

where $r(X)$ is perhaps interpreted as "X is rich." The first three elements of the expansion of b as defined by these rules begins

$$\begin{gathered} r(X)l(X,Y) \\ r(X)k(X,W_0)r(W_0)l(W_0,Y) \\ r(X)k(X,W_0)r(W_0)k(W_0,W_1)r(W_1)l(W_1,Y) \end{gathered} \tag{6}$$

which contains no inter-element or intra-element redundancy.

The predicate c in the first definition in the previous example is what was called a *recursively redundant predicate* in Naughton (1986d). (The word "recursively" is included because neither rule r_1 nor r_2 contain any predicates that are redundant when the rule bodies are viewed as relational expressions.) Briefly, a predicate instance p in a recursive definition D is recursively redundant if and only if there is an equivalent definition D' in which that instance of p appears in no recursive rule. Recursively redundant predicates will produce intra-element redundancy.

This definition of redundancy is similar but not identical to that used in Sagiv (1987), which states that a predicate instance p in a recursive rule is redundant if the definition defines the same relation when that predicate instance is removed. One difference is that according to Naughton's definition, a predicate instance is recursively redundant if it can be removed from a recursive rule by changing the nonrecursive rules or adding some new nonrecursive rules; Sagiv's definition, on the other hand, considers only removing a predicate instance without changing anything else in the definition. Because of this difference the redundancy removing algorithms in Sagiv (1987) are able to handle more general recursions than those given in Naughton (1986c).

The two definitions happen to coincide on the previous example, but this is not always the case. In the recursion consisting of the two rules

$$\begin{array}{ll} r_4: & b(X,Y) \text{ :- } l(X,Y). \\ r_5: & b(X,Y) \text{ :- } k(X,W), b(W,Y), c(Y). \end{array} \tag{7}$$

the instance of $c(Y)$ in r_4 is recursively redundant by Naughton's definition (and, therefore, will cause intra-element redundancy in the expansion) but not by Sagiv's. In this case an optimized equivalent program is

$$\begin{array}{ll} r_6: & b(X,Y) \text{ :- } l(X,Y). \\ r_7: & b(X,Y) \text{ :- } b'(X,Y). \\ r_8: & b'(X,Y) \text{ :- } k(X,W), l(W,Y), c(Y). \\ r_9: & b'(X,Y) \text{ :- } k(X,W), b'(W,Y). \end{array} \tag{8}$$

There are also redundant predicates that will be detected by Sagiv's optimizing algorithms but not by Naughton's. These include predicate instances that are redundant in the body of the rule when viewed as a relational expression. In this paper we concentrate on recursively redundant predicates as defined by Naughton, because (as we will show) there is a close relationship between intra-element and inter-element redundancy and predicates of this type.

The following is a recursion that produces inter-element redundancy.

Example 2. Suppose that we introduce a new relation $i(X)$, which stands for "person X is impressionable." Also, suppose now that a person will buy a product if the person likes it, or if the person is impressionable and someone else has bought it. This situation is captured by the following rules:

$$\begin{aligned} r_{10}: \quad & b(X,Y) \text{ :- } l(X,Y). \\ r_{11}: \quad & b(X,Y) \text{ :- } i(X), b(W,Y). \end{aligned} \tag{9}$$

The first three elements of this expansion are

$$\begin{gathered} l(X,Y) \\ i(X)l(W_0,Y) \\ i(X)i(W_0)l(W_1,Y) \end{gathered} \tag{10}$$

One can prove that for any valuation of i and l, the relation returned by any element other than the first will be identical to that returned by the second element, so the recursion produces an expansion with inter-element redundancy. The recursion can be re-written equivalently as

$$\begin{aligned} r_{10}: \quad & b(X,Y) \text{ :- } l(X,Y). \\ r'_{11}: \quad & b(X,Y) \text{ :- } i(X), l(W,Y). \end{aligned} \tag{11}$$

which produces no inter-element redundancy.

In the previous example something interesting has happened. We were able to replace a recursive definition by a nonrecursive definition. Recursions that are replaceable by nonrecursive definitions are called *bounded recursions.* Bounded recursions are a particularly interesting subset of recursions with inter-element redundancy. Early work on sufficient conditions for a recursion to be bounded appeared in Minker and Nicolas (1982). Sagiv (1985) gave necessary and sufficient conditions for strongly typed, single predicate recursions to be bounded. Cosmadakis and Kanellakis (1986) showed that Sagiv's result can be extended to certain untyped recursions. Recently Gaifman *et al.* (1987) have proven that the general problem is undecidable. Here with respect to boundedness we summarize the line of research beginning in Ioannidis (1985) and in Naughton (1986a) and continued in Naughton and Sagiv (1987).

As we will show, all but a finite number of elements in the expansion of a bounded recursive definition are redundant. Thus, boundedness of a recursive definition means inter-element redundancy, but the converse is not necessarily true. Another observation is that boundedness is a special case of intra-element redundancy, as can be seen when we ask "what if every predicate in every recursive rule is recursively redundant?" Then, by definition of recursively redundant, there is an equivalent definition in which no predicate appears in any recursive rule. But this means that there is an equivalent nonrecursive definition, so that the recursion is bounded. The price we pay for this observation is an easy reduction that proves that detecting redundant predicates is undecidable.

The ideal would be an algorithm that, for any recursive definition, outputs an equivalent definition that contains no inter-element or intra-element redundancy in its expansion. The previous undecidabilty results show that this is unachievable, but actually it is worse than that — in Section 4 we prove that there are recursions that are not equivalent to any inter-element redundancy free recursion. However, we can isolate subclasses of recursions for which redundancy eliminating algorithms do exist. Inasmuch as these subclasses of recursions cover most of the recursions we have encountered in actual horn-clause programs, we expect these algorithms will be useful as part of the optimization of horn-clause database query languages.

We now summarize the rest of the paper. Section 2 precisely defines the expansion of an IDB predicate, inter-element redundancy, and intra-element redundancy. Section 3 introduces the *A/V graph*, a graph constructed from a recursive rule that is useful in detecting both types of redundancy. This section also introduces *chains* and *branches*, properties of elements of expansions that determine whether or not redundancy will occur.

Section 4 concentrates on inter-element redundancy. This section relates boundedness to inter-element redundancy, defines two types of boundedness, and reviews current results about the decidability of boundedness. Section 5 concentrates on intra-element redundancy. In it we relate intra-element redundancy to syntactic properties of the elements of an expansion, and show how to detect certain intra-element redundancies from the A/V graph for a recursive rule. We conclude in Section 6 with some open problems and directions for future research.

2. Expansions of IDB Predicates

The expansion of an IDB predicate t is the set of all conjunctions of EDB predicates that can be generated by some sequence of rule applications to t. More precisely, let s be a conjunction of predicate instances. To apply a rule to s, find some rule r such that the head of r unifies with some predicate instance p in s. Then replace that instance of p by the body of r, after applying the most general unifier (of the rule head and p) to both the body of r and the predicate instances in $s - p$. Then the expansion of an IDB predicate t is the set of all conjunctions of EDB predicates such that each conjunction can be produced by some sequence of rule applications beginning with an application of some rule to t. For recursive predicates, the expansion is infinite.

Recall that in this paper, we are dealing with linear recursive rules without constants or repeated variables in the rule head. Procedure ExpandRule (Fig. 1), enumerates the expansion of such definitions consisting of a recursive rule, r_r, and a nonrecursive rule, r_e. The output of ExpandRule is the expansion of the recursively defined predicate, represented by the infinite set S.

```
Give all variables in rules subscript 0;
S := ∅;
CurString := t;
while true do
        S := S ∪{CurString with r_e applied;
        CurString := CurString with r_r applied;
        increment the subscripts of all variables in r_r and r_e;
endwhile;
```

Fig. 1. Procedure ExpandRule.

Throughout the procedure, the string-valued variable CurString will have exactly one occurrence of the recursive predicate t. To "apply" a rule r to CurString, replace that occurrence of t by the right side of r, after the substitutions required to unify it with the head of the rule. In the initialization, we subscript the variables in the rules so that no variable appears in both CurString and one of the rules. On each iteration, we increment the subscripts for the same reason.

Example 3. If e is the edge relation of a digraph, then the following rules define the transitive closure t of the graph.

$$\begin{array}{ll} r_r: & t(X,Y) \text{ :- } e(X,Z), t(Z,Y). \\ r_e: & t(X,Y) \text{ :- } e(X,Y). \end{array} \tag{12}$$

In order to distinguish between an application of the recursive rule and an application of the nonrecursive rule, we let e denote the occurrence of e in the recursive rule, and e' denote the occurrence in the nonrecursive rule. The first four strings in the set S are

$$\begin{gathered} e'(X,Y), \\ e(X,Z_0)e'(Z_0,Y), \\ e(X,Z_0)e(Z_0,Z_1)e'(Z_1,Y), \\ e(X,Z_0)e(Z_0,Z_1)e(Z_1,Z_2)e'(Z_2,Y). \end{gathered} \tag{13}$$

A string in an expansion may contain multiple occurrences of each predicate appearing in the recursive rule. We will use "predicate instance" to refer to occurrences of predicates in the strings of the expansion.

The strings in an expansion are conjunctive queries, a subset of relational expressions. If a variable V appears in the head of the rule, then V is a *distinguished* variable; otherwise, it is *nondistinguished.* The distinguished variables appearing in each element of the expansion are those of the original recursive rule (possibly not all of them.) On the other hand, new nondistinguished variables are created in each element of the expansion. The variable W_i denotes a nondistinguished variable that appears for the first time in the $i+2$ element of the expansion, and it corresponds to the nondistinguished variable W in the body of the recursive rule.

If $V_1, V_2, \ldots, V_i$ are the distinguished variables, and $W_1, W_2, \ldots, W_j$ the nondistinguished variables, then the relation specified by the string $p_1 p_2 \ldots p_n$ is

$$\{(V_1, V_2, \ldots, V_i) | (\exists W_1, W_2, \ldots, W_j)(p_1 \wedge p_2 \wedge \cdots \wedge p_n)\} \tag{14}$$

The recursively defined relation is the union of the relations for the strings in the expansion.

In the next section we will need to decide equivalences between conjunctive queries; to do this, we use techniques developed by Aho *et al.* (1979) and by Chandra and Merlin (1977).

Definition 1. A mapping m from the variables of a string s_1 to the variables of a string s_2 is a *containment mapping* if distinguished variables map to themselves, and if $p(X_1, \ldots, X_n)$ appears in s_1, then $p(m(X_1), \ldots, m(X_n))$ appears in s_2.

The following lemma was proved by Chandra and Merlin (1977). (Aho *et al.* also attributed this to Chandra and Merlin although they proved it independently, because Chandra and Merlin (1977) was published a year earlier.)

Lemma 1. *If a string s_1 maps to a string s_2, then the relation specified by s_2 is contained in the relation specified by s_1.*

Aho et al. (1979) and Chandra and Merlin (1977) defined a conjunctive query c to be minimal if there is no conjunctive query c' such that c' is equivalent to c yet contains fewer predicate instances. Any conjunctive query c has a unique minimal form (up to a renaming of variables), and that minimal form is always a subset of the predicate instances in c.

We are now ready to define inter-element and intra-element redundancy.

Definition 2. An expansion S of an IDB predicate t contains *inter-element redundancy* if there are two elements s_1 and s_2 of S such that s_1 maps to s_2.

If an expansion contains no inter-element dependency we say that the expansion is *containment-free.*

Definition 3. An expansion S of an IDB predicate t contains *intra-element redundancy* if there exists an element s in S such that s is not minimal as a conjunctive query.

3. The A/V Graph and Branches

Inter-element and intra-element redundancies depend on certain properties of the strings. In this section we discuss these properties, and develop tools to detect them. The lemmas and facts cited in this section are proven in Naughton (1986b, 1986c).

3.1. A/V Graphs and Expansions

To relate the patterns of variables appearing in the strings of S to the structure of the rules, we define the *argument/variable (A/V) graph*:

- For each variable appearing in the rules add a variable node.
- For each argument position in each rule body add an argument node.
- Draw an undirected edge from each argument node to the node for the variable that appears in that position in the rule. This kind of edge is called an *identity edge*.
- Draw a directed edge from each argument node corresponding to a position p in the recursive predicate to the node for the distinguished variable that appears in p in the rule head. This kind of edge is called a *unification edge*.

The node for a variable X is labeled X, and the node for argument position i of a predicate p is labeled p^i. A node for a distinguished variable is a distinguished variable node; all other variables nodes are nondistinguished. Because of the one-to-one correspondence between positions in the bodies of rules and the argument nodes in the A/V graph, we use position names to refer to both an argument position and the argument node it is represented by. Similarly, we use variable names to refer to variable nodes.

Many of the subsequent results depend on the existence of certain kinds of paths through the A/V graph. Some nonstandard terminology arises because we allow the directed edges in an A/V graph to be traversed from head to tail as well as from tail to head; thus a path in an A/V graph can contain unification edges traversed in either direction.

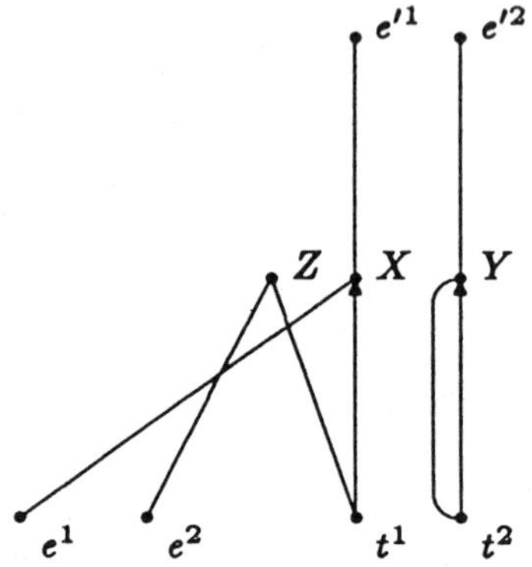

Fig. 2. A/V graph for Example 3.

Example 4. Fig. 2 gives the A/V graph for the rules of Example 3.

There is a close relationship between the A/V graph and procedure ExpandRule of Section 2. If a predicate instance first appears through applying a rule on iteration i, then we say that predicate instance was produced on iteration i. (The first iteration of the while loop is iteration 0.) There are two ways a predicate appearing in a string s of S can be produced on iteration i. It can be added to CurString through applying the recursive rule, or, if s was added to S on iteration i, it can be produced by applying the nonrecursive rule.

Consider iteration i. At line 8 on iteration $i-1$, the variables in the rules were given subscript i. Letting the argument nodes of the A/V graph represent the bodies of the rules, we represent iteration i by subscripting the labels of the variable nodes by i. (Fig. 3(a)).

Because the heads of the rules contain no repeated variables or constants, the unification can be done by replacing the subscripted distinguished variables by the variables appearing in the instance of t in CurString. If we consider the argument nodes for t as representing that instance of t, the variable at the head of a unification edge is replaced by the variable appearing in the argument at the tail. On iteration 0, because of the initialization of CurString, the arguments contain the distinguished variables. On all other iterations, they hold the variables that were put there on the previous iteration — in this case, Z_{i-1} and Y. (Fig. 3(b)).

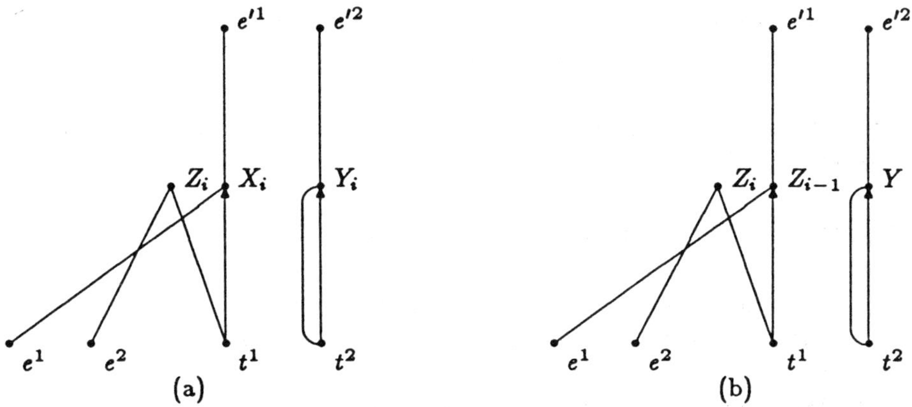

Fig. 3. A/V graph for Example 3.

After the substitution, argument a of a predicate instance produced on iteration i will contain the variable that is the label of the node at the end of its incident identity edge. In our example, the predicate instance added by the nonrecursive rule will be $e'(Z_{i-1}, Y)$ and the predicate instances added by the recursive rule will be $e(Z_{i-1}, Z_i)t(Z_i, Y)$.

The previous two paragraphs show how we can determine what variable appears in any position of any predicate instance in the expansion. The following two facts can be proven by induction:

Fact 1. A nondistinguished variable W_i appears in position p in a predicate instance produced on iteration $i + k$ if and only if there is a path from W to p containing k unification edges, all traversed in the forward direction.

Fact 2. A distinguished variable V appears in position p on iteration i if and only if there is a path from V to p containing i unification edges, all traversed in the forward direction.

Any A/V graph can be divided into two kinds of connected components, those containing nondistinguished variables and those containing only distinguished variables. (Connected components in A/V graphs can require unification edges to be traversed in either direction.) Each type of component has a specific structure.

Lemma 2. *If a connected component in an A/V graph contains a nondistinguished variable W, it is a tree, and W is the only nondistinguished variable in the component.*

Lemma 3. *If a connected component contains no nondistinguished variable, that component must contain a cycle.*

Lemmas 2 and 3 combine with Facts 1 and 2 to prove that

1. Arguments in connected components that contain a cycle will always contain the distinguished variables appearing on the cycle.
2. Arguments in connected components that contain no cycles will eventually contain only subscripted instances of the nondistinguished variable in the component.

In view of point number 1 above, we have the following definition.

Definition 4. A *persistent variable* is a distinguished variable that appears in a cyclic component of the A/V graph.

Example 5. See Fig. 2 for the A/V graph for Example 3. There are two connected components in this graph. The first, $\{t^2, Y, e'^2\}$, contains the cycle $t^2 \to Y \to t^2$. Then Fact 2 implies that Y always appears in e'^2. Y is a persistent variable. The remaining nodes form a tree, with Z at the root. By Fact 1, Z_i appears in e^2 and e'^1 on iteration i, and in e^1 on iteration $i+1$.

In the following subsection, it will be important to know how variables are shared between the predicate instances in the expansion. Things are complicated by the possibility of repeated variables in the rule body. Repeated variables give rise to branches in the paths from variable nodes to argument nodes, and the variable at the root of such paths appears on all branches. Thus to determine when argument positions share variables, we need to follow unification edges backward (toward the argument nodes) as well as forward.

To count the net number of forward unification edges in a path, we introduce weights on the edges of the A/V graph. The weight of an identity edge is 0; the weight of a unification edge traversed in the forward direction is 1, and the weight of a unification edge traversed in the reverse direction is -1. The weight of a path in the A/V graph is the sum of the weights of the edges in the path. With this definition, we have the following lemma:

Lemma 4. *A variable appears in position p^1 of a predicate instance produced on iteration i, and in position p^2 of a predicate instance produced on iteration $i+k$, if and only if there is a path from p^1 to p^2 of weight k such that the weight of the minimal point on the path, relative to p^1, is $\geq -i$.*

3.2. Chains and Branches

In this subsection we discuss some properties of the strings of an expansion that will be useful in determining redundancy. These properties depend only on the recursive rule.

Definition 5. A *chain* is a sequence of predicate instances $p_1, p_2, \ldots, p_n$, such that for $1 \leq i < n$, p_i and p_{i+1} share a nondistinguished variable.

Definition 6. A *branch* is a sequence of predicate instances such that for all p and q in the sequence, there is a chain containing p and q.

Example 6. Assuming that X and Y are the only distinguished variables,

$$p(X, W_0)p(W_0, W_1)p(W_1, Y) \tag{15}$$

is a chain of length three. The sequence

$$p(W_0, X)p(X, Y)p(Y, W_1) \tag{16}$$

contains three chains of length one. The sequence

$$p(X, W_0, Z_0)q(W_0)p(Z_0, W_1, Z_1)q(W_1)p(Z_1, W_2, Z_2)q(W_2) \tag{17}$$

is a branch.

By examining the A/V graph for a recursive rule, we can tell what chains (and therefore, what branches) will appear in the strings of the expansion. Chains depend only on shared nondistinguished variables, so the first task is to eliminate from the A/V graph the nodes for arguments that contain only distinguished variables. By Lemmas 2 and 3, we can do this by removing all connected components that contain cycles.

For the second step, we augment the remaining subgraph by adding predicate edges between adjacent argument nodes of the same nonrecursive predicate. The result is the *augmented* A/V graph for the rule.

Example 7. The following rule illustrates the concepts that will be developed in this section.

$$r_r: \quad t(X,Y) :- t(X,V), p(X,W), q(W,V), r(X,Y). \tag{18}$$

The A/V graph for this rule is given in Fig. 4. X appears in a connected component with a cycle, so we remove the nodes for X, t^1, p^1, and r^1. The augmented A/V graph for the rule is given in Fig. 5.

The following lemma and its two corollaries show the relationship between the augmented A/V graph for a rule and the branches that appear in its expansion. Recall that in paths in an A/V graph, unification edges (arcs) can be traversed in either direction.

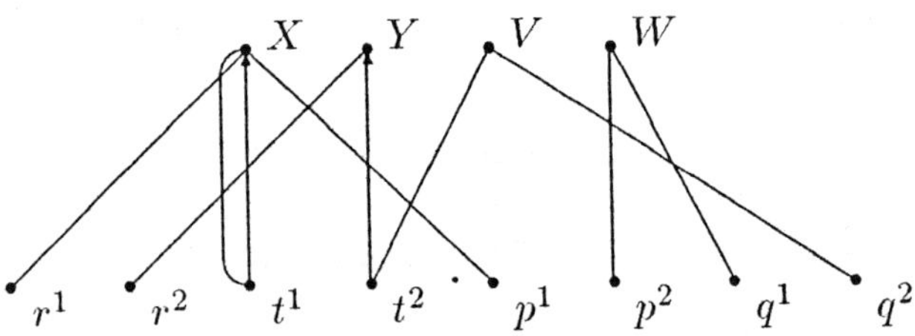

Fig. 4. A/V graph for Example 7.

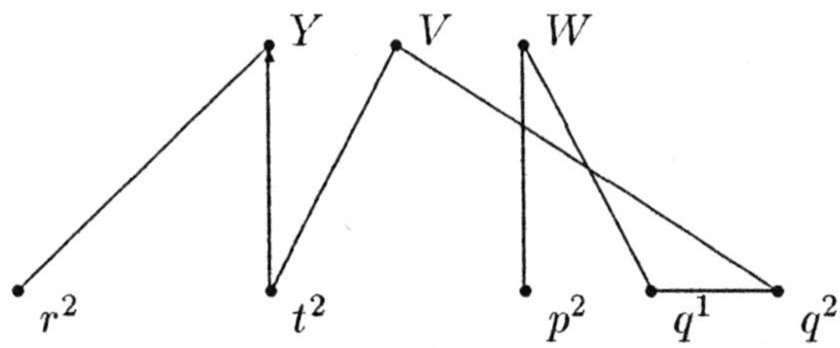

Fig. 5. Augmented A/V graph for Example 7.

Lemma 5. *There is a chain containing an instance of predicate r produced on iteration i, and an instance of s produced on iteration $i+k$, if and only if there is a path of weight k from some argument of r to some argument of s.*

Corollary 1. *Instances of two predicates appear in the same branch if and only if the argument nodes for those predicates appear in the same connected component.*

Definition 7. The *rank* of a predicate is the weight of a maximal path from any variable node to an argument node of the predicate. The *span* of a branch is the rank of its maximal predicate.

If there are cycles the rank of some predicates may be infinite.

Corollary 2. *The span of a branch is the maximum weight of any path in the connected component for the branch.*

Two branches in a string are *instances of the same branch* if the predicates in each appear in the same connected component in the augmented A/V graph. A rule produces multiple branches if there are multiple connected components in the augmented A/V graph for the rule. An instance of each branch is begun on every iteration. If the span of a branch is k, the first complete instance of that branch will be produced by iteration $k+1$. In addition to the complete instances of branches, there will also be incomplete instances of branches produced.

Example 8. In Fig. 5, there is only one connected component, and it contains all the argument nodes. This implies that there is a branch (chain, in this case) containing instances of p, q, and r. There are paths of weight 0 from V and W to the nodes of p and q, thus p and q are of rank 0. There is a path from V to r^2 of weight 1, so r is of rank 1. The span of the branch is 1.

The third string in the expansion,

$$p_2(X,W_1)q_2(W_1,V_1)r_2(X,V_0)p_1(X,W_0)q_1(W_0,V_0)r_1(X,Y), \quad (19)$$

contains a partial instance of the branch produced on iteration 1 ($r_1(X,Y)$), a complete instance produced on iterations 1 and 2

$$p_1(X,W_0)q_1(W_0,V_0)r_2(X,V_0) \quad (20)$$

and another partial instance produced on iteration 2 ($p_2(X, W_1)$ $q_2(W_1, V_1)$). This last partial instance will be completed on iteration 3.

If there is a cycle of nonzero weight in a component of an augmented A/V, then there will be a branch with an infinite span. Such branches will never be completed; in this case we say the expansion contains *unbounded branches.*

Definition 8. A connected component of an augmented A/V graph is a *bounded component* if it contains no cycle of nonzero weight; otherwise it is an *unbounded component.*

Definition 9. A nonzero weight cycle in the augmented A/V graph is called a *chain-generating path.*

Definition 10. Any argument position on a chain-generating path is a *linking position*; the variables that appear in linking positions in the elements of the expansion are called *linking variables.*

Example 9. The augmented A/V graph for the recursive rule of the transitive closure example (Example 3), Fig. 6, contains an unbounded component. The cycle in that component is a chain-generating path. Argument positions e^1 and e^2 are linking positions. The expansion contains unbounded branches — for any n, we can find a string in the expansion containing a branch of at least n instances of e.

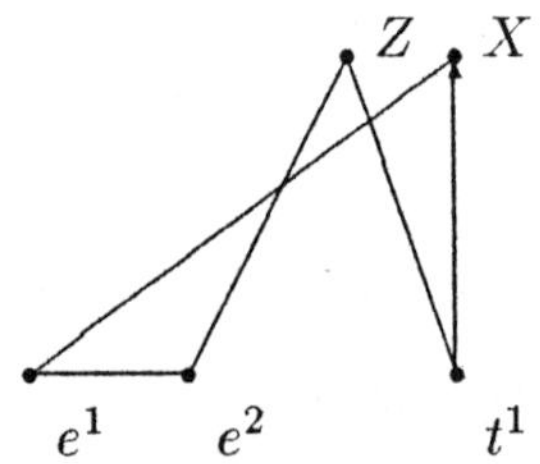

Fig. 6. Augmented A/V graph for the recursive rule of Example 3.

4. Inter-Element Redundancy

In this section we concentrate on inter-element redundancy. We begin by giving a sufficient condition for an expansion to contain no inter-element redundancy.

Theorem 1. *Let D be a recursion with a single, linear recursive rule r. Suppose that the augmented A/V graph for D has some unbounded component and that the body of r contains no repeated predicates. Then the expansion of D contains no inter-element redundancy.*

Proof. Let the augmented A/V graph have a chain-generating path. We prove that if an element s_1 of the expansion maps to an element s_2 of the expansion, then the predicate instance appearing i predicate instances from the start of the chain in s_1 must map to the instance appearing i instances from the start in s_2. The proof is by induction on the predicate instances in the chains.

Let p be the first predicate in the chain. Then, by Fact 4.2 from Naughton (1986b), there is a linking position a in p that contains a distinguished variable. Because there are no repeated predicates in the rule, by Fact 4.1 from Naughton (1986b), this is the only instance of p such that a contains that distinguished variable. Thus if s_1 is to map to s_2, the first predicate instance of the chain in s_1 must map to the first predicate instance in the chain in s_2.

Now suppose that for $i < n$, if s_1 maps to s_2, then the *ith* predicate instance from the beginning of the chain in s_1 maps to the *ith* predicate instance from the beginning of the chain in s_2. The *nth* predicate in from the beginning of the chain in s_1 must, by definition of a chain, share some linking variable V with the instance $n-1$ from the beginning. By induction, any mapping m must map V to some variable, say V', that appears in the same position in the predicate $n-1$ from the beginning in s_2. But again by definition of chain, V' must appear in the same position in the predicate n from the beginning in s_2 as V does in s_1. But because there are no repeated predicates in the body of the recursive rule, this is the only instance in which V appears in that position, the instance in s_1 that is n instances from the beginning must map to the predicate instance n from the beginning in s_2.

But if $s_1 \neq s_2$, then the instance of the chain is longer in one string than in the other. If s_1 is longer than s_2, then we will some predicate instance in the chain in s_1 with no predicate instance in s_2 to map to, and hence s_1 cannot be mapped to s_2. If s_2 is longer than s_1, then the exit predicate t_0 will have to be mapped to some other predicate in the chain. But this is impossible by definition of mapping, so again, s_1 cannot be mapped to s_2.

Because this holds for any distinct pair of strings s_1 and s_2, the expansion must contain no inter-element redundancy. □

We now turn to definitions of bounded recursions. Strictly speaking, because of the class of recursions we are considering, in this paper we are talking about *uniform boundedness.* For a discussion of uniform vs. simple boundedness see Naughton and Sagiv (1987).

There are two natural ways to define boundedness.

Definition 11. A program P is bounded if there is a nonrecursive program P' such that P and P' are equivalent.

Definition 12. A program P is bounded if there is some k such that for any value of the relations, the iterative bottom-up construction of the fixpoint will converge in $\leq k$ iterations.

The following lemma is from Naughton and Sagiv (1987).

Lemma 6. *Definitions 11 and 12 are equivalent.*

The following theorem is from Naughton (1986a). It shows the relationship of boundedness to a particular kind of inter-element redundancy.

Theorem 2. *A set of rules defining a predicate t is bounded if and only if, in the expansion of t, there exists an n_0 such that for all $n > n_0$, s_n is mapped to by some previous string.*

Also in Naughton (1986a) it was shown that the absence of a chain generating path was a sufficient condition for bounded recursion. However, not all programs with chain generating paths are unbounded.

Example 10. The following program

$$\begin{aligned} t(X,Y,Z) &:- t(X,W,Z) \,\&\, e(W,Y) \,\&\\ &\quad\; e(W,Z) \,\&\, e(Z,Z) \,\&\, e(Z,Y).\\ t(X,Y,Z) &:- t_0(X,Y,Z). \end{aligned}$$

has a chain generating path but is bounded.

We now turn to a class C for which a chain-generating path is necessary and sufficient for the recursion to be unbounded. This class was defined in Naughton and Sagiv (1987).

Definition 13. Let r be a linear recursive rule, and let s_i and s_j, where $i < j$, be any pair of strings in the expansion of r. Then r is in the class C if and only if there can be no mapping m from s_i to s_j such that m maps a linking variable in s_i to a persistent variable in s_j.

Theorem 3. *Let P be a recursive rule in class C. Then P is bounded if and only if there is no chain generating path in the A/V graph for P.*

The simplest way to prove that a recursive rule is in the class C is to prove that no persistent variable ever appears in a linking position. This technique allows us to prove that the following four classes are subclasses of C. The following lemmas are from Naughton and Sagiv (1987).

Lemma 7. *If no sequence of argument positions in the recursive predicate in the body of the recursive rule in P contains a permutation of the variables that appear in the same positions in the instance of the recursive predicate in the head, then P is in C.*

Ioannidis first gave a necessary and sufficient condition for boundedness in this class in Ioannidis (1985).

Lemma 8. *If the body of the recursive rule in a recursion P has no repeated predicates, then P is in the class C.*

A chain-generating path was first proven necessary and sufficient for recursions in this class in Naughton (1986a).

Lemma 9. *If no persistent variable appears in a nonrecursive predicate in the body of the recursive rule of P, then P is in C.*

Ioannidis (1986) has shown that his condition from Ioannidis (1985) holds for this class as well. It includes recursions in which no distinguished variable appears both in a nonrecursive predicate and in the recursive predicate in the body.

Lemma 10. *If no persistent variable appears in a linking position in the body of the recursive rule of P, then P is in C.*

The subclasses of C described by Lemmas 7, 8, and 9 are incommensurate and there are linear time membership algorithms for each. The subclass described by Lemma 10 properly includes that described by Lemma 9 and has a polynomial time membership algorithm.

We finish this section with a negative result about eliminating all redundancy from an expansion:

Theorem 4. *There are recursively defined predicates that cannot be defined by any inter-element redundancy free definition.*

Proof. Consider the following definition:

$$\begin{aligned} t(X,Y) &:- e(X,Y), e(Y,Y). \\ t(X,Y) &:- t(X,W), e(W,Y), e(W,W). \end{aligned} \tag{21}$$

The expansion S of this recursion begins

$$\begin{gathered} e(X,Y)e(Y,Y) \\ e(X,W_0)e(W_0,W_0)e(W_0,Y)e(Y,Y) \\ e(X,W_1)e(W_1,W_1)e(W_1,W_0)e(W_0,W_0)e(W_0,Y)e(Y,Y) \end{gathered} \tag{22}$$

It is easy to verify that this expansion has the property that for all $i, j > 0$, if $j \geq i$ then s_j maps to s_i. Call this recursion D.

Now suppose that there is some other definition D' that is equivalent to D, and let S' be the expansion of D'. Furthermore, assume that S' has no inter-element redundancy. We first prove that S' must be infinite.

Assume that S' is finite. This means that S is definable by a finite union of conjunctive queries, and this finite union would then be a nonrecursive program equivalent to S. But by Theorem 3 and Lemma 10, S is not bounded. So S' must be infinite.

Take any element s'_k from S'. By a theorem due to Sagiv and Yannakakis (1980), if S and S' are equivalent, then there is at least one element s_m in S such that s_m maps to s'_k. Similarly, s_m must be mapped to by some s'_n of S', and so s'_n maps to s'_k. Since S' does not have any inter-element redundancy, $k = n$, so each element s'_k of S' has an equivalent element s_m in S.

Now take any two distinct elements of S', say s'_k and s'_l, and let s_m and s_n be their equivalent elements in S. Because there is containment between every pair of elements in S, there is containment between s_m and s_n, hence there must also be containment between s'_k and s'_l, which is a contradiction. □

5. Intra-Element Redundancy

In this section we concentrate on intra-element redundancy. We show that if a predicate p appears in a bounded component of the augmented A/V graph, then p is recursively redundant, that is, the elements of the expansion will contain redundant p instances. For a useful subset of recursive rules, only predicates appearing in bounded components are redundant. These predicates can be found by a linear-time algorithm.

5.1. A Sufficient Condition

The following theorem from Naughton (1986d) relates redundancy to connected components in the augmented A/V graph for the recursive rule. (Recall that in a connected component of an A/V graph, unification edges can be traversed in either direction.)

Theorem 5. *Let a predicate p appear in a rule r, and suppose that no argument of p appears in an unbounded component of the augmented A/V graph for r. Then p is recursively redundant in r.*

Note that if one argument of p appears in an unbounded component of the augmented A/V graph, then all arguments of p are in that component.

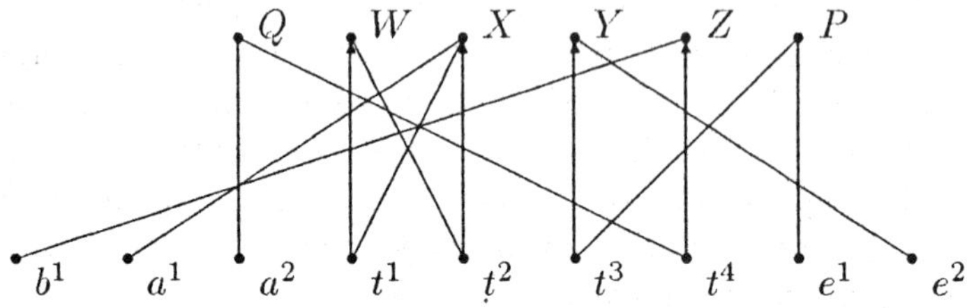

Fig. 7. A/V graph for the recursive rule of Example 11.

Example 11. Consider the following rules:

$$\begin{aligned} &t(W,X,Y,Z) :- t_0(W,X,Y,Z). \\ &t(W,X,Y,Z) :- t(X,W,P,Q), e(P,Y), a(X,Q), b(Z). \end{aligned} \tag{23}$$

In the A/V graph for the recursive rule (Fig. 7), there is a cycle involving X and W of weight 2. The nodes connected to this cycle form an unbounded component.

In the augmented A/V graph for the recursive rule (Fig. 8) there are two components. The argument nodes of e appear in an unbounded component, while all other predicate arguments appear in a component with a maximal path of weight 1. The rank of a is 0, and the rank of b is 1. Consider the string produced on iteration four. We display it below, one branch to a line, one column for each iteration, starting with iteration 0 on the right.

1)	$t_0(W,X,P_3,Q_3)$	$e(P_3,P_2)$	$e(P_2,P_1)$	$e(P_1,P_0)$	$e(P_0,Y)$
2)	$b(Q_3)$	$a(W,Q_3)$			
3)		$b(Q_2)$	$a(X,Q_2)$		
4)			$b(Q_1)$	$a(W,Q_1)$	
5)				$b(Q_0)$	$a(X,Q_0)$
6)					$b(Z)$

The instances of a and b produced on iteration 2 ($a(X,Q_2)$ and $b(Q_2)$) are redundant. $\square$

More detail can be found in Naughton (1986c), which shows how to determine which instances of redundant predicates will be redundant, and gives an algorithm to rewrite recursive definitions to remove redundant predicates.

5.2. A Necessary and Sufficient Condition

In general, the converse of Theorem 5 fails to hold: some predicates appearing in unbounded components are redundant. If the nonrecursive predicates are IDB predicates, their definition may cause redundancy. For example, if we take the standard transitive closure rules,

$$\begin{aligned} t(X,Y) &:- e(X,Y). \\ t(X,Y) &:- t(X,W), e(W,Y). \end{aligned} \tag{24}$$

and add

$$e(X,Y) :- a(X), b(Y). \tag{25}$$

then t is completely defined by the rule

$$t(X,Y) :- a(X), b(Y). \tag{26}$$

and e is redundant in the recursive rule.

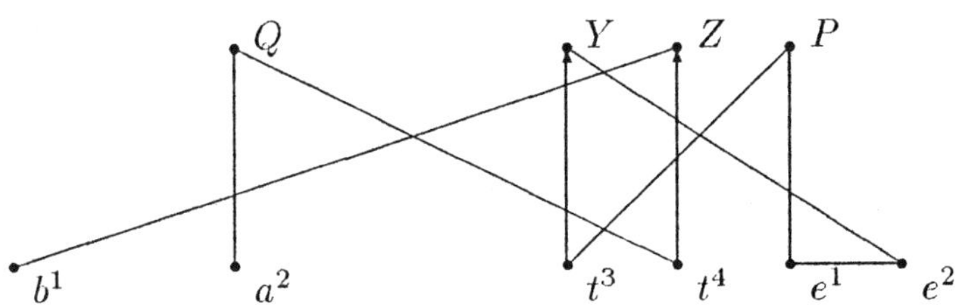

Fig. 8. Augmented A/V graph for the recursive rule of Example 11.

Even if the nonrecursive predicates are EDB predicates, interactions between the recursive and nonrecursive rule can make predicates redundant. The pair of rules

$$\begin{aligned} t(X,Y) &:- b(X), t(X,W), e(W,Y). \\ t(X,Y) &:- b(X), e(W,Y). \end{aligned} \tag{27}$$

can be replaced by the nonrecursive rule alone.

However, we can state the following theorem, again from Naughton (1986d).

Theorem 6. *Let all nonrecursive predicates in r be EDB predicates, and suppose that there are no repeated nonrecursive predicates in r. Then a predicate p is recursively redundant in r if and only if p appears in a bounded component of the augmented A/V graph.*

5.3. Detecting Redundant Predicates

In this subsection we present a linear-time algorithm that detects redundant predicates in an augmented A/V graph. A variant of this algorithm was presented in Ioannidis (1985). There it was used on a different graph, the α-graph, to decide the bounded recursion problem for a restricted class of rules.

An A/V graph can be converted to an augmented A/V graph by a straightforward application of depth-first search. We assume that the augmented A/V graph is presented in adjacency list form. Each edge is represented by a pair of directed edges. Associated with each directed edge is a weight. Both directed edges for a predicate or identity edge have weight zero; the directed edge corresponding to a forward traversed unification edge has weight one, while the directed edge for a reverse unification edge has weight minus one.

A linear-time algorithm to detect redundant predicates is given in Figure 9.

6. Conclusion

Removing redundancy is only one part of efficient evaluation of queries on recursively defined relations. An evaluation algorithm should attempt to use selection constants to restrict the EDB tuples looked at during evaluation, and attempt to reuse the results from partial evaluation of one element of an expansion when evaluating another. These issues are ignored here because they are orthogonal to the issue of removing redundancy from a recursion.

This paper focussed on recursions containing one, linear recursive rule. We are currently extending the redundancy and boundedness detection procedures for more general recursions.

Input. An augmented A/V graph $G = (V, E)$ represented by adjacency lists $L[v]$, for $v \in V$. The function $w(e(v, u))$ returns the weight for each edge $e(v, u)$.
Output. A list of the bounded components in the graph.

```
cycle:   boolean;            /* true if current component contains a cycle */
weight:  array of integer;  /* weight[u] holds weight of u if visited */

Procedure SearchComp(v, cycle);

begin
    mark v old;
    for each vertex u on L[v] do
      if u is marked "old" and weight[u] ≠ weight[v] + w(e(v, u)) then
        cycle := true;
        return;
      else
        weight[u] := weight[v] + w(e(v, u));
        SearchComp(v);
      endif;
    endfor;
end;

begin
  mark all vertices "new";
  while there exists a vertex v in V marked "new" do
    cycle := false;
    SearchComp(v);
    if not cycle then
      list bounded component containing v;
    endif;
  endwhile;
end.
```

Fig. 9. An algorithm to detect redundant predicates.

Another, more specific question is: is testing for containment-freedom (no inter-element redundancy) decidable? The general techniques given in Gaifman et al. (1987) do not apply here, because although containment-freedom is strongly non-trivial, it is not semantic — there exist equivalent recursions D_1 and D_2 such that D_1 is containment free whereas D_2 is not.

Finally, the condition for redundancy given in Theorem 5 is similar to the condition for boundedness given by Theorem 3. Yet the class of recursions for which the condition of Theorem 5 is complete (no repeated predicates) is much more restrictive than the class C for which the condition of Theorem 3 is complete. Can the class of recursions for which Theorem 5 is complete be extended?

Acknowledgement

The importance of containment-free expansions in minimizing recursions first arose in discussions with Jeff Ullman and Allen Van Gelder.

References

Aho, Alfred V., Sagiv, Yehoshua, and Ullman, Jeffrey D. (1979). "Equivalence of relational expressions," *SIAM Journal of Computing*, **8**(2), 218–246.

Bancilhon, François and Ramakrishnan, Raghu (1986). "An amateur's introduction to recursive query processing strategies," In *Proceedings of the ACM SIGMOD International Conference on Management of Data*, 16–52, Washington, D.C.

Cosmadakis, Stavros S. and Kanellakis, Paris C. (1986). "Parallel evaluation of recursive rule queries," In *Proceedings of the ACM SIGACT-SIGMOD Symposium on Principles of Database Systems*, 280–293, Cambridge, MA.

Chandra, Ashok K. and Merlin, Phillip M. (1977). "Optimal implementation of conjunctive queries in relational data base," In *Conference Record of the Ninth Annual ACM Symposium on Theory of Computing*, 77–90, Boulder, CO, May 1977.

Gaifman, Haim, Mairson, Harry, Sagiv, Yehoshua, and Vardi, Moshe Y. (1987). "Undecidable optimization problems for database logic programs," In *Proceedings of the Second IEEE Symposium on Logic in Computer Science*, 106–115, Ithaca, NY, June 1987.

Ioannidis, Yannis E. (1985). *Bounded recursion in deductive databases.* Technical Report UCB/ERL M85/6, UC Berkeley, Berkeley, CA, February 1985.

Ioannidis, Yannis E. (1986). "A time bound on the materialization of some recursively defined views," In *Proceedings of the 11th International VLDB Conference*, 219–226, Stockholm, Sweden, August 1985.

Minker, Jack, and Nicolas, Jean M. (1982). "On recursive axioms in relational databases," *Information Systems*, **8**(1), 1–13.

Naughton, Jeffrey F. (1986a). "Data independent recursion in deductive databases," In *Proceedings of the ACM SIGACT-SIGMOD Symposium on Principles of Database Systems*, 267–279, Cambridge, MA, March 1986. Selected for special issue of JCSS.

Naughton, Jeffrey F. (1986b). *Data independent recursion in deductive databases.* Technical Report STAN-CS-86-1114, Stanford, CA, May 1986.

Naughton, Jeffrey F. (1986c). *Optimizing Function-free Recursive Inference Rules.* Technical Report STAN-CS-86-1114, Stanford University, Stanford, CA, May 1986.

Naughton, Jeffrey F. (1986d). "Redundancy in function-free recursive inference rules," In *Proceedings of the IEEE Symposium on Logic Programming*, 236–245, Salt Lake City, Utah.

Naughton, Jeffrey F. and Sagiv, Yehoshua (1987). "A decidable class of bounded recursions," In *Proceedings of the ACM SIGACT-SIGMOD-SIGART Symposium on Principles of Database Systems*, 227–236. Selected for special issue of the Journal of Logic Programming.

Sagiv, Yehoshua (1985). "On computing restricted projections of representative instances," In *Proceedings of the ACM SIGACT-SIGMOD Symposium on Principles of Database Systems*, 171–180, Portland, OR.

Sagiv, Yehoshua (1987). "Optimizing datalog programs," In *Proceedings of the ACM SIGACT-SIGART-SIGMOD Symposium on Principles of Database Systems*, 349–362, Austin, TX, March 1987.

Sagiv, Yehoshua and Yannakakis, Mihalas (1980). "Equivalences among relational expressions with the union and difference operators," *JACM*, **27**(4), 633–655, October, 1980.

10 Tree Monoids and Recognizability of Sets of Finite Trees

MAURICE NIVAT
ANDREAS PODELSKI
LITP
Université Paris VII
Paris, France

1. Introduction

The new structure which is introduced in this paper is the monoid of trees with a pointed border node, that is the set $\Sigma^{(\#)}$ of pairs (t, f) where t is a binary tree on Σ and f is a border node of t, equipped with the associative composition law

$$(t, f) + (t', f') = (t + ft', ff'). \tag{1}$$

The notations are borrowed from Nivat (1986a). A border node f of t is different from a leaf. It is a node, itself not belonging to the tree, which is the son of a node of the tree (formally, of the form gi, where $i \in \{1, 2\}$, g is a node of t and gi is not).

The sum $t + ft'$ in the tree is obtained by adding to t the tree t' rooted in f so that the support $\text{Supp}(t+ft')$ is the (disjoint) union of the support $\text{Supp}(t)$ of t and $f\text{Supp}(t') = \{fh \mid h \in \text{Supp}(t')\}$ which is the translation of $\text{Supp}(t')$ by f.

It is shown that the transition monoid of a deterministic finite ascending tree automaton (dfata) is very naturally a homomorphic image of the monoid $\Sigma^{(\#)}$.

Resolution of Equations in
Algebraic Structures
Volume 1

ISBN 0-12-046370-9

This transition monoid is, as in the case of word automata, a monoid of functions from the set of states into itself under the ordinary composition law.

In the case of a dfata the function associated to the tree t with pointed border node f maps the state q on the state which results of the computation of the automaton when all the border nodes but f are initially in the initial state q_0, and f is given the state q. If $T_{(t,f)}$ is the function, the composition product of $T_{(t,f)}$ and $T_{(s,g)}$ is clearly given by

$$T_{(t,f)}T_{(s,g)} = T_{(t+fs,fg)} = T_{(t,f)+(s,g)}. \tag{2}$$

This fact is the major property on which all the results presented below rely. These results constitute the beginning of an algebraic theory of tree automata (deterministic, finite and ascending) since they relate what can be called the syntactic monoid of a recognizable finitary language L to the transition monoid of the dfata recognizing L (in fact they are proved to be equal).

Not surprisingly the set of states of this dfata can be identified with the set of equivalence classes modulo a root (or left) invariant equivalence relation associated with L, which has finite index and is the natural extension to trees of the classical Nerode equivalence for words.

And the syntactic monoid of L is the quotient of $\Sigma^{(\#)}$ for the root and leaf (left and right) invariant equivalence relation which can be associated to a tree language L.

2. Binary Trees and Ascending Finite Automata

Given is a finite alphabet Σ, and $\mathcal{D}$ denotes the free monoid generated by {1,2}.

A binary *tree* on Σ denoted t is the pair formed by

- its support Supp(t) which is a subset of $\mathcal{D}$ satisfying the condition
$$FG(\text{Supp}(t)) = \text{Supp}(t) \tag{3}$$
- its valuation or labeling $\hat{t}$ which maps Supp(t) into Σ.

We denote by $FG(\text{Supp}(t))$ the set of left factors (or prefixes) of the words in Supp(t).

A set which satisfies $FG(\mathrm{Supp}(t)) = \mathrm{Supp}(t)$ is also said to be closed by left factors or factorial.

Only finite binary trees will be considered in the sequel, a tree being finite iff its support is finite. The set of finite binary trees on Σ is denoted by $\Sigma^{\#}$.

There exists a large wealth of standard terminological designations:

- the condition $FG(\mathrm{Supp}(t)) = \mathrm{Supp}(t)$ implies $\varepsilon \in \mathrm{Supp}(t)$, this node ε of t is called the root of t,
- for all $f \in \mathrm{Supp}(t)$, $f1$ is called the left son of f and $f2$ the right son of the node f of the tree,
- each node $f \in \mathrm{Supp}(t) \setminus \{\varepsilon\}$ can be written in a unique way: $f = gi$, with $g \in \mathrm{Supp}(t)$ and $i \in \{1,2\}$, and then g is called the father of f,
- the leaves of the tree t are the elements of the set $\mathrm{Max}(\mathrm{Supp}(t))$ which can be written as $\{f \in \mathrm{Supp}(t) | \; f\mathcal{D} \cap \mathrm{Supp}(t) = \{f\}\}$.

Less standard denotations are the following:

The *border* of the tree t is the set

$$\mathcal{B}(t) = \{fi | \; f \in \mathrm{Supp}(t), i \in \{1,2\}, fi \notin \mathrm{Supp}(t)\} \tag{4}$$

which can also be written as $\mathrm{Supp}(t)\{1,2\}\backslash\mathrm{Supp}(t)$, and each element $fi \in \mathcal{B}(t)$ is a border node of t.

We denote by Ω the *empty tree* whose support $\mathrm{Supp}(\Omega) = \Phi$ is empty and its border is conveniently defined as $\mathcal{B}(\Omega) = \{\varepsilon\}$.

For all $a \in \Sigma$, the *punctual tree* whose support is equal to $\{\varepsilon\}$ and whose labelling of its only node is a is also denoted by a.

This notation leads to $\mathrm{Supp}(a) = \{\varepsilon\}$ and $\hat{a}(\varepsilon) = a$. It allows for using $\hat{t}(\varepsilon)$ as a letter from Σ and as a denotation of the corresponding punctual tree.

The following two basic notions will play a major role. For $t \in \Sigma^{\#}$ and $f \in \mathcal{D}$, the *terminal subtree of t rooted in f*,

$$t' = f^{-1}t \tag{5}$$

is the tree given by

$$\mathrm{Supp}(t') = f^{-1}(\mathrm{Supp}(t)) = \{g \in \mathcal{D} | fg \in \mathrm{Supp}(t)\} \tag{6}$$

and for all $g \in \text{Supp}(t')$

$$\widehat{t'}(g) = \hat{t}(fg). \tag{7}$$

Clearly $s = f^{-1}t$ is a tree since $FG(\text{Supp}(t')) = \text{Supp}(t')$.

For all $t, t' \in \Sigma^{\#}$ and $f \in \mathcal{B}(t)$, the *sum*

$$s = t + ft' \tag{8}$$

is the tree given by

$$\text{Supp}(s) = \text{Supp}(t) \cup f\text{Supp}(t') \tag{9}$$

and for all $g \in \text{Supp}(s)$

$$\hat{s}(g) = \begin{cases} \hat{t}(g) & \text{if } g \in \text{Supp}(t) \\ \widehat{t'}(f^{-1}g) & \text{if } g \in f\,\text{Supp}(t'). \end{cases} \tag{10}$$

Obvious remarks are that $\text{Supp}(t)$ and $f\text{Supp}(t')$ are disjoint sets and that their union $\text{Supp}(t) \cup f\,\text{Supp}(t')$ is closed by left factor, this implying that $t + ft'$ is indeed a tree.

If $f \notin \mathcal{B}(t)$, then one of these two conditions is not fullfilled, and the sum $t + ft'$ is undefined.

Extending this notion of sum, every tree $t \in \Sigma^{\#}$ can be written in a unique way as the sum

$$t = a + 1t_1 + 2t_2 \tag{11}$$

where $a = \hat{t}(\varepsilon)$ and $t_1 = 1^{-1}(t)$ and $t_2 = 2^{-1}(t)$.

Indeed we can identify the set of trees $\Sigma^{\#}$ with the smallest set of expressions which

- contains the expression Ω
- contains the expression $a + 1t_1 + 2t_2$ whenever t_1, t_2 belong to $\Sigma^{\#}$ and a is a letter from Σ.

The border can be defined recursively by

- $\mathcal{B}(\Omega) = \{\varepsilon\}$,
- $\mathcal{B}(t + ft') = \mathcal{B}(t) \setminus \{f\} \cup f\mathcal{B}(t')$.

The definition of a finite ascending tree automaton can be rephrased using this set of notations.

A *finite ascending tree automaton (fata)* on Σ

$$A = (Q, Q_{\text{init}}, Q_{\text{fin}}, \lambda) \tag{12}$$

is given by

- its finite set of states Q
- the subsets Q_{init} and Q_{fin} of Q
- a mapping λ of $Q \times Q \times \Sigma$ into $\mathcal{P}(Q)$, the powerset of Q.

A *successful computation* of A on a finite tree t on Σ is a finite tree S on Q, the *state tree* $S = (\text{Supp}(S), \hat{S})$ which satisfies the four following conditions:

- $\text{Supp}(S) = \text{Supp}(t) \cup \mathcal{B}(t)$
- $\forall g \in \mathcal{B}(t) : \hat{S}(g) \in Q_{\text{init}}$
- $\forall f \in \text{Supp}(t) : \hat{S}(f) \in \lambda(\hat{S}(f1), \hat{S}(f2), \hat{t}(f))$
- $\hat{S}(\varepsilon) \in Q_{\text{fin}}$.

The tree language L *recognized by* A, $L = \mathcal{L}(A)$, is the subset of $\Sigma^{\#}$ defined as the set of all finite trees t for which there exists a successful computation of A on t.

It is well known that an fata can be made deterministic:

A *deterministic fata* or *dfata* is an fata such that the transition function λ maps $Q \times Q \times \Sigma$ into Q and Q_{init} is a singleton $\{q_0\}$.

The construction of a dfata recognizing the same tree language as a given fata is exactly the same as the construction of a deterministic finite automaton (for words) recognizing the same word language as a given finite automaton.

Let $A = (Q, q_o, Q_{\text{fin}}, \lambda)$ be a dfata with $q_o \in Q$ and $\lambda : Q \times Q \times A \mapsto Q$.

We can define the mapping

$$\tilde{\lambda} : \Sigma^{\#} \mapsto Q \tag{13}$$

in the following recursive way.

$$\begin{aligned} \tilde{\lambda}(\Omega) &= q_0 \\ \tilde{\lambda}(a + 1t_1 + 2t_2) &= \lambda(\tilde{\lambda}(t_1), \tilde{\lambda}(t_2), a) \end{aligned} \tag{14}$$

It should be clear that with this definition

$$\mathcal{L}(A) = \{t|\, \tilde{\lambda}(t) \in Q_{\text{fin}}\} = \tilde{\lambda}^{-1}(Q_{\text{fin}}) \tag{15}$$

for the only successful computation of A on t is the tree S on Q such that

$$\text{Supp}(S) = \text{Supp}(t) + \mathcal{B}(t) \tag{16}$$

and for all $t \in \text{Supp}(S)$

$$\hat{S}(f) = \tilde{\lambda}(f^{-1}(t)) \tag{17}$$

The family $\text{Rec}(\Sigma^{\#})$ of all sets of finite trees which are recognized by an fata is the family of *recognizable tree languages* .

We shall constantly use the fact that a tree language $L \subseteq \Sigma^{\#}$ is recognizable iff it can be recognized by a dfata.

3. The Nerode Equivalence for Trees

Let $L \subseteq \Sigma^{\#}$ be a tree language. We associate with L a relation on $\Sigma^{\#}$ denoted $\sim_L$ and given by

$$t_1 \sim_L t_2 \text{ iff } \quad \forall t \in \Sigma^{\#}\, \forall f \in \mathcal{B}(t) :\ (t + ft_1 \in L \Leftrightarrow t + ft_2 \in L). \tag{18}$$

This is obviously an equivalence relation. It is also *root invariant* which means that

$$t_1 \sim_L t_2 \quad \Rightarrow \quad \forall t' \in \Sigma^{\#}\, \forall f' \in \mathcal{B}(t') :\ t' + f't_1 \sim_L t' + f't_2. \tag{19}$$

The root invariance is fairly obvious from the fact that

$$t + f(t' + f't_1) = (t + ft') + ff't_1. \tag{20}$$

This root invariant equivalence relation is called the *Nerode equivalence* modulo L (by analogy with the equivalence associated by Nerode to word language).

An equivalence relation on any set E is said to have *finite index* iff the set of equivalence classes $\{[t]_{\sim_L} \mid t \in \Sigma^{\#}\}$ is finite.

In order to establish the major result of this section, we shall use the following.

Lemma 1. $\forall t, t' \in \Sigma^{\#}\ \forall f \in \mathcal{B}(t) : \ \hat{\lambda}(t + ft') = \hat{\lambda}(t, f, \hat{\lambda}(t'))$.

Proof. by induction on the length of f:
$|f| = 0$, then $f = \varepsilon$ and $\varepsilon \in \mathcal{B}(t) \Rightarrow t = \Omega$, and then

$$\hat{\lambda}(\Omega + \varepsilon t') = \hat{\lambda}(t') = \hat{\lambda}(\Omega, \varepsilon, \hat{\lambda}(t')) \tag{21}$$

$|f| = n + 1$, then $f = if'$ with $i = 1$ or $i = 2$ and we assume the lemma is true for f' of length n.

Then $t = a + 1t_1 + 2t_2$ where a $= \hat{t}(\varepsilon)$ and $f' \in \mathcal{B}(t_i)$.

Assume $i = 1$ (the symmetric case $i = 2$ is treated in the same way).

$$\begin{aligned} \hat{\lambda}(a + 1t_1 + 2t_2 + 1f't') &= \hat{\lambda}(a + 1(t_1 + f't') + 2t_2) \\ &= \lambda(\hat{\lambda}(t_1 + f't'), \hat{\lambda}(t_2), a) \\ &= \hat{\lambda}(a + 1t_1 + 2t_2, 1f', \lambda(t')) \end{aligned} \tag{22}$$

□

Theorem 1. *The tree language L is recognizable iff $\sim_L$ has finite index.*

Proof. Assume card $\{[t]_{\sim_L} \mid t \in \Sigma^{\#}\}$ is finite. We build a dfata

$$A = (Q, q_0, Q_{\text{fin}}, \lambda) \tag{23}$$

which recognizes L in the following way:

- $Q = \{[t]_{\sim_L} \mid t \in \Sigma^{\#}\}$
- $q_0 = [\Omega]_{\sim_L}$
- $Q_{\text{fin}} = \{[t]_{\sim_L} \mid t \in L\}$
- $\lambda([t_1]_{\sim_L}, [t_2]_{\sim_L}, a) = [a + 1t_1 + 2t_2]_{\sim_L}$

Clearly

$$t \in L \Leftrightarrow \Omega + \varepsilon t \in L \tag{24}$$

and this implies

$$\forall t \in L : \ [t]_{\sim_L} \subseteq L \tag{25}$$

(for $t \in L$ and $t' \sim_L t$ implies $\Omega + \varepsilon t' = t' \in L$).

The definition of λ is well-founded because of the root invariance of $\sim_L$:

$t_1 \sim_L t_1'$ and $t_2 \sim t_2'$ imply

$$\begin{aligned} a + 1t_1 + 2t_2 &\sim_L (a + 1t_1) + 2t_2' \\ &\sim_L (a + 2t_2') + 1t_1 \\ &\sim_L (a + 2t_2') + 1t_1' \\ &\sim_L a + 1t_1' + 2t_2'. \end{aligned} \tag{26}$$

To end the first part of the proof of Theorem 1 it suffices to see that

$$\forall t : \tilde{\lambda}(t) = [t]_{\sim_L} \tag{27}$$

but this is immediate from the definition, and that $[t]_{\sim_L} = [t']_{\sim_L}$ for some $t' \in L$ implies $t \in L$, which has been proved.

Now consider a dfata $A = (Q, q_0, Q_{\text{fin}}, \lambda)$ which recognizes $L \subseteq \Sigma^{\#}$. We can attach to A the relation $\sim_A$ on $\Sigma^{\#}$ defined by

$$t_1 \sim_A t_2 \;\Leftrightarrow\; \hat{\lambda}(t_1) = \hat{\lambda}(t_2) \tag{28}$$

This relation is obviously an equivalence relation and it is root invariant for $t_1 \sim_A t_2$ implies $\hat{\lambda}(t + ft_1) = \hat{\lambda}(t + ft_2)$ for all $t \in \Sigma^{\#}$ and $f \in \mathcal{B}(t)$.

This last implication is a straightforward consequence of the following definition and lemma.

We denote, for all $t \in \Sigma$, $f \in \mathcal{B}(t)$, $q \in Q$, by

$$\hat{\lambda}(t, f, q) \tag{29}$$

the state defined recursively in the following way.

$$\begin{aligned} \hat{\lambda}(\Omega, \varepsilon, q) &= q \\ \hat{\lambda}(a + 1t_1 + 2t_2, 1f_1, q) &= \lambda(\hat{\lambda}(t_1, f_1, q), \hat{\lambda}(t_2), a) \\ \hat{\lambda}(a + 1t_1 + 2t_2, 2f_2, q) &= \lambda(\hat{\lambda}(t_1), \hat{\lambda}(t_2, f_2, q), a) \end{aligned} \tag{30}$$

Clearly for all $f \in \mathcal{B}(t)$:

$$\hat{\lambda}(t, f, q_0) = \hat{\lambda}(t) \tag{31}$$

Indeed $\hat{\lambda}(t, f, q)$ is the state at the root of t which results of the computation of A on t when all border nodes but f are given the initial state q_0 and f is initially given the state q.

The second part of the proof of Theorem 1 is then easily completed. We remark that

$$t_1 \sim_A t_2 \Rightarrow t_1 \sim_L t_2$$

for $\hat{\lambda}(t_1) = \hat{\lambda}(t_2)$ implies $\hat{\lambda}(t + ft_1) = \hat{\lambda}(t + ft_2)$ for all $t \in \Sigma^{\#}$ and $f \in \mathcal{B}(t)$, which implies $t + ft_1 \in L \Leftrightarrow t + ft_2 \in L$. Whence the equivalence relation $\sim_A$ is coarser than $\sim_L$, and since card $\{[t]_{\sim_A} | t \in \Sigma^{\#}\} \leq$ card (Q), the equivalence $\sim_L$ has finite index. $\square$

A next remark, which is essential in the rest of this paper, is

Lemma 2. $\forall t, s \in \Sigma^{\#}\ \forall f \in \mathcal{B}(t)\ \forall g \in \mathcal{B}(s)\ \forall q \in Q:$

$$\hat{\lambda}(t, f, \hat{\lambda}(s, g, q)) = \hat{\lambda}(t + fs, fg, q) \tag{32}$$

Proof. by induction on the length of f:
$|f| = 0$, then $f = \varepsilon$ and $\varepsilon \in \mathcal{B}(t) \Rightarrow t = \Omega$. Thus

$$\hat{\lambda}(\Omega, \varepsilon, \hat{\lambda}(s, g, q)) = \hat{\lambda}(s, g, q) = \hat{\lambda}(\Omega + \varepsilon s, \varepsilon g, q) \tag{33}$$

$|f| = n + 1$, then $f = if'$ with $i = 1$ or $i = 2$ and $|f'| = n$.
One can write $t = a + 1t_1 + 2t_2$ where $a = \hat{t}(\varepsilon)$ and $f' \in \mathcal{B}(ti)$.
Assume i = 1 (the case i = 2 is completely symetric).

$$\begin{aligned} \hat{\lambda}(a + 1t_1 + 2t_2, 1f', \hat{\lambda}(s, g, q)) &= \hat{\lambda}(\hat{\lambda}(t_1, f', \hat{\lambda}(s, g, q), \hat{\lambda}(t_2), a) \\ &= \hat{\lambda}(\hat{\lambda}(t_1 + f's, f'g, q), \hat{\lambda}(t_2), a) \\ &= \hat{\lambda}(a + 1(t_1 + f's) + 2t_2, 1f'g, q) \\ &= \hat{\lambda}(t + fs, fg, q) \end{aligned} \tag{34}$$

$\square$

An immediate corollary of Theorem 1 is

Corollary 1. *If L is saturated by a root invariant equivalence relation $\sim$ which has finite index then L is recognizable.*

Proof. L is saturated by $\sim$ iff $L = \cup\{[t]_\sim | \ t \in L\}$. The construction of the dfata associated to $\sim_L$ is valid for $\sim$ as well. $\square$

4. Transition Monoid of a Tree Automaton

Consider a dfata $A = (Q, q_0, Q_{\text{fin}}, \lambda)$. We can define for all $t \in \Sigma^{\#}$ and $f \in b(t)$ the mapping

$$T_{(t,f)} : Q \mapsto Q \tag{35}$$

by

$$T_{(t,f)}(q) = \hat{\lambda}(t, f, q). \tag{36}$$

The interesting phenomenon lies in the composition of such mappings. Let us denote

$$T_{(t,f)}T_{(s,g)} \tag{37}$$

the mapping whose value for a given q is

$$T_{(t,f)}T_{(s,g)}(q) = T_{(t,f)}(T_{(s,g)}(q)). \tag{38}$$

We have, by Lemma 2, for all $q \in Q$

$$\begin{aligned} T_{(t,f)}T_{(s,g)}(q) &= T_{(t,f)}(\hat{\lambda}(s, g, q) \\ &= \hat{\lambda}(t, f, \hat{\lambda}(s, g, q)) \\ &= \hat{\lambda}(t + fs, fg, q). \end{aligned} \tag{39}$$

Thus, if we define

$$(t, f) + (s, g) = (t + fs, fg) \tag{40}$$

we have

$$T_{(t,f)}T_{(s,g)} = T_{(t,f)+(s,g)}. \tag{41}$$

This leads to two definitions.

$$\Sigma^{(\#)} = \{(t, f)|\ t \epsilon \Sigma^{\#}, f \epsilon \mathcal{B}(t)\} \tag{42}$$

with the following composition law.

$$(t, f) + (s, g) = (t + fs, fg) \tag{43}$$

This composition law is clearly associative.

$$\begin{aligned} ((t_1, f_1) + (t_2, f_2)) + (t_3, f_3)) &= (t_1 + f_1t_2, f_1f_2) + (t_3, f_3) \\ &= (t_1 + f_1t_1 + f_1f_2t_3, f_1f_2f_3) \\ &= (t_1 + f_1(t_1 + f_2t_3), f_2f_3) \\ &= (t_1, f_1t_1) + ((t_2, f_2) + (t_3, f_3)). \end{aligned} \tag{44}$$

This composition law has also a neutral element, namely (Ω, ε).

$$\begin{aligned}(\Omega, \varepsilon) + (t, f) &= (\Omega + \varepsilon t, \varepsilon f) \\ &= (t, f) \\ &= (t + f\Omega, f\varepsilon) \\ &= (t, f) + (\Omega, \varepsilon).\end{aligned} \tag{45}$$

The *transition monoid* of the dfata A is the set $M(A)$ of all mappings $T_{(t,f)}$ with the ordinary composition law as above:

$$M(A) = \{T_{(t,f)} : Q \mapsto Q | \; (t, f) \in \Sigma^{(\#)}\}. \tag{46}$$

Since Q is finite, the cardinality of mappings of Q into Q is finite, and the transition monoid $M(A)$ is also finite:

$$\phi_A : \Sigma^{(\#)} \mapsto M(A). \tag{47}$$

Our next result is a clear extension of well known results on dfa for words.

Theorem 2. *Every recognizable set of trees $L \subseteq \Sigma^{\#}$ is recognized by a minimal automaton, whose transition monoid is the syntactic monoid of L.*

Proof. We first ought define the notions of minimal automaton and syntactic monoid in the context of tree languages.

1) A dfata A is called minimal if every dfata A' recognizing the same tree language can be mapped onto A by a morphism π.
2) A morphism π from $A = (Q, q_0, Q_{\text{fin}}, \lambda)$ on $A' = (Q', q_0', Q_{\text{fin}}', \lambda')$ is a mapping of Q onto Q' which satisfies the following conditions:

$$\begin{aligned}\pi(\lambda(q_1, q_2, a)) &= \lambda'(\pi(q_1), \pi(q_2), a) \\ \pi(q_0) &= q_0' \\ \pi(Q_{\text{fin}}) &= Q_{\text{fin}}'.\end{aligned} \tag{48}$$

The *syntactic monoid* of L, $M(L)$ can only be defined as a quotient monoid of $\Sigma^{(\#)}$ using an equivalence relation which is both root invariant and leaf invariant (i.e., left invariant and right invariant referring to the sum operation of the monoid $\Sigma^{(\#)}$), $\equiv_L$, defined on $\Sigma^{(\#)}$ by $(t_1, f_1) \equiv_L (t_2, f_2)$ iff $\forall (t, f), (t', f') \in \Sigma^{(\#)}$:

$$(t,f)+(t_1,f_1)+(t',f') \in \tilde{L} \Leftrightarrow (t,f)+(t_2,f_2)+(t',f') \in \tilde{L} \tag{49}$$

where we define

$$\tilde{L} = \{(t,f) |\; t \in L, f \in b(t)\} \tag{50}$$

The root and leaf invariance stems straightforwardly from the definition, and

$$M(L) = \Sigma^{(\#)}_{/\equiv_L} \tag{51}$$

is thus a monoid.

Already defined using the Nerode eqivalence,

$$A_{\text{min}} = (Q_{\text{min}}, q_0, Q_{\text{fin}}, \lambda) \tag{52}$$

is given by

$$\begin{aligned} &— Q_{\text{min}} = \{[t]_{\sim_L} : t \in \Sigma^{\#}\} = \Sigma^{(\#)}_{/\sim_L} \\ &— q_0 = [\Omega]_{\sim_L} \\ &— Q_{\text{fin}} = \{[t]_{\sim_L} | t \in L\} \\ &— \lambda([t_1]_{\sim_L}, [t_2]_{\sim_L}, a) = [a + 1t_1 + 2t_2]_{\sim_L}. \end{aligned} \tag{53}$$

Its transition monoid $M(A_{\text{min}})$ is the set of mappings

$$\{T_{(t,f)} : Q_{\text{min}} \mapsto Q_{\text{min}} |\; \forall q \in Q_{\text{min}} : T_{(t,f)} = \hat{\lambda}(t,f,q)\} \tag{54}$$

with the ordinary composition of functions as monoid composition law.

$$\phi : M(A_{\text{min}}) \mapsto \Sigma^{(\#)}_{/\equiv_L} \tag{55}$$

which is given by

$$\phi(T_{(t,f)}) = [(t,f)]_{\equiv_L}. \tag{56}$$

The proof that ϕ is well-defined amounts to check that

$$T_{(t_1,f_1)} = T_{(t_2,f_2)}. \Rightarrow [t_1,f_1]_{\equiv_L} = [t_2,f_2]_{\equiv_L} \tag{57}$$

And this is clear from

$$\begin{aligned} (t,f)+(t_1,f_1)+(t',f') \in \tilde{L} &\Leftrightarrow T_{(t,f)+(t_1,f_1)+(t',f')}(q_0) \in Q_{\text{fin}} \\ &\Leftrightarrow T_{(t,f)}T_{(t_1,f_1)}T_{(t',f')}(q_0) \in Q_{\text{fin}} \\ &\Leftrightarrow T_{(t,f)}T_{(t_2,f_2)}T_{(t',f')}(q_0) \in Q_{\text{fin}} \\ &\Leftrightarrow (t,f)+(t_2,f_2)+(t',f') \in \tilde{L}. \end{aligned}$$

The morphism is obviously surjective. To prove that it is injective, consider two elements of $M(A_{\min})$, say $T_{(t_1,f_1)}$, $T_{(t_2,f_2)}$, having the same image under ϕ, that is $(t_1, f_1) \equiv_L (t_2, f_2)$, and any element $[t']_{\sim_L} \in Q_{\min}$. Then

$$\begin{aligned} T_{(t_1,f_1)}([t']_{\sim_L}) &= \hat{\lambda}(t_1, f_1, [t']_{\sim_L}) \\ &= \hat{\lambda}(t_1, f_1, \hat{\lambda}(t')) \\ &= \hat{\lambda}(t_1 + f_1 t') \\ &= [t_1 + f_1 t']_{\sim_L}. \end{aligned} \tag{58}$$

But

$$(t, f) + (t_1, f_1) + (t', f') \in \tilde{L} \Leftrightarrow (t, f) + (t_2, f_2) + (t', f') \in \tilde{L} \tag{59}$$

implies

$$[t_1 + f_1 t']_{\sim_L} = [t_2 + f_2 t']_{\sim_L} \tag{60}$$

for

$$(t, f) + (t_1, f_1) + (t', f') = (t + f(t_1 + f_1 t'), f f_1 f') \tag{61}$$

and this element belongs to $\tilde{L}$ iff

$$t + f(t_1 + f_1 t') \in L \tag{62}$$

Similarly

$$(t, f) + (t_2, f_2) + (t', f') \in \tilde{L} \Leftrightarrow t + f(t_2 + f_2 t') \in L \tag{63}$$

whence

$$t_1 + f_1 t' \sim_L t_2 + f_2 t' \tag{64}$$

by the very definition of $\sim_L$. We conclude that

$$\forall t' : T_{(t_1,f_1)}([t']_{\sim_L}) = T_{(t_2,f_2)}([t']_{\sim_L}) \tag{65}$$

that is $T_{(t_1,f_1)} = T_{(t_2,f_2)}$. $\square$

5. Monoid Recognizability

We say that the tree language $L \subseteq \Sigma^{\#}$ is recognized by the finite monoid M iff there exists a morphism

$$\phi : \Sigma^{(\#)} \mapsto M \tag{66}$$

and a subset P of M such that

$$\tilde{L} = \phi^{-1}(P) \tag{67}$$

Theorem 3. *A tree language is recognizable iff it is recognized by a finite monoid.*

Proof. If $L = L(A)$ where $A = (Q, q_0, Q_{\text{fin}}, \lambda)$, we clearly have a morphism ϕ of $\Sigma^{(\#)}$ onto $M(A)$ which satisfies $\tilde{L} = \phi^{-1}(P)$ for some subset of $M(A)$ given by

$$P = \{T_{(t,f)} | \ T_{(t,f)}(q_0) \in Q_{\text{fin}}\} \tag{68}$$

Indeed, $\tilde{L} \supseteq \phi^{-1}(P)$ follows from

$$\begin{aligned} \phi(t, f) \in P &\Rightarrow T_{(t,f)}(q_0) \in Q_{\text{fin}} \\ &\Rightarrow \hat{\lambda}(t, f, q_0) = \hat{\lambda}(t) \in Q_{\text{fin}} \\ &\Rightarrow t \in L \\ &\Rightarrow (t, f) \in \tilde{L} \end{aligned} \tag{69}$$

and $\tilde{L} \subseteq \phi^{-1}(P)$ follows from

$$\begin{aligned} (t, f) \in \tilde{L} \Rightarrow t \in L = L(M) & \\ \Rightarrow \hat{\lambda}(t) = \hat{\lambda}(t, f, q_0) &= T_{(t,f)}(q_0) \in Q_{\text{fin}} \\ \Rightarrow \phi(t, f) = T_{(t,f)} &\in P \\ \Rightarrow (t, f) \in \phi^{-1}(P) & \end{aligned} \tag{70}$$

Assume conversely the existence of M, ϕ and P such that $\tilde{L} = \phi^{-1}(P)$. We define a dfata $A = (Q, q_0, Q_{\text{fin}}, \lambda)$, which recognizes L, i.e $L = \tilde{\lambda}^{-1}(Q_{\text{fin}})$, by: $Q = M$, $Q_{\text{fin}} = P$, $q_0 = 1 \in M$, and

$$\lambda(m_1, m_2, a) = \phi(a + 1t_1 + 2t_2, 1f_1) \tag{71}$$

for arbitrary $(t_1, f_1), (t_2, f_2) \in \Sigma^{(\#)}$ such that $\phi(t_1, f_1) = m_1$, $\phi(t_2, f_2) = m_2$. Then

$$\hat{\lambda}(t) \in Q_{\text{fin}} \Leftrightarrow \phi(t, f) \in P \tag{72}$$

where f is any border node of t.
$\hat{\lambda}(\Omega) = q_0 = 1 = \phi(\Omega, \varepsilon)$.

We take any $m \in M$, suppose $m = \phi(t', f')$, and any $f \in b(t)$. Then

$$\begin{aligned}
P \ni m\tilde{\lambda}(t) &= \phi(t', f')\lambda(\tilde{\lambda}(t_1), \tilde{\lambda}(t_2), a) \\
&= \phi(t', f')\phi(a + 1\bar{t_1} + 2\bar{t_2}, 1\bar{f_1}) && (73) \\
&= \phi(t' + f'a + f'1\bar{t_1} + f'2\bar{t_2}, f'1\bar{f_1}) \\
&= \phi(t' + f'a + f'2\bar{t_2}, f'1)\phi(\bar{t_1}, \bar{f_1}) \\
&= \phi(t' + f'a + f'2\bar{t_2}, f'1)\tilde{\lambda}(t_1) \ \in P \\
\Leftrightarrow \ & \phi(t' + f'a + f'2\bar{t_2}, f'1)\phi(t_1, f_1) \ \in P && (74) \\
&= \phi(t' + f'a + f'1t_1 + f'2\bar{t_2}, f'1f_1) \ \in P \\
\Leftrightarrow \ & (t' + f'a + f'1t_1 + f'2\bar{t_2}, f'1f_1) \ \in \tilde{L} \\
\Leftrightarrow \ & (t' + f'a + f'1t_1 + f'2\bar{t_2}, f'2\bar{f_2}) \ \in \tilde{L} \\
\Leftrightarrow \ & \phi(t' + f'a + f'1t_1 + f'2\bar{t_2}, f'2\bar{f_2}) \ \in P \\
&= \phi(t' + f'a + f'1t_1, f'2)\phi(\bar{t_2}, \bar{f_2}) \\
&= \phi(t' + f'a + f'1t_1, f'2)\tilde{\lambda}(t_2) \ \in P \\
\Leftrightarrow \ & \phi(t' + f'a + f'1t_1, f'2)\phi(t_2, f_2) \ \in P && (75) \\
&= \phi(t' + f'a + f'1t_1 + f'2t_2, f'2f_2) \ \in P \\
\Leftrightarrow \ & (t' + f'a + f'1t_1 + f'2t_2, f'2f_2) \ \in \tilde{L} \\
\Leftrightarrow \ & (t' + f'a + f'1t_1 + f'2t_2, f'f) \ \in \tilde{L} \\
\Leftrightarrow \ & \phi(t' + f'a + f'1t_1 + f'2t_2, f'f) \ \in P \\
&= \phi(t', f')\phi(a + 1t_1 + 2t_2, f) \\
&= m\phi(t, f) \ \in P
\end{aligned}$$

where in (73), by definition of λ, $(\bar{t_1}, \bar{f_1})$ and $(\bar{t_2}, \bar{f_2})$ are arbitrary such that $\phi(\bar{t_1}, \bar{f_1}) = \tilde{\lambda}(t_1)$ and $\phi(\bar{t_2}, \bar{f_2}) = \tilde{\lambda}(t_2)$, and (74) and (75) are valid by induction hypothesis. □

The similarity between syntactic monoids of recognizable subsets of $\Sigma^{(\#)}$ and the syntactic monoid of recognizable subsets of the free monoid Σ^* leads to several natural questions:

1. Is it possible to define a family of rational subsets of $\Sigma^{(\#)}$, or, equivalently, a class of rational tree expressions denoting subsets of $\Sigma^{(\#)}$ in such a way that every rational subset would be recognizable and conversely every rational subset would be rational?
2. Can one define right linear systems of equations with tree variables whose solution would be recognizable subsets of $\Sigma^{(\#)}$ and such that every recognizable subset of $\Sigma^{(\#)}$ would be a component of the solution of such a system? This second problem is obviously closely linked with the first one.
3. The second problem is a subproblem of the general problem of tree equations which can be written using the sum of trees with pointed border nodes (i.e., the basic operation used in the present paper), the union of two compatible trees, set union and the iterated sum playing the role of Kleene's star. Obviously the theory of tree equations has to be built from nothing or almost nothing since the operations already considered are very different. (See Aït-Kaci, 1986; Aït-Kaci, 1984; Engelfriet, 1978; Huet, 1976; Makanin, 1979.)

References

Aït-Kaci, H. (1986). "An algebraic semantics approach to the effective resolution of type equations," *Theor. Comp. Sci.* **45**, 293–351.

Aït-Kaci, H. (1984). "A lattice-theoretic approach to computation based on a calculus of partially ordered type structures," Ph.D. Thesis, University of Pennsylvania, Philadelphia, Pennsylvania.

Beauquier, D., and Nivat, M. (1985). "About rational subsets of algebras of infinite words" *Automata, Languages and Programming* (W. Brauer, ed.), *LNCS* **194**, Springer-Verlag, Berlin, 33–42.

Blanchard, P., and Hansel, G. (1986). "Systèmes codés," *Theor. Comp. Sci.* **45**, 165–176.

Doner, J. E. (1970). "Tree acceptors and some of their applications," *J. Comp. System. Sci.* **4**(2), 406–451.

Eilenberg, S. (1974). *Automata, Languages and Machines, vol. A.* Academic Press, New York.

Engelfriet, J. and Schmidt, E. M. (1978). "IO and OI". *JCSS* **15** 328–353, and *JCSS* **16**, 67–99.

Huet, G. (1976). "Résolution d'équations dans des languages d'ordre $1, 2, \ldots, \omega$," Thèse d'Etat, Université Paris VII, Paris, France.

Nivat, M. (1986a). "Eléments d'une théorie générale des codes d'arbres, partie I." Rapport LITP 86-61.

Nivat, M. (1986b). "Eléments d'une théorie générale des codes d'arbres, partie II". Rapport LITP 86-70.

Makanin, M. (1979). "The theory of equations in a free semigroup," *Math. Sb.* **103** (145), 147–236, Engl. transl. in: *Math. USSR Sb.* **32** (1977).

Thomas, W. (1984). "Logical aspects in the study of tree languages," In *Proc. Ninth Colloquium on Trees in Algebra and in Programming* (B. Courcelle, ed.), Cambridge University Press, Cambridge, England, 31–49.

11 Recursively Defined Types in Constructive Type Theory

PRAKASH PANANGADEN

Computer Science Department
Cornell University
Ithaca, New York

PAUL MENDLER

Computer Science Department
Manchester University
Manchester, England

MICHAEL I. SCHWARTZBACH

Computer Science Department
Aarhus University
Aarhus, Denmark

1. Introduction

In this paper we give an account of recursive definitions in a constructive type theory. This is the subject of a recent thesis by Mendler (1988). Here we shall concentrate on explaining these definitions from the viewpoint of category theory. In Mendler (1988) there is a detailed discussion of the proof theory and a demonstration of the consistency of type theory with the recursive definition facility. Here we will not try to be particularly formal as there now exist several books that discuss these topics in detail. Excellent introductory articles on category theory and categorical logic are contained in a

Resolution of Equations in
Algebraic Structures
Volume 1

ISBN 0-12-046370-9

recent book, "Category Theory and Computer Programming"(Pitt et al., 1986).

It is by now well-known that there is an intimate connection between logic and category theory. The work of Lawvere (1963, 1964, 1969), which initiated this study, was quickly followed up by Freyd, Tierney, Fourman, Kock, Lambek and many others. This connection is the subject of a number of recent books of which two particularly interesting ones are "Introduction to Higher Order Categorical Logic" by Lambek and Scott (1986) and "First Order Categorical Logic" by Makkai and Reyes (1977). This work resulted in the development of topos theory as a generalization of set theory. One can associate an internal logic with a topos and it turns out that, in general, this will be intuitionistic. Thus, there is a strong connection between constructive principles and categorical logic. In Lambek and Scott's book (1986), this connection between an intuitionistic type theory and topos theory is discussed in detail.

In the present paper we will examine the relationship between Martin-Löf's Intuitionistic Type Theory (henceforth, ITT) (Martin-Löf, 1975) and categorical logic. This theory forms the basis of the Nuprl system designed and developed at Cornell by R. Constable and J. Bates (1986). The type system of Nuprl is discussed in the next section. The recursive type constructors defined by Mendler (Constable and Mendler, 1985; Mendler et al., 1986; Mendler, 1988; Mendler, 1987) are expressed as additions to the Nuprl type theory. Thus, to understand our categorical presentation of recursive type definitions it is first necessary to understand the relationship between ITT and categorical logic. This connection is discussed by Seely for ITT without the hierarchy of universes (Seely, 1984).

The organization of this paper is as follows. In Section 2 we summarize some basic definitions from category theory. Section 3 reviews ITT, Section 4 discusses the connection between category theory and type theory. Section 5 describes recursive type constructors in Nuprl and Section 6 explains how these are expressed categorically. We conclude with a discussion of the construction of recursively defined domains and compare that theory with our construction. Sections 2 and 3 can be skipped if the reader knows a little bit of category theory and intuitionistic logic. We have tried to repeat the important definitions from these sections later on as they are needed so that one can do this without losing the main thread of the argument.

2. Basic Category Theory

This section contains a quick review of the definitions of category theory. It is not intended to be a first introduction to the subject but rather as an orientation for the reader who is not very familiar with category theory. There are several good textbooks on the subject we particularly recommend MacLanes's classic "Categories for the Working Mathematician" (MacLane, 1971). A gentle introduction is contained in "Topoi" by R. Goldblatt (1986).

Definition 1. A *category*, $\mathcal{C}$, consists of two collections called *objects*, $\mathcal{O}$, and *morphisms*, $\mathcal{M}$ (or *arrows*), two functions from $\mathcal{M}$ to $\mathcal{O}$ called *dom* and *cod* and a partial binary operation on $\mathcal{M}$ called *composition* and written in infix form $\circ$. Two morphisms, f, g are called *composable* if $\text{dom}(g) = \text{cod}(f)$, whenever f and g are composable then there exists a morphism, written $g \circ f$ with

$$\text{dom}(g \circ f) = \text{dom}(f) \text{ and } \text{cod}(g \circ f) = \text{cod}(g). \tag{1}$$

Composition of morphisms is associative and for every object, A, there is a unique morphism I_A that serves as a two-sided identity for composition.

If f is a morphism with $\text{dom}(f) = A$ and $\text{cod}(f) = B$ then this is written in the suggestive notation $f : A \to B$.

We have been deliberately vague about foundational issues. In particular, we have not said whether the collections $\mathcal{O}$ and $\mathcal{M}$ form sets. If $\mathcal{O}$ is a set then the category $\mathcal{C}$ is said to be *small.* If the collection of all morphisms between two fixed objects A and B (written $\text{hom}(A, B)$) is a set then $\mathcal{C}$ is said to be *locally small.* Locally small categories are quite common, unfortunately, there are common constructions on categories that produce categories that are not locally small from categories that are locally small. Rather than distract attention with a discussion of such issues, we refer the interested reader to MacLane's book or to the book by Herrlich and Straeker (1973). The reader may also be interested in the following quote from the preface to "Toposes, Theories and Triples" by Barr and Wells (1985)

> The well-known set theorist Andreas Blass gave a talk on the interaction between category theory and set theory, in which, among other things, he

offered three set-theoretic foundations for category theory. One was ...; but his first suggestion, and one that he clearly thought at least reasonable, was: *None*. This is the point of view that we shall adopt.

Examples of categories are **Set**, **Group** and **Poset**. The category **Set** has, as objects, all sets and, as morphisms, all functions between sets. The category **Group** has, as objects, all groups and, as morphisms, all group homomorphisms. Similarly, **Poset** has, as objects, all partially ordered sets and, as morphisms, all monotonic functions. Many categories are of this nature, but not all. There are categories in which the morphisms are not functions. An example is **Rel**, the category in which the objects are sets and the arrows are binary relations. This shows that two entirely different categories may have the same collection of objects. Another example is provided by considering a specific poset $\langle P, \leq \rangle$ as a category. The objects are the elements of P and a morphism is said to exist from x to y (where $x, y \in P$) iff $x \leq y$. There is at most one morphism between two objects. Posets are thus rather "poor" in categorical structure, i.e., there are few morphisms. The viewpoint that category theory brings out is that "structure preserving maps are as important as structures". The structure need not be algebraic, as in the examples above, but could be topological, as in the category **Top** where the objects are topological spaces and the morphisms are continuous functions.

In categories one can define notions that are analogues of the notions of "one-to-one" and "onto" for ordinary functions. We have to do this purely in terms of morphisms, not in terms of elements, since, in a general category, morphisms are not functions. Consider the following fact about one-to-one functions. If $f : A \to B$ is a one-to-one function, then given any two functions, p and q, from C to A, $f \circ p = f \circ q$ iff $p = q$. It is easy to verify this fact. We can make this the definition of a property of morphisms

Definition 2. A *monic* is a morphism from A to B such that, given any two morphisms, p and q, from C to A we have $f \circ p = f \circ q$ implies $p = q$. We also say that a monic is *left-cancelable.* Monics are sometimes called *monomorphisms.*

There is a similar definition for morphisms that mimic the behavior of onto functions.

Definition 3. An *epic* is a morphism from A to B such that, given any two morphisms, p and q, from B to C we have $p \circ f = q \circ f$ implies $p = q$. We also say that an epic is *right-cancelable*. Epics are sometimes called *epimorphisms*.

One can check that onto functions have this property. There are common examples of categories where one has epics that are not onto maps. Consider the category **Top** of topological spaces and continuous functions. Consider the space R of the real numbers with their usual topology and the space Q of the rationals with the subspace topology that they inherit from R. The inclusion map is an epic but is obviously not onto. If the last example made no sense it can be ignored with no loss of continuity.

Categories sometimes contain certain special objects, called initial and terminal objects.

Definition 4. An object **1** in a category $\mathcal{C}$ is said to be *terminal* if there is exactly one arrow from every object in $\mathcal{C}$ to **1**. An object **0** in a category $\mathcal{C}$ is said to be initial if there is exactly one arrow from **0** to every object in $\mathcal{C}$.

Initial and terminal (sometimes called final) objects are fairly common. In the category **Set** the initial object is $\emptyset$, the empty set and any singleton set is a terminal object. In the category **Group** the trivial group with a single element is both initial and terminal. Note that there are potentially many distinct initial or terminal objects, however, they are all "essentially equivalent". In order to make this precise we need to define the category theoretic notion of isomorphism.

Definition 5. A morphism $h : A \to B$ is said to be an *isomorphism* if there exists a unique morphism $k : B \to A$ such that $h \circ k = I_B$ and $k \circ h = I_A$. The objects A and B are said to be isomorphic.

In **Set** isomorphisms are bijections. In any category the identities are isomorphisms. In category theory, isomorphic objects are regarded as virtually identical and many important concepts are defined "upto isomorphism". It is an easy exercise to check that all terminal (or initial) objects are identical.

One of the advantages of category theory is that one can give very general definitions of standard constructions. Initially these definitions may seem a bit contrived because they are expressed entirely in terms of morphisms, however, the resulting definition is more widely applicable than the corresponding definitions given in terms of sets and membership. We illustrate this with a couple of simple but important examples.

Definition 6. Given two objects A and B in a category $\mathcal{C}$, a *product* of A and B, written $A \times B$, is an object in $\mathcal{C}$ together with morphisms $\pi_A : (A \times B) \to A$ and $\pi_B : (A \times B) \to B$ such that, given any other object C and morphisms $f : C \to A$ and $g : C \to B$ there is a unique morphism $k : C \to (A \times B)$ such that $f = \pi_A \circ k$ and $g = \pi_B \circ k$.

It is routine to check that Cartesian product of sets is a product in this sense. Once again we do not say "the" product, but one can check easily that if there are two or more products for a given pair of objects then the products are all isomorphic. In a poset regarded as a category the product of two objects is their greatest lower bound. It is important to realize that there may not be products for any two objects in a category. A category in which every pair of objects has a product is said to "have products".

One can use the definition of product of objects to define a pairing operation on morphisms. Suppose that $f : A \to C$ and $g : B \to D$ are morphisms in a category $\mathcal{C}$. Then one can define a paired morphism, written $\langle f, g \rangle$, from $A \times B$ to $C \times D$. First one has morphisms, $f \circ \pi_A : A \times B \to C$ and $g \circ \pi_B : A \times B \to D$. Now the morphism $\langle f, g \rangle$ is the unique arrow induced from $A \times B$ to $C \times D$ required by the fact that $C \times D$ is the product. The situation is illustrated in Fig. 1.

Given a particular concept in category theory one can obtain a dual concept by reversing the sense of the morphisms. Thus, for example, the concepts of initial object and terminal object are dual. One can define a concept dual to the product, it is called the *co-product.*

Definition 7. Given two objects A and B in a category $\mathcal{C}$, a *co-product* of A and B, written $A + B$, is an object in $\mathcal{C}$ together with morphisms $\iota_A : A \to A + B$ and $\iota_B : B \to A + B$ such that, given any other object C and morphisms $f : A \to C$ and $g : B \to C$ there

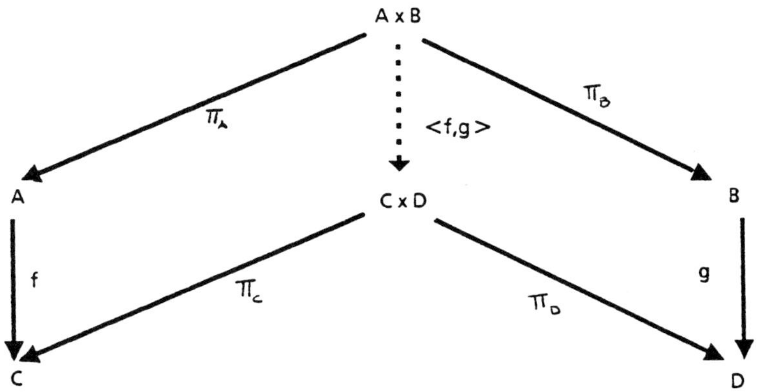

Fig. 1. Pairing morphisms.

is a unique morphism $k : (A + B) \to C$ such that $f = k \circ \iota_A$ and $g = k \circ \iota_B$.

An example of a co-product is the disjoint sum construction on sets. In a poset viewed as a category the co-product corresponds to the least upper bound. Both products and co-products are examples of type constructions as we shall see in the later sections.

Two very general constructions are *limits* and *co-limits*. Indeed the product and co-product defined above are examples of limits and co-limits respectively. We shall explain limits and leave it to the reader to formulate the dual definition of co-limit.

Definition 8. A *diagram* in a category $\mathcal{C}$ is a collection of of objects and arrows between these objects. Neither collection has to be non-empty.

Limits are defined for various diagrams.

Definition 9. Given a diagram D in a category $\mathcal{C}$, a *cone* over D is an object C together with morphisms from C to every object in the diagram, such that if $f : A \to B$ is a morphism in D and if the morphisms from C to A and B respectively are called μ_A and μ_B then $\mu_B = f \circ \mu_A$.

We shall usually use the name of the object in the cone to name the entire cone. A schematic picture of a cone for a general diagram is shown in Fig. 2. Note that we do not require the morphisms in D to commute. A cone is almost a limit except that, as with products, we need to identify a special cone which encodes all the information in any other cone. The following defines what is often called a "universal property".

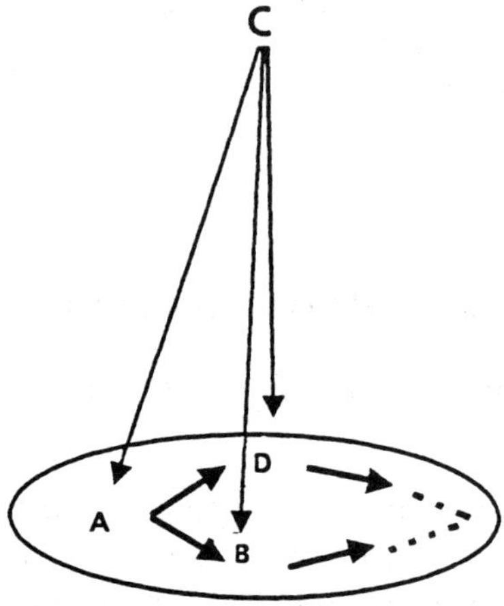

Fig. 2. A cone.

Definition 10. A *limit* for a diagram D is a cone C such that if C' is any other cone over D there is a unique morphism $h : C' \to C$ such that for all morphisms $\mu'_A : C' \to A$ in the cone C' we have

$$\mu'_A = \mu_A \circ h \tag{2}$$

The morphism h is called the mediating morphism.

Once again, a given category may or may not have limits for any particular type of diagram. A category that has all limits is said to be complete. A category that has all co-limits is said to be co-complete. The dual of the universal property is called, somewhat surprisingly, a "co-universal" property.

Another example of a limit is a *pullback*. A pullback is illustrated in Fig. 3. Given objects A, B and C and morphisms $f : A \to C$ and $g : B \to C$ as shown, a pullback is an object D together with morphisms $p : D \to A$ and $q : D \to B$ such that the diagram

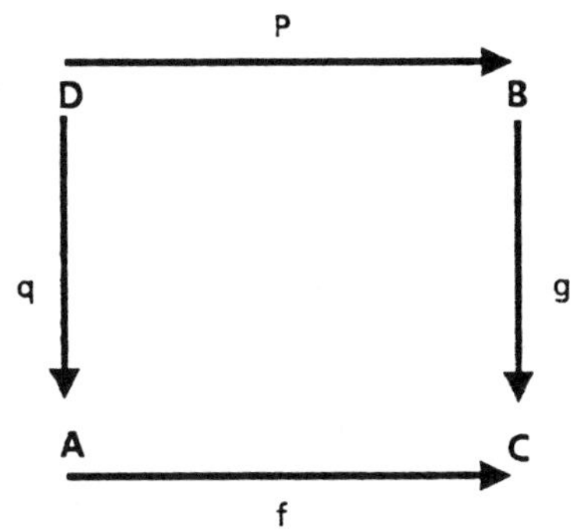

Fig. 3. A pullback diagram.

commutes. Furthermore, if there is another object E and morphisms $u : E \to A$, $v : E \to B$ such that $f \circ u = g \circ v$ then there is a *unique* morphism $k : E \to D$ such that $u = p \circ k$ and $v = q \circ k$. In the case where A and B are subsets of a set C and f and g are inclusion maps, the pullback is the intersection of A and B.

One important construction that is not a limit or a co-limit is the exponent.

Definition 11. An exponent for a pair of objects A, B in a category $\mathcal{C}$ is an object, written B^A, and a morphism $apply : B^A \times A \to B$ such that whenever there is a morphism $f : C \times A \to B$ there is a unique morphism

$$\langle \text{curry}(f), I_A \rangle : C \times A \to B^A \times A \tag{3}$$

such that $f = apply \circ \langle \text{curry}(f) I_A \rangle$.

This expresses the concept of function space. The names *apply* and curry are intended to suggest the similarity with the corresponding λ-calculus operations. Note that we needed to assume that binary products of the appropriate sort exist. The availability of exponents says that, roughly speaking, the collection of morphisms between two objects (called a "hom-set") forms an object. A category that has finite products, a terminal object and exponents is called a *Cartesian closed category*. An example of a Cartesian closed category is the category **Set**. A category that is not Cartesian closed is **Group**, this is because the group homomorphisms between two groups do not form a group.

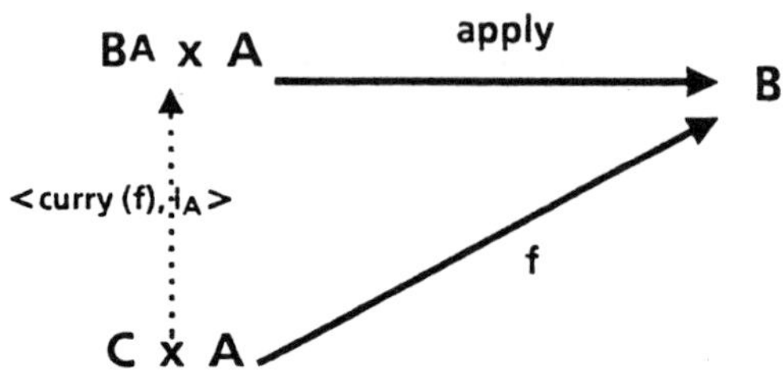

Fig. 4. The definition of Exponent.

The diagram in Fig. 4 illustrates exponentiation.

One can recover many set theoretic ideas in category theory. The resulting theory of "set-like" categories is called *topos theory* and arose from considerations in algebraic geometry. Topos theory comes with its own logic and turns out to define an intuitionistic set theory. We shall describe some of the basic definitions of topos theory.

The most fundamental notion from set theory that we have lost is the concept of an *element.* How does one express the idea of an element using morphisms. The category **Set** is a fairly good (but not completely reliable!) guide. In **Set** one has a one element set, the terminal object 1. The elements of any set A can be identified with the morphisms from 1 to A. Accordingly we define an element as follows.

Definition 12. Given an object A in a category, an *element* (more precisely, a *global element*) of A is a morphism from 1, the terminal object, to A.

These analogues are well known by topos theorists, see, for example, Lambek and Scott (1986) or Barr and Wells (1985).

We need to have some mechanism analogous to the comprehension principle in set theory that allows us to form subsets of a given set. First we define an analogue of subset.

Definition 13. A *subobject* of an object A is a monic $f : B \to A$.

Strictly speaking, we should define equivalence classes of such monics under isomorphism and refer to the equivalence classes as the subobjects. Now in order to classify subobjects of a given object we define a *subobject classifier.*

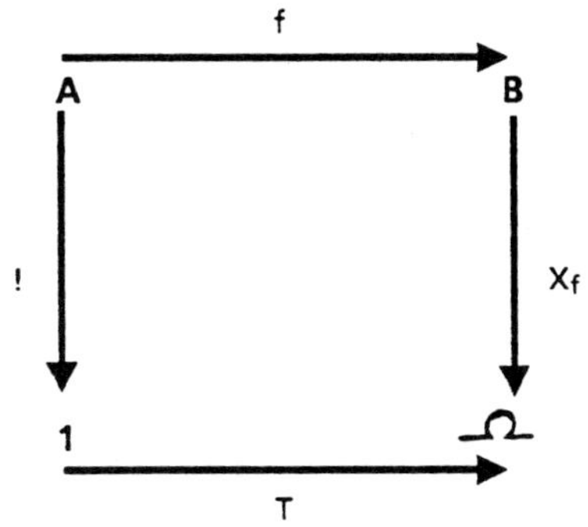

Fig. 5. The definition of subobject classifier.

Definition 14. A *subobject classifier*, Ω in a category $\mathcal{C}$, is an object such that there exists a distinguished element of Ω called T, with the property that if $f : A \to B$ is a subobject of B, then there exists a unique morphism $\chi_f : B \to \Omega$ that makes the diagram in Fig. 5 a pullback.

This rather complex definition can be understood easily once we point out what all these entities are in **Set**. The subobject classifier is any two element set, let us for definiteness say that it is the set $\{T, F\}$, and the morphism T is the element T of this set. Now if A is a subset of B and i is the inclusion map of A into B, then the morphism χ_i is the characteristic function of A. A *topos* is a Cartesian closed category with a subobject classifier.

Finally we introduce the concept of functor to relate two categories.

Definition 15. A functor F from a category $\mathcal{C}$ to a category $\mathcal{D}$ is a mapping F_o from the objects of $\mathcal{C}$ to the objects of $\mathcal{D}$ and a mapping F_m from the morphisms of $\mathcal{C}$ to the morphisms of $\mathcal{D}$ such that

1. if $f : A \to B$ is a morphism in $\mathcal{C}$ then $F_m(f)$ has domain $F_o(A)$ and co-domain $F_o(B)$,
2. $F_m(I_A) = I_{F_o(A)}$, i.e., identities are preserved,
3. $F_m(f \circ g) = F_m(f) \circ F_m(g)$, i.e., compositions are preserved.

We usually use the same symbol F to stand for both the mappings. Functors allow one to relate very different categories. They were first identified by algebraic topologists who were expressing topological properties in terms of algebraic structures like groups. A standard example of a functor is the *forgetful* functor from **Group** to **Set** that associates with every group its set of elements and with every group homomorphism, the same function now viewed as just a function. Sometimes one has functors that reverse the directions of the morphisms but still preserve compositions and identities. Such functors are called *contravariant* functors. If one wishes to emphasize that functors, as we have defined them, preserve the directions of morphisms, they are called *covariant* functors.

3. Constructive Type Theories

In this section we introduce constructive type theories and explain their semantics. We begin with a quick discussion of natural deduction due originally to Gentzen (1934). An excellent discussion of these topics and of the philosophical underpinnings of intuitionism is contained in "Elements of Intuitionism" by Dummett (1977). A good account of natural deduction, proof theory, and model theory of intuitionistic logic is contained in "Logic and Structure" by van Dalen (1980).

The basic unit of inference in natural deduction is the *sequent*. A sequent is written schematically as

$$\text{set of formulas} \vdash \text{formula} \tag{4}$$

or symbolically as

$$\Gamma \vdash \phi \tag{5}$$

where the formulas are from whatever logic is being formulated. We assume that the formulas are specified by an inductive definition and contain the connectives $\wedge$ for **and**, $\vee$ for **or**, $\neg$ for **negation**, $\Rightarrow$ for **implication** and $\perp$ for the constant **false**. There are no quantifiers for the moment. Intuitively, one understands a sequent $\Gamma \vdash \phi$ as asserting that if all the formulas in Γ are proved then the formula ϕ is also proved. The idea behind natural deduction is that one understands the logical connectives through their use in proofs.

The presentation of natural deduction is organized around the idea that rules of inference are fundamental rather than axioms. There is only one axiom scheme, namely that any sequent of the form

$\Gamma \vdash \phi$ where $\phi \in \Gamma$ is true. The rules of inference are organized into *introduction* and *elimination* rules; so called because they introduce or remove the various logical connectives.

Connective	**Introduction**	**Elimination**
$\wedge$	$\dfrac{\Gamma \vdash \phi \quad \Delta \vdash \psi}{\Gamma, \Delta \vdash \phi \wedge \psi}$	$\dfrac{\Gamma \vdash \phi \wedge \psi}{\Gamma \vdash \phi}$
$\vee$	$\dfrac{\Gamma \vdash \phi}{\Gamma \vdash \phi \vee \psi}$	$\dfrac{\Gamma \vdash \phi \vee \psi \quad \Gamma, \phi \vdash \xi \quad \Gamma, \psi \vdash \xi}{\Gamma \vdash \xi}$
$\Rightarrow$	$\dfrac{\Gamma, \phi \vdash \psi}{\Gamma \vdash \phi \Rightarrow \psi}$	$\dfrac{\Gamma \vdash \phi \Rightarrow \psi \quad \Delta \vdash \phi}{\Gamma, \Delta \vdash \psi}$
$\perp$	(no intro rule)	$\dfrac{\Gamma \vdash \perp}{\Gamma \vdash \phi}$

The elimination rule for **and** and the introduction rule for **or** each have another similar rule, not shown in the table above, which are the reversed versions of the rules shown. In addition, there is a "structural" rule called *thinning*

$$\frac{\Gamma \vdash \phi}{\Gamma, \psi \vdash \phi}. \tag{6}$$

The logical connective $\neg$ is regarded as a defined connective, thus $\neg\phi$ is viewed as shorthand for $\phi \Rightarrow \perp$. Sets of formulas are combined using ordinary union, thus Γ, Δ stands for $\Gamma \cup \Delta$ and Γ, ϕ stands for $\Gamma \cup \{\phi\}$. Note that in the above rules the formulas on the left of the $\vdash$ are never touched; in other formalisms like the sequent calculus there are rules that modify the formulas on either side of the turnstile.

The above rules define intuitionistic logic. Proofs are constructed by building a finite tree, called a *proof-tree*, whose nodes are sequents whose root node at the bottom. At each node in the proof tree we either have no nodes above it, in which case the node has to be an axiom or there are nodes immediately above it, in which case these nodes and the node below must form a substitution instance of one of the rules. A formula ϕ is said to be a *theorem* if there is a proof of the sequent $\vdash \phi$. The following proof-tree establishes $\vdash \phi \Rightarrow (\psi \Rightarrow \phi)$:

$$(\Rightarrow\text{-intro})\ \frac{(\Rightarrow\text{-intro})\ \frac{\phi,\psi \vdash \phi}{\phi \vdash \psi \Rightarrow \phi}}{\vdash \phi \Rightarrow (\psi \Rightarrow \phi)}. \tag{7}$$

Proofs in intuitionistic logic have an important property; every proof of a disjunct $\phi \vee \psi$ contains a proof of either ϕ or of ψ. Thus to prove a disjunct one must prove either one or the other of the formulas in the disjunction. This means, for example, that in order to prove $\phi \vee \neg\phi$ one must prove either ϕ or $\neg\phi$. This means that if $P(x)$ is a predicate in order to establish $\forall x.P(x) \vee \neg P(x)$ one must exhibit a decision procedure for P. In classical logic, of course, $\phi \vee \neg\phi$ is an axiom scheme.

Beginners find this the hardest aspect of intuitionistic logic to accept. The classical notion of truth values seems very natural and, after an early struggle with the truth table for implication, a student finds it hard to understand how anyone can reject the law of the excluded middle. It is important to note that what is being rejected is not just the classical truth table for $\vee$ but rather the entire classical semantic account of the meaning of the logical connectives. In intuitionistic logic the logical connectives acquire their meaning through the rules for introducing them and eliminating them from proofs. In other words, the semantics is expressed in terms of methods for constructing proofs of formulas.

This does not mean that one cannot construct models of intuitionistic logic, indeed it has a very much richer model theory than classical logic precisely because fewer theorems hold, but that the basic understanding of the connectives does not come from the model theory. One point worth mentioning is that it is impossible to model intuitionistic logic with finitely many truth values. One can argue this as follows. Suppose that there are n truth values. Let $p_1, \ldots, p_{n+1}$ be distinct propositional symbols. Construct the formula

$$P \equiv \bigvee_{0<i,j\leq n+1, i\neq j} (p_i \Leftrightarrow p_j). \tag{8}$$

The formula P cannot be proven in intuitionistic logic because one would have to prove one of the disjuncts and none of them are provable even classically. With only n truth values, however, one must assign the same truth value to one pair of distinct propositional symbols p_k, p_l and thus the disjunct $p_k \Leftrightarrow p_l$ is true and thus P is true. So no finite set of truth values characterizes intuitionistic provability.

One can understand the semantics of the connectives in terms of rules that state what counts as a proof of a formula. The rules of inference state this precisely. Here we shall give the so called "propositions-as-types" interpretation (Howard, 1980; Martin-Löf, 1982; Constable and Mendler, 1985) which embodies this semantic account of intuitionistic logic. One thinks of a formula in intuitionistic logic as defining a collection of all its proofs. Let us agree to call such a collection a "type".

What counts as a proof of a complex formula is explained by defining operations on the corresponding types. Consider the case of conjunction. Let us say that the type of all proofs of ϕ is A and the type of all proofs of ψ is B. Let a be a member of A, i.e. a proof of ϕ and let b be a member of B i.e. a proof of ψ. Then a proof of $\phi \wedge \psi$ is a pair, written $\langle a, b\rangle$ where a is a proof of A and b is a proof of B. Implicit in the notation $\langle a, b\rangle$ is the idea that the pair can be decomposed into its constituents. Thus from a proof of $\phi \wedge \psi$ one can recover proofs of ϕ and ψ. The collection of proofs of $\phi \wedge \psi$ define a new type, written $A \times B$, consisting of all pairs of the form $\langle a, b\rangle$. This new type is easily recognized as arising from the familiar type construction *binary product*.

This observation that the connectives in intuitionistic logic are akin to type constructions was made long ago by Kolmogorov (1932) and was put on a more rigorous footing by Howard in 1969[1] (1980). Before explaining type theory we will explain disjunction, implication and negation using a naive conception of type and type construction as we did for conjunction. A proof of $\phi \vee \psi$ consists of a proof of either ϕ or ψ, hence, the type of proofs of $\phi \vee \psi$ contains all proofs of ϕ as well as all proofs of ψ, furthermore the proofs are labeled with information that allows one to recognize which disjunct it is a proof of. In terms of type constructions, if the type of all proofs of ϕ is A and the type of all proofs of ψ is B, the type of all proofs of $\phi \vee \psi$ is $A + B$ where $+$ stands for disjoint union. Note that the constructive reading of disjunction is essential for this correspondence.

What should count as a proof of an implication? Once again, let the type of all proofs of ϕ be A and the type of all proofs of ψ be B. A proof of $\phi \Rightarrow \psi$ should be an effective method for producing a proof of ψ from a proof of ϕ. In other words, it should be a function

[1]A published account did not appear until 1980.

from A to B, thus the type of proofs of $\phi \Rightarrow \psi$ is the function type $A \Rightarrow B$. An absurd statement like $0 = 1$ should not have any proofs at all, accordingly, introduce the empty type **void**, to stand for the type of proofs of absurd propositions. Negation is just a notation for functions into **void**, thus a proof of $\neg\phi$ is a function from A to **void**.

At this stage one may wonder what constitutes a proof of a quantified formula. In order to explain this we need to introduce new type constructions. Rather than continue with the naive understanding of types, we shall begin a more formal explanation of Martin-Löf's Intuitionistic Type Theory (1975). The previous discussion indicates that one can embed intuitionistic logic into a constructive theory of types. We will only give an explanation of some aspects of the theory, we will not delve into issues like "higher universes" or "equality types" which are needed for a complete understanding of the theory, one should read Martin-Löf's original paper (1975) or "Foundations of Constructive Mathematics" by Beeson (1985) or "Implementing Mathematics" by R. Constable et al. (1986) for a complete discussion of the theory.

Intuitionistic Type Theory is designed to provide a constructive handle on the fundamental mathematical operation of forming collections. In this sense, it can be regarded as an alternative to classical set theory which is also organized around the idea of forming collections.

The theory ITT consists of three things; an untyped calculus of *terms*, a system of organizing terms into *types* and a collection of rules of inference for reasoning about equality of terms, membership of terms in types and equality of types. The logic is higher-order in that the types are included in the ensemble of terms and one may compute with types. The terms are essentially lambda terms together with a few special constants (with their own reduction rules) and some constructor functions. The reduction relation is defined to be lazy in the sense that a term is considered to be in normal form as soon as an outer constructor function is visible. Thus, for example, $\langle\_,\_\rangle$ is a function representing pairing; a term of the form $\langle a, b\rangle$ is deemed to be *canonical* (i.e. in normal form) regardless of whether a or b by itself is in normal form.

Following Bishop's conception of sets (Bishop, 1967), a type consists of *elements* and an *equality relation* between elements. To define a type, one gives a way of recognizing its members and a way of determining when two elements are equal. Elements are represented by *terms*, which are divided into two classes: *canonical* and *non-canonical*. Canonical terms are the irreducible elements, they name the values, whereas non-canonical terms are expressions that denote values. A term is deemed *typeable* when it can be reduced to a canonical term in some type; it is possible for a typeable term to contain untypeable subterms. The reduction relation respects the equality.

Types may be built through *type constructors*. These are presented in a very uniform scheme defining the *formation* of a new type, the *introduction* of canonical terms, the *elimination* (or *analysis*) of terms, and the *equality* between terms. We can illustrate these concepts through a simple example: the *product type* constructor. If A and B are types we posit that we can form the product type $A \times B$. If a is a term of type A and b is a term of type B we introduce $\langle a, b\rangle$ as a canonical term of type $A \times B$. If t is a term of type $A \times B$ we give the two elimination forms $\text{left}(t)$ and $\text{right}(t)$ as non-canonical terms of type A respectively type B. These may be transformed into canonical terms through the two reductions: $\text{left}(\langle a, b\rangle) = a$ and $\text{right}(\langle a, b\rangle) = b$. Finally, we say that two canonical terms $\langle a, b\rangle$ and $\langle a', b'\rangle$ are equal exactly when a and a' are equal terms of type A and b and b' are equal terms of type B.

In order to write down the rules of inference, we need to introduce the various kinds of sequents[2] that one uses in ITT. They take the following four forms:

$$\begin{gathered}\Gamma \vdash A \textbf{ type} \\ \Gamma \vdash A = B \textbf{ as types} \\ \Gamma \vdash a \textbf{ in } A \\ \Gamma \vdash a = b \textbf{ in } A \end{gathered} \tag{9}$$

In the above sequents, Γ stands for a sequence of hypotheses of the form $x : A$ where x is a variable and A is a type. Thus, one may think of Γ as a collection of type assignments to variables, or a context. The last rule shows that each type comes with its own notion of equality. One can have two distinct types with the same canonical terms in both of them but with different equality relations defined on them; this arises, for example, when one forms a new type by taking

[2]These are called *judgements* by Martin-Löf.

the quotient of an existing type with respect to some equivalence relation. In Nuprl there are only forms of sequents namely the ones involving equality above, the first form of sequent above is regarded as shorthand for

$$\Gamma \vdash A = A \ \ \textbf{as types} \tag{10}$$

while the third form of sequent above is shorthand for

$$\Gamma \vdash a = a \ \ \textbf{in } A. \tag{11}$$

The rules of inference for the product constructor are:

Formation rule: $\dfrac{\Gamma \vdash A\ \textbf{type} \quad \Delta \vdash B\ \textbf{type}}{\Gamma, \Delta \vdash A \times B\ \textbf{type}}$

Introduction rule: $\dfrac{\Gamma \vdash a\ \textbf{in}\ A \quad \Delta \vdash b\ \textbf{in}\ B}{\Gamma, \Delta \vdash \langle a, b\rangle\ \textbf{in}\ A \times B}$

Elimination rule: $\dfrac{\Gamma \vdash c\ \textbf{in}\ A \times B}{\Gamma \vdash \mathrm{left}(c)\ \textbf{in}\ A}$

with a similar rule involving right(c).

The above rule introduced new combinators, left and right. It turns out that it is convenient to have a single, more general, term for doing this. Such a term is written sprea $d(t; u, v.e)$, where t is a term, e is an expression containing the free variables u, v and the u and v bind their free occurrences in e. This term is a redex when t is a pair $\langle a, b\rangle$. The reduction is

$$\mathrm{spread}(\langle a, b\rangle; u, v.e) \rightarrow e[u \mapsto a, v \mapsto v] \tag{12}$$

where the notation $e[x \mapsto a]$ means "replace free occurrences of x in e with a with appropriate renaming of bound variables to avoid capture". One can express left as spread$(c; u, v.u)$ and right as spread$(c; u, v.v)$. This allows us to write a single elimination rule

$$\frac{\Gamma \vdash c\ \textbf{in}\ \ A \times B \quad \Gamma, u : A, v : B \vdash e\ \textbf{in}\ \ T}{\Gamma \vdash \mathrm{spread}(t; u, v.e)\ \textbf{in}\ \ T}. \tag{13}$$

There are similar rules for disjoint union. The canonical elements of a type $A+B$ are of the form $\mathrm{inl}(a)$ where a is in A or $\mathrm{inr}(b)$ where b is in B. The notation inl, inr stands for "inject left" and "inject right" respectively. Analogously to spread above one can introduce a new term for computing with members of $A + B$. This term is written $\mathrm{decide}(t; u.e; v.f)$. The computation rules for this are

$$\mathrm{decide}(\mathrm{inl}(a); u.e; v.f) \to e[u \mapsto a]$$

and

$$\mathrm{decide}(\mathrm{inr}(b); u.e; v.f) \to f[v \mapsto b]. \tag{14}$$

There are similar rules one can write down for function spaces, the notation used for writing down the members of function types is ordinary lambda notation. The Nuprl logic contains many other type constructors and the structure of sequents are much more complicated than is suggested in the discussion above. We now introduce two type constructors that the reader may not have come across before. These are the *dependent product* and the *dependent sum.* In the Nuprl logic these are called dependent function space and dependent product respectively; despite the confusion caused, it is worth being aware that both terminologies make good sense as we will explain in a later section. We will continue to say dependent product and dependent sum as these are more common but, in our opinion, the Nuprl terminology is more accurate.

Intuitively, these two type constructors allow one to take products and sums of families of types and not just of finite collections. The notation for the dependent product is $\prod x : A.B(x)$ and for the dependent sum is $\sum x : A.B(x)$. The notation $B(x)$ is intended to suggest an expression that contains a free variable x, for any a in A, $B[x \mapsto a]$ is a type. Thus, in a vulgarized set-theoretic notation, $\{B(a) \mid a \in A\}$ is an $A-$ *indexed* family of types. The dependent product, $\prod x : A.B(x)$ consists of A-indexed tuples where the element in the position indexed by a is of type $B(a)$. A more realistic way of thinking of the elements of the dependent product type is as functions from A to a type that depends on the value of the argument, hence the use of the name "dependent function space". An example of such a dependent function is as follows. Consider a language with arrays. Let us assume that each array size is a separate type. Consider a function $\mathrm{alloc}(n)$ that allocates an empty array of size n. The type of this function is from *integer* to an array type that depends on the argument to alloc.

The dependent sum, $\sum x : A.B(x)$ can be understood similarly. The members of $\sum x : A.B(x)$ are pairs of the form $\langle a, b \rangle$, where a is from type A and b is from type $B(a)$. When $B(x)$ does not actually contain x, i.e. in the case where there is no real dependence, the dependent product reduces to the ordinary function space (*not* to the ordinary product) while the dependent sum reduces to the ordinary product. These type constructors were introduced by De Bruijn (1970) in the AUTOMATH project. Both these constructors have begun to make an impact on type systems for programming languages. It was realized by several people that abstract data types correspond to dependent sums, this insight appeared in print for the first time in a paper by Mitchell and Plotkin (1985). McQueen (1986) has proposed that a type system for modules can be developed using dependent types.

From the propositions-as-types point of view, the dependent types are what are needed to handle quantified formulas. What should one understand as the constructive meaning of $\forall x : A.B(x)$? It should be a procedure that takes as argument any element a of A and constructs a proof of $B(a)$. Such a proof object is precisely an inhabitant of $\prod x : A.B(x)$. A constructive proof of $\exists x : A.B(x)$ is should be a putative witness a of type A to the existentially quantified assertion together with a proof that it satisfies $B(a)$. This is just what is in the type $\sum x : A.B(x)$.

To summarize, a constructive type theory is organized around rules for reasoning about inhabitation of types. The rules are such that from a proof that a given type is inhabited one can effectively construct an inhabitant of the type. Ordinary intuitionistic logic can be encoded using the propositions-as-types correspondence in which a formula of intuitionistic logic is identified with the type of all its constructive proofs; to coin another slogan "truth-is-inhabitation".

The other slogan often heard in connection with constructive type theory is "proofs are programs" (Bates and Constable, 1985). Implicit in constructive proofs are effective procedures. The Nuprl system is designed to exploit this connection by automatically extracting executable terms from proofs developed using the system. In order to understand this connection a bit more precisely let us return to the fragment of simple intuitionistic logic obtained by using

$\Rightarrow$ as the only connective. The proof trees that one constructs can be viewed as simply-typed lambda terms.[3] First consider all the propositional symbols in the proof as types. Each sequent of the fringe of the tree has the form $\phi_1, \ldots, \phi_n \vdash \phi_i$. We rewrite this by introducing fresh variables to stand for inhabitants of the types obtaining

$$x_1 : \phi_1, \ldots \vdash x_i : \phi_i. \tag{15}$$

Every time one encounters an introduction of an implication, perform a lambda abstraction. Thus, for example, if at some point in the proof tree one has

$$\frac{p, q \vdash p}{p \vdash q \Rightarrow p} \tag{16}$$

we rewrite this to

$$\frac{x : p, y : q \vdash x : p}{x : p \vdash \lambda y.x : q \Rightarrow p}. \tag{17}$$

another implication introduction would yield

$$\vdash \lambda x.\lambda y.x : p \Rightarrow (q \Rightarrow p). \tag{18}$$

The original proof tree that proved the formula $p \Rightarrow (q \Rightarrow p)$ has become the λ-term $\lambda x.\lambda y.x$. Implication eliminations are represented by applications of the corresponding λ-terms.

The proof tree, above, is also exactly what one would get if one were trying to assign types to the lambda term above using one of the well-known type inference formalisms (Cardelli, 1987). The fact that proof trees give rise to λ-terms is another way of realizing that proofs in intuitionistic logic are to be understood as methods for constructing evidence. One may wonder what the role of β-reduction is in this analogy. Consider a proof with an implication introduction immediately followed by implication elimination. The corresponding λ-term has an abstraction followed by an application, i.e., a β-redex. This proof tree is redundant and can be simplified. The simplification effected by the removal of the unnecessary implication introduction corresponds to carrying out the β-reduction. The strong normalization theorem for simply typed λ-calculus corresponds to the fact that proofs in intuitionistic logic can be written in a standard form

[3]For a reader familiar with polymorphic type-checking (Milner, 1978), the connection is probably already clear.

in which all eliminations appear before all the introductions. This is the basis of a decision procedure for intuitionistic logic, Statman has shown that this decision problem is P-space complete. A thorough exposition of the basic ideas in proof theory is to be found in a review article called "Ideas and results in proof theory" by Prawitz (1970).

The correspondence is much more subtle in a larger logical theory like Nuprl which was designed to be expressive enough to formalize large portions of constructive mathematics. Our interest in the rest of the paper will be in showing that constructive type theory closely corresponds to certain kinds of categories.

4. Type Theory Expressed Categorically

In this section we give an account of the connections between category theory and type theory. We will concentrate on those type constructors that correspond to propositional connectives. At the end of this section we will discuss, briefly, how to express dependent types as internal limits or co-limits.

An important goal of our program is to mimic the propositions-as-types correspondence described in the last section. In intuitionistic type theory the concept of "truth of a formula" is replaced by the concept of "inhabitation of a type." It is this that makes the theory constructive. A proof contains a prescription for constructing a term. The category theoretic reading of the basic type constructors has this constructive principle built into it in a natural way.

Our mimicry of type theory within category theory is carried out using the category-theoretic analogues for the notions of element and inhabitation or membership. We say an object is inhabited if it has an element. Since all the categories we ever talk about are in fact also toposes the above definition is always sensible.

Types are viewed as objects in a category. Simple type constructors can be thought of as limits and co-limits in the category. The terms are viewed as elements (in the categorical sense) of the types. In other words, a term is a morphism and a *closed term* is a morphism from 1. Given this view of inhabitation, the introduction and elimination rules associated with type constructors can be easily shown to correspond to the universality (co-universality) properties of certain limits (co-limits). To illustrate this we will demonstrate this connection for the product and sum type constructors. The thesis by Schwartzbach (1987) demonstrates this connection for many of the type constructors used in the Nuprl system (Constable *et al.*, 1986). In the following we use uppercase letters $A, B, \ldots$ to stand for types and also for the objects, in a suitable category, that they denote.

The product of two types A and B is written $A \times B$. We model this type constructor as the limit of the diagram consisting of just A and B, in short, by the category theoretic product, $A \times B$. Thinking in terms of inhabitation, the correspondence is quite straightforward. When is $A \times B$ inhabited? Exactly when both A and B are inhabited. To see this note that inhabitation of A and B implies that there is a morphism from 1 into both A and B. But then universality of products induces an morphism from 1 into $A \times B$. This is exactly what the introduction rule for products says. The elimination rule says that inhabitation of $A \times B$ immediately implies inhabitation of A and of B. The category theoretic definition of product includes the existence of morphisms, called projections, to both A and B. Thus, composing the morphism from 1 to $A \times B$ which establishes that the product type is indeed inhabited, with the projections produces the inhabitants of A and B. Note that inhabitation is established by actually exhibiting an inhabitant.

The sum type constructor written $A + B$ is modeled by the co-product (which is a co-limit). In order to show that this is indeed the right construct, we need to show how reasoning about morphisms from 1 using elementary category theoretic definitions leads to the type-theoretic concept of the disjoint sum. Intuitively, $A + B$ is inhabited exactly when either A is inhabited or B is inhabited. This is essentially the introduction rule for $A + B$. Recall that the co-product $A + B$ is defined as an object together with morphisms ι_A and ι_B from A and B respectively into $A + B$. Furthermore, given any object C such that A and B have morphisms f and g (respectively) into C there must exist a *unique* morphism h from $A + B$ into C such that $\iota_A \circ h = f$ and $\iota_B \circ h = g$. Suppose that we have an morphism from 1 into either A or B, then we immediately get an morphism from 1 into $A + B$ by composing with the appropriate ι. The elimination rule is a little more interesting in this case. Suppose we have an inhabitant of $A + B$. Now the usual way of presenting an elimination rule for disjunction in logic is to provide a mechanism for case-split. This is done by introducing an *elimination form*, i.e., an addition to the syntax of terms which allows one to express conditional constructs. In the Nuprl logic the syntax used is $\mathbf{decide}(x; u.y; v.z)$. This is understood as a term whose evaluation (roughly speaking) proceeds as follows. First x is evaluated, if it is a member of A then $(\lambda u.y)x$ is evaluated otherwise if it is a member of B $\lambda v.z$ is evaluated. In the context of category theory, terms are modeled by morphisms and application is composition of morphisms. Now suppose that k is an element of $A + B$, in other words, k is a morphism from 1 to $A + B$. The term

$$\mathbf{decide}(x; u.f \circ u; v.g \circ v) \tag{19}$$

is exactly the morphism h induced by co-universality. The relevant morphisms are shown in Fig. 6.

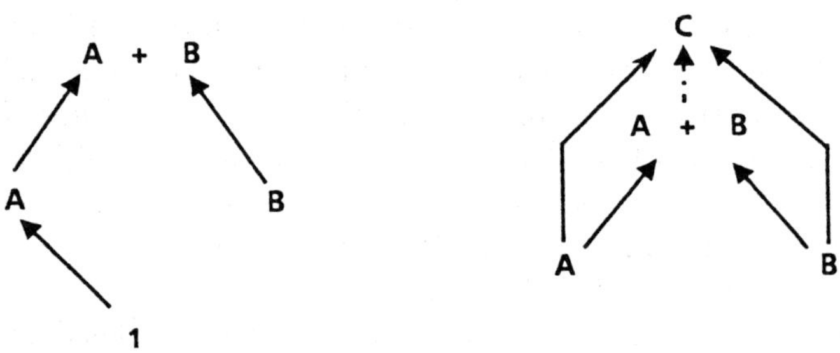

Fig. 6. The co-product as disjoint sum.

The function space type constructor, written $A \rightarrow B$, does not correspond to a limit or a co-limit. It corresponds to exponentiation in categories. The connection between exponentiation in categories and the typed λ-calculus are by now well known (see for example the recent book by Lambek and Scott (1986)) so we will not repeat that discussion in detail here. The following rule describes abstraction.

$$(\rightarrow\text{-intro}) \qquad \frac{H \vdash \lambda x.f \in A \rightarrow B}{H, x : A \vdash f(x) \in B}. \tag{20}$$

We notice that the antecedent of this rule tells us that f can be abstracted as a morphism, so we define $\lambda x.f = \text{curry}(f)$. Fig. 7 shows the defining diagram for exponentiation with the morphisms named suggestively. One again, one can develop an intuitive understanding by thinking about the situation in the category **Set**. Given a function $f : C \times A \to B$ in **Set**, we can obtain a function from A to B by fixing an element of C. A function from A to B is exactly a member of B^A. Thus we get a map, written curry(f), from C to B^A using lambda-notation we can write curry$(f)(c) = \lambda x : A.f(c, x)$. The fact that the diagram commutes just says that eval$(\lambda x : A.f(c, x))(a) = f(c, a)$; this, of course, justifies the name *eval*.

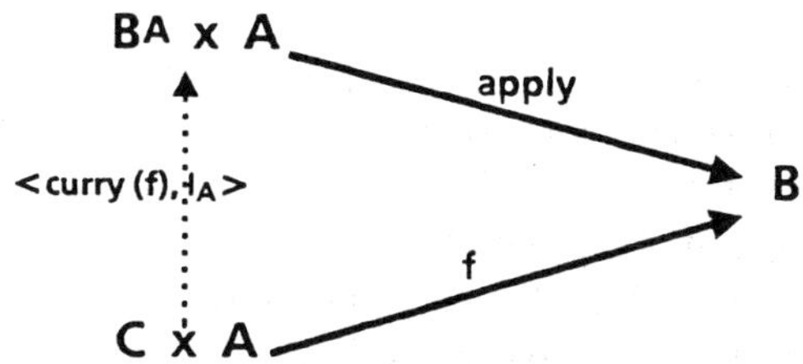

Fig. 7. The Exponent as function space.

This does not show how to model the dependent types. The problem is that we need a way of modeling indexed families in purely category theoretic terms. It turns out that the "right" way to model an A-indexed family of objects is as a morphism from B to A. To see this, consider the usual set theoretic definition of an A-indexed family of sets; $\{B(a)|a \in A\}$. From this we can define the set B as $B = \{\langle a, b\rangle \mid b \in B(a)\}$ and we can define a function $p : B \to A$ by $p(\langle a, b\rangle) = a$. One can recover the $B(a)$ as $p^{-1}(a)$. In category theory then, we can take $p : B \to A$ as the definition of an A-indexed family. Now we need to define a category of A-indexed objects. This is provided by the following definition.

Definition 16. Given a category $\mathcal{C}$ and an object A in $\mathcal{C}$, we define a category $\mathcal{C} \setminus A$ which has, as objects, morphisms from objects of $\mathcal{C}$ to A. The morphisms of $\mathcal{C} \setminus A$ are morphisms of $\mathcal{C}$ that make the triangles shown in Fig. 8 commute.

This defines what is called the "slice" category or "comma" category.

The dependent type $\Pi(A, B)$ can now be defined as an "internal limit" in this slice category. The picture depicting this is shown in Fig. 9. Intuitively, we think of a product like $\Pi(A, B)$ as having

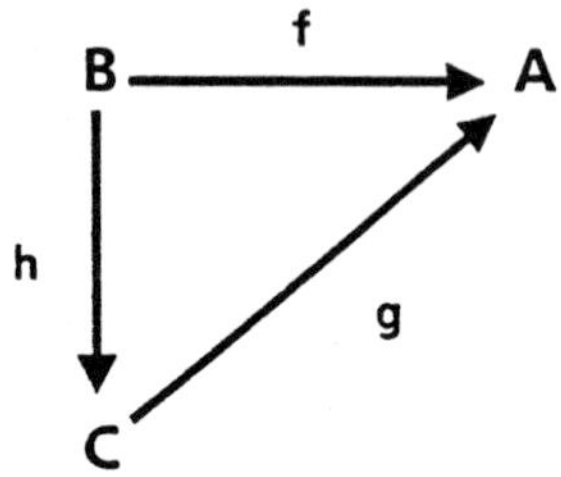

Fig. 8. Morphisms in the Slice Category.

morphisms from itself to each component; we do not have these components we have the composite B instead. Instead of projections to each component we can think of "packaging" all the projections together with tags stating which component of B they are mapped to. The morphism labeled π in Fig. 9 represents this bundle of individual projections, it now goes from $\Pi(A, B) \times A$ to B instead of from $\Pi(A, B)$ to $B(a)$. We need to be assured that π sends each element into the right component; this is guaranteed by the fact that $\pi \circ p = \mathit{right}$. Finally, we would like $\Pi(A, B)$ to be universal; in other words, if C is any other object with morphisms from $C \times A$ to A and B such that the diagram commutes then there must be a unique morphism, shown by the dotted line, from $C \times A$ to $\Pi(A, B) \times A$. The interesting point about this picture is that one can view it as the exponent in the slice category $\mathcal{C} \setminus A$ if one looks at the triangle formed by π f and the dotted arrow and compares this with the diagram in Fig. 4. This is one reason why this type constructor is called the dependent function space. Showing that this actually models the introduction and elimination rules for the Π type is a little more difficult and we will not discuss it further here. This is discussed in Schwartzbach's thesis in detail (Schwartzbach, 1987).

5. Recursive and Infinite Types in Nuprl

In this section we present simple versions of the recursive type and infinite type constructors as they have been given in extensions to the Nuprl type theory (Constable and Mendler, 1985; Mendler *et al.*, 1986). In order for the reader to compare our presentation of the rules with those used by the PRL group we will use their conventions

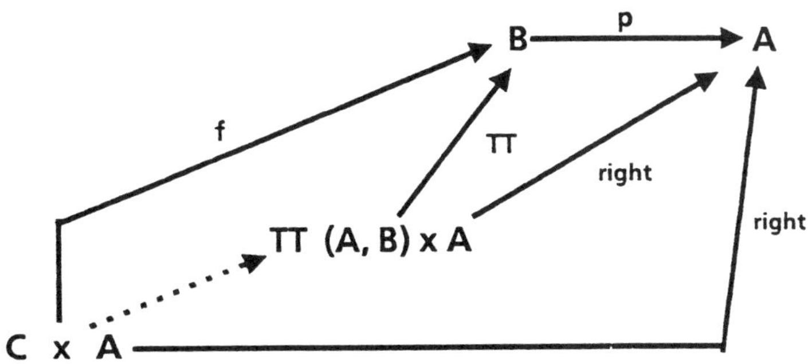

Fig. 9. The Internal Limit of an Indexed Family.

in the rest of the paper. The most important difference is that the rules are written "upside-down" with respect to our presentation so far. In other words, one writes the sequent that the rule establishes on top, and the sequents that must be proved in order for the first sequent to be proven below. This is useful for implementation purposes, one uses the rules as generators of "subgoals" and proceeds until all the subgoals are trivial. From the proof development point of view it is more natural to proceed from desired goal to subgoals.

For convenience we write the rule

$$\frac{H \vdash G}{H, H_1 \vdash G_1; \ldots ;\ H, H_n \vdash G_n} \tag{21}$$

in the form

$$\begin{array}{ll} H & \vdash G \\ & H_1 \vdash G_1 \\ & \vdots \\ & H_n \vdash G_n. \end{array} \tag{22}$$

As usual, H stands for a sequence of hypotheses.

5.1. Recursive Types

The motivation for "recursive types", as we call them, is to include general schemes of inductively defined types into the logic.

We present here the rules for a simple class of inductive definitions. These are denoted $rec(g.c)$, which can be thought of as the type inductively defined by the recursive equation $g = c$. For example we might think of the type of finite integer lists as the solution to the equation

$$\text{int-lists} = NIL + \text{int} \times \text{int-lists} \tag{23}$$

where NIL is a type containing just the empty list. In our notation we would write this as $\text{rec}(L.NIL + \text{int} \times L)$.

In the thesis by Mendler (1988) other, more complex and general, schemes are defined. These include simultaneously defined families of recursive types, and types defined simultaneously with functions from them. These types allow one to express recursively defined predicates and partial functions in the logic. We will not be presenting their rules here since all the essential ideas are contained in the simpler recursive type definitions.

The key intuition behind the recursive type definitions is that one needs a rule for "unwinding" a recursive type definition, the introduction rule will have this character, and a rule that allows one to define functions by induction, the elimination rule will permit this. The elimination form, i.e. the syntactic term one needs to express the elimination rule, will be a term that permits inductive definitions. It is called a "rec-ind" form.

To simplify the statement of the rules, we use the abbreviations

$$\begin{aligned} \Phi &= \text{rec}(g.c) \\ \omega_t &= \text{rec-ind}(t; h, z.d) \\ \Phi' &= \{v : \Phi \mid p(v)\}. \end{aligned} \tag{24}$$

The term ω_t is the form chosen for the elimination form of the recursive type. It is a redex that evaluates in one step to $d[\lambda z.\omega_z, t/h, z]$. The formation rule for this type constructor is

$$\begin{aligned} H \quad &\vdash \Phi \in U_i \\ &g : U_i \vdash c \in U_i \end{aligned} \tag{25}$$

with the syntactic restriction that g occurs positively in c. In a semantic account of these recursive definitions one can define a lattice whose elements model types and the recursive types can be defined as least fixed points of functions on this lattice. The restriction to positivity guarantees the monotonicity of $\lambda g.c$ when interpreted as a function on this lattice. This semantic account is described in Constable and Mendler (1985) and Mendler *et al.* (1986) and in more detail in the thesis (Mendler, 1988). In our account we will not elaborate this view any further since we are offering a category theoretic semantics. It will turn out that positivity will ensure that a certain type construction will define a functor.

The introduction rule for the type is

$$\begin{array}{ll} H & \vdash r \in \Phi \\ \quad H & \qquad \vdash r \in c[\Phi/g] \\ \quad H & \qquad \vdash \Phi \in U_k \end{array} \tag{26}$$

this captures the most direct notion that elements of the recursive type are the elements of the unfolding of the type.

The elimination rule asserts that the recursive type is the least, inductively closed type with respect to $\lambda g.c$:

$$\begin{array}{ll} H & \vdash G[t/v] \\ & H, p : \Phi \to U_k,\ h : (v : \Phi' \to G),\ z : c[\Phi'/g] \vdash G[z/v] \\ & H \vdash t \in \Phi \end{array} \tag{27}$$

The computation rule defines how the elim form computes:

$$\begin{array}{ll} H & \vdash \omega_t = u \in G \\ & \vdash d[\lambda z.\omega_z, t/h, z] = u \in G. \end{array} \tag{28}$$

As an example of how the rec-ind form can be used consider the type *Tree* of binary trees with integers at the leaves. This is given by the type expression

$$\mathrm{Tree} = \mathrm{rec}(T.\,\mathrm{int} + T \times T). \tag{29}$$

An expression that sums the values in the leaves is

$$\begin{gathered} \text{rec-ind}(t; \mathrm{sum}, \mathrm{tree}\,.\,\mathrm{decide}(\mathrm{tree}; \\ \mathrm{leaf}\,.\,\mathrm{leaf}; z.\,\mathrm{spread}(z; \mathrm{left}, \mathrm{right}\,.\,\mathrm{sum}(\mathrm{left}) + \mathrm{sum}(\mathrm{right})))). \end{gathered} \tag{30}$$

In functional programming notation, one would write

```
sum(tree) = if leaf(tree) then tree
                          else sum(left(tree))
                               +sum(right(tree)).
```

5.2. Infinite Types

Sometimes not the inductive, least solution but the greatest solution is sought for a recursively defined type. An example is provided by our familiar recursive type equation for integer lists,

$$\text{int-lists} = NIL + \text{int} \times \text{int-lists}\,. \tag{31}$$

In the description we have given so far the solution of this equation produces the collection of all finite lists of integers. Suppose we wished to have the collection of all finite and infinite lists of integers. This is used for modeling streams, for example. Then we want not the least (speaking set theoretically) solution of the above equation, but, rather, the greatest solution. It is possible to have interesting solutions of the above equation that are different from the least or greatest solution. An example is the collection of lists in which there are only a finite number of zeros. The type constructors that we develop cannot model this example. We can, however, introduce a new type constructor to express the greatest solution.

We provide a type constructor for this dual situation: $\text{inf}(g.c)$. Since the elements of this type may be infinite in nature, for example, a stream, we analyze them through a natural limiting process: by viewing them as elements to types that approximate the infinite type. The grossest approximation is the atomic type *Term* which is inhabited by all terms and equates them all; thus it serves us as a convenient terminal type. This approach allows us to avoid notions of partial objects and approximation orderings (Mendler et al., 1986).

The formation rule is as before:

$$\begin{aligned} H \quad & \vdash \text{inf}(g.c) \in U_i \\ & g : U_i \vdash c \in U_i \end{aligned} \tag{32}$$

Introduction shows how we accept elements in the infinite type through a limiting process:

$$\begin{aligned} H \quad & \vdash e \in \text{inf}(g.c) \\ & n : \text{int} \vdash e \in (\lambda g.c)^n(\text{Term}) \\ & \vdash \text{inf}(g.c) \in U_k. \end{aligned} \tag{33}$$

Elimination allows us to use the type by unfolding it:

$$\frac{H, z : \Phi, H' \vdash G}{H, z : c[\Phi/g], H' \vdash G}. \tag{34}$$

This sort of recursive type definition is also possible using domain theory (Smyth and Plotkin, 1982). Our type definition also uses the intuition that infinite objects, like infinite lists, should be understood via approximations of finite entities, but our approximation is expressed between the types. Thus, in our theory, we do not have partial objects.

6. Recursive and Infinite Types Expressed Categorically

In this section we shall show how to view the recursive types and infinite types as co-limits and limits, respectively, of appropriate diagrams. The basic idea is to define the meaning of a recursively defined type as the fixed point of a function from types to types. Under our interpretation of types as objects in a category, such a function is a functor from the category to itself. The fixed-point is constructed by a successive approximation technique, just as fixed points of functions are constructed.

It is important to point out that we cannot describe how to express all recursive types categorically. One example of a type that we cannot define is the W-type used by Martin-Löf to represent the Brouwer ordinals (well-founded trees).

The basic plan is to define a recursive type, written rec(F), given a functor F. Similarly, we define another type, inf(F) to represent the type of infinite objects. In order for the functors to have sensible fixed-points we need to ensure that they preserve certain limits or co-limits. The following definitions make this precise. Note how they generalize the ordinary definitions of least upper bound in a poset and of continuous function between complete partial orders.

Definition 17. An ω-chain in a category $\mathcal{C}$ is a diagram of the form

$$O_1 \xrightarrow{f_1} O_2 \xrightarrow{f_2} \dots O_n \xrightarrow{f_n} \dots \tag{35}$$

where the O_i are objects and the f_i are morphisms. The same diagram with arrows reversed is called an *ω-cochain.*

The next definition is the categorical analogue of a complete partial order.

Definition 18. A category with an initial object and co-limits for ω-chains is called an ω-complete category. Dually, a category with a terminal object and limits of ω-cochains is said to be ω*-cocomplete* category.

Finally, we have the analogue of a continuous function.

Definition 19. Given a category $\mathcal{C}$, a functor F from $\mathcal{C}$ to $\mathcal{C}$ is said to be ω*-continuous* if it preserves colimits of ω-chains. A functor is said to be ω*-cocontinuous* if it preserves limits of ω-cochains.

The nomenclature is, unfortunately, quite confusing but is now standard (Smyth and Plotkin, 1982).

The next critical point is to determine when categories have limits or co-limits of ω-diagrams and when functors preserve them. The question of defining the fixed points of functors was considered first by Adamek and Koubek (1979). The applications of these ideas to recursively defined types was carried out by Wand (1979), and by Smyth and Plotkin (1982). They considered categories of ordered sets and showed that the common functors had the requisite continuity properties. We shall discuss the situation without reference to order structure, but our constructions will be less general, for example, we will not be able to define types that are isomorphic to their own function spaces.

The types *rec*(F) and *inf*(F) are constructed by a limiting process, the continuity or cocontinuity restriction will guarantee that the limit construction will exist after iterating through all "finite" approximations. The type W requires transfinite iteration. Thus it can be expressed through the recursive definition facility described in the last section but not by the categorical construction of the present section.

We will define a little language to express functors. These are written as expressions over one variable using constants, $+$, $\times$, and $\rightarrow$. If we insist that the variable does not occur in the domain of any $\rightarrow$-type, i.e., that the expressions must be *strongly positive*, we are guaranteed to get a covariant and continuous functor. The proofs of these statements are straightforward, one explicitly checks all the conditions. To illustrate, we sketch the proof that the product functor is continuous and cocontinuous. Consider an object A of the category $\mathcal{C}$. Define the functor F from $\mathcal{C}$ to $\mathcal{C}$ by

$$F(B) = A \times B \tag{36}$$

on objects B of $\mathcal{C}$, and

$$F(f : B \to C) = \langle I_A, f \rangle \tag{37}$$

where the notation $\langle f, g \rangle$ on arrows is defined by using the universality of the product. Now consider an ω-diagram Δ, let the objects in Δ be $D_1, \ldots, D_n, \ldots$ and the morphisms be $f_i : D_i \to D_{i+1}$. Suppose that it has a co-limit, D, let the morphisms from the D_i to D be g_i. The claim is that when the functor F is applied to all these objects and morphisms, the object $F(D)$ is the co-limit of the ω-diagram consisting of the $F(D_i)$ as objects and the $F(f_i)$ as morphisms; furthermore, the cocone formed by the $F(g_i)$ is the co-universal cocone. These conditions amount to several equations between the morphisms. The fact that they hold follows from the following easy lemma.

Lemma 1. *If $p : X \to Y$, $q : Y \to Z$ and $r : X \to Z$ are morphisms in $\mathcal{C}$ and the equation $r = q \circ p$ holds, then the equation $F(r) = F(q) \circ F(p)$ also holds.*

Proof. Consider $F(q) \circ F(p)$, using the morphism part of F we get $\langle I_A, q \rangle \circ \langle I_A, p \rangle$ which is $\langle I_A \circ I_A, q \circ p \rangle$ or $\langle I_A, q \circ p \rangle$. This latter morphism is exactly $F(q \circ p)$.

The functor $x \mapsto (x \to A)$ is contravariant and neither continuous nor cocontinuous. All the strongly positive expressions, however, are continuous. The $\to$-type constructor is, however, not cocontinuous in its second argument in general categories. Thus we arrive at similar restrictions as Constable *et al.* (1986), but with a completely different motivation. In Mendler (1988) the requirements have been weakened to just *positive* expressions of which $x \mapsto ((x \to A) \to B)$ is an example. These expressions are all covariant but demonstrably *not* continuous or cocontinuous; only the property of *monotonicity* is used by Mendler (1988).

Let $\mathcal{U}$ be the category of types. Let $F : \mathcal{U} \to \mathcal{U}$ be a covariant functor. Then *cochain*(F) is the diagram obtained by iterating F

on the unique induced morphism $couniv(F(0), \emptyset, \emptyset) : 0 \rightarrow F(0)$, and $chain(F)$ is the diagram obtained by iterating F on the unique induced morphism $univ(F(1), \emptyset, \emptyset) : F(1) \rightarrow 1$.

- rec(F) is the colimit of cochain(F).
- inf(F) is the limit of chain(F).

These constructions are illustrated in Fig. 10 and 11.

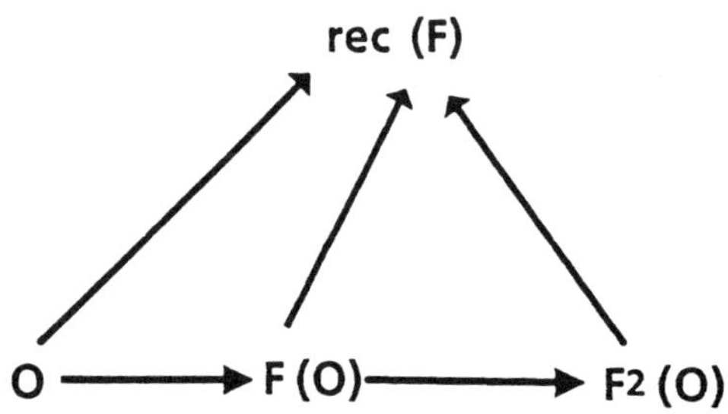

Fig. 10. Cochain and Colimit.

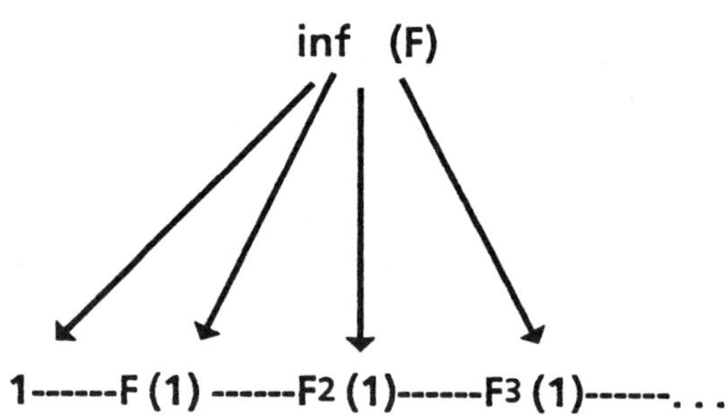

Fig. 11. Chain and Limit.

It is now clear why we must insist on covariant functors; a contravariant functor would alternate the direction of morphisms in (co)chains. When the functor is cocontinuous it follows that the colimit of F(cochain(F)) is F(rec(F)); but cochain(F) and F(cochain(F)) is essentially the same diagram, so co-universality tells us that $\text{rec}(F) \cong F(\text{rec}(F))$. Similarly, continuity yields $\text{inf}(F) \cong F(\text{inf}(F))$. However, (co)continuity is not necessary to derive sensible induction rules, so it is conceivable that we could introduce a notion of monotonicity and define inductive types as in Mendler (1988). We can think of the recursive type as the *smallest* solution to $x \cong F(x)$ and correspondingly of the infinite type as the *largest* solution (the ordering being intuitive rather than formal). For example, if $F(x) = 1 + A \times x$ then $\text{rec}(F)$ defines *finite lists* and $\text{inf}(F)$ defines *streams.*

We get some very pleasing rules.

$$\text{(rec-intro)} \qquad \frac{H \vdash \text{inc}(n, x) \in \text{rec}(F)}{H, n : \text{Int} \vdash x \in F^n(0)}. \tag{38}$$

Here $F^n(0)$ abbreviates $\text{iter}(F, 0)(n)$. The rule states that we can obtain an element of the recursive type by including an element from any unfolding. We can merely define $\text{inc}(n, x) = C(\text{cochain}(F), F^n(0)) \cdot x$. This introduction rules is illustrated in Fig. 12.

$$\text{(rec-elim)} \qquad \frac{H \vdash \text{def}(f) \in \text{rec}(F) \to B}{H \vdash f \in F(B) \to B}. \tag{39}$$

This rule tells us how to define recursive functions on recursive types. Soundness is a little complicated to demonstrate. We clearly have the unique morphism from 0 to B; applying F yields a morphism from $F(0)$ to $F(B)$; composing with f we get a morphism from $F(0)$ to B. By iterating this we get a morphism from each $F^n(0)$ to B. But then we have a cocone p over $\text{cochain}(F)$ and we can define $\text{def}(f) = \text{couniv}(B, p, \text{cochain}(F))$. This elimination rule is illustrated in Fig. 13.

To see how this rule works we shall define a length function on lists, i.e. a morphism $\text{length} : \text{rec}(F) \to \text{Int}$, where $F(x) = 1 + A \times x$. Our rule tells us to define a morphism $1 + A \times \text{Int} \to \text{Int}$; let $0 : 1 \to Int$ be the zero morphism and let $S : \text{Int} \to \text{Int}$ be the successor morphism. Then we can define $\text{length} = \text{def}(\text{decide}(0, S \cdot \text{right}))$.

The rules for infinite types are dual in nature.

$$\text{(inf-intro)} \qquad \frac{H \vdash \text{gen}(f) \in B \to \text{inf}(F)}{H \vdash f \in B \to F(B)} \tag{40}$$

We see that elements on an infinite type must be generated inductively. As before we start with the unique morphism from B to 1;

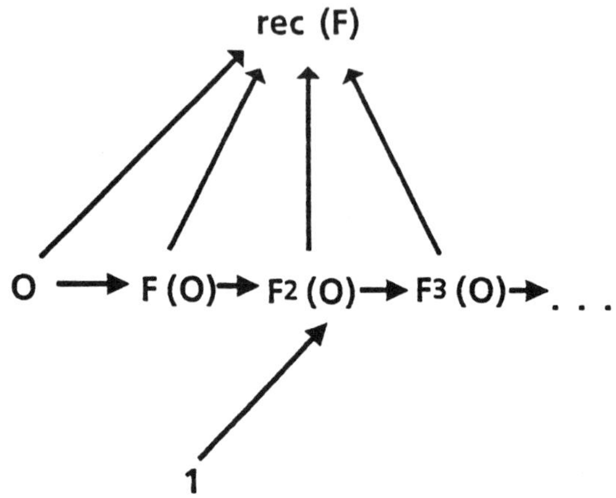

Fig. 12. Introduction Rule for Recursive Types.

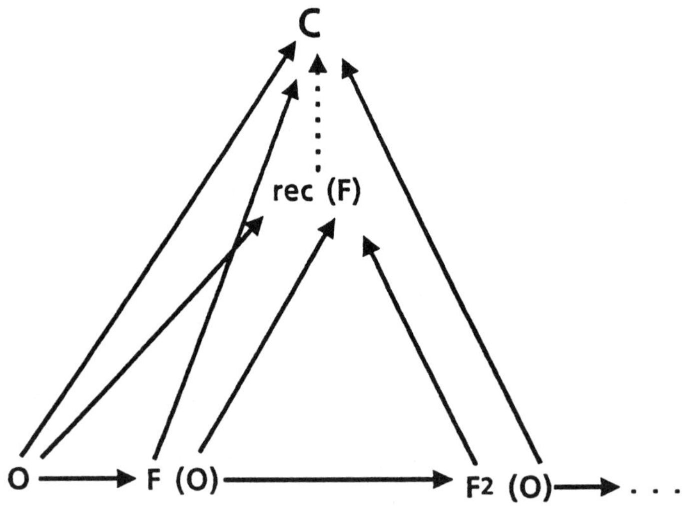

Fig. 13. Elimination Rule for Recursive Types.

applying F we get a morphism from $F(B)$ to $F(1)$; composing with f we get a morphism from B to $F(1)$. By iterating we get a morphism from B to each $F^n(1)$. But then we have a cone p over chain(F) and we can define gen$(f) = $ univ$(B, p, $ chain$(F))$. This introduction rule is illustrated in Fig. 14.

$$\text{(inf-elim)} \qquad \frac{H, n : \text{Int} \vdash \text{app}(n, x) \in F^n(1)}{H \vdash x \in \inf(F)} \tag{41}$$

The rule states that from an element of the infinite type we can get an element of any approximant. We simply define $\text{app}(n, x) = L(\text{chain}(F), F^n(1)) \cdot x$. This elimination rule is illustrated in Fig. 15.

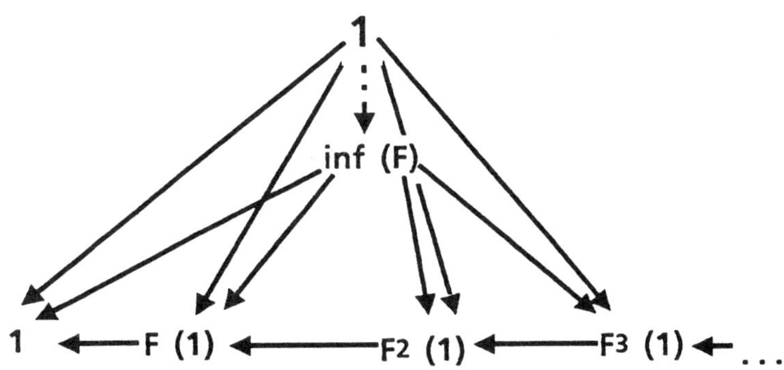

Fig. 14. Introduction Rule for Infinite Types.

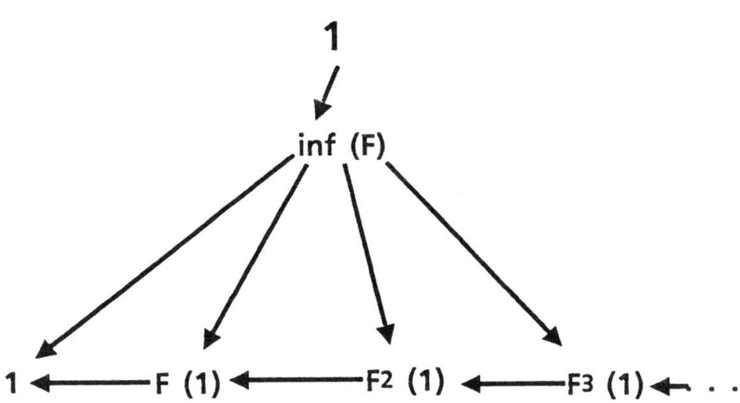

Fig. 15. Elimination Rule for Infinite Types.

It is tempting to speculate on what happens if we form the limit of a cochain or the colimit of a chain. This turns out not to be very interesting; regardless of what F is, we find that the colimit of chain(F) is 1 and the limit of cochain(F) is 0.

It is interesting to compare the recursive and infinite type constructors defined here, with the category theoretic solution of recursive domain equations. This is discussed in detail in the classic paper of Smyth and Plotkin (1982). In domain theory it is possible to solve equations of the form $D = D \to D$ as is needed, for example, to obtain a model of the untyped lambda calculus. This is rather surprising since the function space constructor cannot ordinarily be viewed as a functor.

This is because it is covariant in its second argument and contravariant in its first argument. To see this, consider the morphism part of the functor that takes X, an object of $\mathcal{C}$, to $\hom(A, X)$, the set of morphisms from A to X. Let us call this functor H. What should H do to $f : X \to Y$? It should produce a function from $\hom(A, X)$ to $\hom(A, Y)$. The obvious such function is composition with f, more precisely, the morphism part of H should be $H(f) = \lambda u : \hom(A, X).f \circ u$. Thus we have a covariant functor. Now consider the functor that takes an object X to the set of morphisms $\hom(X, A)$, where A is a fixed object of the category $\mathcal{C}$. The morphism part of this functor has to take a morphism $f : X \to Y$ into a function between hom-sets. This time the natural composition operation leads to a contravariant functor. To see this, consider an element of $\hom(X, A)$, there is no way to compose it with $f : X \to Y$; on the other hand, an element of $\hom(Y, A)$ can be composed with $f : X \to Y$ to give an element of $\hom(X, A)$.

Thus we need to set up a situation where we have something that can be both contravariant and covariant at the same time. Furthermore, the limit that gets constructed has to be both a co-limit and a limit at the same time. Such a phenomenon is called *limit-colimit* coincidence. It was first used by Scott in the construction of the D_∞ model of the untyped λ-calculus. He used it in the category of complete lattices. Subsequently, Smyth and Plotkin (1982) showed how one could do this in a more general categorical setting.

First consider the category in which one solves the equation $D = D \to D$, normally one supposes this to be s Dom, the category that has, say, Scott domains as objects and continuous functions as the as the morphisms. This gives a Cartesian closed category as one expects. The functor $L(X) = X \rightarrow X$ is not really a endofunctor on this category for the reason discussed above. Furthermore, there is no initial object in this category with which one could begin the construction.

There is, however, a closely related category on which L is a *continuous* endofunctor and this category does have an initial object. This category is obtained in the following way. Take the same objects as in $SDom$ but a morphism from A to B is a pair of continuous maps $\langle e, p\rangle$ with $e : A \rightarrow B$ and $p : B \rightarrow A$ and satisfying the conditions $e \circ p \sqsubseteq I_B$ and $p \circ e \sqsubseteq I_A$. Such a pair is called an *embedding-projection* pair. One member of an embedding-projection pair determines the other.

Now we have a category, called $SDom^E$, in which there is an initial object, namely the domain containing just $\perp$. This category has exactly the structure that one needs, the arrows are really pairs of arrows and thus one can view an arrow as "going in either direction" at will. Now it is possible to show that L is a functor on $SDom^E$. In fact, the functor L is a *continuous* functor on $SDom^E$; this is the main contribution of the paper by Smyth and Plotkin (1982).

One can now solve the recursive domain equation in $SDom^E$ and one gets an object of $SDom$ as the limit. The category $SDom^E$ is not Cartesian closed so by itself is not a suitable candidate for the category of domain needed for denotational semantics. The limit that is constructed in $SDom^E$ is also a co-limit.

7. Conclusions

In this paper we have drawn a connection between category theory and type theory and have shown how one can explain recursive type definitions in this framework. The existing domain theoretic constructions Smyth and Plotkin (1982) for recursive type definitions also define recursive types as limits (co-limits) of chain diagrams in suitable categories. Our treatment is rather different in details and philosophy. In particular we are interested in preserving the relation between type theory and logic via the so called propositions-as-types principle. Thus we would like to have, for example, an empty type. In domain theory empty types are not possible since all domains are

The connection between category theory and type theory is not new. Our presentation is intended to present this in a pedestrian manner. Seely (1984) has shown how one can mimic the mechanics of the type constructors in *locally Cartesian closed categories.* There, collections forming diagrams are taken as given and the definition of *internal* limit and co-limit is used to define the type constructors. It turns out that the internal limit is in fact just the exponent in a suitable slice category.

We have found the connection between category theory and type theory useful in the development of the rules for the recursive type constructors but it must be admitted that the connection does not explain Martin-Löf's ITT or Nuprl in detail, rather it shows how the rules for type constructors arise in natural way by viewing them as universal constructions. We have not satisfactorily explained the hierarchy of universes nor have we modeled equality types. There has been a lot of recent work on relating category theory to impredicative type theories.

Acknowledgements

We have benefited from discussions with Robert Constable, Thierry Coquand, Carl Gunter, Edmund Robinson, Pino Rosolini, Philip Scott, and Robert Seely. We gratefully acknowledge funding support from the National Science Foundation under grant DCR-86-02072.

References

Adamek, J. and Koubek, V. (1979). "Least fixed-point of a functor," *JCSS* **19**, 163–178.

Barr, Michael and Wells, Charles (1985). *Toposes, Triples and Theories.* Springer-Verlag, New York.

Bates, J. and Constable, R. L. (1985). "Proofs as programs," *Transactions on Programming Languages and Systems* **7**(1), 113–136.

Beeson, Michael J. (1985). *Foundations of Constructive Mathematics.* Springer-Verlag, New York.

Bishop, E. (1967). *Foundations of Constructive Analysis.* McGraw-Hill, New York.

Cardelli, Luca (1987). "Basic polymorphic typechecking," *Science of Computer Programming* **8**, 147–172.

Constable, R. L., and Mendler, N. P. (1985). "Recursive definitions in type theory," In *Proceedings of Logics of Programs '85*, Springer-Verlag, New York.

Constable, R. L., Allen, S. F., Bromley, H. M., Cleaveland, W. R., Cremer, J. F., Harper, R. W., Howe, D. J., Knoblock, T. B., Mendler, N. P., Panangaden, P., Sasaki, J. T., and Smith, S. F. (1986). *Implementing Mathematics with the Nuprl Proof Development System.* Prentice-Hall, Englewood Cliffs, New Jersey.

DeBruijn, N. G. (1970). "A survey of the project automath," In *Symposium on Automatic Demonstration*, Springer-Verlag, New York, 589–606.

Dummett, M. (1977). *Elements of Intuitionism. Oxford Logic Series*, Clarendon Press, Oxford.

Gentzen, Gerhard (1934). "Untersuchungen über das logische Schliessen," *Math. Zeitschrift* **39**, 176–210.

Goldblatt, R. (1986). *Topoi: the categorical Analysis of Logic.* Volume 98 of *Studies in Logic and the Foundations of Mathematics*, North-Holland, Amsterdam.

Gunter, Carl (1985). "Comparing categories of domains," In *Proceedings of the Symposium on Mathematical Foundations of Programming Language Semantics; Lecture Notes in Computer Science* **239**, Springer-Verlag, New York, 101–121.

Herrlich, Horst, and Straeker, George E. (1973). *Category Theory.* Allyn and Bacon, Boston, Massachusetts.

Howard, W. "The formulas-as-types notion of construction," *To H. B. Curry: Essays in Combinatory Logic, Lambda-Calculus and Formalism* (J. P. Seldin and J. R. Hindley, eds.), Academic Press, New York, 479–490.

Kolmogorov, A. N. (1932). "Zur Deutung der intuitionistischen Logik," *Mathematische Zeitschrift*, **35**, 58–65.

Lambek, J., and Scott, P. J. (1986). *Introduction to Higher Order Categorical Logic.* Volume 7 of *Cambridge Studies in Advanced Mathematics*, Cambridge University Press, Cambridge.

Lawvere, F. W. (1963). "Functorial semantics of algebraic theories," *Proc. Nat. Acad. Sci. U.S.A.*, **50**, 869–872.

Lawvere, F. W. (1964). "An elementary theory of the category of sets," *Proc. Nat. Acad. Sci. U.S.A.*, **51**, 1506–1510.

Lawvere, F. W. (1969). "Adjointness in foundations," *Dialectica*, **23**, 281–296

MacLane, Saunders (1971). *Categories for the Working Mathematician.* Volume 5 of *Graduate texts in Mathematics*, Springer-Verlag, New York.

McQueen, David (1986). "Using dependent types to express modular structure," In *Proceedings of the 13th Annual ACM Symposium on Principles of Programming Languages*, 277–286.

Makkai, Michael, and Reyes, Gonzalo (1977). *First Order categorical Logic, Lecture Notes in Mathematics* **611**. Springer-Verlag, New York.

Martin-Löf, Per (1975). "An intuitionistic theory of types: predicative part," In *Proceedings of Logic Colloquium '73*, North-Holland, Amsterdam.

Martin-Löf, Per (1982). "Constructive mathematics and computer programming," In *Sixth International Congress for Logic, Methodology and Philosophy of Science*, North-Holland, Amsterdam, 153–175.

Mendler, N. P. (1987). "Recursive types and type constraints in second-order lambda calculus," In *Proceedings of the Second Symposium on Logic in Computer Science*, 30–36.

Mendler, N. P. (1988). *Inductive Definitions in Type Theory.* PhD Thesis, Cornell, Ithaca, New York.

Mendler, N. P., Panangaden, P., and Constable, R. L. (1986). "Infinite objects in type theory," In *Proceedings of Symposium on Logic in Computer Science '86*, IEEE Computing Society Press.

Milner, Robin (1978). "A theory of type polymorphism in programming," *J. Comput. System Sci.*, **17**, 348–375.

Mitchell, John C., and Plotkin, Gordon D. (1985). "Abstract types have existential types," In *Proceedings of the 12th Annual ACM Symposium on Principles of Programming Languages*, 37–51.

Pitt, David, Abramsky, Samson, Poigne, Axel, and Rydeheard, David, editors. (1986). *Category Theory and Computer Programming.* Volume 240 of *Lecture Notes in Computer Science*, Springer-Verlag, New York.

Prawitz, Dag (1970). "Ideas and results in proof theory," In *Proceedings of the Second Scandanavian Logic Symposium* (J. E. Fenstad, ed.), 235–308, North-Holland, Amsterdam.

Schwartzbach, Michael I. (1987). *A Category Theoretic Analysis of Predicative Type Theory.* PhD Thesis, Cornell University, Ithaca, New York. Available as CS Technical Report TR-CS-86-793.

Seely, R. A. G. (1984). "Locally cartesian closed categories and type theory," *Mathematical Proceedings of Cambridge Philosophical Society*, **95**.

Smyth , M. B. and Plotkin, G. D. (1982). "The category theoretic solution of recursive domain equations," *Siam Journal of Computing*, **11**(4).

Statman, R. (1979). "Intuitionistic propositional logic is PSPACE-complete," *Theoretical Computer Science*, **9**, 73–81.

Taylor, Paul (1987). *Indexed Category Theory, Recursive Domains and Polymorphism.* PhD Thesis, University of Cambridge, Cambridge.

vanDalen, Dirk (1980). *Logic and Structure*, Second Edition. Springer-Verlag, Berlin.

Wand, M. (1979). "Fixed-point constructions in order-enriched categories," *Theoretical Computer Science*, **8**, 13–30.

12 Rule Transformation Methods in the Implementation of Logic Based Languages

DOMENICO SACCA[1]
University of Calabria
Rende, Italy

CARLO ZANIOLO
MCC
Austin, Texas

1. Introduction

Several novel techniques have been developed to support the efficient implementation of Logic based languages oriented towards data intensive applications, such as LDL (Tsur and Zaniolo, 1986; Beeri et al., 1987) and NAIL! (Morris et al., 1987) and several others (Zaniolo, 1987). A crucial problem in the implementation of these languages is the efficient support for recursion. Among the most significant techniques developed for the solution of this problem, we find the Differential (Seminaive) Fixpoint Method (Balbin and Ramamohanarao, 1987; Saccà and Zaniolo, 1988), the Magic Set Method (Bancilhon et al., 1986), the Minimagic Method (Saccà and Zaniolo, 1987), the Counting Method (Saccà and Zaniolo, To appear) and the Magic Counting Method (Saccà and Zaniolo, 1986; 1987).

[1]Part of this work was done while the author was visiting at MCC. This author was also supported by a grant of C.N.R. (project "Sistemi evoluti per Basi di Dati").

ISBN 0-12-046370-9

These techniques were used in the implementation of the LDL system that was designed to support and manage efficiently large data and rule sets (Chimenti et al., 1987), and construct all-answer solutions to queries more efficiently than current Prolog implementations. Thus, LDL is a powerful database languages, which removes the "impedance mismatch" between the query language and the host programming language, currently besetting the development of database applications.

Take for instance the recursive predicate SG of Fig. 1,

$$r_o{:}SG(x,y) : -P(x,x1), SG(x_1,y_1), P(y,y_1)$$
$$r_1{:}SG(x,x) : -H(x).$$

Fig. 1. The same-generation example.

where $P(x,x_1)$ is a previous parent predicate (either defined as a *parent* database relation or as the union of two database relations *mother* and *father*) and $H(x)$ is a database predicate denoting all humans. Non recursive rules over a fact base can be supported efficiently in a bottom up fashion by an extended relational algebra (Zaniolo, 1985) implementing the *immediate consequence* operator T_P (Lloyd, 1984). In conformity with the bottom up, matching-based execution strategy and the operator-based approach we want to implement recursive queries, by a least Fixpoint operator (Van Emden and Konolski, 1976).

For instance, for a program P consisting of the SG rules and the query,

$$G1\text{: } ?SG(x,y) \tag{1}$$

one will start by setting the initial value of a variable relation SG to empty and computing a new value, say

$$\mathbf{SG}' = \mathbf{SG} \cup T_P(\mathbf{SG}) \tag{2}$$

where T_P denotes, e.g., the relational algebra equivalent to the SG rules (Ullman, 1982):

$$T_P(\mathbf{SG}) = \pi_{1,1}\mathbf{H} \cup \pi_{1,5}((\mathbf{P} \bowtie_{2=1} \mathbf{SG}) \bowtie_{4=2} \mathbf{P})). \tag{3}$$

Thus, after replacing $\mathbf{SG}$ by $\mathbf{SG}'$, the computation iterates and terminates when $\mathbf{SG}' = \mathbf{SG}$. Since our rules contain only positive goals, the function T_P is monotonic and continuous in the lattice of relations defined by set containment, and this procedure computes a unique least solution of the Fixpoint equation $\mathbf{SG} = T_P(\mathbf{SG})$ (Tarski, 1955).

This naive Fixpoint approach suffers from two problems. One is the fact that some computation is repeated unnecessarily— for instance the naive Fixpoint algorithm will recompute the first level ancestors of a pair (a, b) at the first step of the Fixpoint iteration and then again at each step that follows. Differential techniques (Paige and Koenig, 1982) used to solve this problem, yielding the differential (alias seminaive) Fixpoint method discussed in the next section. The second problem occurs when a recursive predicate is called with some arguments bound. For instance, for the query,

$$G2\text{: } ?SG(adam, y) \tag{4}$$

the Fixpoint approach will still compute all possible pairs of humans that are of the same generation, only to later discard those that do not have "adam" as their first argument. Clearly, we need a strategy for taking advantage of the constants present in the query. Unfortunately, the simple approach of specializing the rules by substituting the constants in place of the variables does not work for recursive predicates. For instance, if we replace the occurrence of the variable x by "adam" we obtain rules that do not produce humans of the same generation as adam. (Of course, there are also cases in which the substitution trick works (Aho and Ullman, 1979); but recognizing those cases is, in general, undecidable (Beeri et al., 1987).) Thus, we need novel techniques for taking advantage of constants in recursive queries. Such a need is underscored by the safety issues discussed next.

To illustrate a first aspect of the safety issue let us assume that the only goal (i.e., $H(x)$) is removed from rule r_1 in Fig. 1. Then, assuming an infinite underlying universe, the query $G1$ becomes unsafe, since any pair (a, a), with a an arbitrary constant would satisfy it. The query $G2$ is safe, per se, since only a finite number of people can be of the same generation as adam; but its evaluation using the Fixpoint approach is still unsafe and, therefore, unfeasible. To solve this problem we need an execution plan capable of taking advantage of instantiated arguments in the recursive goal.

The safety issue is even more of a concern for "computational" predicates such as that of Fig. 2, which are normally intended for use only with certain input bindings. The potential sources of unsafe behavior in the rules of Fig. 2 are two. One is that the non-recursive rules r_2 and r_3 define MG to be satisfied by infinitely many pairs of x values—this is the same problem as in the previous example. However, even if the x-values range over a finite set, the recursive rules r_0 and r_1 generate longer and longer lists at each step in the Fixpoint computation, which, therefore, never ends (unsafe computation).

$$r_0\colon MG(x \bullet y, x_1 \bullet y_1, x \bullet w) : -MG(y, x_1 \bullet y_1, w), x \geq x_1$$

$$r_1\colon MG(x \bullet y, x_1 \bullet y_1, x_1 \bullet w) : -MG(x \bullet y, y_1, w), x < x_1$$

$$r_2\colon MG(\mathbf{nil}, x, x)$$

$$r_3\colon MG(x, \mathbf{nil}, x)$$

Fig. 2. Merging two sorted lists.

In conclusion, an effective usage of the instantiated arguments in a recursive goal is vital for performance reasons and to avoid the non-termination pitfall. Therefore, it should be of no surprise that a number of interesting approaches were proposed to deal with this problem, including (Bancilhon et al., 1986; Beeri and Ramakrishnan, 1987; Gardarin and De Maindreville, 1986; Chandra and Harel, 1982; Henschen and Naqvi, 1984; Kifer and Lozinskii, 1986; McKay and Shapiro, 1981; Saccà and Zaniolo, 1986; Ullman, 1985; Vieille, 1986). The reader is referred to Bancilhon and Ramakrishnan (1986) for an overview and a comparison of these approaches. As described in Bancilhon and Ramakrishnan (1986), some of these approaches lack generality—i.e., the realm in which they work is either limited or not clearly understood—and they lack robustness—i.e., they are formulated as a sequential algorithm for main memory based execution —hence their compatibility with different execution models and architectures is unclear.

In this paper, we propose a comprehensive rule rewriting approach and supporting algorithms on which the actual implementation of LDL is based (Chimenti et al., 1988). The outline of the paper is as follows:

In Section 2 we present the differential (alias seminaive) Fixpoint algorithm.

In Sections 3 and 4 we turn to the implementation of recursive predicates when some arguments in the recursive goal are instantiated. We describe two methods, *the magic set method and the counting method.* These rewrite the original rules, that, as such, cannot be implemented safely and efficiently as a single Fixpoint, into equivalent ones that can be implemented by a pair of (seminaive) Fixpoint computations safely and efficiently.

In Section 5, we consider the *specialized single Fixpoint method* that handles various cases of practical import where one of the two Fixpoint computations can be eliminated.

In Section 6, we compare the pros and cons of various methods and study their computational complexity.

In Section 7 we introduce the magic counting method that combines the strong features of the previous methods.

2. Seminaive Fixpoint Algorithm

In a logic program LP, a predicate P is said to *imply* a predicate Q, written $P \rightarrow Q$, if there is a rule in LP having predicate Q as its head and predicate P in its body, or there exists a P' where $P \rightarrow P'$ and $P' \rightarrow Q$ (transitivity). Then, a predicate P will be called *recursive* when $P \rightarrow P$. Two predicates P, and Q are called *mutually recursive* if $P \rightarrow Q$ and $Q \rightarrow P$. Then, the sets of all predicates in LP can be divided into recursive predicates and non-recursive ones (such as database predicates). This implication relationship can also be used to partition the recursive predicates into disjoint subclasses of mutually recursive predicates, which we will call *recursive cliques* (having the graph representation of $\rightarrow$, in mind). All predicates in the same recursive clique will have to be computed together, they cannot be computed separately. Say that $\mathbf{R}$ denotes the vector of predicates in our recursive clique. Then, a rule defining a predicate in $\mathbf{R}$ will be called *recursive* if it contains a predicate in $\mathbf{R}$ among its goals, and it will be called an *exit rule* otherwise (e.g., in Fig. 2, r_0 and r_1 are recursive rules, while r_2 and r_3 are exit rules). A recursive rule that only contains one occurrence of a recursive predicate in $\mathbf{R}$ in its body is called *linear.* The recursive rules in Figs. 1 and 2 are linear.

Say now that $\mathbf{E}$ defines the value of the relations in $\mathbf{R}$ according to the exit rules, and $T_P(\mathbf{R})$ is the mapping defined by the recursive rules (e.g., the relational algebra equivalent of the recursive rules). Then, the Fixpoint algorithm can be formulated as follows:

```
Naive Fixpoint Algorithm
  begin
  j =: 0;
  R := φ;
  R' := E;
  while R' ≠ R do
        begin
        R := R';
        R' := T_P(R) ∪ R;
        j := j + 1
        end
  end.
```

Fig. 3. A naive formulation of the Fixpoint algorithm.

The value **R** returned by this new algorithm is the least Fixpoint solution of the recursive equation $\mathbf{R} := T_P(\mathbf{R}) \cup \mathbf{E}$, which, in turn, defines the minimum model for the given recursive rules and exit rules defining **R**. In the algorithm of Fig. 3, we also have include a step counter j, which, although not needed at this point, will become useful later.

The problem with the naive Fixpoint algorithm is that tuples are produced over and over again. For instance, in the same generation example of Fig. 1, the parents of a given human generated in the first iteration step are generated again in the second step, where grandparents are first computed. Differential techniques have been proposed by several authors to solve the problem, including Bancilhon (1985) and Balbin and Ramanohanarao (1987). Our approach, similar to that proposed in Balbin and Ramanohanarao (1987), has the advantage of being general and not requiring differentiation of relational algebra expressions.

Assume that $\delta\mathbf{R}$ denotes the "new" tuples obtained at each step of the algorithm of Fig. 3. Then the previous algorithm can be modified into the one of Fig. 4. The final answer of the seminaive Fixpoint algorithm, i.e., **R** after a step where that $\delta\mathbf{R} = \phi$, is obviously identical to the answer **R** produced by the naive Fixpoint algorithm. This second algorithm is, however, more efficient than the previous one since we will next recast the computation $\delta\mathbf{R} := T_P(\mathbf{R}) - \mathbf{R}$ into a form that avoids the recomputation of old results.

Seminaive Fixpoint Algorithm

begin
$j := 0;$
$\delta\mathbf{R} := \phi;$
$\mathbf{R} := \mathbf{E};$
while $\delta\mathbf{R}_j \neq \phi$**do**
  begin
  $\delta\mathbf{R} := T_P(\mathbf{R}) - \mathbf{R};$
  $\mathbf{R} := \mathbf{R} + \delta\mathbf{R};$
  $j := j + 1$
  end
end.

Fig. 4. The seminaive formulation for the Fixpoint algorithm.

While previous authors have proposed rather cumbersome transformations on the relational algebra expression for $F(\mathbf{R})$, we will propose here a more direct and expressive rule rewriting approach.

For the Same Generation example the above seminaive Fixpoint algorithm could be expressed as follows:

begin
$j := 0;$
$SG(0, x, x) : -H(x);$
while $SG(j, \_, \_)$ **do**
$SG(j+1, x, y) : -P(x, x_1), SG(j, x_1, y_1), P(y, y_1),$ not $(SG(\_, x, y));$
  $j := j + 1$
  end
end.

Fig. 5. Seminaive Fixpoint for the same generation example with an Unimproved Iterative Step.

Some remarks about the notation are needed here. First of all, observe that $J+1$ is just a short-hand notation of convenience, inasmuch as in a proper Prolog program $J+1$ would be replaced by $J1$ and the goal "$J1$ *is* $J+1$" would be added to the rule. Also observe that we are now treating our rules as those of production systems, inasmuch as we use them to updated our current computation state. Indeed, we have a three column relation SG to which we initially assign a value derived according to the exit rule, and, at each step of the Fixpoint iteration, we augment the current SG according to the recursive rule. Note that $SG(\_, x_1, y_1)$ will return the x_1, y_1 projection of all the tuples produced so far in the Fixpoint iteration—i.e., the **R** relation is updated implicitly, thus eliminating the need for the explicit step $\mathbf{R} := \mathbf{R} + \delta\mathbf{R}$ in the algorithm of Fig. 4. Thus, $not(SG(\_, x, y))$ amounts to a set difference—i.e., to the $-\mathbf{R}$ operation above. Finally, $\delta\mathbf{R} \neq \phi$ holds whenever $SG(j, \_, \_)$ is true. Thus, Fig. 5 is in fact a faithful restatement of the seminaive Fixpoint algorithm in the production rule notation.

Furthermore, one need not write the algorithm of Fig. 5. explicitly, since the application of the Naive Fixpoint of Fig. 3 to the pair of rules that follows tantamounts to the algorithm of Fig. 5.

$$\begin{aligned} &SG(0, x, x) : -H(x). \\ &SG(J+1, x, y) : -P(x, x_1), SG(J, x_1, y_1), P(y, y_1), \\ &\qquad \text{not } (SG(\_, x, y)). \end{aligned} \tag{5}$$

By now, our reader has probably begun to grasp how the rule rewriting approach works. We have a program P that we would like to implement with a seminaive algorithm when only the basic naive algorithm is at hand. Then, we transform P into an *equivalent* P' such that the naive algorithm on P' behaves as the seminaive on P.

The program above is only a first step, inasmuch as it suffers from two of problems. The first is that, because of the presence of negation it does not have a unique minimal model; furthermore, the result of the Fixpoint algorithm may not yield least solution of the Fixpoint equation $R = T_P(R)$—which in fact may not have a least solution (Kolaitis and Papadimitriou, 1988). Without a clear Fixpoint-based or model-theoretic semantics a claim of equivalence preserving methods would have no basis. To solve this problem we observe that we can rewrite these rules as

$$\begin{aligned} &G(0, x, x) : -H(x). \\ &SG(J+1, x, y) : -P(x, x_1), SG(J, x_1, y_1), P(y, y_1), \\ &\qquad \text{not } (SG(I, x, y), I \leq J). \end{aligned} \tag{6}$$

The program so transformed is *locally stratified* (Przymusinski, 1987). A program P is said to be locally stratified when its Herbrand Base H can be partitioned into strata $H_0, H_1, \ldots$ such that for every instantiation r of each rule in P, the negated goals of r

belong to strata strictly lower than that of the head of r, while the positive goals of r belong to strata not higher than that of the head of r. Programs that are locally stratified have a well-defined model theoretic semantics, based on the concept of *perfect model* (Przymusinski, 1987). Furthermore, the locally stratified programs produced by our seminaive methods, also have a least-Fixpoint based semantics (Saccà and Zaniolo, 1988). This provides a formal basis to the claim that our transformed program is equivalent to the original one.

Let us now turn to the second problem, which is that of performance. To derive the seminaive improvement we can observe, that, at step j, every value of $SG(J, x_1, y_1)$ with $J < j$ had already been used at previous steps of the Fixpoint computation and can be dropped. Thus, the previous rules can now be simplified into those of Fig. 6.

$$\begin{aligned} &SG(0, x, x) \ :- \ H(x). \\ &SG(j+1, x, y) :- P(x, x_1), SG(j, x_1, y_1), P(y, y_1), \\ &\qquad \text{not } (SG(I, x, y), I \leq j). \end{aligned}$$

Fig. 6. Seminaive Fixpoint for the same generation example with an Improved Iterative Step.

where the j value to be used (in the relational algebra based implementation of these rules) is that of the running index of the Naive Fixpoint of Fig. 3. The improvement of the iterative step in the seminaive computation is thus straightforward for linear rules.

Non-linear rules require a more complex rewriting, which is discussed next. In general, each non-linear rule in a recursive clique has the form

$$r\text{: } P :- D_0, P_1, D_1, P_2, \ldots, P_n, D_n, \tag{7}$$

where $P_1, \ldots, P_n$ denote the *occurrences of the recursive predicates* from the clique, $(0 \leq i \leq n)$, and $D_0, \ldots, D_n$ denote non-recursive predicates or (possibly empty) conjunctions thereof. Let $P[j+1]$, $P[j]$ and $P[I_i]$, respectively, denote the predicate P expanded by the addition of a first argument $j+1$, j and "I_i," then our rule r is expanded into a disjunction of n rules to be used in the seminaive algorithm template as follows:

$$
\begin{aligned}
r_1\colon\ & P[j+1] : -D_0,\ P_1[j], D_1, P_2[I_2], \ldots, P_n[I_n], D_n \\
& \qquad I_2 \leq j, \ldots, I_n \leq j, \text{not}\,(P[I], I \leq j). \\
r_2\colon\ & P[j+1] : -D_0, P_1[I_1], D_1, P_2[j], \ldots, P_n[j], D_n \\
& \qquad I_1 \leq j, \ldots, I_n \leq j, \text{not}\,(P[I], I \leq j). \\
& \vdots \\
r_2\colon\ & P[j+1] : -D_0, P_1[I_1], D_1, P_2[I_2], \ldots, P_n[j], D_n \\
& \qquad I_2 \leq j, \ldots, I_{n-1} \leq j, \text{not}\,(P[I], I \leq j).
\end{aligned}
\tag{8}
$$

For instance, for the non-linear formulation of ancestors, of Fig. 7,

$$
\begin{aligned}
& Anc(x,y) : -P(x,y). \\
& Anc(x,z) : -Anc(x,y), Anc(y,z).
\end{aligned}
$$

Fig. 7. Non-linear formulation for ancestors.

The rules resulting from the seminaive method are given in Fig. 8.

$$
\begin{aligned}
& Anc(0,x,y) : -P(x,y); \\
& Anc(j+1,x,z) : -Anc(j,x,y), Anc(I_2,y,z), I_2 \leq j, \\
& \qquad not(Anc(I,x,z), I \leq j). \\
& Anc(j+1,x,z) : -Anc(I_1,x,y), Anc(j,y,z), I_1 \leq j, \\
& \qquad not(Anc(I,x,z), I \leq j).
\end{aligned}
$$

Fig. 8. Seminaive improvement for the example of Fig. 7.

3. The Magic Set Method and the Minimagic Method

As previously discussed, it is often the case that constants cannot be migrated into recursive rules. For these cases however, a small set of values, called a *magic set*, can be derived from these constants and used to restrict the Fixpoint computation and make it either safe or more efficient. This observation is at the basis of the *magic set method* introduced in Bancihon er al. (1986). For example, in the query of Fig. 9,

$$G5{:}?SG(adam, y)$$
$$r_0{:}SG(x, y) : -P(x, x_1), SG(x_1, y_1), P(y, y_1)$$
$$r_1{:}SG(x, x) : -H(x).$$

Fig. 9. Find persons of the same generation as adam.

we can restrict our search to the values of x in r_1 and to the values of x and x_1 in r_0 denoting persons that are ancestors of adam. Thus, the set of ancestors of adam is the magic set for the query at hand. From an operational viewpoint, magic set values are those instantiated by Prolog during the goal expansion phase (first phase) of the SLD resolution. During this phase, r_0 is invoked first and the x value is bound to "adam" while x_1 is bound to a parent of adam. As rule r_0 is called again, with x now bound to a parent of adam, x_1 becomes instantiated to a grandparent of adam, and so on. Also observe that y_1 and y are not instantiated during this phase; these variables become instantiated during the second (tree reduction) phase, which begins after new ancestors of adam run out and rule r_1 is used. Therefore, the application of the magic set method involves three steps.

Step 1: The *binding passing analysis* determines which arguments in the recursive predicates will be instantiated during the first phase of the SLD resolution,

Step 2: The *rule rewriting step* adds rules for computing the magic sets and modifies the original rules to take advantage of these magic sets, and

Step 3: The *Fixpoint-based execution* of the rules produced in Step 2 constructs the answer.

Let us illustrate these steps on the example of Fig. 9. The binding passing analysis will determine that

(a) the first argument in the recursive predicate is bound by the query, and
(b) given that the first argument (i.e., x) is instantiated in the head of the recursive rule, then, the first argument in the recursive predicate of the body (i.e., x_1) can also be instantiated via the goal $P(x, x_1)$.

Thus, we have a cyclic situation illustrated by the graph of Fig. 10 (also known as a *binding graph* (Saccà and Zaniolo, 1987)), where SG with a bound first argument (denoted by SG^1) in turn binds the first argument of SG again.

$$SG^1$$

Fig. 10. Binding passing graph for the example of Fig. 9.

Therefore, for the example in Fig. 9, the first argument of SG is always instantiated during the first phase of the SLD resolution, and thus the magic set will collect the various values taken by this argument during this phase.

The binding passing analysis (Step 1) can be done at query compilation time since it only depends on the intentional information (query and rules). Step 2 can also be performed at compilation time since it operates only on the intentional database by adding rules defining a magic predicate and modifying the original recursive rules with the addition of the magic predicate as a goal to act as a filter. However, the magic set computation depends on the extensional information and must be done at execution time.

For the Same Generation example, for instance, the magic rules are given in Fig. 11.

%magic rules—ancestors of adam
$m.SG(adam)$
$m.SG(x_1) : -P(x, x_1), m.SG(x).$

%modified rules
$SG'(x, y) : -m.SG(x), H(x).$
$SG'(x, y) : -m.SG(x), P(x, x_1), SG'(x_1, y_1), P(y, y_1).$

%modified query
$?SG'(adam, y)$

Fig. 11. Magic rules and modified rules for the query of Fig. 9.

Note that $m.SG(x)$ contains the set of all ancestors of adam, as expected. Therefore, the addition of the goal $m.SG(x)$ to the original rules yields a much more constrained (i.e., more efficient) computation for SG'. The starting point of this computation (as per the exit rule) no longer consists of the set of all humans, but rather the set of ancestors of adam—thus, the goal $H(x)$ could be removed without making the query unsafe. Moreover, owing to the $m.SG$ goal in the modified recursive rule, persons who are not ancestors of adam are discarded at each step of the Fixpoint computation of SG'.

Therefore, the result of the magic set method is a transformed program featuring (i) a new recursive clique containing the magic predicates and (ii) the old recursive clique containing the old recursive predicates with modified rules. Every recursive clique is translated into a Fixpoint computation; therefore, the original program, that is unsuitable to implementation via a single Fixpoint computation, has been transformed into an equivalent one that is amenable to implementation by a pair of Fixpoint computations. These two Fixpoint computations are then further optimized using the seminaive improvement described in the last section.

The reader can refer to Saccà and Zaniolo (1987) for a more complete and formal description of the magic set algorithm, including the binding passing analysis which determines the arguments in the recursive rules that are instantiated by the query constants. Of more direct interest in our discussion are the rewriting rules used to implement the magic set method.

3.1. Rewriting Rules for the Magic Set Method

3.1.1. Exit Magic Rule

There is exactly one such a rule and it consists of a unit clause constructed from the query goal by eliminating non-constant arguments and adding the distinguished prefix "*m.*" to the predicate name.

3.1.2. Recursive Magic Rules

They are derived from the recursive rules as follows:

Case 1. The given rule is linear. Then do the following:

(a) drop the goals with no instantiated arguments(e.g., $P(y, y_1)$ is dropped for the example of Fig. 9),

(b) drop the non-instantiated arguments from the recursive goals, (y and y_1 are dropped for the example of Fig. 9)

(c) exchange the recursive predicate in the head with the recursive predicate in the tail; these predicates are also renamed with the prefix "$m.$" (see Fig. 11 for the case at hand).

Case 2. The given rule is non linear, with $n > 1$ occurrences of recursive predicates in the body. Then we do the following:

(a) Construct n linear rules; each such a rule is obtained by deleting all occurrences of the recursive predicates but one (obviously, a different one for each rule).
(b) Construct a magic rule from each of the n rules so obtained, using the transformation indicated in Case 1.

3.1.3. Building the Modified Rules

Each original rule is modified by the addition of the corresponding magic predicate. This has the form $m.P(X)$, with P the predicate symbol in the head of the rule and X the bound arguments in said head (For the example of Fig. 10, we add $m.SG(x)$).

Observe that there is nothing magic about the particular value adam appearing in the query of Fig. 9, since the solution just described applies to any goal $SG(z, y)$ provided that z was instantiated, say, by previous goals in the rule. All that is needed is the ability to pass this instantiated value in the exit magic rule of Fig. 11. Thus, the magic set method—and, in general all the methods described in this paper—can also be used to support sideways information passing into recursive predicates.

The magic set method was first proposed in Bancilhon et al. (1986) for linear rules; it was then generalized in Saccà and Zaniolo (1986) to deal with function symbols and certain non-linear rules and further extended in Beeri and Ramakrishnan (1987) with a notion of sideway information passing making it applicable to arbitrary non-linear rules.

The magic set method is elegant and general, but suffers from two drawbacks. Concerning the first drawback, observe that the natural way to support the modified rule of Fig. 11 consists of joining the current relation $SG'(x_1, y_1)$ with $P(x, x_1)$ and then with $m.SG(x)$. However, if we assume that $P(x, x_1)$ is conditionally safe, i.e., can be evaluated only when the first argument is instantiated (typically, arithmetic predicates and recursive predicates are conditionally safe) then, the only way to support the modified rules of Fig. 11, is to join (the relation corresponding to) $m.SG(x)$ with $P(x, x_1)$ first, and then join the result with $SG'(x_1, y_1)$ and, finally, with $P(y, y_1)$. The second drawback is that the join of $m.SG(x)$ with $P(x, x_1)$ is exactly the computation performed in building the magic set. Thus, rather than repeating this computation at each step of the second Fixpoint iteration, we can save the result of the first Fixpoint computation and use it in performing the second one. This observation is at the basis of the *minimagic method,* (Saccà and Zaniolo, 1986) where all the results of the first Fixpoint computation are saved in a "magic relation". This relation is then used in the modified rule computation in lieu of the original predicates and solved goals. Thus, in the example in question, the two goals $m.SG(x)$ and $P(x, x_1)$ would be replaced by a goal $m.P(x, x_1)$ which is basically the restriction of the parent relation over the ancestors of "adam". In general, the minimagic method avoids duplicate computation at the price of storing larger data sets.

The magic set transformation for the list-merge example of Fig. 2 and the query

$$?MG(L_1, L_2, w) \tag{9}$$

is given in Fig. 12.

$$m.MG(L1, L2).$$

$$m.MG(y, x_1 \bullet y_1) : -m.MG(x \bullet y, x_1 \bullet y_1), x \geq x_1$$
$$m.MG(x \bullet y, y_1) : -m.MG(x \bullet y, x_1 \bullet y_1), x < x_1$$

$$MG'(x \bullet y, x_1 \bullet y_1, x \bullet w) : -m.MG(x \bullet y, x_1 \bullet y_1), MG'(y, x_1 \bullet y_1, w), x \geq x_1$$
$$MG'(x \bullet y, x_1 \bullet y_1, x_1 \bullet w) : -m.MG(x \bullet y, x_1 \bullet y_1), MG'(x \bullet y, y_1, w), x < x_1$$
$$MG'(\mathbf{nil}, x, x) : -m.MG(\mathbf{nil}, x).$$
$$MG'(x, \mathbf{nil}, x) : -m.MG(x, \mathbf{nil}).$$
$$?MG'(L_1, L_2, w).$$

Fig. 12. The magic set method for the list merge example.

Fig. 12 illustrates the problem inflicted upon the magic set method by predicates that are only conditionally safe. Here the two comparison predicates, $x \geq x_1$ and $x < x_1$, which were used for computing the magic set, can not be used in the modified rules to derive new values of x or x_1. Thus we have to start the computation of the modified rule from the $m.MG$ relation rather than from the smaller MG'.

For this particular example, the minimagic method would use no additional storage, with respect to the magic set method, since it does not require saving the values of any additional variables. Moreover, the minimagic method would avoid duplication of computation by dropping the goal, $x \geq x_1$ ($x < x_1$) from the first (second) modified rule and using two distinct magic relations—one for the first rule and one for the second.

In summary, we have presented the magic set method and also discussed the minimagic method that improves on the former by avoiding the repetition of computation. However, we will not discuss the minimagic method any further since we want to present next the counting method, that eliminates duplication of computation even further. As an historical note, however, we would like to mention the similarity of the minimagic method (Saccà and Zaniolo, to appear) with the Alexander method described in Rohmer et al. (1986). While the Alexander method predates both the magic method and the minimagic method, it was, at first, overlooked in the research community and in topical surveys (an unfortunate event attributable, at least in part, to inadequate documentation). More recently, the method has become better known through the works of Beeri and Ramakrishnan (1987) and Demo et al. (1986).

4. The Counting Method

The Counting Method eliminates the duplicate computation of the magic set method by using counting indices. For instance, the humans of the same generation as adam can be found by first computing the ancestors of adam and then the descendants of these ancestors—provided that the levels of these ancestors and their descendants are also computed and compared as follows. Starting from adam, who is a zero-level ancestor of himself, we increase the count by one every time the recursive rule is used to generate a set of new ancestors. Symmetrically, the descendants of these ancestors are computed, by decreasing the count at each step in the recursive computation; once we obtain a descendant of level zero we can add this name to the query answer. Figure 13 gives the counting method code for the query of Fig. 9 ($j+1$ and $j-1$ are again shorthand notations of convenience similar to that used in Section 2).

> %counting rules—counting up on ancestors
> $cnt.SG(0, adam)$
> $cnt.SG(j+1, x_1) : -P(x, x_1), cnt.SG(j, x).$
>
> %modified rules—counting down on descendants
> $SG'(j, x) : -cnt.SG(j, x), H(x).$
> $SG'(j-1, y) : -SG'(j, y_1), P(y, y_1).$
>
> %modified query
> $?SG'(0, y)$

Fig. 13. Counting rules and modified rules for example of Fig. 9.

The predicate *cnt.SG* computes the ancestors of adam and counts the levels up. The predicate SG' computes the descendants of these ancestors and counts the levels down. The modified query selects the descendants whose associated index is zero and, thus, are of the same generation as adam.

Figure 13 also illustrates the rewriting rules used to implement the counting method. The counting sets are generated in a fashion similar to that of the magic rules; the main differences lie in the introduction of a counting index to record the level. Since a new level is introduced at each step, the seminaive Fixpoint computation of the counting sets can be simplified by removing the duplicate elimination check. Moreover, the modified rules are much simpler than those of the magic set method, since they do not contain the goals previously solved in computing the counting sets. Thus, the counting method avoids duplication of computation.

The idea of counting was first introduced in Bancilhon et al. (1986) and formalized in Saccà and Zaniolo (1986) for a simple class of linear queries. The method was then generalized to more complex situations in Saccà and Zaniolo (1987), using the rule-rewriting approach discussed next.

The *Generalized Counting Method* (Saccà and Zaniolo, 1987) makes use of additional indices and *supplementary counting sets.* The additional indices are required when there is more than one recursive rule defining a recursive predicate or when there are non-linear rules. The supplementary counting set is used to save values that will then be needed during the second phase. Figure 14, for instance, gives the generalized counting method for the query $?MG(L_1, L_2, w)$.

%Counting Rules:
$?cnt.MG(0, 0, L1, L2)$.

$cnt.MG(j+1, 2*k+0, y, x_1 \bullet y_1) : -cnt.MG(j, k, x \bullet y, x_1 \bullet y_1),$
$\quad x \geq x_1$
$cnt.MG(j+1, 2*k+1, x \bullet y) : -cnt.MG(j, k, x \bullet y, x_1 \bullet y_1),$
$\quad x < x_1$

%Supplementary Counting Rules:
$spcnt.MG.1(j, k, x) : -cnt.MG(j, k, x \bullet y, x_1 \bullet y_1), x \geq x_1$
$spcnt.MG.2(j, k, x_1) : -cnt.MG(j, k, x \bullet y, x_1 \bullet y_1), x < x_1$

%Modified Rules:
$MG'(j-1, (k-0)/2, x \bullet w) : -spcnt.MG.1(j-1, k/2, x),$
$\quad MG'(j, k, w)$
$MG'(j-1, (k-1)/2, x_1 \bullet w) : -spcnt.MG.2(j-1, (k-1)/2, x_1),$
$\quad MG'(j, k, w)$
$MG'(j, k, x) : -cnt.MG(j, k, \mathbf{nil}, x).$
$MG'(j, k, x) : -cnt.MG(j, k, x, \mathbf{nil}).$

%Modified Query:
$?MG'(0, 0, w)$.

Fig. 14. The counting method on the list merge example.

The integer arithmetic performed on the second index k is such that the complete history of the rules used in the counting phase are preserved and can be recovered in the second phase.

The counting method has two strong points and a weak one. The first plus for the method is represented by its performance that has been shown to be superior to that of other methods, both in terms of typical behavior (Bancilhon and Ramakrishnan, 1986) and worst case behavior (Marchetti-Spaccamela et al., 1987). The second plus is that it provides a framework to identify many cases of practical interest where one the two Fixpoint computations becomes unnecessary and the method can be further improved. On the minus side there is the potentially unsafe behavior of the method in the presence of cycles in the database, and problems with managing the supplementary indices needed for arbitrary depths of recursion. We will discuss these three issues next, beginning with the cases in which the counting method reduces to a single Fixpoint computation.

5. Specialized Single Fixpoint

In many practical situations, the computation plan prescribed by the magic set method or the counting method is unnecessarily complex. These are the cases in which the given recursive query with constants can be supported efficiently via a single Fixpoint computation. A typical situation is the computation of a transitive closure, such as for the linear ancestor example of Fig. 15:

$$\begin{aligned}&anc(x,x).\\&anc(x,z):-anc(x,y),P(y,z).\end{aligned}$$

Fig. 15. A right linear formulation of ancestor.

$$\begin{aligned}&?anc(john,x).\\&cnt.anc(0,john).\\&cnt.anc(j+1,x):-cnt.anc(j,x).\\&anc'(j,x):-cnt.anc(j,x).\\&anc'(j-1,z):-anc'(j,y),P(y,z).\\&?anc'(0,x).\end{aligned}$$

Fig. 16. Counting method for query (5.1) on rules of Fig. 15.

Since the indices are no longer needed in the modified rules as shown in Fig. 16, the counting rules become useless and can be eliminated. In short, the computation reduces to the program of Fig. 17:

$$\begin{aligned}&anc'(john).\\&anc'(z):-anc'(x),P(y,z).\\&-?anc'(x).\end{aligned}$$

Fig. 17. Specialized single Fixpoint program for query (5.1).

Thus the computation reduces to a single Fixpoint on a unary relation.

Consider now the second query (5.2) on the same set of rules of Fig. 15.

$$?anc(x, mark) \tag{10}$$

For this query, the counting method yields the program in Fig. 18:

$$cnt.anc(0, mark).$$
$$cnt.anc(j+1, y) : -cnt.anc(j, z), P(y, z).$$
$$anc'(j, x) : -cnt.anc(j, x).$$
$$anc'(j-1, x) : -anc'(j, x).$$
$$?anc'(0, x).$$

Fig. 18. Counting method for query (5.2) on rules of Fig. 15.

Observe now that the modified recursive rule performs no function other than decreasing the index by one, and copying the old values for x. But, since the initial value of j established by the exit rule is positive, j eventually becomes zero and x becomes an answer to the query. Therefore, we can dispense with the modified rules, and, hence, with the indices from the counting rules (which then look much as magic rules). The whole computation thus reduces to that of Fig. 19. Symmetric solutions work for the second formulation of the ancestor rule.

$$cnt.anc(mark)$$
$$cnt.anc(y) : -cnt.anc(z), P(y, z).$$
$$anc'(x) : -cnt.anc(x).$$
$$?anc'(x).$$

Fig. 19. Specialized single Fixpoint program for query (5.2).

In summary, the counting method can be specialized to detect and eliminate a trivial first phase or trivial second phase, and produce a set of equivalent rules where (1) constants have been pushed into recursion, and (2) the constant arguments have been eliminated from the recursive predicates. Thus we have a solution that frees the user from having to customize the recursive rules to the particular pattern of bound/free argument in the goal, and also leads to a very efficient implementation.

While the domain of applicability of the techniques just proposed is limited, they supply good heuristics for cases such as transitive closures that are common in applications. In general, the question of whether the given query can be supported by a single Fixpoint remains undecidable (Beeri et al., 1987).

6. Computational Complexity of Methods

Let us consider the program of Fig. 20.

$$?S(a,\ Y).$$
$$S(X,\ Y)\ :-\ E(X,\ Y).$$
$$S(X,\ Y)\ :-\ L(X,\ X_1),\ S(X_1,\ Y_1),\ R(Y,\ Y_1).$$

Fig. 20. The general template of a strongly linear program.

We can view this program as an abstraction of a rather general class of programs known as *strongly linear programs* (Saccà and Zaniolo, 1986). S is the recursive predicate E, L and R are *database predicates* or predicates derived from these, or arbitrary conjunctions of these. Moreover, X, X_1, Y, Y_1 need not represent single variables but could be vectors of variables. The same generation program discussed in the previous sections obviously falls in this class. For notational convenience, we will denote the magic set by MS and the counting set by CS. A simple graph based interpretation will be assigned to the program of Fig. 20 in order to characterize the worst case behavior of our methods. A graph G is associated with the previous query, in the following way:

a) For each value in the domains underlying L or R, there is a corresponding node in G. If the same value appears both in L and R then there will be two distinct nodes in G (one may think they have some label to distinguish them). Therefore, the nodes of G are partitioned into L-nodes and R-nodes.
b) For each pair (b, c) in L, there is an arc (b, c) in G, where both b and c are L-nodes.
c) For each pair (b, c) in E such that b and c are also database constants of L and R, respectively, there is an arc (b, c) in G where b is an L-node and c is an R-node.
d) For each pair (b, c) in R, there is an arc (c, b) in G, where both b and c are R-nodes.

The *query graph* $G_Q = \langle N,\ A\rangle$ is the subgraph of G, induced by all nodes that are reachable from a (recall that a denotes the constant or vector of constants in the query goal). We call a the *source node* of G_Q. The query graph G_Q, in turn, is composed by the three subgraphs $G_L = \langle N_L, A_L\rangle$, $G_E = \langle N_E, A_E\rangle$ and $G_R = \langle N_R, A_R\rangle$, such that A_L, A_E and A_R are all the arcs in A corresponding to pairs in L, E and R, respectively. It is easy to see that N_L and N_R are disjoint and contain L-nodes and R-nodes, respectively, and $N_L \cup N_R = N$. On the other hand, G_E is a bipartite graph having arcs from L-nodes to R-nodes. Finally, A_L, A_R and A_E are disjoint and $A_L \cup A_R \cup A_E = A$. The number of respective nodes of G_Q, G_L, G_R and G_E will be denoted by n, n_L, n_R and n_E, while the number of respective arcs is denoted by m, m_L, m_R and m_E.

Let b and c be two nodes in the query graph. If there is a path from b to c with length k, we say that c has a distance k from b.

Consider the graph G_L. It follows directly from the definitions that the nodes of this graph are the magic set values; $N_L = MS$. Thus we will call G_L the *Magic Graph* and refer to magic graph nodes and magic set values as synonyms. The counting set CS consists of pairs (j, b) where b is a node in the magic graph and j is its distance from the source node a, as it will be shown below. The set of values obtained from CS by projecting the indices out will be denoted CS_{-i}; obviously $CS_{-i} = MS$. Moreover, let I_b denote the set of indices j such that (j, b) is in CS. Then a node b will, respectively, be called

(a) *single* if I_b is a singleton set,
(b) *multiple* if I_b has a finite cardinality greater than one,
(c) *recurring* if I_b is infinite.

The magic graph of a query will be called *regular*, when all its nodes are single and non-regular otherwise.

For the query graph shown in Fig. 21, G_L is the subgraph induced by the nodes a, $a_1, \ldots, a_5$, while G_R is the subgraph, represented by darker arcs, induced by the nodes $b_1, \ldots, b_{10}$. The graph G_E is composed by the dashed arcs in Fig. 21. The magic graph is regular since all nodes in G_L are single, i.e., they have a unique distance from a. If we add the tuple $\langle a_2, a_5\rangle$ to the relation L then the query and the node a_5 become multiple; instead, if we add the tuple $\langle a_5, a_2\rangle$, then the query becomes cyclic and the nodes a_2, a_3 and a_5 become recurring.

Proposition 1. *Let Q be a query and G_Q be the query graph. Then $MS = CS_{-i} = N_L$. In addition, given a node b in G_L,*

a) b is single if and only if all directed paths from the source node a to b in G_L have the same distance,
b) b is multiple if and only if there at least two directed paths from a to b in G_L with different length,
c) b is recurring if and only if there is a cyclic directed path from a to b in G_L, and
d) the set I_b of indices associated with b coincides with the set of all distances of b from a.

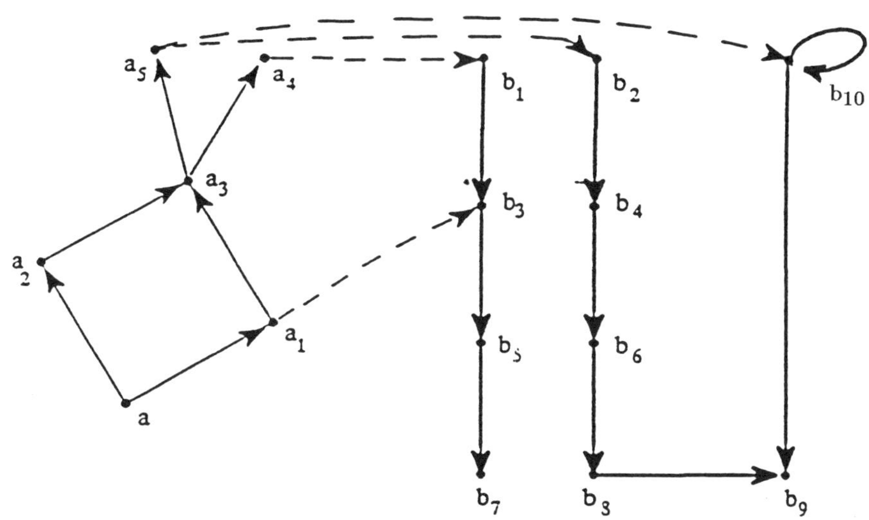

Fig. 21. Query Graph

As shown in Saccà and Zaniolo (1986) and Marchetti-Spaccamela et al. (1987), there is also a simple graph based interpretation of the query answer:

Fact 1. A node b is in the answer of Q if there is a (possibly cyclic) directed path from the source node a to b in G_Q such that this path is composed by exactly k arcs from A_L, one arc from A_E and k arcs from A_R, where k is any non-negative integer. □

Consider the query instance whose graph is shown in Fig. 21. Then b_5 is in the answer because of the path a, a_1, b_3, b_5. The other answer nodes are b_7, b_8, b_9 and b_{10}. Note that the latter node is in the answer because of the cyclic path

$$a, a_1, a_3, a_5, b_{10}, b_{10}, b_{10}, b_{10}. \tag{11}$$

The path from a to b_9 is cyclic as well.

Within this graph formalism, interesting complexity results about the magic set and the counting methods have been obtained in Marchetti-Spaccamela et al. (1987). The costs of the two methods are summarized in Table 1 for the different kinds of Magic Graphs (MG) The basic cost unit is the cost of retrieving a tuple in a database relation. We use the notations O and Θ for describing asymptotic time complexity. If the cost function of an algorithm is $f(n)$, where n is the problem size, and $g(n)$ is another function of n, then

a) $f(n) = O(g(n))$, if there exists a constant d such that $f(n) \leq d \times g(n)$ for all but some finite (possibly empty) set of non-negative values for n, and
b) $f(n) = \Theta(g(n))$, if both $f(n) = O(g(n))$ and $g(n) = O(f(n))$.

Table 1. Costs of the counting and magic set methods.

MG	Counting	MagicSet
Regular	$\Theta(m_L + n_L \times m_R)$	$\Theta(m_L \times m_R)$
Acyclic	$\Theta(n_L \times m_L + n_L \times m_R)$	$\Theta(m_L \times m_R)$
Cyclic	*unsafe*†	$\Theta(m_L \times m_R)$

† In Marchetti-Spaccamela et al. (1987) it has been shown that the counting method can be extended to deal with cyclic graphs and its cost is $\Theta(m \times n^2)$. Also note that the costs for the magic set method are actually higher than those given in Table 1, since arcs not in G_Q can be developed at each step of Fixpoint computation using the modified rules. For simplicity, these costs can be neglected since they simply reinforce the superiority of magic counting methods.

Proposition 2. *Let C and Ms be the costs of the counting method and of the magic set method, respectively. Then*

a) If the magic graphs are regular then $C = O(Ms)$.
b) If the magic graphs are acyclic and $m_L = O(m_R)$, then $C = O(Ms)$.

Proposition 2 says that the counting method always works better than the magic set method when the magic graphs are regular. In addition, since it is realistic to assume that m_R is, on the average, of the same order of m_L, it is fair to say that the counting method, on the average, works better than the magic set method when there is no cycle. In fact, in the average, we have that $C = \Theta(n \times m)$ and $Ms = \Theta(m \times m)$. Note that having $m_L \ll m_R$ is not sufficient for the magic set method to work better than the counting method.

Thus the counting method is superior to the magic set method in terms of worst case behavior. This superiority is even more dramatic when typical behavior is considered; in the comparative study presented in Brancilhon and Ramakrishnan (1986) the counting method was shown to be more efficient than all other methods by an order of magnitude (including the magic set method but excluding Prolog and the Henschen and Naqvi (1984) methods that often deliver comparable performance).

Unfortunately, the potential presence of cycles in the database compromises the applicability of the counting method in many situations. Note that, a database being logically acyclic (e.g., a non-incestuous family tree for the same generation example) does not guarantee that the physical database is cycle free, since checking acyclicity upon updates is very expensive and not often done in practice—thus there could be accidental cycles that throw the counting method astray. Therefore, a method that combines the performance of the counting set method with the safety of the magic set method is highly desirable.

7. Magic Counting Methods

We now propose a family of methods that combine the magic set and the counting methods and that are, therefore, called *magic counting* methods. All methods in the family make use of a *reduced magic set* and a *reduced counting set*. A *reduced* magic set, denoted by RM, is any (possibly empty and not necessarily proper) subset of the magic set MS. Likewise, RC will denote any (possibly empty and not necessarily proper) subset of the counting set CS, while RC_{-i} denotes the set of values in RC without their indices. In addition, for each b in RC_{-i}, RI_b is the set of all indices associated with b in RC (obviously $RI_b \subseteq I_b$).

The general structure of the magic counting methods consists of two steps. In the first step, a reduced magic set RM and a reduced counting set RC is constructed; in the second step, both the magic counting method and the magic set method are applied using the reduced sets. This second step is implemented as shown in Fig. 22.

Modified Rules & Query For MC Methods

$$S_C(J,\ Y)\ :-\ RC(J,\ X),\ E(X,\ Y). \tag{1}$$
$$S_C(J-1,\ Y)\ :-\ S_C(J,\ Y_1),\ R(Y,\ Y_1). \tag{2}$$

$$S_CM(X,\ Y)\ :-\ RM(X),\ E(X,\ Y). \tag{3}$$
$$S_M(X,\ Y)\ :-\ MS(X),\ L(X,\ X_1),\ S_M(X_1,\ Y_1),\ R(Y,\ Y_1). \tag{4}$$

$$Answer(X)\ :-\ S_C(0,\ X). \tag{5}$$
$$Answer(X)\ :-\ S_M(a, X). \tag{6}$$
$$?Answer(X)$$

Fig. 22. The independent magic counting method.

Observe that the modified rules of a magic counting method are basically the juxtaposition of the modified rules for the counting method (Rules 1 and 2) and those for the magic set method (Rules 3 and 4). However, they use reduced counting sets and magic sets. Thus, the predicate RC has replaced CS in the exit rule of S_C (Rule 1), while RM has replaced the original MS in the exit rule for S_M (Rule 3). Since Rules 1 and 2 operate independently from Rules 3 and 4, the methods conforming to the script of Fig. 22 are called independent magic counting methods. The following theorem characterizes the domain of correctness for these methods (Saccà and Zaniolo, 1987).

Theorem 1. *An independent magic counting method is correct when the following two conditions hold:*

a) $RM \cup RC_{-i} = MS$, *and*
b) for each b in $RC_{-i} - RM$, $RI_b = I_b$.

This theorem allows us to divide the nodes of the magic graph into the set RC that uses the counting method and the set RM that uses the magic set method. Since the counting method is better than the magic method for all nodes but the recurring ones, the ideal solution would assign the recurring nodes to RM and all others to RC. However, this ultimate goal is not easy to reach, since the existence of recurring nodes is not known at compile time and must be detected at run time at the price of some computational complexity. Thus, we present three alternative methods that approximate the ultimate goal with solutions that offer practical advantages of their own. Since detecting non-regular graphs is easier than detecting cyclic ones, these methods use the regularity of the magic graph as their decision criterion.

The simplest method to implement is the basic method, as follows:

a) *Basic Method.* If the graph G_L is regular then $RM = \emptyset$ and $RC = CS$, otherwise $RM = MS$ and $RC = \emptyset$. The basic method coincides with the counting method in the former case and with the magic set method in the latter case.

For instance, the graph G_L of Fig. 23 is not regular, thus $RM = MS = \{a, b, \ldots, l\}$ and $RC = \emptyset$.

While the basic method removes the compile-time dilemma of having to choose between counting and magic sets, it is clearly suboptimal since it does not, as it should, use the counting method for the parts of the graph which do not contain any multiple or recurring nodes. The next method accomplishes that by recording the level at which non-regular nodes are first found:

b) *Single Method.* Let k be the maximum index such that all nodes in CS_{-i} having an index less than k are single. Then, RC_{-i} is the set of all (single) nodes with index less than k, and $RM = MS - RC_{-i}$.

In Fig. 23, for example, we have $k = 2$, $RC_{-i} = \{a,\ b,\ c,\ d\}$ and $RM = \{e,\ f, \ldots,\ l\}$.

Using an index to partition the graph horizontally represents too coarse a criterion, since nodes in different vertical branches of the graph are smeared together. For the example of Fig. 23, for instance, the nodes e and f are assigned to RM, although they are single. The next method solves this problem:

c) *Multiple Method.* RC_{-i} is the set of all single nodes and $RM = MS - RC_{-i}$ (i.e., RM contains all multiple and recurring nodes).

For the example of Fig. 24, we have $RC_{-i} = \{a,\ b,\ c,\ d,\ e,\ f\}$ and $RM = \{g,\ h,\ i,\ j,\ k,\ l\}$.

Our final method uses counting for both single and multiple nodes.

d) *Recurring Method.* RC_{-i} is the set of all single and multiple nodes and $RM = MS - RC_{-i}$ (thus, RM contains all recurring nodes).

For the magic graph G_L in Fig. 24, the Recurring Method will produce, $RC_{-i} = \{a,\ b,\ c,\ d,\ e,\ f,\ h,\ k\}$ and $RM = \{g,\ i,\ j,\ l\}$.

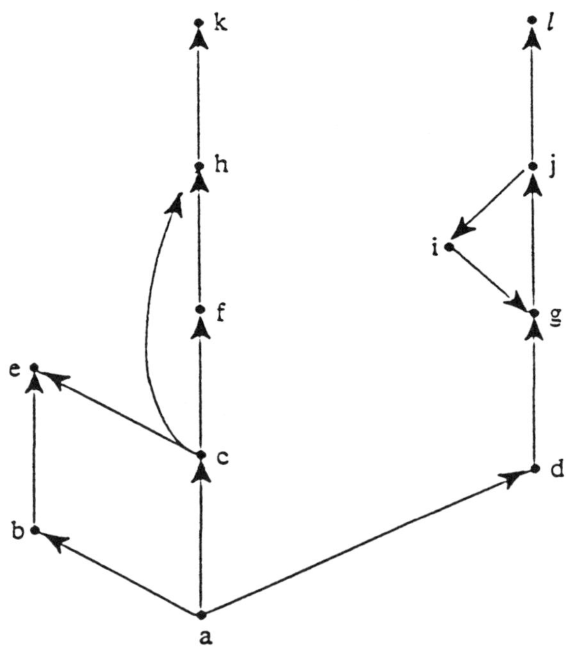

Fig. 23. Magic Graph

Before turning to the actual computation of the reduced set RC and RM, let us observe how, in the last three methods, the RM nodes have been relegated to the part of the graph most remote from the source—i.e., to the upper part of Fig. 24. As the magic set computation for these nodes progresses, it moves to the lower part of the graph (i.e, closer to the source node) where no recurring node exists—thus it can be improved by using the counting method. This idea has given birth to a further refinement of the magic counting method called an *integrated* magic counting method (Saccà and Zaniolo, 1987) (as opposed to the independent method just discussed). A discussion of the integrated counting method is outside the scope of this paper.

The computation of the sets RC and RM must be performed together, using a modification of the seminaive Fixpoint algorithm. The key idea is that the seminaive computation of the magic set and that of the counting set are very similar and, thus, can be combined together. Consider for instance the problem of implementing the Multiple Magic Counting method; here we must detect all single nodes and include them in the reduced counting set. To this end, we can use a temporary relation im, where the second column records whether we are dealing with the first occurrence of a node or the second one. Then, both first and second occurrences are used to generate other nodes in order to identify all multiple or recurring nodes (thus, we may need to use the same path twice) (see Fig. 24).

$$
\begin{aligned}
&\overline{MS}(0,1,a).\\
&\overline{MS}(j+1,1,X_1) : -\overline{MS}(j,1,X), L(X,X_1),\\
&\qquad \text{not } (\overline{MS}(I,\_,X_1), I \le j).\\
&\overline{MS}(j+1,2,X_1) : -\overline{MS}(j,\_,X), L(X,X_1), \overline{MS}(K,1,X_1), K \le j,\\
&\qquad \text{not } (\overline{MS}(I,2,X_1), I \le j).
\end{aligned}
$$

Fig. 24. Reduced sets computation for the multiple magic counting method.

At the end of the above Fixpoint computation, the multiple methods compute RC_{-i} as the set of all single nodes and RM as the set of all multiple/recurring nodes in the following way:

$$
\begin{aligned}
&MS(Y) : -\overline{MS}(\_,1,Y).\\
&RM(Y) : -\overline{MS}(\_,2,Y). \qquad\qquad (12)\\
&RC(I,Y) : -\overline{MS}(I,1,Y), \text{ not } (RM(Y)).
\end{aligned}
$$

If RC happens to be empty (i.e., there are no single nodes), then the integrated method adds the pair $(0, a)$.

The implementation of the recurring method is more complex, since the determination of cyclic structures in a graph is more suitable to depth-first search than to the breath-first expansion of the Fixpoint algorithm. An effective algorithm for detecting all recurring nodes consists in generating first the magic graph

$$ML(X, Y) : -MS(X), L(X, Y). \tag{13}$$

and then use the following algorithm to detect the non-recurring nodes $\overline{NR}(Y)$, with the help of an auxiliary predicate $UP(J, Y)$, which, at the end of each step J, contains all Y's that have an unsolved predecessor, i.e., there is an edge from some node in $\overline{NR}$ into Y.

$$\begin{aligned}
&\overline{NR}(0, a) : - \text{ not } (ML(_, a)).\\
&UP(0, Y) : -ML(X, Y), X \neq a.\\
&\overline{NR}(j+1, Y) : -\overline{NR}(j, X), ML(X, Y), \text{ not } (UP(I, Y), I \leq J).\\
&UP(j+1, Y) : -UP(j, Y), \text{ not } (\overline{NR}(j, Y)).
\end{aligned}$$

Fig. 25. Computation of non-recurring nodes.

The program of Fig. 25 is locally stratified (Przymusinski, 1987); its perfect model can be computed efficiently using the Naive algorithm of Fig. 3 (Saccà and Zaniolo, 1988). At the end of this algorithm we have:

$$RC(X) : -\overline{NR}(\_, X). \tag{14}$$

Thus, all is left is to compute the indices for the set RC as shown in Fig. 26:

$$\begin{aligned}
&RC(0, a).\\
&RC(j+1, Y) : -RC(j, X), ML(X, Y), RC(Y).
\end{aligned}$$

Fig. 26. Computation of indices for non-recurring nodes.

Consider now the computational complexity of performing the seminaive computation on the rules of Fig. 25. Observe that for node Y is added to $\overline{NR}$ after n_m attempts, at most. Moreover, the size of UP is always less than n_L and the basic Fixpoint is performed m_m times. Thus, if n_m and m_m denote the number of non-recurring nodes and arcs between these, respectively, then the worst case complexity of computing ML and $\overline{NR}$ is $\Theta(m_L + n_m \times m_m + n_L \times m_m)$. (This upper bound is the actual one for complete graphs.) For the computation of Fig. 26, observe that for each node, there are at most m_m paths of different length between the node and the origin a. Thus, the complexity of the Fixpoint computation on the rules of Fig. 26 is $\Theta(n_m \times m_m)$. Thus the cost of computing RC is dominated by that of computing $\overline{NR}$.

It also easy to see how the computation of UP in Fig. 25 can be improved by deleting tuples from a UP table, rather than copying the entire table at each step. Then the complexity of the overall computation reduces to $\Theta(m_L + n_m \times m_m)$.

All magic counting methods are safe in the presence of cycles; moreover, they work better than the magic set method, and they coincide with the counting method when the query is regular (Saccà and Zaniolo, 1987). Furthermore, the recurring magic counting method behaves as the counting method, i.e., with complexity $\Theta(n_L \times m_L + n_L \times m_R)$, for acyclic graphs. In the integrated form discussed in Saccà and Zaniolo (1987), the recurring magic counting method has, for cyclic graphs, complexity $\Theta(m_L + n_m \times m_m) + (m_L - n_m) \times m_R + n_m \times m_R)$—thus it behaves as the counting method for all non-recurring nodes and as the magic set method for the recurring ones.

8. Conclusion

In this paper, we have presented a comprehensive solution that uses a Fixpoint based implementation for recursive predicates and compile-time rewriting techniques to make it safe and efficient. The first issue discussed was the application, via rule rewriting scripts, of symbolic finite differencing to improve the basic Fixpoint algorithm by removing redundant computation. Then, we tackled the key problem of taking advantage of bindings existing in recursive goals. To this end, the magic set method and the counting methods were developed which recast the original recursive rules into equivalent ones that are amenable to a safe and efficient implementation by two Fixpoint computations. A novel treatment was then proposed for dealing with those important cases, where one of these two fixed-points can be eliminated. A performance comparison between the magic set and the counting method was also presented; the results of this comparison motivated the introduction of a method, called the magic counting method, that integrates the strengths of the previous methods.

Many approaches were proposed in the past for supporting recursion in logic based languages; all these approaches have strengths and weaknesses (e.g., limited generality). The contribution of this paper consists in integrating some of the more powerful methods into a comprehensive framework which (i) provides a unified comparative treatment of the various methods, and (ii) supplies the basis for a general and robust system implementation where the various methods are chosen and combined to maximize their strengths and compensate for their weaknesses (Chimenti et al., 1988).

Acknowledgments

The authors are grateful to Catriel Beeri, Danette Chimenti, and Raghu Ramakrishnan for their comments and suggested improvements.

References

Aho, A. V., and Ullman, J. (1979). "Universality of Data Retrieval Languages," *Proc. POPL Conference*, San Antonio, Texas.

Balbin, I., and Ramamohanarao, K. (1987). "A Differential Approach to Query Optimization in Recursive Deductive Databases," *Journal of Logic Programming* **4**, No. 2, Sept. 1987, 259–262.

Bancilhon, F. (1985). "Naive Evaluation of Recursively defined Relations," *On Knowledge Base Management Systems* (M. Brodie and J. Mylopoulos, eds.), Springer-Verlag.

Bancilhon, F., and Ramakrishnan, R. (1986). "An Amateur's Introduction to Recursive Query Processing Strategies," *Proc. ACM SIGMOD Int. Conference on Management of Data*, Washington, D.C.

Bancilhon, F., Maier, D., Sagiv, Y., and Ullman, J. (1986). "Magic sets and other strange ways to implement logic programs," *Proc. 5th ACM SIGMOD-SIGACT Symp. on Principles of Database Systems.*

Beeri, C., and Ramakrishnan, R. (1987). "On the Power of Magic," *Proc. 6th ACM SIGMOD-SIGACT Symp. on Principles of Database Systems.*

Beeri, Nagvi, S., Ramakrishnan, R., Shmueli, O., and Tsur, S. (1987). "Sets and Negation in a Logic Data Language (LDL1)," *Proc. 6th ACM SIGMOD-SIGACT Symp. on Principles of Database Systems*, 269–283.

Beeri, C., Kanellakis,P., Bancilhon, F., and Ramakrishnan, R. (1987). "Bound on the Propagation of Selection into Logic Programs," *Proc. 6th ACM SIGMOD-SIGACT Symp. on Principles of Database Systems.*

Chimenti D., O'Hare, T., Krishnamurthy, R., Nagvi, S., Tsur, S., Wert, C., and Zaniolo, C. (1988). "An Overview of the LDL System," *Database Engineering Bulletin.*

Chandra, A. K., and Harel, D. (1982). "Horn clauses and the Fixed-point hierarchy," *Proc. ACM SIGMOD-SIGACT Symp. on Principles of Database Systems*, 158–163.

Demo B., Porta, M., and Sapino, M. L. (1986). "A Stack Machine Semantics for Rewriting Rule Methods in Logical Databases." submitted for publication.

Gardarin, G., and De Maindreville, C. (1986). "Evaluation of Database Recursive Logic Programs as Recursive Function Series," *Proc. ACM SIGMOD Int. Conference on Management of Data*, Washington, D.C.

Gallaire, H., Minker, J., and Nicolas, J. M. (1984). "Logic and Databases: a Deductive Approach," *Computer Surveys* **16**, No. 2.

Henschen, L. J., and Naqvi, S. A. (1984). "On compiling queries in recursive first-order databases," *JACM 31*, **1**, 47–85.

Kifer, M., and Lozinskii, E. L. (1986). "Filtering Data Flow in Deductive Databases," *ICDT '86*, Rome, Italy.

Kolaitis, G. P., and Papadimitriou, C. H. (1988). "Why Not Negation by Fixed-point?" *Proc. 7th ACM SIGMOD-SIGACT Symp. on Principles of Database Systems.*

Lloyd, J. W. (1984). *Foundations of Logic Programming.* Springer-Verlag.

Morris, K. et al. (1987). "YAWN! (Yet Another Window on Nail!)," *Data Engineering* **10**, No. 4, 28–44.

Marchetti-Spaccamela, A., Pelaggi, A., and Saccà, D. (1987). "Worst-case complexity analysis of methods for logic query implementation," *Proc. 6th ACM SIGMOD-SIGACT Symp. on Principles of Database Systems.*

McKay, D., and Shapiro, S. (1981). "Using active connection graphs for reasoning with recursive rules," *Proc. 7th IJCAI*, 368–374.

Paige, R., and Koenig, S. (1982). "Finite Differencing of Computable Expressions," *ACM TOPLAS* **4**, No. 3, 402–454.

Przymusinski, T. (1987). "On the Semantics of Stratified Deductive Databases and Logic Programs." In *Foundations of Deductive Databases and Logic Programming* (J. Minker, ed.), Morgan Kaufman, Los Altos.

Reiter, R. (1978). "On closed world databases." In *Logic and Databases* (H. Gallaire and J. Minker, eds.), Plenum, New York, 55–76.

Rohmer, J., Lescouer, R., and Kerisit, J. M. (1986). "The Alexander Method – A technique for the Processing of Recursive Axioms in Deductive Databases," *New Generation Computing* **4**, No. 3, 273–287.

Saccà, D., and Zaniolo, C. (1986). "On the implementation of a simple class of logic queries for databases," *Proc. 5th ACM SIGMOD-SIGACT Symp. on Principles of Database Systems.*

Saccà, D., and Zaniolo, C. (1987). "Implementation of Recursive Queries for a Data Language based on Pure Horn Logic," *Proc. Fourth Int. Conference on Logic Programming,* Melbourne, Australia.

Saccà, D., and Zaniolo, C. "The Generalized Counting Method for Recursive Logic Queries," *JTC.* To appear. (Also *Proc. ICDT '86.*)

Saccà, D., and Zaniolo, C. (1987). "Magic Counting Methods," *ACM SIGMOD Proceedings.*

Saccà, D., and Zaniolo, C. "Differential Fixed-point Methods and Stratification of Logic Programs." Submitted for publication.

Tarski, A. (1955). "A Lattice Theoretical Fixed-point Theorem and its Application," *Pacific Journal of Mathematics No. 5*, 285–309.

Tsur, S., and Zaniolo, C. (1986). "LDL: A Logic-Based Data Language," *Proc. of 12th VLDB*, Tokyo, Japan.

Ullman, J. D. (1982). *Principles of Database Systems.* Computer Science Press, Rockville, Md.

Ullman, J. D. (1985). "Implementation of logical query languages for databases," *TODS* **10**, (3), 289–321.

van Emden, M. H., and Kowalski, R. (1976). "The semantics of Predicate Logic as a Programming Language," *JACM* **23** (4), 733–742.

Van Gelder, A. (1986). "A Message Passing Framework for Logical Query Evaluation," *Proc. ACM SIGMOD Int. Conference on Management of Data*, Washington, D.C.

Vieille, L. (1986). "Recursive Axioms in Deductive Databases: the Query-Subquery Approach," *Proc. First Int. Conference on Expert Database Systems*, Charleston, S.C.

Zaniolo, C. (1986). "Prolog: a database query language for all seasons." In *Expert Database Systems* (L. Kerschberg, ed.) Benjamin/Cummings.

Zaniolo, C. (1985). "The Representation and Deductive Retrieval of Complex Objects," *Proc. 11-th VLDB*, 459–469.

Zaniolo, C. "Safety and Compilation of Non-Recursive Horn Clauses," *Proc. First Int. Conference on Expert Database Systems.*

Zaniolo, C. (ed.), (1987). "Special Issue on Databases and Logic," *Data Engineering* **10**, No. 4.

Index